AF598607

OCEANOGRAPHY
and
MARINE BIOLOGY

AN ANNUAL REVIEW

Volume 36

OCEANOGRAPHY and MARINE BIOLOGY

AN ANNUAL REVIEW

Volume 36

Editors

A. D. Ansell

R. N. Gibson

Margaret Barnes

The Dunstaffnage Marine Laboratory
Oban, Argyll, Scotland

Founded by Harold Barnes

First published in 1998 by UCL Press

UCL Press Limited
1 Gunpowder Square
London EC4A 3DE
UK

and

325 Chestnut Street
Philadelphia
PA 19106
USA

British Library Cataloguing-in-Publication Data
A catalogue record for this book is available from the British Library.

Library of Congress Cataloging-in-Publication Data
are available.

ISBN:
1-85728-984-6 HB

Typeset by Santype, Salisbury, UK.
Printed and bound in
Great Britain by
T. J. International Ltd,
Padstow

CONTENTS

CONTENTS

PREFACE

Volume 36 of this series of annual reviews contains eight articles with authors contributing from six countries. It is particularly gratifying that these include two contributions from Russian institutions since much Russian literature remains relatively inaccessible. This year's volume lacks a contribution in physical oceanography, but covers both botanical and zoological interests, while two reviews deal in part with interactions between algal and animal communities. There is some emphasis on the effects of natural and man-made disturbances, including the spread of introduced species and the effects of eutrophication and aggregate dredging. Aspects of coral reef communities are explored in two reviews and others deal with symbiosis among the polychaetes and with the eggs and egg masses of cephalopods.

The year has seen our new relationship with publishers Taylor & Francis move to a more settled phase after some initial teething problems. We are grateful to all members of their staff who have contributed to this smooth transition. Editorial policy remains one of maintaining a high standard of authoritative review, both by soliciting articles on subjects that appear ready for such treatment and by accepting suitable reviews that are offered to us. Suggestions of possible subjects from the marine science community are always welcome. We are, as always, grateful to our contributors for agreeing to prepare such comprehensive reviews and for their cooperation and patience in dealing with editorial comments and queries.

Oceanography and Marine Biology: an Annual Review 1998, **36**, 1–64

UCL Press

ECOLOGY OF THE GREEN MACROALGA *CODIUM FRAGILE* (SURINGAR) HARIOT 1889: INVASIVE AND NON-INVASIVE SUBSPECIES

CYNTHIA D. TROWBRIDGE
Gatty Marine Laboratory, School of Environmental & Evolutionary Biology, University of St Andrews, St Andrews, Fife KY16 8LB, Scotland
and Department of Zoology, Hatfield Marine Science Center, Oregon State University, Newport, OR 97365, USA

Abstract The large, green, branching macroalga *Codium fragile* (Suringar) Hariot 1889 (Chlorophyta: Codiaceae) is one of the most abundant and widespread species in the morphologically and taxonomically diverse genus of *Codium*. Six distinct subspecies of *C. fragile* have been recognized, in addition to morphologically heterogeneous populations (with no subspecific name) on temperate and boreal shores throughout the world. Three of the subspecies appear to occur primarily as introduced forms: ssp. *atlanticum* and ssp. *tomentosoides* originated from Japan and ssp. *scandinavicum* originated from Siberia. Ssp. *tomentosoides* is one of the most invasive seaweeds in the world, with extensive transoceanic and interoceanic spread this century; the alga is a serious ecological and economic pest on NW Atlantic shores. Despite the high abundance and broad distribution of *C. fragile*, a disproportionate amount of study has focused on ssp. *tomentosoides*, in a narrow part of its invaded range, namely NW Atlantic shores; results from this region are not necessarily applicable to the alga in other temperate and boreal regions. Furthermore, much of the work on ssp. *tomentosoides* is unrelated to the invasion ecology of this alga, and many authors remain unaware of its exotic origins. In this review, I examine the ecological differences among and within subspecies and evaluate their relative invasiveness. Variation among subspecies of *C. fragile* occurs in the following attributes: (a) sexual reproduction *v.* parthenogenesis, (b) apparent ploidy level of the macroscopic adult thallus, (c) salinity tolerance, and (d) thallus buoyancy in terms both of tissue density and propensity to trap gases. There is little reported evidence, however, that subspecies vary substantially in length of their reproductive period, growth, phenology, vegetative propagation, physiological ecology, herbivore palatability, competitive ability, host–epiphyte interactions, or natural products production. Comparative studies are needed to understand the variable invasiveness of the three introduced subspecies and the non-invasiveness of indigenous forms as well as geographic variation in ecological attributes of ssp. *tomentosoides*.

Introduction

The green macroalgal genus *Codium* (Chlorophyta: Codiaceae) contains about 100 described species, making it one of the most diverse genera of marine algae (Silva 1992). The most abundant and widely distributed species in the genus is *C. fragile* (Suringar) Hariot, generally known as sea staghorn, oyster thief, or dead man's fingers (Table 1). This species has a pan-temperate, bipolar, and antarctic circumpolar distribution (Silva & Womersley 1956, Silva 1979): it occurs from 29° to 54 °S latitude and

Table 1 Common names of *Codium fragile* in different geographic regions.

Common name	Region	References
Sea staghorn	Korea, NE Pacific, NW Atlantic	Loosanoff (1975), Arasaki & Arasaki (1986), Lee (1986), Ahn et al. (1993), Cho et al. (1995), Kim et al. (1995)
Oyster thief	NW Atlantic	Wassman & Ramus (1973b), Hanisak (1980), Lee (1986)
Dead man's fingers	NE Pacific, NW Atlantic	Wassman & Ramus (1973b), Loosanoff (1975), Ramus et al. (1976a,b), Dawson & Foster (1982), Coleman (1996)
Sputnik weed	NW Atlantic	Wassman & Ramus (1973b)
Spaghetti grass	NW Atlantic	Wassman & Ramus (1973b), Loosanoff (1975)
Green fleece or fleece	NE Pacific, NW Atlantic	Madlener (1977), Gosner (1978)
Spongeweed or sponge tang	NE Pacific, NW Atlantic	Madlener (1977), Dawson & Foster (1982)
Green sea fingers	UK	Eno et al. (1997)
Miru	China, Japan	Chapman & Chapman (1980), Arasaki & Arasaki (1986)
Shuisong (water pine)	China	Meiling & Tseng (1984)
Blomstertang (blossom seaweed)	Rogaland	Printz (1952)
Chunggak/chonggak (sea staghorn)	Korea	Madlener (1977), Ahn et al. (1993), Cho et al. (1995), Kim et al. (1995)

33° to 70 °N latitude (Bright 1938, Lund 1940, Fægri & Moss 1952, Printz 1952, Stellander 1969, Boraso & Piriz 1975, Searles et al. 1984). The species is a common to locally common component of many rocky shores throughout the world (e.g. Williams 1925, Isaac 1937, Bright 1938, Eyre 1939, Stephenson 1948, Printz 1952, Silva &

Table 2 Different subspecies of *Codium fragile* in different regions of the world. Data based on Silva (1951, 1955, 1957, 1959), Silva & Womersley (1956), Scagel (1966).

Subspecies of *C. fragile*	Status	Potential origin	Geographic range
ssp. *atlanticum* (Cotton) Silva	introduced	Japan	British Isles and Norway*
ssp. *scandinavicum* Silva	introduced	Siberia	Scandinavia*
ssp. *tomentosoides* (van Goor) Silva	introduced	Japan	N Atlantic, Mediterranean, NE Pacific, New Zealand, SE Australia*
ssp. *capense* Silva	native	—	South Africa (Swakopmund, SW Africa, around Cape of Good Hope to Robberg, Cape Province)
ssp. *novae-zelandiae* (J. Ag.) Silva	native	—	New Zealand, SE Australia (Robe, S Australia to Ballina, New South Wales; Tasmania)
ssp. *tasmanicum* (J. Ag.) Silva	native	—	S Australia (Victor Harbour, S Australia to Walkerville, Victoria; Tasmania)

* Introduced subspecies have a potential origin in the NW Pacific; the origin is not included within the subspecific ranges because distinct subspecies are not recognised there.

Womersley 1956, Lewis 1964, Thorne-Miller et al. 1983, Meiling & Tseng 1984, Kim 1988a, 1991, Bird et al. 1993, Harlin & Rines 1993, Trowbridge 1993, 1995, 1996, Liu et al. 1995, references therein).

The species is currently recognized as a single species with morphologically (a) homogeneous populations (morphological plateaux) considered to be subspecies (Table 2) and (b) heterogeneous populations referred to just as *C. fragile*. Two of the six described subspecies are currently recognized as introduced and a third subspecies may be introduced (Silva 1955, 1957). One of these subspecies (ssp. *tomentosoides*) is among the most invasive seaweeds in the world, with extensive transoceanic and interoceanic spread this century (reviewed by Carlton & Scanlon 1985, Ribera 1994, Verlaque 1994, Trowbridge 1995).

C. fragile is a model study system for many fields from algal physiology, endosymbiosis, and heavy metal accumulation to invasion ecology, algal genetics, and natural products. The alga also has economic value: it is cultivated for human consumption in Asia, used as invertebrate food by the mariculture industry, is a pest of natural and cultivated shellfish beds, is a source of bioactive compounds (antibiotic, anticarcinogenic, immuno-suppressive, anti-insect, and antihelminthic activity), and accumulates heavy metals, thus providing a model indicator of pollution. Because research on *C. fragile* has included both basic and applied aspects of diverse subject areas involving various subspecies in various localities, a review of the literature is warranted.

Scope of review

In this review, I synthesize a broad spectrum of areas, from autecology to community ecology of *C. fragile*. I start by highlighting the complexities within the literature to clarify past issues and to emphasize changes required to prevent future confusion about the species. Next, I cover the following 11 general topics about the alga:

(1) Ecological significance of the alga's macro- and microstructure.
(2) Subspecific diversity.
(3) Patterns of growth and phenology.
(4) Ssexual and asexual reproduction.
(5) Differentiated thalli *v.* undifferentiated "vaucherioid" mats.
(6) Ecophysiological responses to environmental factors.
(7) Invasion ecology of the introduced subspecies
(8) Herbivory patterns for generalist and specialist grazers.
(9) Interspecific and inter-subspecific competition,
(10) epiphytes on *C. fragile*.
(11) Natural products of *C. fragile* and their potential ecological relevance.

For each topic, I focus on inter-subspecific comparisons and inter-regional, intra-subspecific comparisons. I cover the extensive literature on *C. fragile*, highlight major results of unpublished dissertations (>10 years old), and include some original data. Finally, I conclude with future research priorities to address pressing ecological concerns.

I dedicate this paper to Professor Paul C. Silva whose extensive taxonomic work distinguishing different species of *Codium* and subspecies of *C. fragile* has made the present review possible. He identified and clarified the three introduced subspecies of

C. fragile on European shores (Silva 1955, 1957) and recorded the subsequent appearance of ssp. *tomentosoides* in San Francisco Bay, California (Silva 1979). He has generously assisted other researchers tracking the continued spread of ssp. *tomentosoides* throughout the world and those studying *Codium* biology.

Complexities within the literature

In contrast to the wealth of taxonomic information on *C. fragile* (e.g. Lund 1940, Silva 1951, 1952a,b, 1954, 1955, 1957, 1959, 1960, 1962, 1979, 1992, Dellow 1952, Silva & Womersley 1956, Silva & Irvine 1960, and references therein), ecological information on the alga, particularly on the native subspecies, is meagre and almost exclusively descriptive. Silva & Irvine (1960) made the astute observation that the "amount of information available about . . . [*Codium*] is not commensurate with the number of published reports" (p. 632). Although the authors were referring to "*C. amphibium*", the point is clearly applicable to much of the ecological literature on *C. fragile*.

Understanding the ecology of *C. fragile* from the published literature is complicated by a multitude of factors. I mention the following six points to clarify past issues and to emphasize changes required to prevent future confusion.

Wealth of unpublished research

There is a large body of unpublished dissertations (PhD, MSc, MA, and Honours theses) on *C. fragile* (e.g. Gibby 1971, Boerner 1972, Meimer 1972, Perretti 1972, Simon 1972, Lopez 1973, Persson 1973, Thomas 1974, Wilson 1978, Ames 1979, Lewis 1982, Davies 1983, Theis 1985). Consequently, valuable experimental and observational ecological research is generally inaccessible; I highlight their important contributions herein.

Omission of relevant citations

The common omission of citations (historical and contemporary) of directly relevant research leads to repetitious research and failure to acknowledge originators of ideas. As an example, recent papers by Fletcher et al. (1989) and Yang et al. (1997) on the occurrence, growth, and development of juvenile vaucherioid stages of *Codium* spp. omit mention of any of the following pertinent works on the subject (Arasaki et al. 1956, Silva & Irvine 1960, Moeller 1969, Ramus 1972, Lopez 1973, Thomas 1974, Steele 1975, Hanisak 1977, 1979b, Ames 1979). Because of the broad geographic range of *C. fragile*, a consideration of the worldwide literature on the species is essential.

Omission of subspecific names and alga's status

A large percentage of the studies on *C. fragile* omit mention of the subspecies, even when it has been well described. In some geographic regions (e.g. NW Atlantic), the omission will not be problematic; where the distributions of subspecies overlap, authors should be as explicit as possible. In this review, I ascertain the subspecies used in previous studies by communication with authors and/or by deduction from known distributions. For example, the extensive studies on *C. fragile* from Bembridge, Isle of Wight, England by Cobb, Trench, Hinde, and colleagues were conducted with ssp. *tomentosoides* (A. H. Cobb, pers. comm.). In several molecular studies, neither the subspecies nor the locality where it was collected is specified so that the identity of the sequenced material is not clear.

Many authors do not mention the status of their alga (introduced *v.* native subspecies); whether this omission is due to the fact that it is not pertinent to the questions addressed or whether the authors are not aware of the status of their study organisms cannot be distinguished (Ribera & Boudouresque 1995). In many cases, however, the status is directly relevant. For example, the high degree of "specialization" between the sea slug *Elysia viridis* and the introduced *Codium fragile* ssp. *tomentosoides* has been well studied, but there is no explicit acknowledgement that the association is a recent one (only a few decades old).

It is crucial for authors to specify the subspecies being studied. "If the subspecies is not specified, those who are not aware of the taxonomic situation will link the entire body of information resulting from that piece of research to *Codium fragile* ssp. *fragile*, while those who are aware of the taxonomic situation will hold this body of information in abeyance unless the identity of the subspecies can be deduced." (P. C. Silva, pers. comm.).

Dilemma of naturalization

Some authors explicitly refer to introduced seaweeds as if they were natives, even in the phycological literature (e.g. *C. fragile* ssp. *tomentosoides* as "British" in Hornsey & Hide 1974, 1976a,b, even though the alga is not native to the North Atlantic). Farnham (1994) considers *C. fragile* to be naturalized on British shores, and Hanisak (1977) implies that the alga is firmly ensconced in New England coastal communities. At what point do we consider an introduced alga "naturalized"? What should be the operational classification criteria for species' naturalization?

Intraspecific extrapolations

Extrapolating from an individual subspecies to other conspecific subspecies or to the NW Pacific stock requires considerable caution. The implicit assumption of such extrapolations is that subspecies are similar; however, it has not yet been demonstrated that the different populations (particularly introduced subspecies and assumed parental stock) are fundamentally similar ecologically or biochemically. For example, although indigenous populations of *C. fragile* on Japanese, Korean, and Chinese shores are edible (e.g. Chapman & Chapman 1980, Arasaki & Arasaki 1986, Cho et al. 1995), Eno et al. (1997) imply that ssp. *atlanticum* and ssp. *tomentosoides* are also consumed by people in Asia (where individual subspecies are not recognized).

Nature and quality of ecological evidence

Because the study of *C. fragile* spans so many different fields (ecology, morphology, reproduction, systematics, physiology, etc.), there are understandably different approaches to these studies and different accepted norms of what constitutes "evidence". Much of the intriguing ecological information on *C. fragile* is partially or totally unsubstantiated. Studies are not necessarily wrong, but rather insufficient quantitative data are presented to evaluate authors' interpretations and assertions. The study of *Codium* biology would progress much more rapidly if researchers would quantify observed patterns and test hypothesized causal processes, rather than asserting subjective opinions.

Morphology

A brief review of the morphology of *C. fragile* is necessary because it is rather unusual, varies subtly among subspecies, and directly affects the alga's ecology. Excellent monographs on the taxonomy and morphology of *Codium* spp. have been written by Lund (1940), Silva (1951, 1952a,b, 1954, 1955, 1957, 1959, 1960, 1962, 1979, 1992), Silva & Womersley (1956), Silva & Irvine (1960), and references therein. My review complements these classical works by emphasizing the ecological significance of morphological attributes. For example, chloroplast structure, cell wall composition, and utricle hairs are integral aspects of the alga's intimate association with ascoglossan (=sacoglossan) sea slugs that feed on *Codium*.

Macrostructure

The genus *Codium* is characterized by extremely high intrageneric diversity in thallus form, including species with prostrate, globose, procumbent, cylindrical, rope-like, erect and flattened, and erect and dichotomously branching thalli (Silva 1954, 1992). *C. fragile* (Fig. 1A), the topic of this review, is a large branching species (up to ~1 m long, 3.5 kg ww). The alga has up to ten orders of dichotomous branching; the cylindrical or terete branches are 3–10 (–14) mm in diameter (Silva 1951, 1955, Wood 1962, Scagel 1966, Boraso & Piriz 1975, Stegenga et al. 1997). One to many fronds arise either out of the holdfast – a broad, spongy basal disk (Lund 1940, Silva 1951, 1955, Scagel 1966, Meiling & Tseng 1984, Oh et al. 1987) – or from a vaucherioid mat (*sensu* Silva & Irvine 1960). When submerged, the spongy, uncalcified thalli are buoyant due to gas trapped within the thallus (Moeller 1969, Dromgoole 1982). The composition of the gas (by weight) for ssp. *tomentosoides* is 94.3% nitrogen, 4.1% oxygen, 1.6% argon, and sometimes carbon dioxide (Moeller 1969).

The gross morphology of *C. fragile* may be ecologically important for several reasons.

(1) Intermediate-sized, coarsely branching algal species (like *C. fragile*) are difficult for generalist grazers to grasp (Steneck & Watling 1982); thus, the alga's gross morphology may limit herbivory. The alga's morphology may also affect physiological attributes such as photosynthesis and nutrient uptake; many processes such as light absorption, gas exchange, and nutrient uptake are directly related to the alga's surface area to volume ratio (Ramus 1978, Arnold & Murray 1980, Littler &

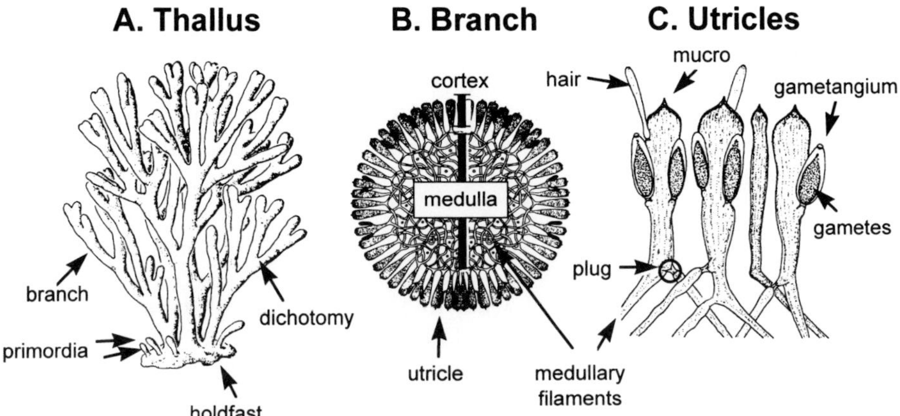

Figure 1 Macro- and microstructure of *Codium fragile*, and terms used in text. Drawings reproduced from Smith (1944, 1955) and Silva (1959) with permission from Stanford University Press, McGraw-Hill, and *Journal of Plant Research*. (A) Entire thallus, (B) cross-section of branch, and (C) close-up of utricles (swollen cortical tips of interwoven filaments).

Arnold 1982). *C. fragile* has a low ratio (10.8 : 1) compared with foliose algae such as *Ulva lactuca* (286 : 1) (Ramus 1978).

(2) The degree of branching varies spatially and temporally (ssp. *tomentosoides*: Benson et al. 1983, Williams et al. 1984; ssp. *novae-zelandiae*: Trowbridge 1996) and this variation may affect algal – sea slug interactions. Sea slugs that feed on *Codium* generally lay their egg masses at branch dichotomies (Macnae 1954), and slug herbivory is often concentrated at the dichotomies and "axillary" areas (Clark 1975, Trowbridge 1993).

(3) The high degree of branching may render thalli prone to wave dislodgement (see Dromgoole 1979). The number of branches and their regularity increase with decreased wave exposure (Dellow 1952).

Microstructure

General features

Irrespective of the gross macrostructure, adult thalli of *Codium* species are formed by the interweaving of siphonous filaments that compose two regions of the thallus. The central region (medulla) is composed of long, colourless filaments (medullary filaments) that run longitudinally within the thallus and are densely intertwined (Fig. 1B). They vary in size among different subspecies (Table 3) but are typically 26 to 68 (–112) μm in diameter (e.g. Silva 1951, 1955, 1957, Scagel 1966). The peripheral region (cortex) is formed of cylindrical or club-shaped siphonous swellings called utricles that are tightly compressed together (Fig. 1C).

The thallus has both extracellular and intracellular spaces (among and within utricles, respectively). The space among the utricles provides a micro-environment where large populations of cyanobacteria and heterotrophic bacteria occur (Dromgoole et al. 1978, Rosenberg & Paerl 1981, Gerard et al. 1990). Of the intracellular space, 93–95% of the utricle volume is occupied by vacuoles (Dromgoole 1979).

The intracellular fluid varies in ion composition and density between two subspecies on New Zealand shores (Dromgoole 1982) and perhaps among other subspecies as well. Cytoplasm containing nuclei, chloroplasts, and other organelles forms a thin parietal lining around the vacuoles (see Hawes 1979, Sealey et al. 1990, Williams & Cobb 1992). Cytoplasmic streaming occurs within each utricle, but the chloroplasts remain concentrated at the apex.

Utricles

Utricle diameter varies from about 130 μm to 350 μm among subspecies of *C. fragile*; utricle length varies from about 480 μm to 1500 μm or more (Table 3) (Silva 1951, 1955, 1957, 1959, Silva & Womersley 1956, Scagel 1966, Kim 1988b). Lateral walls of the utricles are typically 1–3 μm thick whereas apical walls are considerably thicker. Some forms have a rounded apex with laminated walls; other forms have pointed apices with a mucronate tip to 120 μm long (Lund 1940, Scagel 1966).

Although these morphological variations are extremely useful taxonomic attributes, the potential adaptive significance of the variation has not been extensively discussed. Herbivory by sea slugs may be a factor involved in the evolution of the apparent buttressing of the exposed utricle tips. The ascoglossan sea slugs puncture utricle apices with a long, pointed radular tooth (> 100 μm long) and then suck out the algal cytoplasm. Structural buttressing (thickened walls, mucronate tips, and, in other *Codium* species, trabeculae within the utricle tips) may be an antiherbivore defence. Measurements of the puncture resistance of different types of utricle walls would be needed to evaluate this hypothesis.

Wilson (1978) remarked on the propensity of red algal spores to become trapped by the "cobblestone" surface relief formed by the utricle surfaces. Inter-subspecific variation in surface relief may lead to different communities of epiphytes, even on sympatric thalli. As different subspecies of *C. fragile* overlap in distribution (see Table 2, p. 2), subtle structural differences may lead to unexpectedly large ecological differences. Measurements of spore settlement patterns on different surfaces of *C. fragile* would be needed to test this epiphyte hypothesis.

The chemical composition of cell walls affects slug herbivory: different species of ascoglossan sea slugs have radular teeth effective at puncturing different types of algal polysaccharides (Jensen 1980, 1993). The utricle walls of *Codium* are composed primarily of mannan (Iriki & Miwa 1960, Percival & McDowell 1967), and slugs with blade-shaped radular teeth with median denticles are able to puncture mannan cell walls (Jensen 1980, 1993). Slugs with other types of teeth are not able to feed on *Codium*. This constraint is ecologically important when evaluating the frequency with which indigenous species of slugs will switch to feed on newly introduced subspecies of *Codium*.

Utricle hairs

In all subspecies of *C. fragile*, 1–3 hairs extend from the upper region of each utricle into the surrounding medium (water). Hairs are 25–60 μm wide and 50–650 (–1600 μm) long (Table 3) (Hurd 1916, Dellow 1952, Silva & Womersley 1956, Oh et al. 1987). Although many authors refer to them as "colourless", the hairs in fact contain not only cytoplasm but also a few chloroplasts (Lund 1940, C. D. Trowbridge pers. obs.).

Certain subspecies are more hairy (i.e. tomentose) than others; in particular, ssp. *tomentosoides* is typically covered with dense hairs.

The function of utricle hairs is controversial: suggestions include protection of *C. fragile* against intense light, dislodgement of epiphytes, and facilitation of nutrient uptake (Hurd 1916, Dellow 1953, Moeller 1969, Head & Carpenter 1975, Benson et al. 1983). Head & Carpenter (1975) found that utricle hairs persisted in cultures of *C. fragile* ssp. *tomentosoides* with low nitrogen concentrations but were lost in cultures with high nitrogen, indicating that hairs may enhance the alga's nitrogen assimilation.

Because hairs may serve multiple functions, I have explored whether hairs influence herbivore–algal interactions. To investigate the possibility that utricle hairs deter small herbivores, I removed hairs from certain thalli. The presence or absence of hairs did not influence the preference of small suctorial sea slugs for the alga (ssp. *tomentosoides*) (Trowbridge 1995). In recent laboratory experiments (Fig. 2A), competent larvae of sea slugs (*Elysia viridis*) settled and metamorphosed more frequently on pieces of *C. fragile* with utricle hairs present than on pieces with hairs removed; the growth of the recently metamorphosed juveniles (post-larvae), however, did not differ between treatments (Fig. 2B) (C. D. Trowbridge & C. D. Todd, unpubl. data). Thus, spatial and temporal variation in abundance of hairs within (and presumably among) subspecies can contribute to differences in grazer attack.

Gametangia

Gametangia (where gametes are formed) are attached to the lateral walls of the utricles on short pedicels (Scagel 1966) and their tips are below the apical surface of the utricles

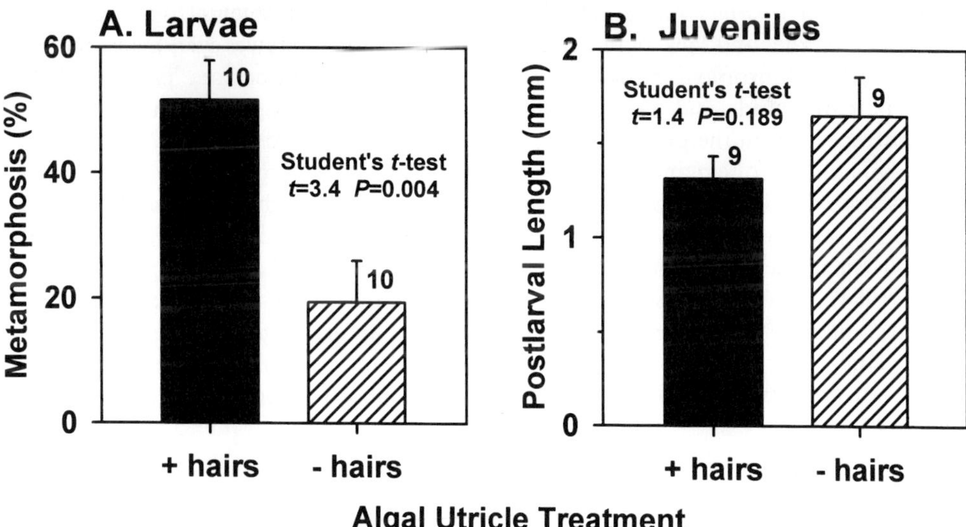

Figure 2 Larval metamorphosis and post-larval (=juvenile) growth of herbivorous sea slugs (*Elysia viridis*) on *Codium fragile* ssp. *tomentosoides* collected from Loch Sween, Argyll, Scotland. Utricle hairs were gently removed with a scalpel under a microscope for − hair treatment; hairs were left intact for the + hair control. In A, replication was 10 dishes per treatment, each with 15 competent larvae. In B, replication was nine dishes per treatment, each with one slug per dish. Error bars indicate 1 SE.

(Fig. 1B,C). Gametangia are ovate to clavate and are generally 70–170 μm in diameter and 185–450 μm long (Silva 1951, 1955, 1957, Silva & Womersley 1956, Scagel 1966, Oh et al. 1987). Utricles of *C. fragile* each bear 1–3 gametangia that mature consecutively (Silva 1951, Scagel 1966). Male gametangia produce numerous tiny yellow–green gametes whereas female gametangia produce a smaller number of larger, dark green gametes (Williams 1925, Smith 1955, Arasaki et al. 1956, Scagel 1966). Both male and female gametes have two terminally inserted flagella. After gametes are released, gametangia are shed, leaving a persistent scar (Williams 1925). Most thalli of *Codium* are dioecious; monecious thalli are assumed to arise from male and female germlings intertwining to form a single macroscopic thallus (Arasaki et al. 1956).

The coenocytic thalli of *Codium* contain no transverse walls or septa *per se*. Yet, there are internal plugs or pad-shaped thickenings that (a) hold the cell contents in place and (b) prevent the loss of an entire filament when injury occurs (Hurd 1916, Scagel 1966). These plugs can be complete or partial (with a persistent small opening). In *C. fragile*, plugs occur between utricles and medullary filaments, at the base of gametangia and hairs, and periodically within medullary filaments (Fig. 1C). When mechanical damage occurs in *C. fragile*, the partial plugs are rapidly blocked and sealed by cytoplasmic organelles, thus preventing extensive loss of cytoplasm (Wassman & Ramus 1973a). When the highly specialized sea slugs attack *C. fragile*, they puncture individual utricles and drain them; thus, the slugs have to puncture every utricle on which they feed. In contrast, conspecific slugs feeding on *Bryopsis* can readily drain the entire coenocytic thallus (C. D. Trowbridge pers. obs.).

Chloroplasts

Chloroplasts in *Codium fragile* are small (~2–3 μm long), ellipsoidal, bounded by a double membrane, and lack pyrenoids (Hori & Ueda 1975, Hawes 1979). The chloroplasts contain chlorophyll *a* and *b* as well as a variety of accessory photosynthetic pigments (carotenoids and xanthophylls) that increase the light-harvesting ability of the alga, especially in the green wavelengths (Benson & Cobb 1981, 1983). The xanthophyll pigments siphonein and siphonaxanthin are ecologically noteworthy as they have been used to validate purported trophic interactions with herbivorous sea slugs (e.g. Greene 1970b). The chloroplast membrane in *C. fragile* is unusually resistant to mechanical and chemical stress (Trench et al. 1973a,b, Cobb & Rott 1978, Grant & Borowitzka 1984). One consequence of this resistance is that the chloroplasts retain their structural integrity even after ingestion and retention (in digestive diverticula) by suctorial sea slugs. Some of the species of slugs that feed on *Codium* spp., including on *C. fragile*, not only retain the chloroplasts but derive photosynthetic benefit from them (see pp. 43–44).

What selective pressures caused the plastids to become so durable? Numerous explanations have been suggested, but the most convincing one is that tough chloroplasts evolved in response to the alga's coenocytic condition. In the absence of internal walls, the risk of lysosomes breaking within the cytoplasm and destroying organelles (such as chloroplasts) increases. However, other authors, such as Hinde (1983), suggest that slug grazing may have selected for chloroplasts capable of extended survival, which may result in less herbivory than would ephemeral chloroplasts. An unexplored question is whether the chloroplasts of different subspecies of *C. fragile* are equally stable.

Subspecific diversity

Subspecific status

C. fragile is currently recognized as a single species with six described subspecies (Tables 2, 3). Silva (1951, 1955, 1957, 1959) suggested that *C. fragile* is composed of a geographically discrete series of (a) morphologically uniform populations (morphological plateaux) in areas distant from the centre of distribution and (b) heterogeneous populations near the centre of distribution (N Pacific shores). In the latter regions, it is difficult to recognize distinct forms so the populations are referred to as *C. fragile* (i.e. no subspecific name). Silva (1951, 1955, 1957) and Silva & Womersley (1956) discuss the rationale for considering the populations subspecies rather than species. The crucial point is that "almost all subspecies (geographically discrete series of morphologically uniform populations) can be matched by one or more Japanese collections" (p. 573, Silva 1957).

Scagel (1966) suggested there are at least two distinct native forms of *C. fragile* on NE Pacific shores, one of which has utricles with sharply pointed apices. This form was described as *C. mucronatum* var. *californicum* by J. Agardh in 1883 but was not recognized by Silva (1951). Goff et al. (1992) referred to it as *C. fragile* ssp. *fragile* but the typical California form cannot be matched by any Japanese population (P. C. Silva, pers. comm.) so referring the California specimens to the Japanese type specimen (ssp. *fragile*) is not appropriate. Scagel (1966) thought a second form, which has utricles with blunt apices, resembled ssp. *novae-zelandiae* but should be referred to as a new variety. This latter form occurs in Oregon (C. D. Trowbridge, pers. obs.) and was the alga studied by Trowbridge (1989, 1992, 1993) and Wheeler & Björnsäter (1992). I will adopt the conservative approach used by Silva (1951) and refer to the NE Pacific populations as indigenous *C. fragile* (with no subspecific identity).

Distinguishing subspecies

Subspecific differences (Table 3) are based primarily on morphological variation in the size and shape of utricles (Silva 1951, 1955, 1957, 1959, Silva & Womersley 1956, Gibby 1971, Dromgoole 1975, Burrows 1991). The number of fronds per thallus and branch width may also be useful characters indicative of subspecies (Trowbridge 1996), but microscopic examination of utricle size and shape is imperative for correct subspecific identification. Because of the high intrathallus variation in utricle attributes, the basis of inter-subspecific comparisons are from utricles collected 2 cm from branch tips (Silva 1957, Dromgoole 1975, 1979); *C. fragile* has apical growth (Hurd 1916, Silva 1954) so the utricles are most consistent in the apical tip region. Silva (1957) reported that "an individual utricle is mature by the time the tip has grown about 5 mm beyond it". Subspecific differences based on morphology are supported by recent electrophoretic and polymerase chain reaction (PCR) results (Malinowski 1974, Goff et al. 1992).

Gross morphological differences among the subspecies of *C. fragile* are not always obvious, particularly where they coexist (generally due to human-assisted introductions). Apparent intermediates exist although whether they are genetic hybrids or environmentally produced phenotypes has not been ascertained (Silva 1955, 1957,

Table 3 Comparison of macro- and microstructural features of different subspecies and regional forms of *Codium fragile* (data from Lund 1940, Silva 1951, 1955, 1957, Dellow 1952, Fægri & Moss 1952, Scagel 1966, Meiling & Tseng 1984, Oh et al. 1987, Burrows 1991).

Subspecies	Thallus length (cm)	Utricle diameter (μm)	Utricle length (μm)	Lateral wall thickness (μm)	Apical wall thickness (μm)	Utricle apex	Hairs (no./utricle)	Medullary filament diameter (μm)
C. fragile ssp. *scandinavicum*	10–20 (–30)	190–420	480–850	<4?	4–12	rounded or thecate tip, sometimes with mucron to 20 μm	1–3	34–68
C. fragile ssp. *atlanticum*	15–25 (–50)	170–330	780–1100	1.5	12	slightly rounded, with blunt point or umbo	1–2	26–68
C. fragile ssp. *tomentosoides*	to 90	165–325	550–1050	1.5	12–65	rounded-apiculate, point prolonged into mucron to 68 μm	1–2	26–68
C. fragile ssp. *novae-zelandiae*	to 25	30–350 (–580)	750–1500 (–2980)	2?	8–40	rounded or bluntly mucronate	1–3	40–60
C. fragile ssp. *tasmanicum*	10–30	(70–) 130–330 (–390)	(719–) 1000–1450 (–1750)	2	thickened	pointed, prolonged into mucro to 77 μm	1–2	26–46
C. fragile ssp. *capense*	10–30	(80–) 130–270 (–355)	730–1100	1.5–2	thickened	broadly rounded, prolonged into mucro to 80 μm	1–2 (–3)	26–46
C. fragile (NE Pacific)	10–30 (–40)	150–350 (–650)	(600–) 675–1125 (–1500)	2–3	?	usually mucronate; mucro to 120 μm	2	26–112
C. fragile (China)	10–30	215–350	900–1025	?	?	cylindrical clavate, more or less distinct, mucro	numerous	?
C. fragile (Korea)	15–30	150–410	1000–1300	?	?	ovate, fusiform	1–2	40–70

Silva & Womersley 1956). On Australasian shores, the native ssp. *novae-zelandiae* is generally macrostructurally distinct from the introduced ssp. *tomentosoides* although not completely so (Trowbridge 1996); utricle features, however, are very distinct. On British shores, ssp. *atlanticum* and ssp. *tomentosoides* are distinct in thallus length, branch width, tidal level, wave exposure, utricle morphology, and other anatomical features (Gibby 1971, Burrows 1991, C. D. Trowbridge & C. D. Todd unpubl. data). Furthermore, ssp. *tomentosoides* differs physiologically from the native ssp. *novae-zelandiae* on New Zealand shores (Dromgoole et al. 1978, Dromgoole 1980, 1982).

In these studies (as well as all inter-subspecific comparisons made to date), a confounding factor is that the subspecies do not yet generally coexist at the same site, even when their geographic ranges overlap (but see Hardy 1990, Burrows 1991). Ideally, comparative measurements of growth, herbivory, or any other ecological or physiological attribute should be made with sympatric individuals. Where the distributions of subspecies do not overlap, it is often necessary to collect individuals from separate locations (e.g. east *v.* west coast of New Zealand, Dromgoole 1980, 1982, Trowbridge 1995, 1996). In such cases, collection location and subspecies are inextricably confounded so significant differences must be viewed cautiously. Ideally, multiple collection localities should be sought for each subspecies.

Past research on subspecies

Codium fragile ssp. *tomentosoides* has received considerably more attention than any of the other described subspecies (Fig. 3A), and most studies have been conducted on NW Atlantic shores. There are at least two reasons for this: (a) Because ssp. *tomentosoides* invaded a geographic region with no existing native species of *Codium* – NW Atlantic north of Cape Hatteras, North Carolina (see Silva 1962) – the alga had higher visibility relative to regions in which ssp. *tomentosoides* invaded areas with morphologically similar (erect dichotomously branching) congeneric species and even resident conspecifics (see Table 1 in Trowbridge 1995). (b) Ssp. *tomentosoides* has negatively affected the shellfish industry and became a nuisance alga ("oyster thief", Table 1, p. 2); these effects will be discussed in detail later (see pp. 39–41).

An unfortunate consequence of this geographic bias is that many marine biologists assume that the population ecology and dynamics of ssp. *tomentosoides* in other parts of its invaded range are similar to that in the NW Atlantic. The validity of this assumption remains to be demonstrated. Ssp. *tomentosoides* occurs from North Carolina to Nova Scotia in the NW Atlantic, from northern Africa to Norway in the NE Atlantic, throughout the Mediterranean, and in NE New Zealand and SE Australia. Seawater and air temperature regimes, salinity ranges, photoperiods, and nutrient levels vary considerably among these different geographic regions (see pp. 25–32) so caution is needed when making broad ecological extrapolations.

The most studied form of *C. fragile* is an introduced alga: ssp. *tomentosoides* is one of the most invasive seaweeds in the world, with ecological and economic effects. Thus, it is ironic that many of the studies are not related to invasion biology (Fig. 3B). The alga has been used as a model system for studying photosynthesis, algal physiology, haemagglutination, coagulation, unusual genetic attributes, etc. Because these topics have been well reviewed elsewhere, I will refer to them only to the extent that they have direct bearing on the ecology of *C. fragile*. Similarly, there is a growing number of

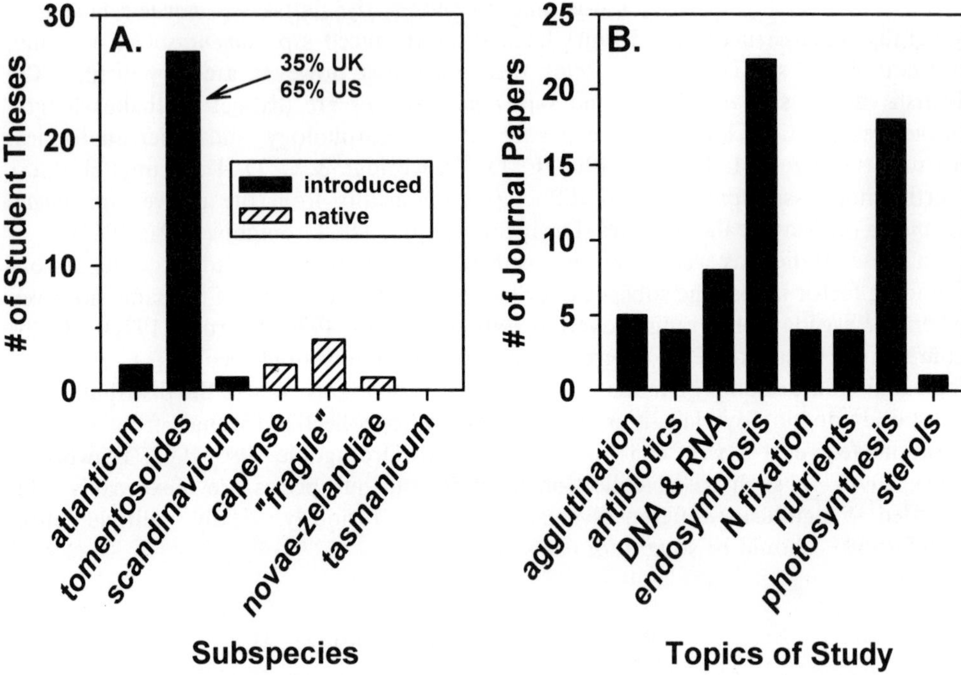

Figure 3 A, Frequency of study of different subspecies of *Codium fragile*. Data based on the number of PhD, MSc, MA, Honours, and Cand. Real. theses found in the literature. B, Frequency of different major research topics of *C. fragile* ssp. *tomentosoides*, other than the alga's invasion ecology; data based on journal papers and book chapters focusing on alga.

studies on ssp. *atlanticum* on its haemagglutination properties but not its invasion biology. This neglect is particularly unfortunate because ssp. *atlanticum* is one of the few marine introduced species that often occurs on wave-swept shores.

Growth and phenology

Mode of growth

The increase in thallus length of *C. fragile* occurs by apical elongation of medullary filaments, primarily at branch tips (Shannon & Altman 1930, as *C. mucronatum*, Silva 1954). Some intercalary growth occurs in intermediate parts of the thallus, and the basal region (main axis or stipe) grows thicker with age (Shannon & Altman 1930, Wilson 1978). New utricles are continually added to the older parts of the thallus (Silva 1954, 1957). Because these newly produced utricles remain slender, older parts of the thallus often appear to have dimorphic or even polymorphic utricles (e.g. see Fig. 8C in Trowbridge 1995). Utricle size variation can influence the alga's vulnerability to grazer attack. For example, Macnae (1954) noted that slug size was directly related to utricle diameter; large slugs could not feed on small utricles and vice versa.

Phenology

Native Codium fragile

The phenology of *C. fragile* varies among geographic regions and habitats within regions, but there is no compelling evidence that introduced subspecies have fundamentally different phenology or growth than native conspecifics. Native populations on NE Pacific shores have been comparatively well studied (Shannon & Altman 1930, Gunnill 1980, 1985, Theis 1985, Trowbridge 1993). In severe winters (e.g. 1995–96), entire thalli including holdfasts are ripped from the substratum; in most years, however, thalli overwinter, and there are a range of frond or axis lengths on each thallus (Trowbridge 1993 and unpubl. data). On Oregon shores, I have recorded spring and summer growth rates of 1–2 cm month^{-1} (C. D. Trowbridge unpubl. data). For Washington shores (Puget Sound), Shannon & Altman (1930) reported summer growth rates of 7.5 cm month^{-1} (i.e. 2.5 mm day^{-1}) for *C. fragile* (as *C. mucronatum*). "Recruitment" occurs from May to November for southern California populations of the alga (Gunnill 1980, 1985, Stewart 1991), although it would be extremely difficult to distinguish among (a) recruitment from settling zygotes, (b) perennation of perennial holdfasts, and (c) growth and differentiation of vaucherioid mats (see pp. 22–24). On Asian shores, Arasaki et al. (1956) and Kim (1988b) both collected *C. fragile* throughout the year. On the Pacific coast of Japan, young thalli appear in early winter and spring and become quite large in summer. In the Sea of Japan, thalli grown on culture ropes in the sea generally die in summer; growth rates are about 1–2 cm month^{-1} (Yotsui & Migita 1989). Only partial phenological data are available on the other indigenous forms of *C. fragile.* Seasonal measurements of thallus length on New Zealand shores indicate that ssp. *novae-zelandiae* is perennial and growth rates are about 1–3 cm month^{-1} (Trowbridge 1996). On Argentinian shores, *C. fragile* is present throughout the year (Rico & Pérez 1993). The phenology of ssp. *tasmanicum* (southern Australia) and ssp. *capense* (South Africa) has not yet been reported.

Introduced Codium fragile

Little has been documented about the phenology of the introduced ssp. *scandinavicum.* Parkes (1975) reported that *C. fragile* ssp. *atlanticum* is perennial to pseudoperennial and exhibits seasonal growth on Irish shores. Thalli die back to a vaucherioid, mossy growth in winter; fronds emerge in February, and maximum development occurs in summer (Parkes 1975). The precise timing varies (Cotton 1912, Parkes 1975, Burrows 1991), undoubtedly due to spatial and temporal variation in environmental factors. On Scottish shores, ssp. *atlanticum* does not necessarily die back in winter (C. D. Trowbridge & C. D. Todd unpubl. data). Yang et al. (1997) reported growth rates for juvenile *Codium* spp., including *C. fragile* ssp. *atlanticum* and ssp. *tomentosoides*, of ~0.6 g month^{-1} (0.22–0.40 g 15 day^{-1}) on British shores. Ramus (1972) reported juvenile growth of 3.6 g month^{-1} (10 g in 12 wk) for ssp. *tomentosoides* on NW Atlantic shores.

In striking contrast to the other subspecies, the phenology of *C. fragile* ssp. *tomentosoides* has been very well studied, particularly on NW Atlantic shores (Moeller 1969, Fralick 1970, Churchill & Moeller 1972, Fralick & Mathieson 1972, 1973, Meimer 1972, Ramus 1972, Lopez 1973, Malinowski & Ramus 1973, Malinowski 1974, Thomas 1974, Hanisak 1977, 1979a,b,c, Hanisak & Harlin 1978, Ames 1979). Growth reportedly occurs for up to nine months per year, generally beginning when seawater temperatures are $>10\,^{\circ}\mathrm{C}$ and salinity >22 ppt (Malinowski & Ramus 1973, Malinowski 1974). The growth of ssp. *tomentosoides* varies from 3–7 cm month^{-1} (Moeller 1969, Thomas 1974) to 1–2 cm day^{-1} during the growing season (Malinowski 1974). The alga often forms high-density populations (up to 20 adult thalli m^{-2}) with a maximum standing crop of 10.8 $\mathrm{kg\,m}^{-2}$ (Malinowski 1974). On NW Atlantic shores, the large thalli die back in winter (Fralick & Mathieson 1972, 1973, Malinowski & Ramus 1973, Hanisak 1977, 1979b) due to environmental and/or ecological factors. The prostrate, filamentous base, however, can persist 2–3 years (Moeller 1969, Fralick 1970). It takes about one month for gametes to settle, germinate, and grow into differentiated axis primordia (Malinowski 1974, Thomas 1974). In some locations, a thick, mossy growth of undifferentiated *Codium* covers much of the shallow infralittoral in winter (Moeller 1969, C. D. Trowbridge pers. obs.).

On New Zealand shores, growth of low intertidal and shallow subtidal thalli of ssp. *tomentosoides* occurs in the spring and summer; in fall, these thalli die back, leaving the perennial holdfast (Dromgoole 1979, Trowbridge 1996). Thalli in high intertidal rock pools continue to grow during the winter ($\sim$1 cm month^{-1}) and do not die back until spring when herbivorous sea slugs (*Placida dendritica*) recruit to their algal hosts and their feeding weakens the thalli (Trowbridge 1996). At present, there is little quantitative data on the alga's dynamics on British shores (Burrows 1991), although studies are in progress at Plymouth, England (A. Albrecht, pers. comm.) and Argyll, Scotland (C. D. Trowbridge & C. D. Todd unpubl. data). Maximum growth of the alga in S England occurs in winter and spring; in summer, the standing crop is "prolific" (Williams et al. 1984). On Scottish shores, *Codium fragile* ssp. *tomentosoides* does not die back, winter fragmentation does not occur, and growth is lush throughout the year (C. D. Trowbridge & C. D. Todd unpubl. data). On Scandinavian shores, thalli fragment in fall or winter (Printz 1952). Thus, the timing of frond senescence varies amongst geographic regions and habitats within regions.

Reproduction

Reproductive structures

Gametangia of *C. fragile* form on sides of utricles (Fig. 1B,C, p. 7). In thalli of the NE Pacific form, gametangia are distributed throughout the thallus but are most abundant 1–5 cm of the tip (Hurd 1916, as *C. mucronatum*). In the introduced ssp. *tomentosoides*, gametangia are also distributed throughout the alga except for the tips when the thallus is actively growing (Moeller 1969, Boerner 1972), but they produce parthenogenetic rather than sexual gametes. For most subspecies, the distribution of gametangia has not been reported. *C. fragile* is considered to be dioecious, although numerous reports of monecious thalli do exist (e.g. Arasaki et al. 1956, Borden & Stein 1969, Kim

1988b). Monecious (bisexual) thalli could develop when coenocytic filaments of male and female germlings intertwine so a morphological individual is a genetic and reproductive mosaic.

There is little information about minimum size of reproductive thalli or the release and settlement of gametes for any subspecies of *C. fragile*. Rico & Pérez (1993) reported that thalli of Argentinian *C. fragile* >300 g ww were all fertile; only 51% of the thalli <50 g were fertile. Comparable information is needed for other subspecies. Gametes are released into sea water, either just before or soon after daybreak (Moeller 1969). For intertidal thalli, gametes are discharged on incoming tides (Smith 1955); the stimulus for subtidal thalli is not known. Gamete concentrations have been reported for only one subspecies (*tomentosoides*): densities range from 1 to 100 gametes ml^{-1} sea water (Moeller 1969). For all subspecies examined, gametes swim "weakly" and generally settle in less than one day (Delépine 1959, Borden & Stein 1969, Moeller 1969, Churchill & Moeller 1972).

Sexual reproduction

Based on detailed cytological work, Williams (1925) reported that (a) *C. fragile* (as *C. tomentosum*) was a diploid gametophyte with 20 chromosomes, (b) meiosis occurred in gametangia, and (c) gametes were haploid with 10 chromosomes. *C. tomentosum* does

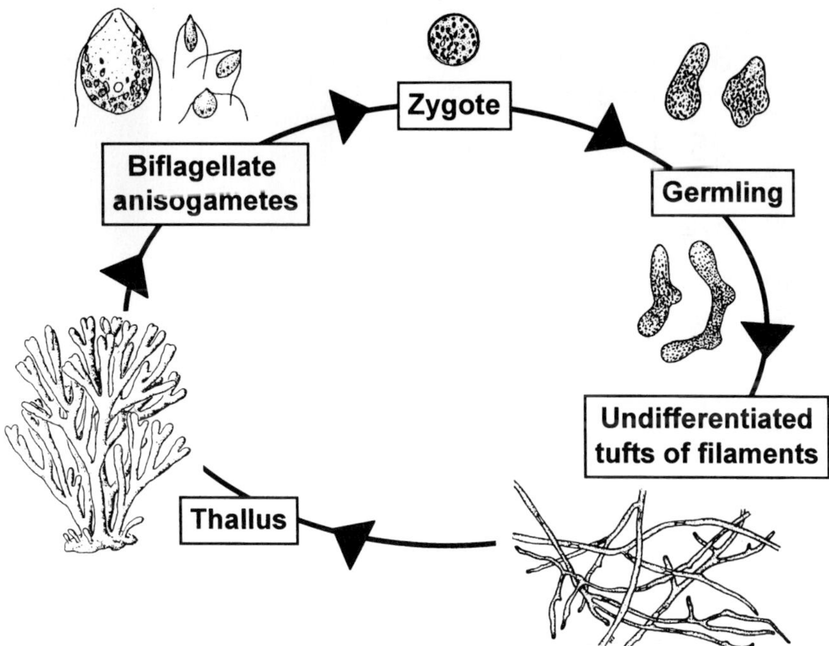

Figure 4 Reproductive life cycle of *Codium fragile*. Female gametes are ~20 μm long; male gametes are 3–4 μm long; zygote is ~10 μm in diameter. The undifferentiated stage or "primary thallus" covers large areas on shore. Figures reproduced from Smith (1944), Arasaki et al. (1956) and Silva & Irvine (1960) with permission from Stanford University Press, Cambridge University Press, and *Journal of Plant Research*.

not occur in Australia (Silva 1955, 1957, Silva & Womersley 1956). Williams's record refers to ssp. *novae-zelandiae* based on utricle morphology and geographic location – low intertidal zone at Bondi Beach, New South Wales, Australia (Silva & Womersley 1956, Trowbridge 1996, P. C. Silva pers. comm.). (The subspecies is not ssp. *tomentosoides*, as suggested by Kapraun & Martin (1987)).

In all subspecies that have been examined (other than ssp. *tomentosoides*), gametangia release anisogamous biflagellate gametes (Fig. 4) (South America: Svedelius 1900; New Zealand: Dellow 1952; Asia: Arasaki et al. 1956, Kim 1988b; Europe: Silva 1955, 1957, Feldmann 1956; NE Pacific: Borden & Stein 1969). For NE Pacific indigenous populations, male gametes are 3–4 μm long and female gametes are 20–25 μm (Borden & Stein 1969); on NW Pacific shores, gametes are slightly larger (Arasaki et al. 1956, Kim 1988b) (Fig. 5). Silva (1957) pointed out that presence of gametes was, by itself, not necessarily indicative of sexual recombination. Although gamete fertilization has not been observed, Borden & Stein (1969) noted that for the NE Pacific *C. fragile*, germination occurred in cultures with fertile male and female branches together but not in unisexual cultures. The apparent zygotes they observed were circular and about 10 μm in diameter (Fig. 4). Thus, for this alga, there is compelling evidence of sexual reproduction.

Germination of zygotes is rapid, and undifferentiated, juvenile thalli develop in 4–7

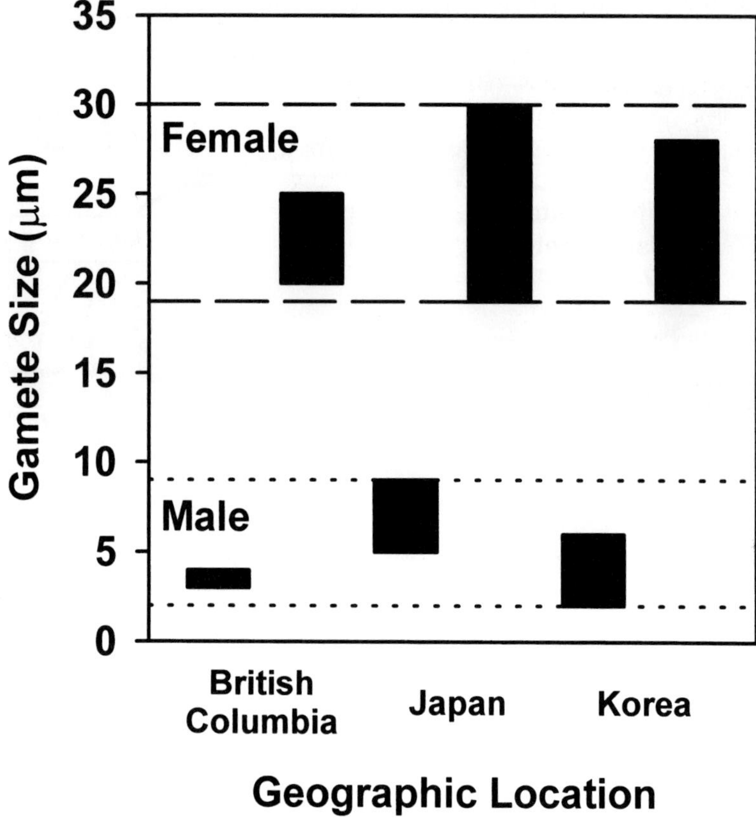

Figure 5 Gamete size from indigenous populations of *Codium fragile*. Data are from Arasaki et al. (1956), Kim (1988b), and Borden & Stein (1969).

days (Fig. 4) (Arasaki et al. 1956, Borden & Stein 1969). This undifferentiated stage is non-septate, lacks utricles, and can persist for months to years without forming utricles (Arasaki et al. 1956, Borden & Stein 1969, Burrows 1991). These juveniles commonly occur on the shore (Silva & Irvine 1960, C. D. Trowbridge pers. obs.). Silva & Irvine (1960) postulated that environmental factors may constrain juvenile stages from maturing and, thus, persist as juveniles. Borden & Stein (1969) further suggested that juvenile stages could serve as a means of vegetative propagation and thallus attachment.

Parthenogenesis

Within Codium fragile

Parthenogenesis occurs in at least three species of *Codium* (Burrows 1991). Although it occurs within *C. fragile* ssp. *tomentosoides*, whether it ever occurs in other subspecies is not known. Hurd (1916) reported a female population of *C. fragile* (as *C. mucronatum*) in Puget Sound, Washington; she speculated that it could be parthenogenetic. Furthermore, Rico & Pérez (1993) reported a female population of Argentinian ssp. *novaezelandiae*. It is an intriguing question whether these reports are the first records of parthenogenesis within subspecies of *C. fragile* other than *tomentosoides*.

Codium fragile *ssp.* tomentosoides

The occurrence of parthenogenesis in *C. fragile* ssp. *tomentosoides* has been reported by numerous authors (Dangeard & Parriaud 1956, Feldmann 1956, Dangeard 1958, Delépine 1959, Boerner 1972, Churchill & Moeller 1972, Ramus 1972, Dromgoole 1979). The biflagellate reproductive cells (also called "swimming cells" or "swarmers") are (18.5–) 20–25 (–30) μm long (Feldmann 1956, Dangeard 1958, Delépine 1959, Boerner 1972, Ramus 1972, Dromgoole 1979), a size comparable to that reported for female gametes in NE Pacific indigenous populations (Borden & Stein 1969).

Male gametes, which in other subspecies of *C. fragile* are much smaller (3–4 μm long) than female gametes (Fig. 5) and are yellowish green rather than dark green, have been reported only twice for ssp. *tomentosoides* (Coffin & Stickney 1967, Prince 1988, 1991). The size of the purported male gametes – 13–14 μm in diameter (Prince 1988, 1991) – is extremely large compared to all the other described male gametes (see Fig. 5), prompting me to question the authenticity of the report. The figures in Prince (1988) clearly show that he was dealing with abnormal development of gametes (P. C. Silva, pers. comm.). Coffin & Stickney (1967) reported bisexual thalli in a population from Boothbay Harbor, Maine, but they did not provide photographs or illustrations. Malinowski (1974) found electrophoretic variation between Maine and Long Island Sound populations of ssp. *tomentosoides*, suggesting to him that sexual reproduction was taking place. The bisexuality of Boothbay Harbor populations remains to be confirmed.

By deploying autoclaved shells on the shore, several authors demonstrated that gamete settlement and germination occur in late summer to early fall (Ramus 1972, Malinowski & Ramus 1973, Thomas 1974). They germinate within a few hours, or at most a few days, following settlement; the early developmental stages have been well illustrated and photographed (Delépine 1959, Moeller 1969, Weber 1969, Churchill & Moeller 1972, Ames 1979). Germlings produce an unconsolidated mass of coenocytic filaments that may cover large areas on the shore or on floating structures (Moeller 1969, Fletcher 1980, Fletcher et al. 1989, C. D. Trowbridge pers. obs.).

Feldmann (1956) suggested that production of parthenogenetic gametes was due to failure of meiosis so that diploid gametes germinate to form diploid thalli. Kapraun & Martin (1987), however, demonstrated that macroscopic thalli of *C. fragile* ssp. *tomentosoides* in North Carolina were haploid and produced haploid reproductive cells. They also suggested that the alga was functionally a tetraploid, based on large chromosome size relative to congeners (Kapraun & Martin 1987, Kapraun et al. 1988). A major weakness of their study is that they made an incomplete comparison between diploid and haploid congeners, rather than diploid and haploid conspecifics. Thus, are sexual diploid thalli of *C. fragile* also functionally tetraploid? The intriguing results and suggestions of Kapraun & Martin (1987) clearly warrant future examination in other subspecies of *C. fragile*.

Irrespective of the type of parthenogenesis, it would appear that ssp. *tomentosoides* comprises a single clone or at most a few clones (Silva 1979). All NW Atlantic individuals analyzed electrophoretically by Malinowski (1974), except those from Boothbay Harbor, Maine, appear to be a single genotype. Goff et al. (1992) and Coleman (1996) reported chloroplast DNA evidence, which indicates that New England and San Francisco Bay populations of ssp. *tomentosoides* are genetically the same. Malinowski (1974) reported that an English population (Bembridge, Isle of Wight) differed electrophoretically from most NW Atlantic populations. To trace whether specific populations are primary *v.* secondary introductions, we will need to explore intra-subspecific genetic variation of *C. fragile* throughout the world.

As the parthenogenetic ssp. *tomentosoides* continues to spread, it becomes locally or regionally sympatric with sexual conspecifics. The risk of interbreeding has not yet been explored. In parthenogenetic kelps, female gametes (from gametophytes derived from parthenogenetic haploid sporophytes) could be fertilized with male gametes (Lewis et al. 1993). Thus, interbreeding is theoretically possible. Another possibility is that asexuality may be facultative and that reversion to sexuality may occur. Given the extremely serious pest-like behaviour of ssp. *tomentosoides*, the potential for the alga to interbreed with native conspecifics (with broader tolerances to wave action) warrants immediate appraisal. Finally, one puzzling point is that ssp. *tomentosoides* grows larger than other subspecies and indigenous forms (Table 3, p. 12). This is the opposite pattern of what is expected, given that ssp. *tomentosoides* is haploid (Kapraun & Martin 1987, Kapraun et al. 1988) and that ssp. *novae-zelandiae* (Williams 1925) and presumably other indigenous forms are diploid. Lewis et al. (1993) found that normal diploid sporophytes were typically longer and wider than parthenogenetic, haploid sporophytes of the kelp *Laminaria japonica*.

Reproductive period

Indigenous Codium fragile

Do introduced subspecies have a longer fertility period than native congeners? For indigenous populations of *C. fragile* in southern California, Gunnill (1980) reported that recruitment occurred from May to November, indicating that the reproductive period was presumably in winter and spring. Gunnill admitted, however, that some of the apparent recruits may have been fronds that regenerated from perennating bases hidden in the coralline algal turf. In more northerly locations (British Columbia, Canada), Scagel (1966) reported mature gametangia from mid-May until September and Borden & Stein (1969) observed gametangia from April to November. On Japanese and Korean shores (Fig. 6), *C. fragile* has gametangia all year (Arasaki et al. 1956,

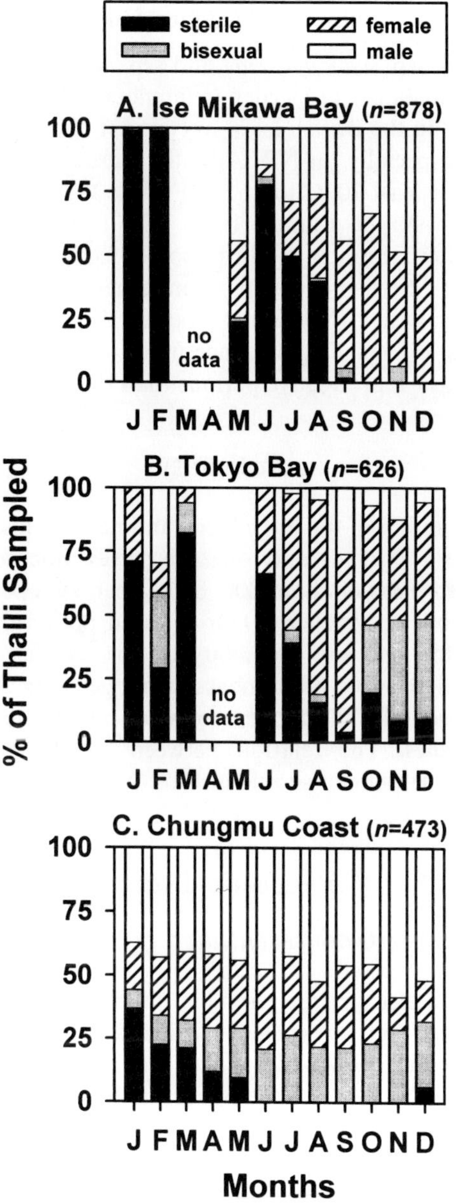

Figure 6 Reproductive phenology of *Codium fragile* on Pacific coast of Honshu, Japan (A, B) and South Korean shores (C). Sample size indicates number of thalli examined during year. Data from Arasaki et al. (1956) and Kim (1988b).

Kim 1988b). On Argentinian shores, most thalli have gametangia from late spring to winter (Rico & Pérez 1993). Information on the reproductive period for the other native forms is not available.

Introduced Codium fragile

Data for the introduced ssp. *scandinavicum* and ssp. *atlanticum* are limited; the former is fertile from July to September in Sweden (references in Lund 1940) whereas the latter is fertile all year around in the British Isles. In contrast, there is a wealth of information on ssp. *tomentosoides*, especially on NW Atlantic shores. Gametangia are produced at water temperatures of at least 10 °C to 16 °C (Fralick 1970, Boerner 1972, Churchill & Moeller 1972, Malinowski & Ramus 1973). Thalli release their gametes in fall (Churchill & Moeller 1972, Fralick & Mathieson 1973, Thomas 1974, Hanisak 1979b) at temperatures >12 °C. Settlement of parthenogenetic gametes and germling establishment occurs from August to November (Ramus 1972, Malinowski & Ramus 1973, Thomas 1974, Hanisak 1979b). This evidence indicates that the three introduced subspecies do not have longer reproductive periods than those of native conspecifics.

Fecundity

Comparisons of the fecundity of the introduced and native subspecies of *C. fragile* have not yet been made. Data exist for only one subspecies. Boerner (1972) reported peak fecundity of ssp. *tomentosoides* of about 300–800 gametangia per cm alga from September to January; Churchill & Moeller (1972) reported peak values of 1200–1600 gametangia per cm alga for one site and 400–800 for another site. Their method of sampling 5–6 cm apical sections of branches, homogenizing them, and counting the gametangia in 1 ml subsamples should be used as a standard technique for future comparative fecundity studies. The relation between gametangial abundance, gamete release (dehiscence), and settlement should also be quantified (see Boerner 1972).

Vaucherioid stage

Undifferentiated growth

The coenocytic filaments of *C. fragile* resemble the coenocytic, mat-forming alga *Vaucheria*; hence the *Codium* mats are called vaucherioid stages (see Arasaki et al. 1956, Silva & Irvine 1960, Moeller 1969). These mats occur on any hard substrata, including rocks, rafts, macroalgae, barnacles, gastropods, and bivalves (Borden & Stein 1969, Moeller 1969, Fletcher et al. 1989).

Using extensive culture studies, Lopez (1973) and Ames (1979) demonstrated that growth of vaucherioid *C. fragile* ssp. *tomentosoides* requires specific types of bacteria. Different forms of nitrogen did not stimulate growth of ssp. *tomentosoides* in axenic culture; stimulation requires exogenous compounds produced by specific types of bacteria (*Pseudomonas* type II) (Lopez 1973, Ames 1979). The implication is that inter-subspecific variation and intra-subspecific variation in growth of *Codium fragile* may be determined by the indirect effects of the microbial assemblage rather than the direct effects of environmental factors *per se*.

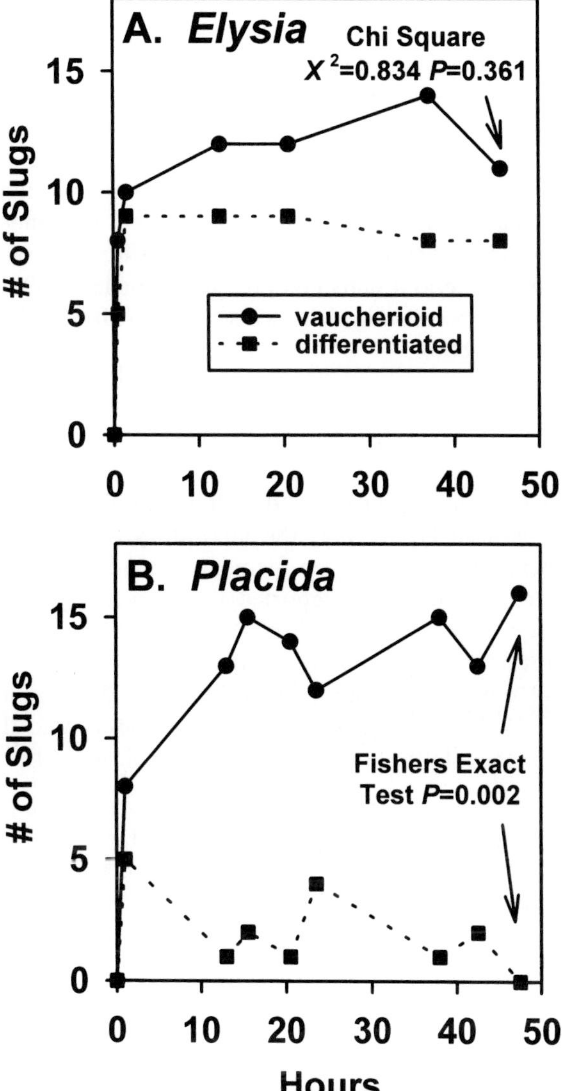

Figure 7 Feeding preferences of individual herbivorous slugs, A, *Elysia viridis* and B, *Placida dendritica* when offered pairwise choices of *Codium fragile* ssp. *tomentosoides* with vaucherioid (undifferentiated) and non-vaucherioid (differentiated) sections. Statistical tests were done only at last time period.

Differentiation

The undifferentiated juvenile stage generally requires wave action or current flow to produce utricles and develop into a macroscopic alga (Ramus 1972, Lewis 1982). Steele

(1975) reported that thallus differentiation occurs when sea water is not enriched with nutrients or trace metals; germlings and branch tips grown in enriched sea water exhibit vaucherioid growth. Hanisak (1977, 1979b) and Yang et al. (1997) reported that the vaucherioid stage of *Codium* spp. is tolerant of a wide range of environmental conditions (temperature, irradiance, and salinity); while Hanisak (1977, 1979b) concluded that adult thalli were more tolerant than germlings, Yang et al. suggested the opposite. However, Yang et al. (1997) presented few data, no summary or inferential statistics, and their species and collection localities were confounded; consequently, their conclusions should be viewed cautiously. "Green bumps" of differentiated axis primordia arise from vaucherioid mats (Wood 1962, Moeller 1969, Ramus 1971, 1972, Thomas 1974). Buds can be dislodged by ice scouring (Moeller 1969) and may be a form of vegetative reproduction.

De-differentiation

Codium fragile also has the capacity to "de-differentiate" (*sensu* Ramus 1972). The development has been seen typically in the laboratory (Moeller 1969, Ramus 1972, C. D. Trowbridge pers. obs.) where green filaments extend out from the alga's surface by up to a centimetre. Ramus (1972) reported that utricles "sprout narrow apically-growing filaments from their distal ends" (p. 48). Populations of ssp. *tomentosoides* in Scottish sea lochs not only form extensive undifferentiated carpets in the low intertidal and shallow subtidal zone but also adult thalli often "de-differentiate" at tips, centre, or base of fronds (C. D. Trowbridge pers. obs.). I hypothesize that the undifferentiated growth may be due to the relative absence of water turbulence in the lochs.

Ecological relevance

I investigated experimentally the ecological significance of these vaucherioid filaments, focusing on the primary grazers on Scottish shores – the stenophagous ascoglossan (= sacoglossan) sea slugs *Elysia viridis* and *Placida dendritica* that consume *Codium fragile* (see pp. 43–4). Individual slugs were offered a pairwise choice of differentiated and partially undifferentiated fronds of ssp. *tomentosoides*, and slug preferences were recorded for a two-day period (Fig. 7). Adult *Elysia viridis* showed no preference between algal fronds, but the smaller slug *Placida dendritica* strongly preferred the loose, dissociated filaments (Fig. 7). These experimental results (C. D. Trowbridge & C. D. Todd unpubl. data) are consistent with my observations of the slug distributions on the shore.

Vegetative propagation

Thallus fragments, vegetative buds, individual utricles, and medullary filaments can all de-differentiate (*sensu* Ramus 1972) and regenerate (Borden & Stein 1969, Fralick 1970, Yotsui & Migita 1989). Although Williams (1948) and Hanisak (1979b) questioned the ecological role of actual reattachment of branch fragments in the field, regeneration of attached utricle and medullary filaments has been demonstrated (Yotsui & Migita

1989). These modes of propagation occur not only in various subspecies of *Codium fragile* but also in different species of *Codium* (e.g. Lund 1940, Williams 1948).

Thallus fragmentation occurs in several subspecies and at different times in different areas. In NW Atlantic populations of ssp. *tomentosoides*, fragmentation occurs in winter or early spring (Fralick 1970, Churchill & Moeller 1972, Fralick & Mathieson 1972, 1973, Clark 1975, Hanisak 1977, 1979b). However, in New Zealand populations of ssp. *tomentosoides*, the time of fragmentation varies with habitat: thalli in the low intertidal and shallow subtidal zones fragment in late summer to autumn (Dromgoole 1975, Trowbridge 1996) while those in high intertidal pools fragment in spring when slugs recruit to hosts (Trowbridge 1996). On Oregon shores, fragmentation in native *C. fragile* occurs in spring (Trowbridge 1993). On Korean shores, fragmentation occurs in winter (Oh et al. 1987).

Despite the fact that many authors cite the "low temperature" fragmentation of ssp. *tomentosoides* reported by Fralick (1970) and Fralick & Mathieson (1972, 1973), the original data were based on only two "representative" thalli: a large thallus that fragmented and a small one that did not. Furthermore, the authors showed only an association between low temperature and fragmentation; they did not present any correlation coefficients or experimental evidence supporting the inferred causal agent.

Hanisak (1977) also investigated thallus fragmentation of ssp. *tomentosoides*. He reported that 35–58% of the standing crop of the alga was lost over the winter; this represents 45–58% of the alga's annual production. Hanisak (1977) also noted that fragmentation (a) occurs more frequently in older thalli than younger thalli and (b) decreases with increased water depth. He considers the introduced alga a "valuable part of the coastal ecosystem" (Hanisak 1977, p. 17) because of its large contribution of tissue to the detrital food chain in winter. An implicit assumption, however, is that decomposing thalli of *C. fragile* ssp. *tomentosoides* are not toxic to detritivores; living thalli do contain a variety of toxic compounds (see pp. 48–53).

Ecophysiology

Silva (1955) suggested that the wide geographic distribution of *C. fragile* (e.g. 29°–70° latitudes) and the broad range of habitats occupied by the alga indicated that the species (in its entirety) has high physiological tolerances. Other authors have extended this hypothesis to ssp. *tomentosoides*: namely, the introduced subspecies is such a successful invader because of its physiological flexibility and its tolerance to a range of environmental conditions. Although this hypothesis and supporting experimental data are consistent with the attributes contributing to invasion success, an implicit assumption is that the non-invasive conspecifics are less physiologically tolerant. To evaluate the assumption, I review published data on the species' tolerances.

Salinity

Codium fragile ssp. *tomentosoides* inhabits marine and estuarine shores. For example, on NW Atlantic shores, the alga occurs in Long Island Sound, New York at salinities of 18–30 ppt (Malinowski 1974) and in Great South Bay, New York at salinities of 25–30 ppt (Thomas 1974). Moeller (1969) reported tolerances of 17.5–40 ppt with partial survival at 12.5 ppt and 15 ppt based on survival of pieces of adult thalli and germination of parthenogenetic gametes. Steele (1975) found maximum growth at 30–40 ppt. In contrast, Yang et al. (1997) reported that the dissociated juvenile stage of ssp. *tomentosoides* grew optimally at 23.8 ppt whereas ssp. *atlanticum* grew optimally at 34.0 ppt. Other subspecies of *C. fragile* are presumed to be stenohaline, given their marine distributions. For example, Korean populations studied by Kim (1988b) experienced seasonal salinities of 32.9 ppt to 34.5 ppt. Experiments on the salinity tolerance of indigenous *C. fragile* (adults, gametes, and juveniles) are sorely needed.

Seawater temperature

C. fragile occurs in a wide range of geographic regions at markedly different latitudes: 29 °S to 54 °S (Bright 1938, Boraso & Piriz 1975) and 33 °N to 70 °N (Lund 1940, Fægri & Moss 1952, Printz 1952, Stellander 1969, Searles et al. 1984). Some areas have a broad seawater temperature range whereas others have a narrow range. Ssp. *tomentosoides* in the NW Atlantic encounters the broadest temperature range with reported values of −2 °C to 27.5 °C (Moeller 1969, Malinowski 1974, Thomas 1974, Searles et al. 1984). On New Zealand shores, the temperature range in the Hauraki Gulf where ssp. *tomentosoides* became established was about 12–21 °C. On Scottish shores, the alga is most abundant in Argyll, which has a comparatively mild climate (~7–14 °C) given its high latitude (56 °N).

Numerous authors have referred to ssp. *tomentosoides* as a warm-water alga (e.g. Fralick 1970, Boerner 1972) and predicted that it would not spread north of Boothbay Harbor, Maine and that the Maine population occurred in a "hot spot" or warm-water "pocket" (Carlton & Scanlon 1985). Yet, the alga subsequently spread north to Nova Scotia (Bird et al. 1993). Similar predictions were made for the British Isles; yet, ssp. *tomentosoides* spread north to the Outer Hebrides and Orkney Islands. Fægri & Moss (1952) suggested that warming trends or current shifts may enable warm-water subspecies such as *C. fragile* to penetrate high-latitude areas. Ssp. *atlanticum* and ssp. *scandinavicum* seem to be cold-water algae, given their geographic distributions.

Malinowski (1974) conducted a series of laboratory growth experiments with adult thalli (4 cm pieces) and germlings (1–2 month-old cultures) of ssp. *tomentosoides* (Fig. 8). Adult thalli from the Maine population grew best at 4 °C whereas the other populations of ssp. *tomentosoides* from the NW and NE Atlantic grew best at 24 °C; juvenile stages all grew best at 24 °C (Malinowski 1974). Persson (1973) also compared temperature-specific growth rates of ssp. *tomentosoides* from Maine and Cape Cod; her results were not significant but tended to support those of Malinowski. Yang et al. (1997) also conducted growth experiments for juvenile forms of *C. fragile*; they reported that ssp. *tomentosoides* grew best at 25 °C and ssp. *atlanticum* at 20° C. The lethal high temperature of ssp. *tomentosoides* is 34 °C (Moeller 1969, Persson 1973).

Of the native forms, populations of South Korean *C. fragile* encounter water temperatures of 7.5–25.4 °C (Kim 1988b) and on Chinese shores, 10–22 °C (Liu et al. 1995). Japanese populations in western Kyushu inhabit waters of 13–26 °C (Yotsui & Migita 1989); populations on southern Kyushu shores experience ranges of 15–30 °C (Noro & Nanba 1989). On NE Pacific shores, native *C. fragile* encounters temperatures of 12.5 °C to 21 °C at Santa Catalina (Theis 1985) and 9 °C to 13 °C in Oregon (Trowbridge 1992). On South African shores, ssp. *capense* encounters temperatures of

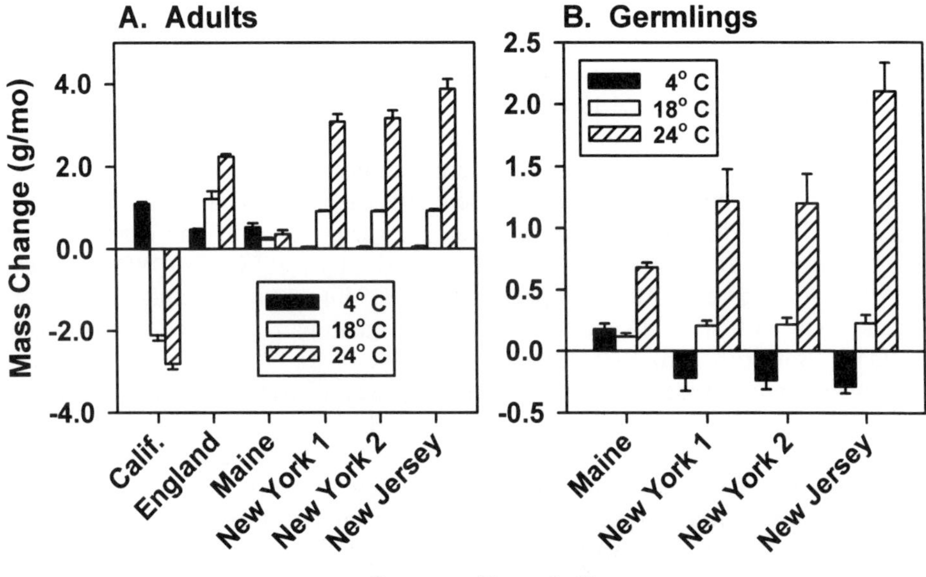

Figure 8 Experimental results of seawater temperature on growth rates of *Codium fragile* from different populations (Malinowski 1974). Although not recognized by Malinowski, the California population at Bodega Head was the indigenous form of *C. fragile*; English and New England populations were ssp. *tomentosoides*.

about 11–19 °C (Isaac 1937, 1949). Thus, non-weedy populations occur in coastal waters ranging from 7 °C to 30 °C.

Malinowski (1974) conducted temperature tolerance experiments with native thalli from northern California (Fig. 8A); maximum growth occurred at 4 °C. At higher temperatures, the native alga lost weight (Fig. 8A); 18 °C and 24 °C are both substantially warmer than seawater temperatures at Bodega Head, California. What has not been realistically assessed is the extent to which acclimation and the dynamics of temperature regime affect temperature tolerance within and among different subspecies of *C. fragile*. Insufficient published information is available to assess whether or not the introduced subspecies have a significantly greater range of temperature tolerance than the indigenous conspecifics.

Tidal heights

The tidal height at which a species lives is a function of the species' tolerance to high and low temperature and dehydration. For most subspecies of *C. fragile*, this type of tolerance information is not available. Regarding dehydration, for most of the subspecies examined, the thallus water content of the thalli is similar – 93–96% of alga's wet weight (e.g. Moeller 1969, Delgado & Duville 1977, Dromgoole 1982, Herbreteau et al. 1997, C. D. Trowbridge, unpubl. data). Thus, water loss would be dictated by the surface area of the thallus, and there are subtle differences in branch thickness, degree of branching, etc. (see Trowbridge 1996).

Indigenous populations of *C. fragile* occur intertidally and subtidally on Japanese, Korean, and Chinese shores (Meiling & Tseng 1984, Kim 1991, Liu et al. 1995). The NE Pacific populations occur in the mid and low intertidal as well as the subtidal (Hurd 1916, Silva 1951, Scagel 1966, Gunnill 1980, 1985, Theis 1985, Trowbridge 1989, 1993, Stewart 1991). Ssp. *novae-zelandiae* on New Zealand and SE Australian shores are common in the low intertidal and the subtidal, occurring in a wide range of wave exposures from protected to exposed (Dellow 1952, Silva & Womersley 1956, Dromgoole 1979, Trowbridge 1995, 1996). Ssp. *capense* occurs in pools and channels of the lower littoral zone and sublittoral fringe (Bright 1938, Eyre 1939, Silva 1959, Lamberth et al. 1995, Stegenga et al. 1997). Stephenson et al. (1940) reported the alga in the lower barnacle zone and in small pools. Finally, on Argentinian shores, *C. fragile* occurs intertidally (Boraso & Piriz 1975, Delgado & Duville 1977).

Are introduced subspecies distributed fundamentally differently than are natives subspecies of *Codium fragile*? Ssp. *tomentosoides* on British shores occurs in pools and rocky surfaces in mid and low intertidal and shallow subtidal (Burrows 1991). On NW Atlantic shores, the alga occurs primarily subtidally (Carlton & Scanlon 1985), although there are several reports of the alga in high intertidal rock pools. Wassman & Ramus (1973a), however, report an upper limit for ssp. *tomentosoides* on vertical rock faces near the mean high water level). On New Zealand shores, the introduced alga occurs in high pools, on emergent low intertidal shores, and in shallow subtidal areas (Trowbridge 1995, 1996, references therein). On Scandinavian shores, ssp. *scandinavicum* occurs subtidally (Lund 1940) but the alga has not been extensively investigated. Finally, on British and Norwegian shores, ssp. *atlanticum* occurs in high, mid, and low intertidal pools as well as in the subtidal (Burrows 1991); thalli on emergent substratum are uncommon. On NE British and NE New Zealand shores, ssp. *tomentosoides* also occurs in high intertidal pools (Hardy 1990, Trowbridge 1995, 1996). Thus, from the published information, I conclude the broad distributions of the different subspecies are quite similar.

In geographic regions where intertidal organisms often freeze in the winter, *C. fragile* occurs primarily subtidally (e.g. NW Atlantic shores); in areas where freezing is unusual (e.g. Argyll, Scotland and North Island, New Zealand), the alga commonly forms dense intertidal stands. The only experimental survivorship data (Fig. 9) come from Moeller (1969) and C. D. Trowbridge & C. D. Todd (unpubl. data). Moeller transplanted ssp. *tomentosoides* at different tidal levels and recorded the number of days thalli survived (Fig. 9A). On New York shores, the thalli survived well only at low intertidal and shallow subtidal levels. In February 1997, I transplanted ssp. *atlanticum* in high tidepools and directly above the waterline of pools at St Andrews Bay, Scotland and monitored their performance after 2.5 weeks (Fig. 9B). The emergent thalli not covered by fucoids rapidly desiccated and deteriorated; emergent thalli under algal canopies performed better but eventually died; submerged thalli grew and persisted for up to six months.

Nutrients and auxins

Nitrogen sources and utilization

There is considerable experimental work on the nutrient relations of *C. fragile* ssp. *tomentosoides* on NW Atlantic shores (Malinowski & Ramus 1973, Malinowski 1974, Head & Carpenter 1975, Steele 1975, Hanisak 1977, 1979a,b,c, Hanisak & Harlin 1978,

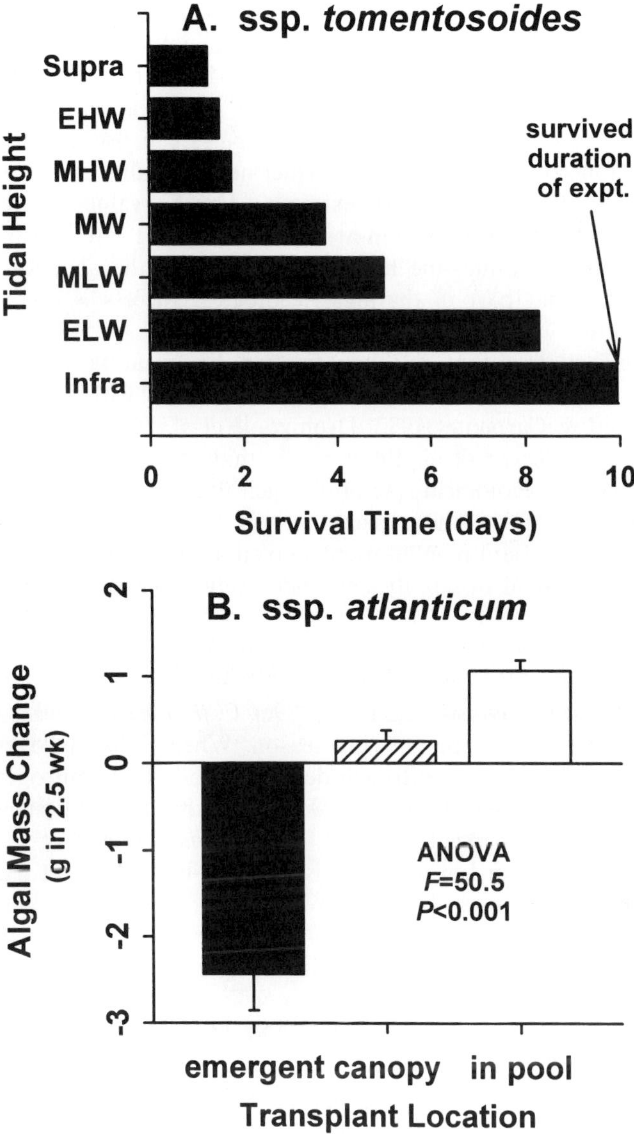

Figure 9 A, Mean survival time of *Codium fragile* ssp. *tomentosoides* transplanted to different tidal levels on shores of New York, USA (Moeller 1969). Thalli transplanted to different infralittoral areas typically survived the duration of the experiments (which was not specified). Data calculated from Table 1 in Moeller (1969). B, Mass change of algal transplants (ssp. *atlanticum*) into high intertidal pools and directly above the pool waterlines (submerged and emergent thalli, respectively). Experiment conducted in winter 1997 at St Andrews Bay, Scotland (C. D. Trowbridge & C. D. Todd, unpubl. data).

Gerard et al. 1990). There is a positive relationship between algal growth and nitrate-nitrite concentrations, and ammonia may also be an important alternative nitrogen source (Malinowski & Ramus 1973). In a series of laboratory experiments, Hanisak (1979b) demonstrated that nitrate, nitrite, ammonia, and urea are equally good sources of nitrogen to the introduced alga; ssp. *tomentosoides* can use all forms of nitrogen simultaneously (Hanisak & Harlin 1978). Furthermore, the alga can compete with phytoplankton for nitrogen because of its extremely low K_s values (Hanisak & Harlin 1978). K_s is the substrate concentration at which the uptake rate is half the maximum value; the lower the K_s value, the higher the affinity of the alga for nitrogen. The authors hypothesize that part of the alga's ecological success is due to its nitrogen acquisition attributes.

Several workers suggest that nitrogen fixation by cyanobacteria and/or heterotrophic bacteria may provide *C. fragile* ssp. *tomentosoides* with some of its nitrogen requirements (Head & Carpenter 1975, Dromgoole et al. 1978, Asare & Harlin 1983) although others (e.g. Gerard et al. 1990) report that little nitrogen is actually translocated to the alga. The ecological role of nitrogen fixation is consequently controversial, particularly as it affects the capacity of ssp. *tomentosoides* to persist through spatial and temporal variation in nutrient depletion. What does seem unequivocal, however, is that the spread of ssp. *tomentosoides* may be enhanced by eutrophication (Ramus 1971).

Nitrogen limitation

Based on nitrogen tissue analysis (Hanisak 1979c), *C. fragile* ssp. *tomentosoides* may be nitrogen-limited for much of the growing season. When tissue nitrogen was $<1.9\%$ nitrogen (DW basis), the alga was nitrogen-deficient; when tissue nitrogen was $>1.9\%$, the alga had nutrient reserves (Hanisak 1979c). Using this critical tissue nitrogen level, other authors have hypothesized that ssp. *tomentosoides* (Head & Carpenter 1975, Asare & Harlin 1983, Gerard et al. 1990) is nitrogen-limited for at least part of the year. Yet, Steele (1975) found that the growth rate of ssp. *tomentosoides* did not increase substantially with increased nitrogen in culture.

Details for indigenous populations are also meagre. *C. fragile* contained 2.1–2.4% tissue nitrogen on Argentinian shores (Delgado & Duville 1977), 0.9–2.6% nitrogen on Oregon shores (Wheeler & Björnsäter 1992), and 2.5% nitrogen on Japanese shores (Arasaki & Arasaki 1986). If the critical tissue nitrogen level from ssp. *tomentosoides* can be extrapolated to these other geographically distinct conspecifics, then the Oregon population is nitrogen-limited during part of the year whereas other populations are not. Before this hypothesis can be accepted, however, the assumption that the critical tissue nitrogen level is spatially and temporally robust needs to be tested.

On NE Atlantic shores, ssp. *tomentosoides* becomes extremely bleached in summer: thalli are frequently lime-green rather than dark green (Williams et al. 1984, C. D. Trowbridge pers. obs.). Such bleaching can be caused by (a) photo-inhibition and photo-oxidative damage or (b) nutrient-induced chlorosis (Williams et al. 1984). Sealey et al. (1990) demonstrated that the former does not occur in intact thalli (see next section), implying that nutrient stress accounts for the pale coloration. Presumably the alga scavenges the accessory photosynthetic pigments for nitrogen. Whether this process is also true for the other two introduced subspecies (ssp. *atlanticum* and ssp. *scandinavicum*) has not been explored though lime-green thalli of ssp. *atlanticum* are common, particularly in spring (e.g. Cotton 1912, Lund 1940).

Other compounds

Nutrients are not the only stimulatory compounds in sea water. Hanisak (1979a) reported that IAA (indole-3-acetic acid) is present in North Carolina coastal sea water, and its concentration varies seasonally. IAA stimulates growth in *C. decorticatum* (Williams 1952) and *C. fragile* (Hanisak 1979a) with optimal response at 10^{-6} M. The ecological significance of growth regulators such as IAA in coastal waters could alter the distribution and abundance of algal growth, particularly of the invasive subspecies of *C. fragile*. Finally, high concentrations of heavy metals may reduce algal growth (Steele 1975).

Light

Irradiance and photoperiod

C. fragile is very optically dense (Ramus et al. 1976a,b, Ramus 1978, Littler & Arnold 1980, 1982). Neither the invasive ssp. *tomentosoides* nor the NE Pacific native alga exhibits photo-inhibition at high irradiance levels (Ramus et al. 1976a, Ramus 1978, Littler & Arnold 1982, Sealey et al. 1990, but see Steele 1975). High light tolerance may be due to the spatial arrangement of chloroplasts around utricular vacuoles (Fig. 3 in Sealey et al. 1990) that results in optical scattering of light within the thallus and self-shading (Yokohama et al. 1977, Ramus 1978). In fact, the alga's light absorption spectrum is typical of deep-water algal species or shade-dwelling species, despite the fact that *C. fragile* typically occurs in direct light in shallow water (Yokohama et al. 1977). At high irradiance, the chloroplasts in the congener *C. decorticatum* migrate from the utricle apices to lateral areas, thus protecting the alga from photo-oxidative damage (Williams 1948). This mechanism may also occur in *C. fragile*.

Photo-inhibition of juveniles does occur (Hanisak 1979b). For example, optimal growth for ssp. *tomentosoides* in culture is $88\,\mu\mathrm{E\,m^{-2}\,s^{-1}}$; optimal growth for ssp. *atlanticum* is $44\,\mu\mathrm{E\,m^{-2}\,s^{-1}}$ (Yang et al. 1997); at higher irradiances, growth declines (although Yang et al. 1997 do not show the results). Hanisak (1979b) demonstrated that the growth rate of ssp. *tomentosoides* is strongly inhibited by high irradiance at high temperatures (24 °C and 30 °C). Furthermore, as photoperiod increases, the degree of inhibition strongly increases (Hanisak 1979c). The implications of these results are that there would be strong latitudinal differences in the growth of ssp. *tomentosoides*, directly related to these physiological effects. Whether there are spatial, temporal, and subspecific variation in the alga's photobiology has not been adequately explored.

Light absorptance

The propensity of *C. fragile* ssp. *tomentosoides* to accumulate gases within its branch tips is an important spectral adaptation, enhancing the alga's capture of light (Ramus 1978). Ramus compared the alga's light absorptance in branch tips with gas to experimental branches where he replaced the gas with sea water. When the alga was gas-filled, absorptance was close to 90–100%; when branches were filled with sea water, absorptance declined, particularly in the 500–600 nm range (Ramus 1978). The generality of this adaptation to other subspecies is not well known. Ssp. *tomentosoides* and ssp. *atlanticum* accumulate gases in their branches but information for other subspecies is meagre.

Photopigment content

Photopigment concentrations of chlorophyll *a* and *b* as well as the major accessory pigments (siphonoxanthin, siphonein, and neoxanthin) peak in the spring (on a biomass basis) in *C. fragile* ssp. *tomentosoides* (Benson & Cobb 1983). Furthermore, pigment concentrations are greater in smaller, younger fronds than larger, older ones (Benson & Cobb 1981). Despite large spatial and temporal variations in chlorophyll content, the ratio of chlorophyll *a* to *b* is relatively constant (1.5–1.8 to 1) (Benson & Cobb 1983).

C. fragile occurs in habitats with markedly different irradiances. Ssp. *tomentosoides* exhibits a sun/shade adaptation: thalli in low irradiance habitats have higher chlorophyll *a* and *b* concentrations than thalli in high irradiance habitats (Ramus et al. 1976b). In algal transplant experiments, ssp. *tomentosoides* changes the amount of chlorophyll but not the ratio of chlorophyll *a* to *b* (Ramus et al. 1976b). Large amounts of accessory pigments (siphonaxanthin, siphonein, etc.) make the alga good at collecting low irradiances (Yokohama et al. 1977).

Invasion biology

The processes required for introduced species to become established and spread are complex (Carlton 1979, Mollison 1986, Williamson & Brown 1986, Ashton & Mitchell 1989, Lodge 1993, Ribera & Boudouresque 1995). Introduction does not always lead to establishment. Other species become established but do not spread whereas some not only spread but also become ecological and/or economic pests. In contrast to the plethora of ecological theory and empirical data investigating all aspects of the process for terrestrial introductions, marine research has focused on the vectors of introduction and attributes rendering species good invaders. Thus, what renders a marine community vulnerable to invasion is poorly known.

The species *C. fragile* has a variety of subspecies ranging from highly invasive to non-invasive. Two subspecies have been introduced to NE Atlantic shores in the last century: ssp. *tomentosoides* and ssp. *scandinavicum* (Silva 1955, 1957). A third subspecies, ssp. *atlanticum*, was probably introduced, being first recorded in the NE Atlantic about two hundred years ago (Silva 1955, 1957). All three subspecies have strong affinities with NW Pacific populations of *C. fragile*. I summarize the introduction, establishment, spread, and persistence of the introduced subspecies and discuss the economic and conservation implications of the aliens. The causal processes that may have facilitated *Codium* establishment and spread are presented in subsequent sections (pp. 42–5 and pp. 45–7).

Introduction

Codium fragile *ssp.* atlanticum

C. fragile ssp. *atlanticum* first appeared on Irish shores about 1808 and, based on herbarium specimens, spread around the British Isles and to Norway (Silva 1955, 1957, Maggs 1986, C. D. Trowbridge & C. D. Todd pers. comm.). Silva (1955) suggested that the apparent source region was Japan, based on morphological similarity of herbarium specimens. The vector of introduction from the Pacific to SW Ireland is not known. Eno et al. (1997) suggested that the alga was introduced with shellfish but the occurrence of ssp. *atlanticum* primarily in high intertidal rock pools indicates that shellfish are a rather unlikely vector. Because the subspecies is not common in harbours, transport via ship hulls and/or ballast water seems extremely unlikely. However, *C. fragile* ssp. *atlanticum* may have been packed around shellfish shipped to Ireland. Given that the introduction occurred so long ago, the vector will probably never be elucidated. Even vectors with a low probability of success may be the source of introduction given a high frequency of transport and/or a large number of propagules per innoculum (Mollison 1986).

The status of *C. fragile* ssp. *atlanticum* is a bit ambiguous. Silva (1955) suggested that the alga may have been introduced; later, he suggested that the alga might be native to Norway (Silva 1957). Subsequent authors have not explicitly acknowledged this uncertainty. Given the criteria for recognizing introduced species (Chapman & Carlton 1991, Boudouresque 1994, Ribera & Boudouresque 1995), what is the evidence that ssp. *atlanticum* was actually introduced? The primary fact (Table 4) is that numerous phycologists reported the appearance of a new species on the shore (e.g. Cotton 1912). The alga's large size and occurrence in high intertidal pools made the alga's appearance and subsequent spread easier to record than those of the other two introduced subspecies. *C. fragile* ssp. *atlanticum* was also highly localized at individual sites. Given that *C. fragile* apparently did not occur in the North Atlantic or Mediterranean at that time (based on archived herbarium specimens), the appearance of a mucronate species (pointed utricle apices) of *Codium* in the British Isles represented a large geographic discontinuity and exotic evolutionary origins (Table 4). Finally, ssp. *atlanticum* is a noteworthy alga as it often occurs on wave-exposed rocky shores; this habitat is unusual for introduced species (but common for the native conspecifics).

Table 4 Criteria for recognizing introduced species (and subspecies) based on Chapman & Carlton (1991), Boudouresque (1994), and Ribera & Boudouresque (1995).

Criteria	ssp. *atlanticum*	ssp. *tomentosoides*	ssp. *scandinavicum*
New species in area	✓	✓	✓
Geographic discontinuity	✓	✓	✓
Localized sites	✓	✓	??
Initial expansion of localized range after introduction	✓	✓	??
Insufficient passive dispersal to account for observed distribution	✓	✓	✓
Rapid growth	✓??	✓	??
Proximity of potential source of introduction; associated with human mechanisms of dispersal	??	✓	??
Prevalence on or restricted to new or artificial habitats	no	✓	??
Associated with or dependent on other introduced species	no	✓	??
Incomplete genetic variability	??	✓	??
Genetic identify between distant populations	??	✓	??
Exotic evolutionary origins	✓	✓	✓

Codium fragile *ssp.* tomentosoides

The second introduced subspecies of *C. fragile* to appear on European shores was ssp. *tomentosoides.* It was first observed in the Netherlands *c.* 1900, Denmark in 1919, Sweden in 1933, England in 1939, Ireland in 1941, Norway in 1946, and Scotland in 1953 (Lund 1940, Fægri & Moss 1952, Printz 1952, Silva 1955, 1957, references therein). *C. fragile* ssp. *tomentosoides* appeared in the Mediterranean in 1950 (Boudouresque 1994, Ribera 1994, Verlaque 1994, Ribera & Boudouresque 1995). Because the alga was introduced to areas with congeners and even conspecifics that were morphologically similar (at least superficially), ssp. *tomentosoides* represented a partially cryptic invader on European and Mediterranean shores. The alga was noted because of its rapid expansion and large standing crop (Printz 1952, Lewis & Powell 1960, Lewis 1964). The differences between subspecies of *C. fragile* and between the sympatric congeners *C. tomentosum* and *C. vermilara* were not clarified until 1955–57 (Silva 1955, 1957). Records before 1955 must, therefore, be viewed cautiously. Ssp. *tomentosoides* currently occurs from northern Africa to northern Norway in the NE Atlantic and throughout the Mediterranean; details of the alga's spread on these shores are reviewed elsewhere (Norton 1978, Maggs 1986, Ribera 1994, Verlaque 1994, Ribera & Boudouresque 1995, C. D. Trowbridge & C. D. Todd pers. comm.).

Ssp. *tomentosoides* next appeared on NW Atlantic shores in 1957, NE Pacific shores in 1977, New Zealand shores in 1973, and Australian shores about 1995 (Bouck & Morgan 1957, Dromgoole 1975, Silva 1979, Carlton & Scanlon 1985, Trowbridge 1995, C. Hewitt, CSIRO, pers. comm.). The alga's appearance was noticed in most of these cases because the areas invaded lacked native conspecifics. For example, there are no native species of *Codium* in the NW Atlantic north of Cape Hatteras (Silva 1962), and there are no native subspecies of *C. fragile* in Auckland Harbour, New Zealand (Dellow 1952, 1953, Dromgoole 1975, 1979, Trowbridge 1995) or San Francisco Bay, USA (Silva 1979). In Australia, however, the alga invaded Port Phillip Bay and environs (C. Hewitt, pers. comm.). According to Silva & Womersley (1956), there are two native subspecies (*tasmanicum* and *novae-zelandiae*) not only in the region, but also in and around Port Phillip Bay itself, rendering this invasion extremely difficult to detect.

The primary vectors of spread of this subspecies are shellfish and ship hulls (Wood 1962, Dromgoole 1975, 1979, Loosanoff 1975, Dromgoole & Foster 1983, Carlton & Scanlon 1985, Ribera 1994, Verlaque 1994). Ballast water introductions are highly unlikely given the short duration of gametes (<1 day). The evidence that *C. fragile* ssp. *tomentosoides* is introduced is extensive (Table 4) but varies with geographic region. Proximity to potential sources of introduction, prevalence on new or artificial substrata, facultative association with bivalves (especially oysters), asexuality, low genetic variability, and genetic identity among distant populations collectively indicate that the alga is introduced. What is not entirely clear is whether the algal introductions are primary introductions from the presumed source region (Japan) or secondary or even tertiary introductions from areas with earlier introductions. The low genetic variability due to parthenogenesis has hitherto made such analyses difficult to evaluate.

Codium fragile *ssp.* scandinavicum

The third introduced subspecies, ssp. *scandinavicum,* has not been extensively studied. It was first noted in Denmark in 1919, Norway in 1929, and then Sweden in 1932

(Lund 1940, Fægri & Moss 1952, Silva 1957). In 1965, the alga was found at Sandvik, Rebbenesøy in Troms, Norway (over 70 °N latitude) (Stellander 1969); this record is the northernmost limit (a) for ssp. *scandinavicum* in Norway and (b) for *C. fragile* in the world. Based on morphological comparisons of specimens, Silva (1957) suggested that the most likely source region of ssp. *scandinavicum* was Vladivostok, Siberia on the Sea of Japan.

The vector of introduction of ssp. *scandinavicum* is not known, although Lund (1940) and earlier workers hypothesize that the alga was imported by a ship (presumably fouling the hull) or by drifting. Since ssp. *scandinavicum* occurs only in Scandinavia (and its Pacific source region), remote dispersal via drifting is not a plausible explanation. Marginal spread (*sensu* Crisp 1958) was probably greatly facilitated by drifting. Given that the subspecies occurs subtidally in harbours (Lund 1940), there is an opportunity for zygotes to attach to ship hulls, anchor lines, etc. and thus be dispersed by human transport.

Identifying *C. fragile* ssp. *scandinavicum* as an introduced alga is not as problematic as distinguishing the subspecies. For example, in Norway ssp. *atlanticum*, ssp. *tomentosoides*, and ssp. *scandinavicum* all occur; in Sweden and Denmark, the latter two subspecies occur. The appearance of ssp. *scandinavicum* in Sweden and Denmark coincided with the appearance of ssp. *tomentosoides*. Differences in branching patterns, utricle shape, and reproduction (sexual *v.* asexual) varied among subspecies. As can be seen from Table 4, there is a great need for further work on ssp. *scandinavicum* as little ecological information has been published about the alga.

Silva (1957) described morphological intergrades between ssp. *scandinavicum* and ssp. *tomentosoides*. Forty years later, we still do not know whether such intergrades are hybrids or just morphological intermediates. The fact that ssp. *tomentosoides* is parthenogenetic (see pp. 19–20) is not necessarily an impediment to interbreeding. Male haploid gametes from the sexual subspecies (ssp. *scandinavicum*) theoretically could fuse with female haploid gametes of ssp. *tomentosoides*. This mechanism of interbreeding has been described for sexual and parthenogenetic strains of the kelp *Laminaria japonica* by Lewis et al. (1993). Thus, the risk of subspecies of *Codium fragile* interbreeding is realistic and should be investigated.

Establishment phase

Species' attributes

The settlement or establishment phase of introduced benthic species depends in part on species attributes (Williamson & Brown 1986, Trowbridge 1995). The types of attributes that render species successful invaders include rapid growth, early maturation, and high dispersal, attributes that render species effective at locating and utilizing available space (Crisp 1958, Williamson & Brown 1986, Lodge 1993). Other attributes include occurrence in habitats in close proximity to vectors of transport: harbours, bays, aquaculture facilities, etc. Thus, the probability of transport may be higher than for species on wave-exposed shores.

When *C. fragile* populations are considered across the species' entire geographic range, they can exhibit rapid growth (up to 7 cm month^{-1}) and high dispersal (up to 65–70 km yr^{-1}). The capacity for utricles, medullary filaments, or even branches to regenerate increases the probability that an individual alga will be dispersed. The parthenogenetic gametes of ssp. *tomentosoides* also enhance the capacity of that subspecies to establish, even at low densities. This alone could account for the alga's apparently-rapid dispersal rates relative to its sexual conspecifics. Furthermore, several subspecies occur in wave-sheltered waters and/or in association with shellfish (e.g. ssp. *tomentosoides*, ssp. *scandinavicum*, ssp. *novae-zelandiae*) and, thus, are predisposed to be dispersed.

Community attributes

The existence of vacant niches has been repeatedly mentioned by invasion ecologists (e.g. Herbold & Moyle 1986, Williamson & Brown 1986). In general, it is difficult to define a vacant niche until it is occupied so this hypothesis has little predictive value, except in cases in which a niche is vacated due to decline of a resident species. The appearance of *C. fragile* ssp. *tomentosoides* on NW Atlantic shores may be such an example: the alga entered estuarine areas where the spatially dominant eelgrass *Zostera marina* had died back due to the wasting disease. On New Zealand shores, the alga also pre-empted space recently occupied by eelgrass (Dromgoole & Foster 1983).

Communities or regions with low species richness tend to be invaded more frequently than areas with higher richness (Diamond & Case 1986, Case 1991, Boudouresque 1994, Ribera 1994). Although this pattern has been observed repeatedly, the precise mechanism (competitive exclusion, intense predation and herbivory, etc.) producing this invasion resistance is not well understood. Ssp. *tomentosoides* invaded the low diversity NW Atlantic shores more readily than the richer NE Atlantic and NE Pacific shores (based on regional comparisons of algal spread). For example, it has taken two decades since the alga was reported in San Francisco Bay for it to appear in Tomales Bay (the only other documented and verified NE Pacific locality).

The alga also invades local sites that have low diversity of space occupiers. For example, on New Zealand shores, ssp. *tomentosoides* inhabits simple communities dominated by red algal *Corallina* turfs and bare space (Fig. 2 in Trowbridge 1995). Quantitative information on the other two introduced subspecies is lacking, and field experiments are sorely needed to ascertain the role of community diversity patterns and invasion success. Information on the communities containing native, non-weedy *Codium fragile* are also meagre. The NE Pacific *C. fragile* occurs in areas of high biotic diversity (Theis 1985). On Oregon shores, the alga occurs in and around intertidal beds of purple sea urchins; these areas have extremely high species richness (>88 species, C. D. Trowbridge, B. A. Menge, & J. Lubchenco unpubl. data). Ssp. *novae-zelandiae* occurs in low-diversity communities such as low intertidal mussel beds on the wave-swept west coast of New Zealand as well as in high-diversity communities such as the wave-sheltered area around Wellington Harbour (C. D. Trowbridge pers. obs.). Although there is the general impression that low diversity marine communities are more susceptible to invaders than high diversity ones, quantitative evidence is lacking.

In terrestrial and marine communities, introduced plant species are most common in areas with low cover (Crawley 1986). On New Zealand shores, *C. fragile* ssp. *tomentosoides* inhabits areas with about 20% bare space (Fig. 2 in Trowbridge 1995). On Scottish shores, communities with high secondary cover (dense canopies of *Ascophyllum nodosum* and *Fucus serratus*) contain the invader less frequently than communities with low cover. In Argyll, Scotland, ssp. *tomentosoides* is most common at sites with gravel, rubble, rocks, and broken shells. Although the species richness is often high, secondary cover is typically low. Also, ssp. *tomentosoides* frequently attaches to *Ascophyllum nodosum, Fucus serratus,* and other macroalgae (Fægri & Moss 1952, C. D. Trowbridge & C. D. Todd, unpubl. data) so interspecific competition is more complex than that of species competing solely for primary substrata. The native ssp. *novae-zelandiae* occurs both where there is high secondary cover (e.g. mussel beds) or low cover (cobble communities in Wellington Harbour). Whether these observations can be extended to other introduced and native subspecies of *Codium fragile* is not known.

Introduced species frequently become established in high-nutrient areas, particularly eutrophic areas (Leppakoski 1994). *C. fragile* ssp. *tomentosoides* on N Atlantic shores is nutrient-limited in summer (Ramus 1971, Head & Carpenter 1975, Hanisak 1979b,c), and Ramus (1971) suggested that nutrient-rich shores would be particularly vulnerable to the invasive alga. Field experiments manipulating nutrient levels *in situ* have never been conducted but observational evidence around the world does support the hypothesis. In contrast, there is no evidence, direct or indirect, regarding any association of ssp. *atlanticum* and ssp. *scandinavicum* with high nutrient concentrations. Furthermore, in contrast to ssp. *tomentosoides* (Head & Carpenter 1975, Dromgoole et al. 1978, Gerard et al. 1990), it is not clear whether the other two introduced subspecies have epiphytic and endophytic bacteria that fix atmospheric nitrogen and whether such microbial assemblages play a role in enabling the macroalgal hosts to become established in new areas.

Spread or expansion phase

Introduced species frequently spread after establishing themselves in new regions; the spread can occur (a) locally and regionally or (b) geographically (Ribera & Boudouresque 1995). The former can occur either as a natural progression or as human-assisted spread. For *C. fragile* ssp. *atlanticum,* the spread from the apparent point source in Ireland to Scotland, Norway, and NE England appears to be in the direction of prevailing currents. It took over a century to disperse around the Scottish coast from an initial appearance in Argyll in 1826 to its arrival on SE Scottish shores in 1949. Although this spread seems to be much slower than that of other invasive macrophytes reported in the literature (Fig. 10) (Moeller 1969, Boudouresque 1994, Ribera & Boudouresque 1995, Trowbridge 1995), in actuality the northward spread was quite rapid and the southward spread slower.

For ssp. *tomentosoides* on New Zealand shores, the early spread documented by Dromgoole (1975, 1979) and Trowbridge (1995) in the Hauraki Gulf appeared to be a natural progression. Later stages of spread (to Whangarei Harbour, Bay of Islands, and Bay of Plenty) were more irregular with no apparent link with direction of currents or distance. The alga's spread within the Mediterranean and on Scottish shores was certainly irregular (Ribera 1994, C. D. Trowbridge & C. D. Todd, pers. comm.). Given the known broad physiological and environmental tolerance of ssp. *tomentosoides* and the known short duration of the parthenogenetic gametes (generally <1 day), the ecological spread of this subspecies is probably due to both natural marginal dispersal coupled with human-assisted spread (Carlton & Scanlon 1985, Ribera & Boudouresque 1995).

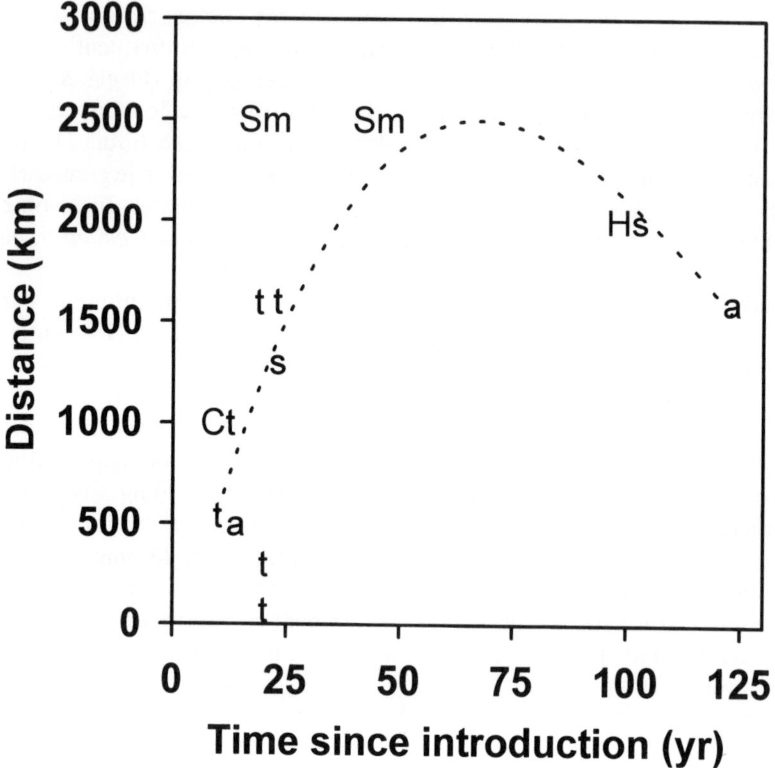

Figure 10 Rates of spread of *Codium fragile* ssp. *atlanticum* (a), ssp. *tomentosoides* (t), ssp. *scandinavicum* (s), and several other introduced macrophytes: *Sargassum muticum* (Sm) *Halophila stipula* (Hs), *Caulerpa taxifolia* (Ct). Data from Moeller (1969), Trowbridge (1995), Ribera & Boudouresque (1995), and calculated from literature. Dashed line is third order polynomial regression line fitted to points.

Geographic expansion (transoceanic and interoceanic) occurs due to human-assisted spread. Even when the vector of spread cannot be specifically pinpointed, the local co-occurrence of a suite of introduced species is strong indirect evidence of human-assisted introduction. Therefore, the appearance of ssp. *atlanticum*, ssp. *tomentosoides*, and ssp. *scandinavicum* (from NW Pacific shores) on NE Atlantic shores was presumably due to human activity.

The most likely vectors for the introduction of ssp. *tomentosoides* to the NW Atlantic shores were ship hulls and/or oysters (Wood 1962, Loosanoff 1975, Carlton & Scanlon 1985). On New Zealand shores, the vector was probably ship hulls because the alga first appeared at the ship container terminal at the Port of Auckland (Dromgoole 1975, 1979, Dromgoole & Foster 1983). For Mediterranean shores, the vector may have been ship hulls, oyster transplants, or marginal dispersal from Atlantic populations (Ribera 1994, Verlaque 1994). On English shores, ssp. *tomentosoides* was first collected from Devon on an oyster, but whether it was introduced with oysters or colonized oysters after its introduction is not clear. Given that the alga was common in the Netherlands, Denmark, and Sweden at that point, the source of the British Isles populations was presumably from mainland Europe, not directly from the Pacific as indicated by

Farnham (1994). On Scottish shores, the vector of introduction was undoubtedly oysters transplanted from Atlantic France (where the alga already occurred) to Loch Sween and Easdale Quarry in the 1940s (C. D. Trowbridge & C. D. Todd pers. comm.). The alga's appearance in Loch Sween cannot be attributable to ballast water introductions or large ship hulls because there are no commercial ports. Although it is possible that ssp. *tomentosoides* arrived via fishing vessels, the region is quite distant from the nearest algal population (English Channel and SW Ireland) and the fishing vessels are primarily local. The long history of oyster introductions for mariculture and scientific purposes in the upper reaches of Loch Sween indicate that oysters were probably the vector in this case (C. D. Trowbridge & C. D. Todd pers. comm.).

Persistence

Introduced species that successfully establish populations in new regions either reach a saturation abundance or actually decline, then stabilize at a lower level (Ribera & Boudouresque 1995). Of the three introduced subspecies of *C. fragile*, information is available only for ssp. *tomentosoides*. On the shores of Mediterranean France, ssp. *tomentosoides* populations peaked in the early 1960s and then regressed (Boudouresque 1994). According to Bird et al. (1993), a similar post-invasion decline in ssp. *tomentosoides* has occurred at Boothbay Harbor, Maine.

Ribera & Boudouresque (1995) suggest five causes of the population decline of introduced species.

(1) Resident consumers may become accustomed to consumption of introduced species ("time since introduction" hypothesis); this may involve changes in digestive physiology of grazers.
(2) Consumers, or
(3) parasites may proliferate and consequently reduce populations of the introduced species.
(4) Adaptation to environmental change or fluctuations may be difficult in introduced species with low genetic variability.
(5) If the introduced species invade areas where native species have temporarily declined in abundance (e.g. eel grass beds infected with the wasting disease), the distribution of introduced species may become more restricted as the native continues to recover.

For *C. fragile*, there are little relevant data to assess the patterns and causal processes of population decline. For ssp. *tomentosoides*, there is some indirect evidence supporting (1) and (5) but rigorous field experiments are now needed. For example, Prince (1988) reported that ssp. *tomentosoides* invaded areas denuded by heavy sea urchin grazing; Prince (1991) and Prince & LeBlanc (1992) now show that sea urchins graze the invasive alga and may help to limit the alga's populations.

Pest

Pests are species that have negative economic or aesthetic effects on human activities. Because the definition focuses on anthropocentric rather than ecocentric criteria, a species can be considered a pest in one area of its geographic range and not in other areas, even when the species exhibits similar attributes in various regions. For example, *C. fragile* ssp. *tomentosoides* is considered a serious pest in many areas on NW Atlantic shores but generally not a pest in other regions (but see Ribera & Boudouresque 1995); other subspecies are not considered pests.

Boudouresque (1994) suggests the major ways in which introduced species can be pests: as a competitor of commercially exploited species; as a nuisance to fishing, aquaculture, tourism, or navigation of vessels; and by fouling.

What is the basis of classification of *C. fragile* ssp. *tomentosoides* as a pest? Basically, the alga negatively affects shellfish industries on New England shores. The alga attaches to living oysters, scallops, surf clams, and mussels in natural and cultivated beds (Galtsoff 1964, Fralick 1970, Ben-Avraham 1971, Ramus 1971, Loosanoff 1975). There are several direct effects of this attachment.

(1) The alga smothers blue mussels and bay scallops by attaching the bivalves' shells shut (Fralick 1970).

(2) When large thalli photosynthesise and trap gas bubbles inside the fronds, their buoyancy lifts shellfish off the beds. These thalli are transported by waves and currents away from shellfish beds and are frequently cast upon the shore as drift, resulting in the death of the shellfish. Consequently, the alga is referred to as the "oyster thief" (Table 1, p. 2).

(3) Oysters (and presumably other shellfish) with attached *Codium* tend to have less meat than comparably sized conspecifics free of the alga (Galtsoff 1962, cited by Moeller 1969). This hypothesis needs to be demonstrated though it seems reasonable that shellfish expend energy in resisting the algal load.

(4) Large thalli increase the frictional drag in water flow, often resulting in dislodgement from the substrata. Hanisak (1977) reported that this occurs in mussels with attached ssp. *tomentosoides.*

(5) The alga can also reduce shellfish mobility and increase their susceptibility to predators (Ramus 1971, Davies 1983).

There are also several indirect effects.

(1) The alga interferes with the harvesting of shellfish: dense beds of ssp. *tomentosoides* clog dredges used in scallop harvesting (Moeller 1969, Fralick 1970). Similarly, collecting clams through thick beds of ssp. *tomentosoides* on Long Island Sound is difficult (Galtsoff 1962, cited by Moeller 1969).

(2) Nets for fin and squid fisheries get fouled by ssp. *tomentosoides* (Carlton & Scanlon 1985).

(3) Because the alga has to be removed from harvested shellfish before shipping to market, ssp. *tomentosoides* increases labour costs by approximately 50% (Carlton & Scanlon 1985).

(4) Scallops fouled by *Codium* were more heavily preyed upon by sea stars and oyster drills than were clean control scallops (Davies 1983). There was not a significant difference in the scallops' response time to predators or distance travelled; Davies suggested that the predators may crawl onto the alga and attack the scallop from above, consequently escaping detection. The presence of the alga on oysters, however, does not influence predation frequency (Perretti 1972).

These deleterious effects are seemingly quite serious, but there is surprisingly little information about the alga's impact. The only quantitative data that I located was in two MSc theses. Fralick (1970) sampled 38 oysters and found a mean of 3.2 thalli of *C. fragile* ssp. *tomentosoides* per oyster, 284 g of algae per oyster, and 3.4 g algae per g oyster. Davies (1983) reported that the alga was on up to 58% of bay scallops surveyed in individual censuses, with an overall value of 35.7%. Various other authors report algal abundance (density values or biomass values) at specific study sites but not in terms of shellfish beds *per se*. Consequently, to evaluate the alga's "pest status", we need spatial and temporal information of ssp. *tomentosoides* on different types of shellfish and shellfish beds as well as values for algal loads caught in scallop dredges and fin and squid fisheries nets.

Because *C. fragile* ssp. *tomentosoides* has not been studied extensively on shores other than the NW Atlantic, it is not clear the extent to which the alga is a pest in other regions. Parkes (1975) reported that the alga was not a pest in Ireland. The alga is not even mentioned within a recent review of the shellfish industry of Great Britain (Spencer 1990). Fletcher (1980) and Fletcher et al. (1989) remark that the alga grows on floating docks and rafts in southern England, but they do not mention whether the alga is a nuisance to human activities. In Loch Sween, Scotland, thalli of ssp. *tomentosoides* frequently grow on introduced Pacific oysters (live and dead) but the frequency of algal fouling at the commercial oyster farms in Argyll is not known. Differences in methods of culturing oysters (e.g. ground *v.* rack culture) may partially explain geographic variation in the alga's pest status. Accumulation of masses of ssp. *tomentosoides* on beaches of the NW Atlantic (Head & Carpenter 1975, Loosanoff 1975, Lee 1986, Coleman 1996), Mediterranean (Ribera & Boudouresque 1995), and New Zealand (C. D. Trowbridge, pers. obs.) is unpleasant. Tractor-drawn harrows remove such algal wrack from beaches in many areas of the world.

Management and conservation

The management options are (a) laissez-faire (no action), (b) interventionist (algal reduction or eradication), and (c) protectionist (prevent invasions) (Fralick 1970, Wassman & Ramus 1973b, Usher 1986, Boudouresque 1994, Ribera & Boudouresque 1995, Eno 1997). To date, the first option has been followed, with little attempt made to slow the alga's spread or to remove existing populations. Potential reduction methods include chemical treatment, hand-picking, or biological controls. Chemical control of *C. fragile* does not seem logistically feasible for intertidal and subtidal algae, given the frequent to continuous submergence in the sea. Quick-lime pellets have been deployed to burn away subtidal algal beds (Wassman & Ramus 1973b), but their effectiveness has never been reported. Dipping oysters with attached algae into a solution of 16.7% Clorox (bleach) for 2 min has proved to be effective but not practical for decontaminating shellfish (Charles Hall, Massachusetts Division of Marine Fisheries, unpubl. ms. cited by Fralick 1970). Manual removal of ssp. *tomentosoides* would be extremely labour intensive (given the large areas, high algal density, and difficulties inherent to removing subtidal thalli). Mechanical control (by dredging or dragging a chain across the benthos to detach attached algae) is possible (Wassman & Ramus 1973b) but this would disturb other species, possible rendering the community even more vulnerable to other invasive species. Furthermore, both manual and mechanical removal would be only temporary as the alga is able to regenerate from vaucherioid mats, medullary threads, and utricles (see Yotsui & Migita 1989). I manually removed all the intertidal thalli of ssp. *tomentosoides* from a beach in New Zealand during the spring and summer; by the following spring, an established algal population had returned (due to regeneration and/or recruitment). Eradication is, for this alga, clearly unrealistic. Wassman & Ramus (1973b) and Ashton & Mitchell (1989) suggest that the direct use of invasive plants for economic benefit (e.g. food, fertilizer) would be a cost-effective way to reduce the algal density. Although the intent is acceptable, such a proposal would undoubtedly lead to the opposite intent: namely, local culture of the pest for economic gain. Thus, I consider a protectionist approach to be the only effective solution to limit future spread. Wassman & Ramus (1973b) report that some oyster farmers erect wire fences around their oyster beds to exclude drift of ssp. *tomentosoides*. While gametes could still enter the area, the immigration of reproductive thalli would be reduced.

Because of the broad distribution of the introduced subspecies of *C. fragile*, populations have entered marine and estuarine conservation areas. For example, on New Zealand shores, the alga occurs in several of the marine reserves in the Hauraki Gulf; on Scottish shores, the alga occurs in Loch Sween, a proposed Nature Conservancy Council reserve. Introduced species make an area "less natural" and may reduce the diversity of other species; thus, they reduce the conservation value of the effected area (Usher 1986). Preventing introductions is more effective than containing established ones (Usher 1986). To quote Bird et al. (1993), "To retard the spread of plants [*C. fragile* ssp. *tomentosoides*] to other areas, measures should be taken to ensure that culture apparatus and shellfish are free of this species before location to new sites" (p.16). Ironically, this advice seems to be generally unheeded. For example, the recent review of the British shellfish industry seems unaware of the presence of the potential pest in its own coastal waters (Spencer 1990).

Herbivory

Herbivory patterns

Generalist grazers

C. fragile is not an attractive food to most generalist grazers (e.g. snails, chitons, sea urchins, isopods), although some species will consume the alga (Wassman & Ramus 1973b, Lubchenco 1978, Nicotri 1980, Knoepffler-Peguy et al. 1987, San Martin 1987, Prince & LeBlanc 1992, Trowbridge 1993, 1995). Although the basis of the alga's unpalatability is not explicitly known, *C. fragile* contains a variety of volatile compounds (see pp. 53–3).

On Oregon shores, the native *C. fragile* is not grazed by common generalist grazers (the snails *Tegula funebralis*, *Littorina scutulata*, and *Lacuna marmorata*; the isopod *Idotea wosnesenskii*; and gammarid amphipods) when alternative seaweeds are available (J. Lubchenco & C. D. Trowbridge, unpubl. data). On New Zealand shores, ssp. *novae-zelandiae* is consumed by at least three common species of generalist invertebrate grazers: the sea urchins *Evechinus chloroticus* and *Centrostephanus rodgersii* and the snail *Cookia sulcata* (Trowbridge 1995). Comparable information on ssp. *capense* (South Africa) and ssp. *tasmanicum* (Tasmania) is lacking.

Codium fragile ssp. *tomentosoides* is consumed by generalist grazers in many areas of its introduction (NW Atlantic: Kitching et al. 1974, Pezzano 1984, Prince & LeBlanc 1992; Mediterranean: Knoepffler-Peguy et al. 1987, San Martin 1987; New Zealand: Trowbridge 1995). In herbivore-inclusion experiments in Massachusetts, the gastropod

Littorina littorea caused slight decreases in abundance of ssp. *tomentosoides*; sea urchins, however, frequently reduced the alga's abundance (Pezzano 1984). In a series of feeding experiments, Prince & LeBlanc (1992) found that the sea urchin *Strongylocentrotus droebachiensis* preferred ssp. *tomentosoides* to *Ulva lactuca*, *Fucus vesiculosus*, and *Zostera marina*. They found no evidence of deterrence: sea urchins consumed ssp. *tomentosoides* when they encountered it. More extensive studies are needed to assess grazing pressure to the alga on the shore.

Little information is available about grazer response to the introduced ssp. *atlanticum* and ssp. *scandinavicum*. On Scottish rocky shores, ssp. *atlanticum* occurs in high pools with high densities of the limpet *Patella vulgata* and moderate densities of *Littorina littorea* (C. D. Trowbridge, pers. obs.). The effects of limpets are unknown, but the littorines frequently graze the vaucherioid mats and the upright fronds of ssp. *atlanticum* in St Andrews Bay (C. D. Trowbridge, pers obs.). Furthermore, in a series of feeding preference experiments, *L. littorea* prefers ssp. *atlanticum* over a variety of different sympatric algal species in tidepools.

Specialist herbivores

Codium fragile is consumed by stenophagous sea slugs (Opisthobranchia: Ascoglossa) that puncture the algal utricles, suck out the cytoplasm, and retain the chloroplasts. Throughout the alga's geographic range there are two genera of slugs that feed on *C. fragile*: *Placida* and *Elysia*. *Placida dendritica* (=*Hermaea dendritica*), as the species is currently recognized, occurs with *Codium fragile* throughout most of the alga's geographic range (Bleakney 1989); thus, the slug consumes both introduced and native subspecies of *C. fragile* (Clark 1975, McLean 1976, Trowbridge 1989, 1993, 1995, 1996, C. D. Trowbridge & C. D. Todd, unpubl. data). The species of *Elysia*, however, are regionally endemic: *E. hedgpethi* in the NE Pacific, *E. atroviridis* in Japan, *E. australis* in Australia, *E. maoria* in Australasia, and *E. viridis* in the NE Atlantic and Mediterranean (Kawaguti & Yamasu 1965, Greene 1970a,b, Greene & Muscatine 1972, Hinde 1983). The latter two slug species consume the introduced *Codium fragile* ssp. *tomentosoides*, although few authors acknowledge it as such (Gallop 1974, Cobb 1978, Gallop et al. 1980, Jensen 1989, 1993; but see Trowbridge 1995).

These ascoglossan species not only ingest chloroplasts from different subspecies of *C. fragile* but also retain them for different lengths of time. In *Placida dendritica*, the chloroplasts remain functional only for an extremely limited period (hours to days) (Taylor 1967, 1968, Greene 1970a,b,c, Greene & Muscatine 1972, McLean 1976, Clark et al. 1990). *Elysia hedgpethi* also retains functional chloroplasts for only a few days (Greene 1970a,b, Greene & Muscatine 1972). At the other extreme, *E. viridis*, *E. australis*, and *E. maoria* retain functional chloroplasts from 6 to 12 weeks (Hinde & Smith 1972, Hinde 1983).

The ascoglossan–chloroplast association is not a true symbiosis because chloroplasts do not divide in the slugs and the animal continues to feed on its hosts. Hence, the terms "kleptoplasty" and "foreign organelle retention" are more appropriate. This association has been studied using ultrastructural and physiological approaches. Because these topics have been well reviewed elsewhere (e.g. Clark et al. 1990, Williams & Cobb 1992, references therein), I mention only two pertinent, though generally overlooked, points.

(1) The wealth of research on the association between *Codium fragile* ssp. *tomentosoides* and *Elysia viridis* on British shores (e.g. Hinde & Smith 1972, 1974, 1975, Gallop 1974, Cobb 1978, Hinde 1978, 1983, Hawes 1979, Hawes & Cobb 1980, Williams & Cobb 1989, 1992) does not allude to the fact that the "specialized" association is quite recent (post-1939 for S England).

(2) The ecological significance of functional kleptoplasty has been extensively reviewed from the slug's perspective; comparatively little is known about the consequences to *Codium fragile* populations (but see Hinde 1983).

If long-term functionality of chloroplasts in ascoglossans reduces slug herbivory on *C. fragile*, then herbivory by non-symbiotic slugs (e.g. *Placida dendritica*) presumably would be more detrimental to host populations than herbivory by endosymbiotic species (e.g. *Elysia viridis*). Trowbridge (1993) demonstrated that slug herbivory by *Placida dendritica* on ssp. *fragile* on Oregon shores was intense in spring and summer; comparable information for other regions and other ascoglossan species is meagre (but see Trowbridge 1995).

Herbivory hypotheses

Escape from consumers

There are several hypotheses to account for the establishment and spread of introduced *C. fragile*. On NW Atlantic shores, numerous authors suggested that the invasive alga was successful because it had "escaped" from its natural consumers presumed to be present in its source range (Ramus 1971, Wassman & Ramus 1973b, Hanisak 1980). This hypothesis, while appealing, has not been rigorously examined. Although ssp. *tomentosoides* is an unattractive food to many generalist grazers, sea urchins do eat it on NW Atlantic shores (Wassman & Ramus 1973b, Pezzano 1984, Prince & LeBlanc 1992) and sea urchins and snails eat the alga on New Zealand shores (Trowbridge 1995). More importantly, the implicit assumption that grazers are ecologically important in reducing the populations of *C. fragile* on Asian shores has not been addressed. In fact, ecological information on ascoglossan populations on *C. fragile* on rocky shores of South Africa, South America, Japan, Korea, and China are generally lacking. Furthermore, there is no evidence that invertebrate grazers regulate the populations of *C. fragile*, introduced or native, anywhere in the alga's geographic range. Thus, demonstrating that ssp. *tomentosoides* is not extensively consumed on NW Atlantic shores is, by itself, not sufficient to test the "escape from consumers" hypothesis.

C. fragile ssp. *atlanticum* may have a spatial escape from ascoglossan slugs. *Elysia viridis* occurs subtidally and in the low intertidal; environmental fluctuations may be too extreme for the slugs to live higher on the shore. Thus, high pool-dwelling thalli of ssp. *atlanticum* escape from slug attack. Although the slug readily consumes the alga in feeding experiments, I have never observed *E. viridis* on ssp. *atlanticum* on the shore despite extensive searches in Berwickshire, Fife, Grampian, Caithness, and Argyll districts of Scotland or on the Orkney Islands. Comparable information on ssp. *scandinavicum* is lacking.

Time since introduction

It is possible that the time since an introduction may influence the extent to which the species is consumed by resident grazers (Southwood 1961, Trowbridge 1995): over time the palatability of the alga and/or the herbivores' preferences, digestive physiology, or abundances could change. This hypothesis would predict that ssp. *tomentosoides* would not be consumed heavily by resident grazers soon after its introduction, but that the alga would be grazed more heavily through time. This hypothesis is difficult to test experimentally. One indirect approach would be to compare the feeding preferences of experienced versus naïve conspecific grazers: animals from where the alga has occurred for decades or centuries versus conspecific grazers in areas where the alga has not yet become established. Another approach would be to compare the palatability of the alga collected from new versus old areas of invasion. A third approach would be to compare herbivory patterns within and among different geographic regions that have been invaded at different times: for example, (a) Nova Scotia v. New England areas and (b) NW v. NE Atlantic shores, NE v. NW Pacific shores.

Coevolved with herbivores

Perhaps *C. fragile* coevolves with its local suite of resident herbivores. If this hypothesis were true, then it should be most apparent in geographically isolated areas with high species endemism (e.g. South Africa, Australasia). Yet, examination of patterns of herbivory by generalist and specialist grazers does not support this hypothesis. On New Zealand shores, six species of grazers (sea urchins, snails, and slugs) exhibit variable preferences between native and introduced subspecies of *C. fragile* (Trowbridge 1995), but the preferences are not related to the life history, geographical range, diet breadth, or mode of feeding of grazers. Furthermore, the fact that *Elysia viridis* feeds on the introduced *Codium fragile* ssp. *tomentosoides* in British Isles and Europe and sequesters functional algal chloroplasts (e.g. Hinde & Smith 1972, 1974, 1975, Trench et al. 1973a,b, Gallop 1974, Trench 1975, Cobb 1978, Hinde 1978, Hawes 1979, Gallop et al. 1980, Hawes & Cobb 1980, Williams & Cobb 1992) does not support the "coevolved with herbivores" hypothesis. Alternatively, the important interactions may occur not at the species level but at the generic level.

Competition

Competition patterns

Quantitative information on interspecific competition and inter-subspecific competition for *C. fragile* is meagre. Malinowski (1974) conducted a field study on NW Atlantic shores, investigating the interactions between *C. fragile* ssp. *tomentosoides* and *Zostera marina*; he reported that the introduced alga was an inferior competitor.

Qualitative information, however, is common though must be considered cautiously. Several British phycologists report that *Codium fragile* ssp. *tomentosoides* has displaced the native congener *C. tomentosum* (Parkes 1975, Norton 1978, Farnham 1980, 1994, Burrows 1991, Eno et al. 1997). Because no data or experiments are presented to support this "displacement" hypothesis, we can infer only that the introduced alga presently occupies some of the habitats previously occupied by the congener; interspecific competition *per se* has not been demonstrated. Similarly, reports that ssp. *tomentosoides* is "displacing" ssp. *atlanticum* on British shores (Hardy 1990, Burrows

1991) are also conjectural. Finally, none of these authors specified whether the type of hypothesized competition was exploitative or interference competition.

A research priority in invasion biology is scientific research rather than relying on subjective impressions (Boudouresque 1994). The only unambiguous way to determine the ecological mechanism driving changes in abundance of *C. fragile* (and other introduced species) is to conduct a series of rigorously conducted competition experiments on the shore. As Kareiva, the editor of the special feature on "predictiveness of invasion ecology" in *Ecology* stated, "Perhaps the most striking feature of these reports on invasion ecology is the absence of manipulative experiments in the tradition of modern community ecology. It is as if researchers had decided invasion is such a long-term and large-scale process that experiments were impractical. One has to wonder whether more progress might have been made if more field research on invasions adopted an experimental approach" (Kareiva 1996: 1652).

Competition hypotheses

Empty niche

As previously mentioned (see p. 36), the existence of vacant niches has been debated by invasion ecologists (e.g. Herbold & Moyle 1986, Williamson & Brown 1986, Lodge 1993). It is difficult to define a vacant niche until it is occupied so this hypothesis has little predictive value, except in cases in which a niche is vacated due to decline of a resident species. The appearance of *C. fragile* ssp. *tomentosoides* on N Atlantic shores may be such an example: the alga entered estuarine areas where the spatially dominant eelgrass *Zostera marina* had died back due to the wasting disease. On New Zealand shores, the alga pre-empted space recently occupied by eelgrass (Dromgoole & Foster 1983). On N Atlantic shores, *Codium fragile* ssp. *tomentosoides* entered habitats in which (a) eelgrass had previously died back due to the wasting disease (Fægri & Moss 1952, Malinowski 1974) and (b) sea urchin grazing had produced a barren area stripped of kelp and other macrophytes (Prince & Le Blanc 1992). In both cases, the alga has apparently filled an empty niche.

Poor competitor

On NW Atlantic and New Zealand shores, the scientific consensus is that ssp. *tomentosoides* is incapable of displacing other algal populations (Malinowski & Ramus 1973, Thomas 1974, Dromgoole 1975, 1979). Once established, however, ssp. *tomentosoides* prevents native species from re-establishing (Malinowski & Ramus 1973, Dromgoole 1975, 1979). The only field experiments testing the "poor competitor" hypothesis were conducted by Malinowski (1974): he demonstrated reduced growth and recruitment of ssp. *tomentosoides* in the presence of the eel grass *Zostera marina. Zostera* outcompetes the introduced alga in two ways: by shading and by causing increased sedimentation (Malinowski & Ramus 1973, Malinowski 1974); the sediment buries not only the mature plants but also surfaces on which the parthenogenetic gametes settle. Other authors have further suggested that juvenile stages of the alga are also poor competitors: (a) that macrophytes such as *Enteromorpha* prevent *Codium* gametes from settling on surfaces (Thomas 1974) and (b) that the vaucherioid mat does not compete well in summer months.

In contrast, British workers consider that ssp. *tomentosoides* outcompetes and displaces the native congener *C. tomentosum* and the introduced ssp. *atlanticum* (see pp. 45–6). This "displacement" could be due to competition but not necessarily so. The increase in ssp. *tomentosoides* and decrease in *C. tomentosum* could be independent events or the two changes could covary with some environmental change (e.g. eutrophication of coastal waters, climate change, etc.). Indeed, the spread of *C. fragile* in Norway occurred simultaneously with other environmental changes (Fægri & Moss 1952); whether they were causal or not is not known.

Epiphytes

Epiphyte-host patterns

C. fragile is frequently covered with epiphytes, particularly the red algae *Ceramium* and *Polysiphonia*, the brown alga *Ectocarpus*, diatoms, cyanobacteria, and heterotrophic bacteria (Silva 1951, Dellow 1953, Boraso & Piriz 1975, Head & Carpenter 1975, Wilson 1977, 1978, Dromgoole et al. 1978, Dromgoole 1979, Lewis 1982, Searles et al. 1984, Trowbridge 1993, 1996). The native *Codium fragile* on British Columbia shores has a diverse assemblage of 30 epiphytic species, 13 of which are common (Lewis 1982). On South African shores, ssp. *capense* has *Ceramium* and the red alga *Pleonosporium* as epiphytes (Stephenson et al. 1940). On New Zealand shores, *Codium* ssp. *novae-zelandiae* is often covered by *Ceramium*, *Polysiphonia*, *Myriogramme*, and hydroids (Dellow 1953, Trowbridge 1996).

There are three general modes of epiphyte attachment: (a) algal species entangle their attachment structures (e.g. rhizoids) around utricle hairs or among the host's utricles, (b) species cement themselves onto the utricles, and (c) species grow within the utricle walls (Silva 1951, Moeller 1969, Lewis 1982). Epiphytes attached to utricle hairs may be sloughed off (by hair abscission) just prior to reproductive maturation (Moeller 1969). Non-specific epiphytes that have rhizoids protruding among the utricles may be dislodged by wave action. The specialized epiphyte *Ceramium codicola*, however, has a bulbous rhizoidal base that is embedded among the utricles and cannot be easily shed.

On NW Atlantic shores, there are at least 17 species of algae epiphytic on *Codium fragile* ssp. *tomentosoides*; the four most common are the brown alga *Ectocarpus*, two species of the red alga *Polysiphonia*, and *Ceramium rubrum* (Moeller 1969). On North Carolina shores, the red algae *Ceramium* and *Audouinella* formed extensive epiphyte loads on the introduced host (Searles et al. 1984). These epiphytes are not host-specific (Wilson 1978), in contrast to the highly specific *Ceramium codicola* on NE Pacific hosts. Epiphytes on *Codium fragile* ssp. *atlanticum* and ssp. *scandinavicum* have not yet been examined.

Ecological consequences

The ecological significance of epiphyte cover on *C. fragile* has been investigated in two areas. On Oregon shores, the presence of the host-specific *Ceramium codicola* on indigenous *Codium fragile* enhances attack by the specialist sea slug *Placida dendritica*

(Trowbridge 1993). Although the slug did not eat the epiphyte, hosts supporting the epiphyte are attacked significantly more frequently than hosts without the epiphyte. When *Ceramium codicola* was experimentally clipped from hosts on the shore, slug recruitment was significantly greater on hosts with intact epiphytes than without (Trowbridge 1993). Several mechanisms (not mutually exclusive) may account for the these results: *C. codicola* may increase metamorphosis of competent slug larvae, enhance larval settlement or post-settlement survival, provide a refuge from predators, facilitate slug feeding, or enhance algal food quality.

On North Carolina shores, epiphytes on the introduced *Codium fragile* ssp. *tomentosoides* modify generalist grazers' preferences (Wahl & Hay 1995). Sea urchins preferred ssp. *tomentosoides* without epiphytes to those with epiphytic *Ectocarpus* (thus, hosts derived "associational resistance"). In contrast, if the red algae *Polysiphonia* or *Audouinella* are epiphytes, their presence led to "shared doom" where sea urchins consumed significantly more *Codium fragile* ssp. *tomentosoides* than of unfouled hosts. Consequently, the presence and identity of epiphytes can modify the fate of *C. fragile* thalli by altering grazing rates of specialist and generalist herbivores (Trowbridge 1993, Wahl & Hay 1995). To understand the ecological importance of these different types of tri-trophic interactions, we need to test how robust the results are to different grazers, different algal subspecies, and a wider range of epiphytic species.

Natural products

There is a wealth of information on the chemical composition and natural products of *C. fragile*, in large part due to the alga's commercial value as human food and for various medicinal and applied applications. I briefly review the major areas of natural product research to emphasize the existing information on the alga; ecological implications, however, are sorely understudied. The next crucial step in understanding the alga's ecological interactions – with consumers, competitors, and symbionts – will be to integrate chemical and ecological details.

Antibiotics

Hornsey & Hide (1974) screened the introduced alga *C. fragile* (probably ssp. *tomentosoides* based on collection details) and the native congener *C. tomentosum* against five types of micro-organisms (Fig. 11A). The algae exhibit moderate bacterial inhibitory action relative to other algal species tested; there is no striking difference in the inhibitory activity of two *Codium* spp., and both are most effective against *Staphylococcus aureus*. In monthly sampling of *Codium fragile* tested against *Staphylococcus aureus*, Hornsey & Hide (1976a) found a spring peak in antimicrobial activity (Fig. 11B). The location of maximal activity was at the growing point below the apical tip (Fig. 11C) (Hornsey & Hide 1976b).

Although Hornsey & Hide did not frame their results in an ecological context, the spatial pattern of antimicrobial activity is inversely related to observed microbial densities. For example, observations by Gerard et al. (1990) indicate that cyanobacteria abundance is highest on older, proximal parts of the thalli and lowest on the younger, distal parts. Furthermore, nitrogen fixation by bacteria on *Codium fragile* ssp. *tomento-*

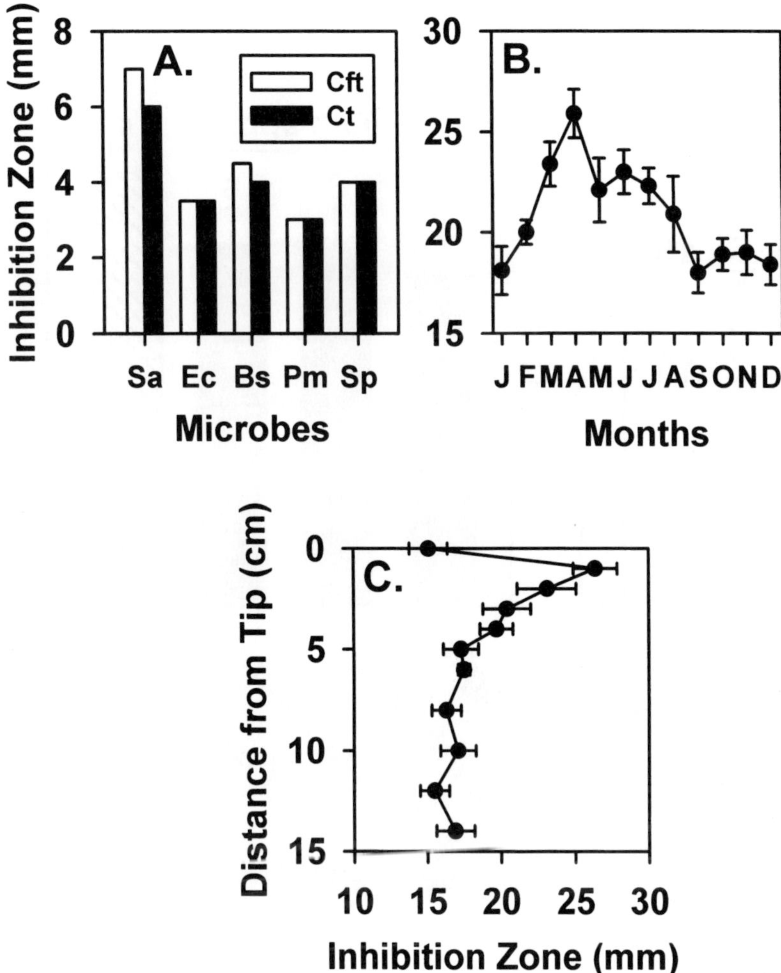

Figure 11 Antimicrobial activity of *Codium*. A, Inhibitory effect of native *C. tomentosum* (Ct) and introduced *C. fragile* ssp. *tomentosoides* (Cft) on four microbial species. Sa = *Staphylococcus aureus*, Bs = *Bacillus subtilis*, Pm = *Proteus morganii*, and Sp = *Streptococcus pyogenes*. Based on spring algal collections. B, Inhibitory effect of *C. fragile* ssp. *tomentosoides* on *S. aureus* in different months. C, Location of inhibitory activity within the thallus of *C. fragile* ssp. *tomentosoides*. Assay microbe was *S. aureus*. Based on spring algal collections. Error bars were standard errors of the mean. Data were from tables in Hornsey & Hide (1974, 1976a,b).

soides is low in young thalli and in the apical tips of older thalli; Head & Carpenter (1975) suggest this pattern may reflect antibacterial activity by the alga. Clearly, the ecological interactions among *C. fragile*, cyanobacteria, and heterotrophic bacteria need to be examined in a more integrated fashion.

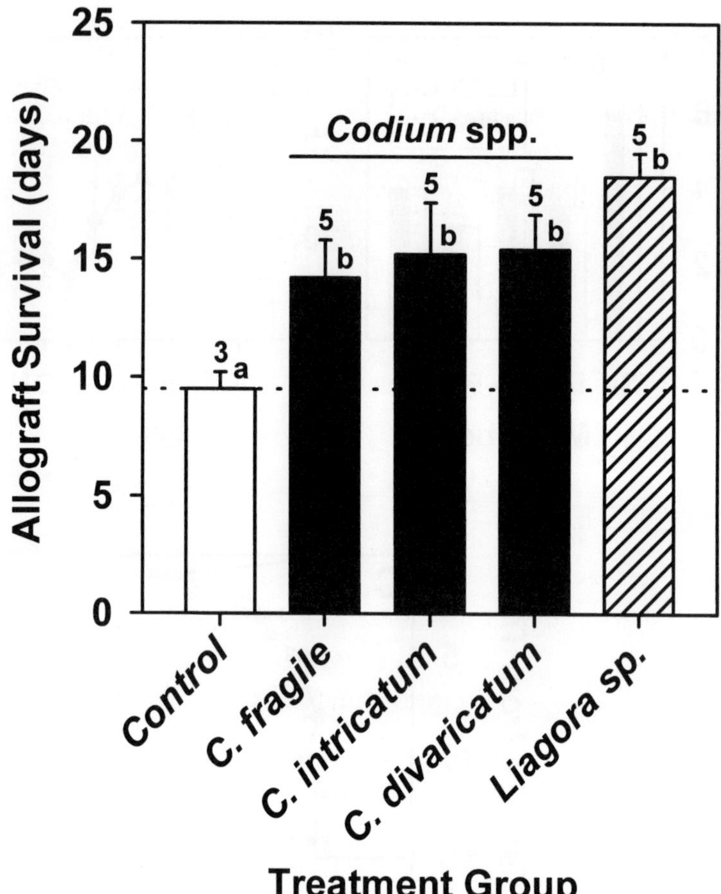

Figure 12 Immuno-suppressive activity of algal extracts on tissue grafts in mice. Data from Table 2 in Mizukoshi et al. (1995). The control was physiological saline 0.9%; algal doses were $50\,\mathrm{mg\,kg^{-1}}$. Error bars denote 1 SD; sample size indicates replication. The dashed horizontal line is the control value; values above the line indicate immuno-suppressive activity.

Toxic compounds

Water-soluble extracts of *Codium* spp., including the native *C. fragile* on Japanese shores, have immuno-suppressive effects (Fig. 12) on the survival of tissue grafts in mice; all four experimental treatments shown had increased survival time relative to saline controls (Mizukoshi et al. 1995). The alga decreases the circulating leukocyte levels when administered for prolonged periods (Cho et al. 1990). Furthermore, Korean *C. fragile* apparently has anticarcinogenic effects (Ahn et al. 1993) and antitumour effects (Cho et al. 1990). Many species of *Codium* that have been studied have inhibitory or even toxic effects (Katayama 1958, Arasaki & Arasaki 1986, Czapla & Lang 1990, Amer et al. 1991, Ballesteros et al. 1992), although the nature of the toxic components is unclear. *C. fragile* has long been consumed by Asians, in part because of its antihelminthic properties: it is effective against nematodes, cestodes, trematodes, and some insects (Katayama 1958, Arasaki & Arasaki 1986, Czapla & Lang 1990). Furthermore, various congeners have antifungal or antiviral properties (Ballesteros et al. 1992); still other *Codium* spp. are highly toxic to insects and mites (Czapla & Lang 1990, Amer at al. 1991).

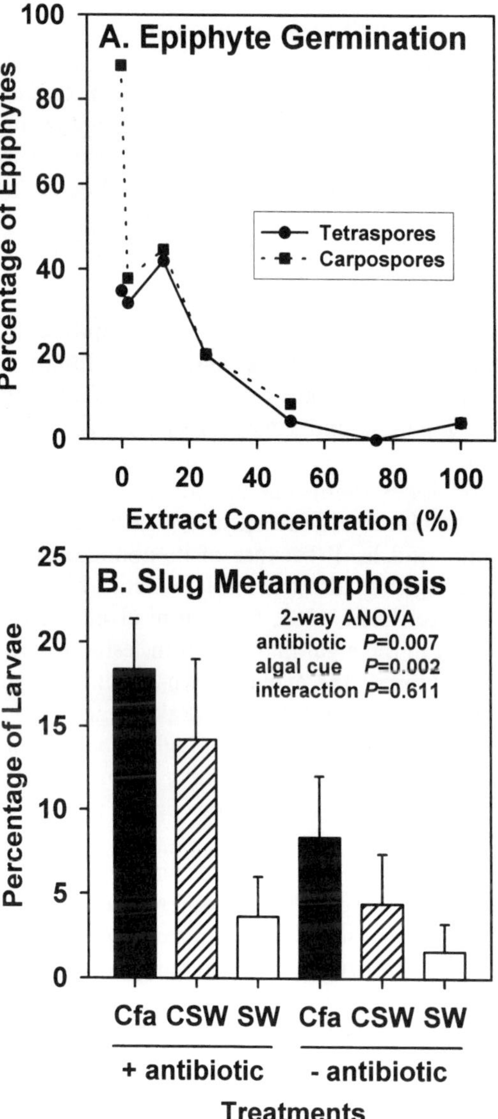

Figure 13 A, Effect of extracts of *Codium fragile* on the germination of the highly specific red algal epiphyte *Ceramium codicola.* Data calculated from Lewis (1982). B, Effect of water soluble compounds from *C. fragile* ssp. *atlanticum* on metamorphosis of competent larvae of the ascoglossan opisthobranch *Elysia viridis.*

The extracts of different subspecies of *C. fragile* have never been tested against the array of natural enemies (herbivores, epibionts, microbes) and competitors (macroalgae and sessile invertebrates) with the exceptions of two studies. Lewis (1982) tested the effect of *C. fragile* extracts against spore germination of the highly host-specific, red algal epiphyte *Ceramium codicola* (Fig. 13A): the host extract inhibited germination of the epiphyte's tetraspores and carpospores. Although he was not sure whether the inhibition was from the host itself or associated contaminants, the highly specific epiphyte was clearly not facilitated by its host. C. D. Trowbridge & C. D. Todd (unpubl. data) found that water-soluble compounds that leach from *Codium fragile* induce larval metamorphosis of host-specific sea slugs (*Elysia viridis*); the induction occurs in the presence and absence of general spectrum antibiotic (gentamicin sulphate) indicating that the active compounds were probably not bacterial (Fig. 13B).

Heavy metals

Indigenous *Codium fragile* (from the Sea of Japan) accumulates heavy metals within its thallus: the lower, older parts of thallus contain substantially more metals than the distal branch tips (Burdin & Poljakova 1984). Ranges of metal concentrations (in $\mu g\, g^{-1}$ DW algae) from Japanese and Korean *C. fragile* are as follows: 6–23 for zinc, 5–15 for copper, 0.3–0.8 for cadmium, 15–17 for nickel, and 5–6 for lead (Pak et al. 1977, Denton & Burdon-Jones 1986). The introduced ssp. *tomentosoides* also accumulates metals and, thus, would be a good species with which to map heavy metal pollution (copper, cadmium, and lead) because of its abundance and broad distribution (Cristiani 1979). Other congeners also accumulate metals (Pak et al. 1977, Castagna et al. 1985, Denton & Burdon-Jones 1986, Costantini et al. 1991). The extent to which heavy metals accumulated in the algal thalli are transferred to grazers and humans that consume the alga is unexplored; the extent to which different subspecies of *C. fragile* accumulate heavy metals is also not known. Pak et al. (1977) consider that metal concentrations in Korean algae are not high enough to be of concern to human health. On some Mediterranean shores, however, metal concentrations in *C. fragile* are considerably higher: up to 14.6 $\mu g\, g^{-1}$ DW algae for copper, 20.4 for cadmium, and 13.3 for lead (Cristiani 1979).

In marked contrast to the studies of metal accumulation, Steele (1975) investigated the effects of trace metals on the growth and differentiation of *C. fragile* ssp. *tomentosoides*. Algal growth was best when trace metals were omitted from the culture medium. The metals also inhibited thallus differentiation. The implication of Steele's results is that metal concentrations in coastal waters may influence whether the species occurs as differentiated thalli or as undifferentiated vaucherioid mats.

Volatile compounds

C. fragile contains an array of volatile compounds (Katayama 1958, Arasaki & Arasaki 1986). On Japanese shores, volatiles compose about 0.034% of the alga's dry mass (Katayama 1958). Aromatic compounds include dimethylsulphide, benzaldehyde, α-methylfurfural, furfural, furfuryl alcohol, 1 : 8-cineol, linalool, terpinolene, geraniol, and eugenol; odorous compounds include formis, propionic, butyric, isovaleric, n-caproic, caprylic, and palmitic acids and p-cresol (Katayama 1958). Linalool, geraniol, and carvone have strong antihelmintic activity (see references in Katayama 1958). In addition, *C. fragile* on Japanese shores contains 0.2% (DW basis) of trimethylamine, a compound that is the source of bad odours in fish (Arasaki & Arasaki 1986).

Whether these aromatic or odorous compounds have any ecological role is not yet known; the extent to which volatile compounds vary among different subspecies of *C. fragile* is also currently not known, although New Zealand ssp. *tomentosoides* is consistently the most odorous *C. fragile* I have ever handled. Preliminary evidence indicates that the unstable, volatile compounds contribute to the species' unpalatability. Freeze-dried tissue reeks for weeks, and when used in feeding experiments, the algal "flour" is more palatable to grazers than the fresh tissue or agar controls.

Nutritional value

Extensive work has been done on natural products of *C. fragile* because it is consumed by human beings and cultured gastropods (e.g. abalone) in Japan, Korea, and China (e.g. Chapman & Chapman 1980, Meiling & Tseng 1984, Arasaki & Arasaki 1986, Pan & Yu 1992, Cho et al. 1995, Kim et al. 1995). There is considerable research on the chemical composition of proteins, fatty acids, sterols, starch, etc. (Wassman & Ramus 1973a, Sato 1975, Arasaki & Arasaki 1986, Delgado & Duville 1977, Aratani et al. 1981, Williams & Cobb 1988, Cho et al. 1990, 1995, Khotimchenko 1993, Kim et al. 1995, Herbreteau et al. 1997). For Korean *C. fragile*, crude protein constitutes 1.6–5.2% (DW) (Cho et al. 1995); for Argentinian *C. fragile*, protein levels are 13.4–15.3% (DW) (Delgado & Duville 1977). Major amino acids include arginine, alanine, aspartic acid, cystine, glutamic acid, glycine, leucine, and valine (Arasaki & Arasaki 1986, Cho et al. 1990).

Lipids compose 1.5–3.7% of *C. fragile* (dry mass basis) (Sato 1975, Herbreteau et al. 1997). Oleic acid is one of the major fatty acids (Khotimchenko 1993). Furthermore, clerosterol is one of the major sterols for *C. fragile* in Japan (Aratani et al. 1981) and for ssp. *tomentosoides* in S England (Rubinstein & Goad 1974). Finally, starch concentrations in ssp. *tomentosoides* are higher in the summer than in winter (Williams et al. 1984, Williams & Cobb 1988); in intertidal thalli than subtidal ones; and in shallow subtidal thalli than deeper ones (Wassman & Ramus 1973a, Ramus et al. 1976a).

Are these natural compounds ecologically important to grazers? There are no experimental data to evaluate this hypothesis. Indirect evidence indicates that there may be an association between slug densities and the amount of palmitic acid in *C. fragile* (Aratani et al. 1981, Trowbridge 1992): both peak in late spring to early summer and decline to a minimum in winter. Furthermore, cultured abalone fed on a diet of *C. fragile* grew at comparable rates to conspecifics fed *Undaria pinnatifida* or *Sargassum muticum* (Pan & Yu 1992).

An important question that has not been addressed, however, is whether the different subspecies are similar in terms of chemical composition and natural products. It is possible that the invasive alga *Codium fragile* ssp. *tomentosoides* may have significantly different amounts or types of natural products than non-weedy conspecifics and that these differences may affect palatability and nutrition. Herbivores exhibit significant differences in preferences between native and introduced subspecies on New Zealand shores (Trowbridge 1995), but whether the differences are due to subspecies *per se* or to algae collected from different sites, different oceans (Tasman Sea *v.* Hauraki Gulf, Pacific Ocean), etc. is not known.

Future directions

C. fragile is a model study system for many fields from algal physiology, endosymbiosis, and heavy metal accumulation to invasion ecology, algal genetics, and natural products. The alga is consumed by humans, used as invertebrate food by the mariculture industry, is a pest of natural and cultivated shellfish beds, is a source of bioactive compounds, and is a potential indicator of heavy metal pollution. Yet, this review demonstrates that ecological information on *C. fragile* is patchy in coverage and quality, particularly for native, non-weedy populations. There are six specific subject areas that warrant future attention.

Phenotypic v. subspecific variation

Given the broad geographic distribution of the species, regional and geographic comparisons are necessary to differentiate between phenotypic and subspecific variation. Williams (1948) investigated this topic for the congener *C. decorticatum*; comparable field transplant experiments are needed for *C. fragile*. Such experiments would be most ethical with non-weedy subspecies; field experiments on ssp. *tomentosoides* should be limited to sites where the alga already occurs. Silva (1957) predicted that environmental variation will have more effect on algal habit than anatomy; forty years later, his hypothesis still has not been tested.

Intermediate phenotypes

The status of "intergrades" between subspecies (Silva 1957) needs to be investigated. Are intergrades environmentally determined intermediates or are they genetic hybrids? If gene flow is occurring, experimental proof and genetic evidence of recombination is needed (Fægri & Moss 1952, Silva 1955, 1957). The risk of interbreeding between introduced and native subspecies warrants immediate appraisal given the rapidly expanding geographic range and pest-like attributes of *C. fragile* ssp. *tomentosoides*.

Genetic basis of ecological differences

Karyological comparisons are necessary to investigate generalizations between ploidy level and mode of reproduction in specific populations of *C. fragile* (see p. 20). Nuclear DNA content can vary not only among species but within them, prompting the question of whether there is a relation between DNA content and algal invasiveness. Polyploid organisms frequently have broader niches than their diploid relatives, and many terrestrial invasive species have uniparental inheritance (parthenogenesis, self-fertilization, or vegetative propagation) (Gray 1986). Briand (1994) reported that comparisons of the genetic profiles of introduced and wild populations are needed to understand whether introduced species (or subspecies) are characterized by specific genetic attributes.

Differentiated thalli v. undifferentiated vaucherioid mats

In tracking the establishment and spread of *C. fragile* ssp. *tomentosoides*, ssp. *scandinavicum*, and ssp. *atlanticum*, phycologists in the past have focused their attention almost exclusively on macroscopic differentiated thalli. Given the capacity of the species to persist as undifferentiated filaments on any hard surface, we need to broaden our perspective to include both life stages. Thus, the microscopic inspection of shellfish, particularly bivalves, is imperative when transplanting maricultural animals between locales. Research should also explore chemical methods to destroy these coenocytic stages on shellfish and ship hulls. Finally, future research should expand on the studies of Borden & Stein (1969), Ramus (1972), and Steele (1975) to address the environmental conditions under which *C. fragile* differentiates from mats of unconsolidated coenocytic filaments to macroscopic thalli.

Source v. introduced populations

A major research priority (Briand 1994) is the comparison of the autecology, population ecology, and community ecology of *C. fragile* in regions where the alga is presumed to be native (NW Pacific) and regions where it is not. This type of information will improve our ability to predict future responses of *C. fragile* in new areas of invasion. Furthermore, we need to determine experimentally what types of marine environments are vulnerable to invasion (Briand 1994) and then to survey such coastal areas to quantify the rate of spread of known exotics (such as *C. fragile* ssp. *tomentosoides*, ssp. *scandinavicum*, and ssp. *atlanticum*).

Relevance of bioactive compounds

Finally, *C. fragile* clearly contains both nutritional and bioactive compounds. The ecological relevance of both classes of compounds should be investigated, particularly as to whether they vary within or among subspecies. There is increasing evidence that *C. fragile* has bioactive compounds that may play multiple roles in the alga's life history and ecology.

Acknowledgements

This research was supported by the Hatfield Marine Science Center, Oregon State University and the generosity of my friends and colleagues. In particular, I thank B. Menge, C. Todd, L. Weber, P. Silva, J. Chapman, and G. Hansen for their support and S. Gilmont, J. Webster, and M. Grundy for their excellent library assistance. Conversations and correspondence with P. Silva, G. Manneveldt, C. Hewitt, G. Hardy, C. Maggs, and K. Van Alstyne provided important phycological information and perspective. Constructive comments and suggestions by P. Silva, G. Hansen, J. Chapman, G. Allison, P. Halpin, and J. Burnaford improved this paper; I thank them for their assistance.

References

Ahn, B. W., Lee, D. H., Yeo, S. G., Kang, J. H., Do, J. R., Kim, S. B. & Park, Y. H. 1993. Inhibitory action of natural food components on the formation of carcinogenic nitrosamine. *Bulletin of the Korean Fisheries Society* **26**, 289–95.

Amer, S. A. A., Dimetry, N. Z. & Reda, A. S. 1991. Toxicity of green marine algae to the two-spotted spider mite *Tetranychus urticae* Koch. *Insect Science and its Application* **12**, 481–5.

Ames, J. A. 1979. *Cultural studies of* Codium fragile. MSc thesis, Adelphi University, New York.

Arasaki, S. & Arasaki, T. 1986. *Low calorie, high nutrition vegetables from the sea to help you look and feel better*. Tokyo: Japan Publications.

Arasaki, S., Tokuda, H. & Fujiyama, K. 1956. The reproduction and morphogeny in *Codium fragile*. *Botanical Magazine Tokyo* **69**, 39–45.

Aratani, T., Okano, M., Funaki, Y. & Mizui, F. 1981. Seasonal variation in sterol, hydrocarbon, and fatty acid fractions in *Codium fragile* (Sur.) Hariot. *Bulletin of the Japanese Society of Scientific Fisheries* **47**, 391–5.

Arnold, K. E. & Murray, S. N. 1980. Relationships between irradiance and photosynthesis for marine benthic green algae (Chlorophyta) of differing morphologies. *Journal of Experimental Marine Biology and Ecology* **43**, 183–92.

Asare, S. O. & Harlin, M. M. 1983. Seasonal fluctuations in tissue nitrogen for five species of perennial macroalgae in Rhode Island Sound. *Journal of Phycology* **19**, 254–7.

Ashton, P. J. & Mitchell, D. S. 1989. Aquatic plants: patterns and modes of invasion, attributes of invading species and assessment of control programmes. In *Biological invasions: a global perspective*, J. A. Drake et al. (eds). John Wiley & Sons, 111–54.

Ballesteros, E., Martín, D. & Uriz, M. J. 1992. Biological activity of extracts from some Mediterranean macrophytes. *Botanica Marina* **35**, 481–5.

Ben-Avraham, Z. 1971. Accumulation of stones on beaches by *Codium fragile*. *Limnology and Oceanography* **16**, 553–4.

Benson, E. E. & Cobb, A. H. 1981. The separation, identification and quantitative determination of photopigments from the siphonaceous marine alga *Codium fragile*. *New Phytologist* **88**, 627–32.

Benson, E. E. & Cobb, A. H. 1983. Pigment/protein complexes of the intertidal alga *Codium fragile* (Suringar) Hariot. *New Phytologist* **95**, 581–94.

Benson, E. E., Rutter, J. C. & Cobb, A. H. 1983. Seasonal variation in frond morphology and chloroplast physiology of the intertidal alga *Codium fragile* (Suringar) Hariot. *New Phytologist* **95**, 569–80.

Bird, C. J., Dadswell, M. J. & Grund, D. W. 1993. First record of the potential nuisance alga *Codium fragile* ssp. *tomentosoides* (Chlorophyta, Caulerpales) in Atlantic Canada. *Proceedings of the Nova Scotian Institute of Science* **40**, 11–17.

Bleakney, J. S. 1989. Morphological variation in the radula of *Placida dendritica* (Alder & Hancock, 1843) (Opisthobranchia: Ascoglossa/Sacoglossa) from Atlantic and Pacific populations. *Veliger* **32**, 171–81.

Boerner, R. E. 1972. *A comparative study of reproductive patterns in populations of* Codium fragile *subspecies* tomentosoides *off the northeast coast of North America.* MSc thesis, Adelphi University, New York.

Boraso, A. L. & Piriz de Núñez de la Rosa, M. L. 1975. Las especies del genero *Codium* (Chlorophycophyta) en la costa Argentina. *Physis Buenos Aires, Sección A* **34**, 245–56.

Borden, C. A. & Stein, J. R. 1969. Reproduction and early development in *Codium fragile* (Suringar) Hariot: Chlorophyceae. *Phycologia* **8**, 91–9.

Bouck, G. B. & Morgan, E. 1957. The occurrence of *Codium* in Long Island waters. *Bulletin of the Torrey Botanical Club* **84**, 384–7.

Boudouresque, C. F. 1994. Les espèces introduites dans les eaux côtières d'Europe et de Méditerranée: etat de la question et conséquences. In *Introduced species in European coastal waters*, C. F. Boudouresque et al. (eds). Luxembourg: European Commission, 8–27.

Briand, F. 1994. Species introductions in coastal waters of Europe: a call for action. In *Introduced species in European coastal waters*, C. F. Boudouresque et al. (eds). Luxembourg: European Commission, 4–7.

Bright, K. M. F. 1938. The South African intertidal zone and its relation to ocean currents. III. – An area on the northern part of the west coast. *Transactions of the Royal Society of South Africa* **26**, 67–88.

Burdin, K. S. & Poljakova, E. E. 1984. Distribution of heavy metals in different parts of thallus of green algae from the Japan Sea. *Vestnik Moskovskogo Universiteta Seriya XVI Biologiya* **39**, 34–9.

Burrows, E. M. 1991. *Seaweeds of the British Isles. Vol. 2: Chlorophyta.* London: Natural History Museum Publications.

Carlton, J. T. 1979. *History, biogeography, and ecology of the introduced marine and estuarine invertebrates of the Pacific coast of North America.* PhD thesis, University of California Davis.

Carlton, J. T. & Scanlon, J. A. 1985. Progression and dispersal of an introduced alga: *Codium fragile* ssp. *tomentosoides* (Chlorophyta) on the Atlantic Coast of North America. *Botanica Marina* **28**, 155–65.

Case, T. J. 1991. Invasion resistance, species build-up and community collapse in metapopulation models with interspecies competition. *Biological Journal of the Linnean Society* **42**, 239–66.

Castagna, A., Sinatra, F., Castagna, G., Stoli, A., Zafarana, S. 1985. Trace element evaluations in marine organisms. *Marine Pollution Bulletin* **16**, 416–18.

Chapman, J. W. & Carlton, J. T. 1991. A test of criteria for introduced species: the global invasion by the isopod *Synidotea laevidorsalis* (Miers, 1881). *Journal of Crustacean Biology* **11**, 386–400.

Chapman, V. J. & Chapman, D. J. 1980. *Seaweeds and their uses.* London: Chapman & Hall.

Cho, D. M., Kim, D. S., Lee, D. S., Kim, H. R. & Pyeun, J. H. 1995. Trace components and functional saccharides in seaweed. I. – Changes in proximate composition and trace elements according to the harvest season and places. *Bulletin of the Korean Fisheries Society* **28**, 49–59.

Cho, K. J., Lee, Y. S. & Ryu, B. H. 1990. Antitumour effect and immunology activity of seaweeds toward sarcoma-180. *Bulletin of the Korean Fisheries Society* **23**, 345–52.

Churchill, A. C. & Moeller, H. W. 1972. Seasonal patterns of reproduction in New York populations of *Codium fragile* (Sur.) Hariot subsp. *tomentosoides* (Van Goor) Silva. *Journal of Phycology* **8**, 147–52.

Clark, K. B. 1975. Nudibranch life cycles in the Northwest Atlantic and their relationship to the ecology of fouling communities. *Helgoländer Wissenschaftliche Meeresuntersuchungen* **27**, 28–69.

Clark, K. B., Jensen, K. R. & Stirts, H. M. 1990. Survey for functional kleptoplasty among West Atlantic Ascoglossa (= Sacoglossa) (Mollusca: Opisthobranchia). *Veliger* **33**, 339–45.

Cobb, A. H. 1978. Inorganic polyphosphate involved in the symbiosis between chloroplasts of alga *Codium fragile* and mollusc *Elysia viridis*. *Nature* **272**, 554–5.

Cobb, A. H. & Rott, J. 1978. The carbon fixation characteristics of isolated *Codium fragile* chloroplasts. Chloroplast intactness, the effect of photosynthetic carbon reduction cycle intermediates and the regulation of $RUBP^+$ carboxylase *in vitro*. *New Phytologist* **81**, 527–41.

Coffin, G. W. & Stickney, A. P. 1967. *Codium* enters Maine waters. *Fishery Bulletin* **66**, 159–61.

Coleman, A. W. 1996. DNA analysis methods for recognizing species invasion: the example of *Codium*, and generally applicable methods for algae. *Hydrobiologia* **326/327**, 29–34.

Costantini, S., Giordano, R., Ciaralli, L. & Beccaloni, E. 1991. Mercury, cadmium and lead evaluation in *Posidonia oceanica* and *Codium tomentosum*. *Marine Pollution Bulletin* **22**, 362–3.

Cotton, A. D. 1912. Clare Island Survey. Marine Algae. *Proceedings of the Royal Irish Academy* **31**, 1–178.

Crawley, M. J. 1986. The population biology of invaders. *Philosophical Transactions of the Royal Society of London, Series B* **314**, 711–31.

Crisp, D. J. 1958. The spread of *Elminius modestus* Darwin in north-west Europe. *Journal of the Marine Biological Association of the United Kingdom* **37**, 483–520.

Cristiani, G. 1979. Espèces indicatrices de pollution littorale par les métaux lourds au débouché d'un émissaire urbain. *Vie Marine* **1979**, 38–51.

Czapla, T. H. & Lang, B. A. 1990. Effect of plant lectins on the larval development of European corn borer (Lepidoptera: Pyralidae) and Southern corn rootworm (Coleoptera: Chrysomelidae). *Journal of Economic Entomology* **83**, 2480–85.

Dangeard, P. 1958. Recherches sur quelques "*Codium*", leur reproduction et leur parthénogenèse. *Le Botaniste* **42**, 65–88.

Dangeard, P. & Parriaud, H. 1956. Sur quelques cas de developpement apogamique chez deux especes de "*Codium*" de la region du Sud-Ouest. *Comptes Rendus de l'Academie des Sciences* **243**, 1981–3.

Davies, T. J. 1983. *The influence of the epibionts* Codium fragile *and* Crepidula fornicata *on the predation of the bay scallop* Agropecten irradians *by the sea star* Asterias forbesi *and the oyster drill* Urosalpinx cinerea. MSc thesis, Adelphi University, New York.

Dawson, E. Y. & Foster, M. S. 1982. *Seashore plants of California. California Natural History Guides: 47.* Berkeley: University of California Press.

Delépine, M. R. 1959. Observations sur quelques *Codium* (Chlorophycées) des côtes françaises. *Revue Generale de Botanique* **66**, 366–94.

Delgado, A. M. de los A. & Duville, C. A. 1977. Estudio de la composición química de *Codium fragile* (Suringar) Hariot (Clorophyta) de Puerto Deseado (Provincia de Santa Cruz, Argentina). Buenos Aires: *Contribución téchnica, Centro de Investigación de Biología Marina* **31**, 1–9.

Dellow, V. 1952. The genus *Codium* in New Zealand. Part 1: – Systematics. *Transactions of the Royal Society of New Zealand* **80**, 119–41.

Dellow, V. 1953. The genus *Codium* in New Zealand. Part II: – Ecology, geographic distribution. *Transactions of the Royal Society of New Zealand* **80**, 237–43.

Denton, G. W. R. & Burdon-Jones, C. 1986. Trace metals in algae from the Great Barrier Reef. *Marine Pollution Bulletin* **17**, 98–107.

Diamond, J. & Case, T. J. 1986. Overview: Introductions, extinctions, exterminations, and invasions. In *Community ecology*, J. Diamond & T. J. Case (eds). New York: Harper & Row, 65–79.

Dromgoole, F. I. 1975. Occurrence of *Codium fragile* subspecies *tomentosoides* in New Zealand waters. *New Zealand Journal of Marine and Freshwater Research* **9**, 257–64.

Dromgoole, F. I. 1979. Establishment of an adventive species of *Codium* in New Zealand waters. *New Zealand Department of Scientific and Industrial Research Information Series* **137**, 411–21.

Dromgoole, F. I. 1980. Desiccation resistance of intertidal and subtidal algae. *Botanica Marina* **23**, 149–59.

Dromgoole, F. I. 1982. The buoyant properties of *Codium. Botanica Marina* **25**, 391–7.

Dromgoole, F. I. & Foster, B. A. 1983. Changes to the marine biota of the Auckland Harbour. *Tane* **29**, 79–96.

Dromgoole, F. I., Silvester, W. B. & Hicks, B. J. 1978. Nitrogenase activity associated with *Codium* species from New Zealand marine habitats. *New Zealand Journal of Marine and Freshwater Research* **12**, 17–22.

Eno, N. C. 1997. Non-native marine species in British waters: effects and controls. *Aquatic Conservation: Marine and Freshwater Ecosystems* **6**, 215–28.

Eno, N. C., Clark, R. A. & Sanderson, W. G. 1997. *Non-native marine species in British waters: a review and directory*. Peterborough: Joint Nature Conservancy Committee.

Eyre, J. 1939. The South African intertidal zone and its relation to ocean currents. VII. An area in False Bay. *Annals of the Natal Museum* **9**, 283–306.

Fægri, K. & Moss, E. 1952. On the occurrence of the genus *Codium* along the Scandinavian coasts. *Blyttia* **10**, 108–13.

Farnham, W. F. 1980. Studies on aliens in the marine flora of southern England. In *The shore environment, Vol. 2. Ecosystems, Systematics Association Special Volume No. 17(b)*, J. H. Price et al. (eds). London: Academic Press, 875–914.

Farnham, W. F. 1994. Introduction of marine benthic algae into Atlantic European waters. In *Introduced species in European coastal waters*, C. F. Boudouresque et al. (eds). European Commission Ecosystem Research Report, 32–6.

Feldmann, J. 1956. Sur la parthénogénèse du *Codium fragile* (Sur.) Hariot dans la Méditerranée. *Comptes Rendus du Congres de la Societe savantes* **243**, 305.

Fletcher, R. L. 1980. The algal communities on floating structures in Portsmouth and Langstone Harbours (south coast of England). In *The shore environment. Vol. 2. Ecosystems, Systematics Association Special Volume No. 17(b)*, J. H. Price et al. (eds). London: Academic Press, 843–74.

Fletcher, R. L., Blunden, G., Smith, B. E., Rogers, D. J. & Fish, B. C. 1989. Occurrence of a fouling, juvenile, stage of *Codium fragile* ssp. *tomentosoides* (Goor) Silva (Chlorophyceae, Codiales). *Journal of Applied Phycology* **1**, 227–37.

Fralick, R. A. 1970. *An ecological study of* Codium fragile *subsp.* tomentosoides *in New England.* MSc thesis, University of New Hampshire.

Fralick, R. A. & Mathieson, A. C. 1972. Winter fragmentation of *Codium fragile* (Suringar) Hariot ssp. *tomentosoides* (van Goor) Silva (Chlorophyceae, Siphonales) in New England. *Phycologia* **11**, 67–70.

Fralick, R. A. & Mathieson, A. C. 1973. Ecological studies of *Codium fragile* in New England, USA. *Marine Biology* **19**, 127–32.

Gallop, A. 1974. Evidence for the presence of a "factor" in *Elysia viridis* which stimulates photosynthate release from its symbiotic chloroplasts. *New Phytologist* **73**, 1111–17.

Gallop, A., Bartrop, J. & Smith, D. C. 1980. The biology of chloroplast acquisition by *Elysia viridis*. *Proceedings of the Royal Society of London, Series B* **207**, 335–49.

Galtsoff, P. S. 1964. The American oyster *Crassostrea virginica* Gmelin. *United States, Fish and Wildlife Service, Fishery Bulletin* **64**, 1–480.

Gerard, V. A., Dunham, S. E. & Rosenberg, G. 1990. Nitrogen-fixation by cyanobacteria associated with *Codium fragile* (Chlorophyta): environmental effects and transfer of fixed nitrogen. *Marine Biology* **105**, 1–8.

Gibby, J. M. 1971. *The present status of* Codium *in Britain.* MSc thesis, Liverpool University.

Goff, L. J., Liddle, L., Silva, P. C., Voytek, M. & Coleman, A. W. 1992. Tracing species invasion in *Codium*, a siphonous green alga, using molecular tools. *American Journal of Botany* **79**, 1279–85.

Gosner, K. L. 1978. *A field guide to the Atlantic seashore. The Peterson Field Guide Series.* Boston: Houghton Mifflin Company.

Grant, B. R. & Borowitzka, M. A. 1984. Changes in structure of isolated chloroplasts of *Codium fragile* and *Caulerpa filiformis* in response to osmotic shock and detergent treatment. *Protoplasma* **120**, 155–64.

Gray, A. J. 1986. Do invading species have definable genetic characteristics? *Philosophical Transactions of the Royal Society of London, Series B* **314**, 655–74.

Greene, R. W. 1970a. Symbiosis in sacoglossan opisthobranchs: functional capacity of symbiotic chloroplasts. *Marine Biology* **7**, 138–42.

Greene, R. W. 1970b. Symbiosis in sacoglossan opisthobranchs: symbiosis with algal chloroplasts. *Malacologia* **10**, 357–68.

Greene, R. W. 1970c. Symbiosis in sacoglossan opisthobranchs: translocation of photosynthetic products from chloroplast to host tissue. *Malacologia* **10**, 369–78.

Greene, R. W. & Muscatine, L. 1972. Symbiosis in sacoglossan opisthobranchs: photosynthetic products of animal-chloroplast associations. *Marine Biology* **14**, 253–9.

Gunnill, F. C. 1980. Recruitment and standing stocks in populations of one green alga and five brown algae in the intertidal zone near La Jolla, California during 1973–1977. *Marine Ecology Progress Series* **3**, 231–43.

Gunnill, F. C. 1985. Population fluctuations of seven macroalgae in Southern California during 1981–1983 including effects of severe storms and an El Niño. *Journal of Experimental Marine Biology and Ecology* **85**, 149–64.

Hanisak, M. D. 1977. *Physiological ecology of* Codium fragile *subsp.* tomentosoides. PhD thesis, University of Rhode Island, Providence.

Hanisak, M. D. 1979a. Effect of indole-3-acetic acid on growth of *Codium fragile* subsp. *tomentosoides* (Chlorophyceae) in culture. *Journal of Phycology* **15**, 124–7.

Hanisak, M. D. 1979b. Growth patterns of *Codium fragile* ssp. *tomentosoides* in response to temperature, irradiance, salinity, and nitrogen source. *Marine Biology* **50**, 319–32.

Hanisak, M. D. 1979c. Nitrogen limitation of *Codium fragile* ssp. *tomentosoides* as determined by tissue analysis. *Marine Biology* **50**, 333–7.

Hanisak, M. D. 1980. *Codium*: an invading seaweed. *Maritimes* **24**, 10–11.

Hanisak, M. D. & Harlin, M. M. 1978. Uptake of inorganic nitrogen by *Codium fragile* subsp. *tomentosoides* (Chlorophyta). *Journal of Phycology* **14**, 450–4.

Hardy, F. G. 1990. The green seaweed *Codium fragile* on the coast of Berwickshire. *History of the Berwickshire Naturalists' Club* **44**, 154–6.

Harlin, M. M. & Rines, H. M. 1993. Spatial cover of eight common macrophytes and three associated invertebrates in Narragansett Bay and contiguous waters, Rhode Island, USA. *Botanica Marina* **36**, 35–45.

Hawes, C. R. 1979. Ultrastructural aspects of the symbiosis between algal chloroplasts and *Elysia viridis*. *New Phytologist* **83**, 445–50.

Hawes, C. R. & Cobb, A. H. 1980. The effects of starvation on the symbiotic chloroplasts in *Elysia viridis*: a fine structural study. *New Phytologist* **84**, 375–9.

Head, W. D. & Carpenter, E. J. 1975. Nitrogen fixation associated with the marine macroalga *Codium fragile*. *Limnology and Oceanography* **20**, 815–23.

Herbold, B. & Moyle, P. B. 1986. Introduced species and vacant niches. *American Naturalist* **128**, 751–60.

Herbreteau, F., Coiffard, L. J. M., Derrien, A. & De Roeck-Holtzhauer, Y. 1997. The fatty acid composition of five species of macroalgae. *Botanica Marina* **40**, 25–27.

Hinde, R. 1978. The metabolism of photosynthetically fixed carbon by isolated chloroplasts from *Codium fragile* (Chlorophyta: Siphonales) and by *Elysia viridis* (Mollusca: Sacoglossa). *Biological Journal of the Linnean Society* **10**, 329–42.

Hinde, R. 1983. Retention of algal chloroplasts by molluscs. In *Algal symbiosis: a continuum of interaction strategies*, L. J. Goff (ed.). Cambridge: Cambridge University Press, 97–107.

Hinde, R. & Smith, D. C. 1972. Persistence of functional chloroplasts in *Elysia viridis* (Opisthobranchia, Sacoglossa). *Nature* **239**, 30–31.

Hinde, R. & Smith, D. C. 1974. "Chloroplast symbiosis" and the extent to which it occurs in Sacoglossa (Gastropoda: Mollusca). *Biological Journal of the Linnean Society* **6**, 349–56.

Hinde, R. & Smith, D. C. 1975. The role of photosynthesis in the nutrition of the mollusc *Elysia viridis*. *Biological Journal of the Linnean Society* **7**, 161–71.

Hori, T. & Ueda, R. 1975. The fine structure of algal chloroplasts and algal phylogeny. In *Advance of phycology in Japan*, J. Tokida & H. Hirose (eds). The Hague: Dr. W. Junk, 11–42.

Hornsey, I. S. & Hide, D. 1974. The production of antimicrobial compounds by British marine algae. I. Antibiotic-producing marine algae. *British Phycological Journal* **9**, 353–61.

Hornsey, I. S. & Hide, D. 1976a. The production of antimicrobial compounds by British marine algae. II. Seasonal variation in production of antibiotics. *British Phycological Journal* **11**, 63–7.

Hornsey, I. S. & Hide, D. 1976b. The production of antimicrobial compounds by British marine algae. III. Distribution of antimicrobial activity within the algal thallus. *British Phycological Journal* **11**, 175–81.

Hurd, A. M. 1916. *Codium mucronatum. Puget Sound Marine Station Publications* **1**, 109–35.

Iriki, Y. & Miwa, T. 1960. Chemical nature of the cell wall of the green algae, *Codium, Acetabularia* and *Halicoryne. Nature* **185**, 178–9.

Isaac, W. E. 1937. Studies of South African seaweed vegetation. I. – West coast from Lamberts Bay to the Cape of Good Hope. *Transactions of the Royal Society of South Africa* **25**, 115–51.

Isaac, W. E. 1949. Studies of South African seaweed vegetation. II. – South coast: Rooi Els to Gansbaai, with special reference to Gansbaai. *Transactions of the Royal Society of South Africa* **32**, 125–60.

Jensen, K. R. 1980. A review of sacoglossan diets, with comparative notes on radular and buccal anatomy. *Malacological Review* **13**, 55–77.

Jensen, K. R. 1989. Learning as a factor in diet selection in *Elysia viridis* (Montagu) (Opisthobranchia). *Journal of Molluscan Studies* **55**, 79–88.

Jensen, K. R. 1993. Morphological adaptations and plasticity of radular teeth of the Sacoglossa (= Ascoglossa) (Mollusca: Opisthobranchia) in relation to their food plants. *Biological Journal of the Linnean Society* **48**, 135–55.

Kapraun, D. F., Gargiulo, M. G. & Tripodi, G. 1988. Nuclear DNA and karyotype variation in species of *Codium* (Codiales, Chlorophyta) from the North Atlantic. *Phycologia* **27**, 273–82.

Kapraun, D. F. & Martin, D. J. 1987. Karyological studies of three species of *Codium* (Codiales, Chlorophyta) from coastal North Carolina. *Phycologia* **26**, 228–34.

Kareiva, P. 1996. Developing a predictive ecology for non-indigenous species and ecological invasions. *Ecology* **77**, 1651–2.

Katayama, T. 1958. Chemical studies on the volatile constituents of seaweed. IX. On the volatile constituents of *Codium fragile. Journal of the Faculty of Fisheries and Animal Husbandry, Hiroshima University* **2**, 67–77.

Kawaguti, S. & Yamasu, T. 1965. Electron microscopy on the symbiosis between an elysioid gastropod and chloroplasts of a green alga. *Biological Journal of the Okayama University* **11**, 57–65.

Khotimchenko, S. V. 1993. Fatty acids of green macrophytic algae from the Sea of Japan. *Phytochemistry* **32**, 1203–7.

Kim, D. S., Lee, D. S., Cho, D. M., Kim, H. R. & Pyeun, J. H. 1995. Trace components and functional saccharides in marine algae. 2. Dietary fiber contents and distribution of the algal polysaccharides. *Journal of the Korean Fisheries Society* **28**, 270–8.

Kim, N. G. 1988a. Benthic marine algal flora of Saryang-Do in the southern coast of Korea. *Bulletin of Tong-Yeong Fisheries Junior College* **23**, 1–4.

Kim, N. G. 1988b. On the morphology of utricles and the producing period of gemetangia in *Codium fragile* (Suringar) Hariot on the coast of Chungmu. *Bulletin of Tong-Yeong Fisheries Junior College* **23**, 77–83.

Kim, N. G. 1991. Benthic marine algal flora of Ch'ungmu area in the southern coast of Korea. *Bulletin of Tong-Yeong Fisheries Technical College* **27**, 118–26.

Kitching, J. A., McLeod, A., Muntz, L., Norton, T., Ebling, F. J., Gamble, J., Hiscock, K. & Hoare, R. 1974. Work in progress at Lough Ine. *Proceedings of the Challenger Society* 1973, **4**, 223 only.

Knoepffler-Peguy, M., Maggiore, F., Boudouresque, C. F. & Dance, C. 1987. Compte rendu d'une experience sur les preferanda alimentaires de *Paracentrotus lividus* (Echinoidea) a Banyuls-sur-Mer. In *Colloque international sur Paracentrotus lividus et leur oursins comestibles*, C. F. Boudoresque (ed.). Marseille, France: GIS Posidonie, 59–64.

Lamberth, S. J., Bennett, B. A., Clark, B. M. & Janssens, P. M. 1995. The impact of beach-seine netting on the benthic flora and fauna of False Bay, South Africa. *South African Journal of Marine Science* **15**, 115–22.

Lee, T. F. 1986. *The seaweed handbook: an illustrated guide to seaweeds from North Carolina to the Arctic.* New York: Dover Publications.

Leppakoski, E. 1994. Non-indigenous species in the Baltic Sea. In *Introduced species in European coastal waters*, C. F. Boudouresque et al. (eds). Luxembourg: European Commission, 67–75.

Lewis, J. R. 1964. *The ecology of rocky shores.* London: English Universities Press.

Lewis, J. R. & Powell, H. T. 1960. Aspects of the intertidal ecology of rocky shores in Argyll, Scotland. I. General description of the area. *Transactions of the Royal Society of Edinburgh* **64**, 45–74.

Lewis, R. J. 1982. *The biology of the host-specific epiphytic red alga* Ceramium codicola *and some other epiphytes of* Codium *spp. in British Columbia*. MSc thesis, University of British Columbia.

Lewis, R. J., Jiang, B. Y., Neushul, M., Fei, X. G. 1993. Haploid parthenogenetic sporophytes of *Laminaria japonica* (Phaeophyceae). *Journal of Phycology* **29**, 363–9.

Littler, M. M. & Arnold, K. E. 1980. Sources of variability in macroalgal primary productivity: sampling and interpretative problems. *Aquatic Botany* **8**, 141–56.

Littler, M. M. & Arnold, K. E. 1982. Primary productivity of marine macroalgal functional-form groups from southwestern North America. *Journal of Phycology* **18**, 307–11.

Liu, J., Zhang, Y. & Yang, Y. 1995. Study of the intertidal benthic algae in the harbour region of Qinhuangdao. *Journal of the Ocean University of Qingdao* **25**, 59–67.

Lodge, D. M. 1993. Species invasions and deletions: community effects and responses to climate and habitat change. In *Biotic interactions and global change*, P. M. Kareiva et al. (eds). Sunderland: Sinauer Associates, 367–87.

Loosanoff, V. L. 1975. Comment. Introduction of *Codium* in New England waters. *Fishery Bulletin* **73**, 215–18.

Lopez, P. J. 1973. *The influence of bacteria on the growth of* Codium fragile *in culture*. MSc thesis, Adelphi University.

Lubchenco, J. 1978. Plant species diversity in a marine intertidal community: importance of herbivore food preference and algal competitive abilities. *American Naturalist* **112**, 23–39.

Lund, S. 1940. On the genus *Codium* Stackh. in Danish waters. *Det Kongelige Danske Videnskabernes Selskab Biologiske Meddelelser* **15**, 1–35.

Macnae, W. 1954. On four sacoglossan molluscs new to South Africa. *Annals of the Natal Museum* **13**, 51–64.

Madlener, J. C. 1977. *The seavegetable book*. New York: Clarkson N. Potter.

Maggs, C. A. 1986. *Scottish marine macroalgae: a distributional checklist, biogeographical analysis and literature abstract*. Nature Conservancy Council Report.

Malinowski, K. C. 1974. Codium fragile – *the ecology and population biology of a colonizing species*. PhD thesis, Yale University.

Malinowski, K. C. & Ramus, J. 1973. Growth of the green alga *Codium fragile* in a Connecticut estuary. *Journal of Phycology* **9**, 102–10.

McLean, N. 1976. Phagocytosis of chloroplasts in *Placida dendritica* (Gastropoda: Sacoglossa). *Journal of Experimental Zoology* **197**, 321–30.

Meiling, D. & Tseng, C. K. 1984. Chlorophyta. In *Common seaweeds of China*, C. K. Tseng (ed.), Beijing and Amsterdam: Science Press/Kugler, 249–301.

Meimer, J. P. 1972. *Morphology and development of* Codium fragile *in Great South Bay, New York*. MSc thesis, Adelphi University, New York.

Mizukoshi, S., Kato, F., Matsuoka, H., Nakamura, K. & Noda, H. 1995. Immunosuppressive effects of marine algae on experimental murine models. *Nippon Suisan Gakkaishi* **61**, 889–94.

Moeller, H. W. 1969. *Ecology and life history of* Codium fragile *subsp.* tomentosoides. PhD thesis, Rutgers University.

Mollison, D. 1986. Modelling biological invasions: chance, explanation, prediction. *Philosophical Transactions of the Royal Society of London, Series B* **314**, 675–93.

Nicotri, M. E. 1980. Factors involved in herbivore food preference. *Journal of Experimental Marine Biology and Ecology* **42**, 13–26.

Noro, T. & Nanba, S. 1989. Distribution and seasonality of seaweeds in Sakurajima Is., Kagoshima, Japan. *Memoirs of the Faculty of Fisheries Kagoshima University* **38**, 69–76.

Norton, T. A. 1978. Mapping species distributions as a tool in marine ecology. *Proceedings of the Royal Society of Edinburgh, Section B* **76**, 201–13.

Oh, Y. S., Lee, Y. P. & Lee, I. K. 1987. A taxonomic study on the genus *Codium*, Chlorophyta, in Cheju Island. *Korean Journal of Phycology* **2**, 61–72.

Pak, C. K., Yang, K. R. & Lee, I. K. 1977. Trace metals in several edible marine algae of Korea. *Journal of Oceanological Society of Korea* **12**, 41–7.

Pan, Z. & Yu, Y. 1992. Experimental studies on abalone, *Haliotis discus hannai* Ino, fed with *Codium fragile*. *Marine Sciences Hai yang Kexue* **5**, 33–6.

Parkes, H. M. 1975. Records of *Codium* species in Ireland. *Proceedings of the Royal Irish Academy Section B* **75**, 125–34.

Percival, E. & McDowell, R. H. 1967. *Chemistry and enzymology of marine algal polysaccharides*. London: Academic Press.

Perretti, C. 1972. *The influence of* Codium fragile *on oyster predation by the oyster drill* Urosalpinx cinerea (*Say*). MSc thesis, Adelphi University, New York.

Persson, N. Y. 1973. *The effect of temperature on the growth and morphology of* Codium fragile (*Sur.*) *Hariot*. MSc thesis, University of Rhode Island.

Pezzano, M. 1984. *Littorina littorea, Arbacia punctulata* and *Strongylocentrotus drobachiensis* as grazers on *Codium fragile*. *Biological Bulletin* **167**, 513 only.

Prince, J. S. 1988. Sexual reproduction in *Codium fragile* ssp. *tomentosoides* (Chlorophyceae) from the northeast coast of North America. *Journal of Phycology* **24**, 112–14.

Prince, J. S. 1991. *Codium fragile* ssp. *tomentosoides* palatability and reproduction. *Journal of Phycology* **27** (suppl.), 60 only.

Prince, J. S. & LeBlanc, W. G. 1992. Comparative feeding preference of *Strongylocentrotus droebachiensis* (Echinoidea) for the invasive seaweed *Codium fragile* ssp. *tomentosoides* (Chlorophyceae) and four other seaweeds. *Marine Biology* **113**, 159–63.

Printz, H. 1952. On some rare or recently immigrated marine algae on the Norwegian coast. *Nytt Magasin for Botanikk* (*Oslo*) **1**, 135–51.

Ramus, J. 1971. *Codium*, the invader. *Discovery* **6**, 59–68.

Ramus, J. 1972. Differentiation of the green alga *Codium fragile*. *American Journal of Botany* **59**, 478–82.

Ramus, J. 1978. Seaweed anatomy and photosynthetic performance: the ecological significance of light guides, heterogeneous absorption and multiple scatter. *Journal of Phycology* **14**, 352–62.

Ramus, J., Beale, S. I. & Mauzerall, D. 1976a. Correlation of changes in pigment content with photosynthetic capacity of seaweeds as a function of water depth. *Marine Biology* **37**, 231–8.

Ramus, J., Beale, S. I., Mauzerall, D. & Howard, K. L. 1976b. Changes in photosynthetic pigment concentration in seaweeds as a function of water depth. *Marine Biology* **37**, 223–9.

Ribera, M. A. 1994. Les macrophytes marins introduits en Méditerranée: biogéographie. In *Introduced species in European coastal waters*, C. F. Boudouresque et al. (eds). Luxembourg: European Commission, 37–43.

Ribera, M. A. & Boudouresque, C. F. 1995. Introduced marine plants, with special reference to macroalgae: mechanisms and impact. *Progress in Phycological Research* **11**, 187–268.

Rico, A. & Pérez, L. 1993. *Codium fragile* var. *novae-zelandiae* (Chlorophyta, Caulerpales) en Punta Borja, Chubut, Argentina: Aspectos reproductivos. *Naturalia patagónica, Ciencias Biológicas* **1**, 1–7.

Rosenberg, G. & Paerl, H. W. 1981. Nitrogen fixation by bluegreen algae associated with the siphonous green seaweed *Codium decorticatum*: effects on ammonium uptake. *Marine Biology* **61**, 151–8.

Rubinstein, I. & Goad, L. J. 1974. Sterols of the siphonous marine alga *Codium fragile*. *Phytochemistry* **13**, 481–4.

San Martin, G. A. 1987. Comportement alimentaire de *Paracentrotus lividus* (Lmk.) (Echinodermata: Echinidae) dans l'Etang de Thau (Herault, France). In *Colloque international sur Paracentrotus lividus et leur oursins comestibles*, C. F. Boudoresque (ed.). Marseille, France: GIS Posidonie, 37–57.

Sato, S. 1975. Fatty acid composition of lipids in some species of marine algae. *Bulletin of the Japanese Society of Scientific Fisheries* **41**, 1177–83.

Scagel, R. F. 1966. Marine algae of British Columbia and northern Washington. Part I: Chlorophyceae (Green algae). *National Museum of Canada*, Bulletin No. 207, Biological Series No. 74, 1–257.

Sealey, R. V., Williams, M. L. & Cobb, A. H. 1990. Adaptations of *Codium fragile* (Suringar) Hariot fronds to photosynthesis at varying flux density. In *Current research in photosynthesis. Vol. II*, M. Baltscheffsky (ed.). The Hague: Kluwer Academic, 455–8.

Searles, R. B., Hommersand, M. H. & Amsler, C. D. 1984. The occurrence of *Codium fragile* subsp. *tomentosoides* and *C. taylorii* (Chlorophyta) in North Carolina. *Botanica Marina* **27**, 185–7.

Shannon, E. L. & Altman, L. C. 1930. Growth in *Codium mucronatum*. *Publications of the Puget Sound Biological Station* **7**, 391–2.

Silva, P. C. 1951. The genus *Codium* in California, with observations on the structure of the walls of the utricles. *University of California Publications in Botany* **25**, 79–114.

Silva, P. C. 1952a. *Codium* Stackhouse. In *An analysis of the siphonous Chlorophycophyta*, L. E. Egerod, *University of California Publications in Botany* **25**, 381–95.

Silva, P. C. 1952b. *Morphotaxonomic studies of the South African representatives of the genus* Codium (*Chlorophycophyta*). PhD thesis, University of California, Berkeley.

Silva, P. C. 1954. Phylogenetic significance of anatomical differences in *Codium*. *Huitième Congrès International de Botanique, Paris* **17**, 102–3.

Silva, P. C. 1955. The dichotomous species of *Codium* in Britain. *Journal of the Marine Biological Association of the United Kingdom* **34**, 565–77.

Silva, P. C. 1957. *Codium* in Scandinavian waters. *Svensk Botanisk Tidskrift* **51**, 117–34.

Silva, P. C. 1959. The genus *Codium* (Chlorophyta) in South Africa. *Journal of South African Botany* **25**, 101–65.

Silva, P. C. 1960. *Codium* (Chlorophyta) in the Tropical Western Atlantic. *Nova Hedwigia* **1**, 497–536.

Silva, P. C. 1962. Comparison of algal floristic patterns in the Pacific with those in the Atlantic and Indian Oceans, with special reference to *Codium*. Proceedings of the 9th Pacific Science Congress, 1957, *Pacific Science* **4**, 201–16.

Silva, P. C. 1979. The benthic algal flora of central San Francisco Bay. In *San Francisco Bay: the urbanized estuary*, T. J. Conomos (ed.). San Francisco, California: Pacific Division of the American Association for the Advancement of Science, 287–311.

Silva, P. C. 1992. Geographic patterns of diversity in benthic marine algae. *Pacific Science* **46**, 429–37.

Silva, P. C. & Irvine, D. E. G. 1960. *Codium amphibium*: a species of doubtful validity. *Journal of the Marine Biological Association of the United Kingdom* **39**, 631–6.

Silva, P. C. & Womersley, H. B. S. 1956. The genus *Codium* (Chlorophyta) in southern Australia. *Australian Journal of Botany* **4**, 261–89.

Simon, H. 1972. *Investigations on the growth of* Codium fragile *subsp.* tomentosoides *in Great South Bay, New York.* MSc thesis, Adelphi University, New York.

Smith, G. M. 1944. *Marine algae of the Monterey Peninsula, California.* Stanford: Stanford University Press.

Smith, G. M. 1955. *Cryptogamic botany, Vol. I. Algae and fungi.* New York: McGraw-Hill.

Southwood, T. R. E. 1961. The number of species of insect associated with various trees. *Journal of Animal Ecology* **30**, 1–8.

Spencer, B. E. 1990. *Cultivation of Pacific Oysters. Laboratory Leaflet No. 63.* Lowestoft: Ministry of Agriculture, Fisheries and Food Directorate of Fisheries Research.

Steele, R. L. 1975. Growth requirements of *Enteromorpha compressa* and *Codium fragile. Proceedings: Biostimulation – nutrient assessment workshop. Ecological Research Series* 660/3-75-034, 213–25.

Stegenga, H., Bolton, J. J. & Anderson, R. J. 1997. *Seaweeds of the South African west coast.* Contributions from the Bolus Herbarium Number 18. Cape Town: Creda Press.

Stellander, O. 1969. Nytt funn av *Codium fragile* (Sur.) Hariot i Nord-Norge. *Blyttia* **27**, 174–7.

Steneck, R. S. & Watling, L. 1982. Feeding capabilities and limitations of herbivorous molluscs: a functional group approach. *Marine Biology* **68**, 299–319.

Stephenson, T. A. 1948. The constitution of the intertidal fauna and flora of South Africa. Part III. *Annals of the Natal Museum* **11**, 207–324.

Stephenson, T. A., Stephenson, A. & Day, J. H. 1940. The South African intertidal zone and its relation to ocean currents. VIII. Lamberts Bay and the west coast. *Annals of the Natal Museum* **9**, 345–80.

Stewart, J. G. 1991. *Marine algae and seagrasses of San Diego County.* San Diego: California Sea Grant College, Report Number T-CSGCP-020.

Svedelius, N. 1900. Algen aus den Ländern der Magellansstrasse und Westpatagonien. I. Chlorophyceae. In *Svenska Expeditionen till Magellandländerna* **3**, 283–316.

Taylor, D. L. 1967. The occurrence and significance of endosymbiotic chloroplasts in the digestive glands of herbivorous opisthobranchs. *Journal of Phycology* **3**, 234–5.

Taylor, D. L. 1968. Chloroplasts as symbiotic organelles in the digestive gland of *Elysia viridis* [Gastropoda: Opisthobranchia]. *Journal of the Marine Biological Association of the United Kingdom* **48**, 1–15.

Theis, C. L. 1985. *Photosynthetic compensation relative to depth in three species of the green alga* Codium *from Santa Catalina Island.* MA thesis, California State University, Fullerton.

Thomas, H. H. 1974. *Investigation on the attachment and early development in a natural population of* Codium fragile *subspecies* tomentosoides. MSc thesis, Adelphi University, New York.

Thorne-Miller, B., Harlin, M. M., Thursby, G. B., Brady-Campbell, M. M. & Dworetzky, B. A. 1983. Variations in the distribution and biomass of submerged macrophytes in five coastal lagoons in Rhode Island, USA. *Botanica Marina* **26**, 231–42.

Trench, R. K. 1975. Of "leaves that crawl". Functional chloroplasts in animal cells. *Symposia of the Society of Experimental Biology*, No. 29, 229–65.

Trench, R. K., Boyle, J. E. & Smith, D. C. 1973a. The association between chloroplasts of *Codium fragile* and the mollusc *Elysia viridis*. I. Characteristics of isolated *Codium* chloroplasts. *Proceedings of the Royal Society of London, Series B* **184**, 51–61.

Trench, R. K., Boyle, J. E. & Smith, D. C. 1973b. The association between chloroplasts of *Codium fragile* and the mollusc *Elysia viridis*. II. Chloroplast ultrastructure and photosynthetic carbon fixation in *E. viridis*.

Proceedings of the Royal Society of London, Series B **184**, 63–81.

Trowbridge, C. D. 1989. Marine herbivore-plant interactions: the feeding ecology of the sea slug *Placida dendritica*. PhD thesis, Oregon State University.

Trowbridge, C. D. 1992. Phenology and demography of a marine specialist herbivore: *Placida dendritica* (Gastropoda: Opisthobranchia) on the central coast of Oregon. *Marine Biology* **114**, 443–52.

Trowbridge, C. D. 1993. Interactions between an ascoglossan sea slug and its green algal host: branch loss and role of epiphytes. *Marine Ecology Progress Series* **101**, 263–72.

Trowbridge, C. D. 1995. Establishment of the green alga *Codium fragile* ssp. *tomentosoides* on New Zealand rocky shores: current distribution and invertebrate grazers. *Journal of Ecology* **83**, 949–65.

Trowbridge, C. D. 1996. Introduced versus native subspecies of *Codium fragile*: how distinctive is the invasive subspecies *tomentosoides*? *Marine Biology* **126**, 193–204.

Usher, M. B. 1986. Invasibility and wildlife conservation: invasive species on nature reserves. *Philosophical Transactions of the Royal Society of London, Series B* **314**, 695–710.

Verlaque, M. 1994. Inventaire des plantes introduites en Méditerranée: origines et répercussions sur l'environnement et les activités humaines. *Oceanologica Acta* **17**, 1–23.

Wahl, M. & Hay, M. E. 1995. Associational resistance and shared doom: effect of epibiosis on herbivory. *Oecologia* **102**, 329–40.

Wassman, E. R. & Ramus, J. 1973a. Primary-production measurements for the green seaweed *Codium fragile* in Long Island Sound. *Marine Biology* **21**, 289–97.

Wassman, R. & Ramus, J. 1973b. Seaweed invasion. *Natural History* **82**, 24–36.

Weber, W. 1969. Morphogenetische und keimungsphysiologische Untersuchungen an einigen Meeresalgen unter besonderer Berücksichtigung der Polarität. *Botanica Marina* **12**, 135–78.

Wheeler, P. A. & Björnsäter, B. R. 1992. Seasonal fluctuations in tissue nitrogen, phosphorus, and N : P for five macroalgal species common to the Pacific Northwest coast. *Journal of Phycology* **28**, 1–6.

Williams, L. G. 1948. The genus *Codium* in North Carolina. *Journal of the Elisha Mitchell Science Society* **64**, 107–15.

Williams, L. G. 1952. Effects of indoleacetic acid on growth in *Codium. American Journal of Botany* **39**, 107–9.

Williams, M. L. & Cobb, A. H. 1988. Observations on chloroplast starch content and synthesis in the intertidal marine alga *Codium fragile* (Suringar) Hariot. *New Phytologist* **108**, 285–90.

Williams, M. L. & Cobb, A. H. 1989. Isolation of functional chloroplasts from the sacoglossan mollusc *Elysia viridis* Montague. *New Phytologist* **113**, 153–60.

Williams, M. L. & Cobb, A. H. 1992. Chloroplasts in animals: photosynthesis and assimilate distribution in chloroplast "symbioses". In *Carbon partitioning within and between organisms*, C. J. Pollock et al. (eds). Cambridge: BIOS Scientific, 27–51.

Williams, M. L., Rutter, J. C., Benson, E. E. & Cobb, A. H. 1984. Photosynthesis by *Codium fragile* in an intertidal zone environment: a growth strategy. In *Advances in Photosynthesis Research, Vol. IV*, C. Sybesma (ed.). Belgium: Martinus Nijhoff/Junk, 287–90.

Williams, M. M. 1925. Contributions to the cytology and phylogeny of the siphonaceous algae. I. The cytology of the gametangia of *Codium tomentosum* (Stack.). *Proceedings of the Linnean Society of New South Wales* **50**, 98–111.

Williamson, M. H. & Brown, K. C. 1986. The analysis and modelling of British invasions. *Philosophical Transactions of the Royal Society of London, Series B* **314**, 505–22.

Wilson, L. 1977. Seasonal distribution of red algal epiphytes on *Codium fragile. Journal of Phycology* **13**, 74 only.

Wilson, L. 1978. *Epiphytes on* Codium fragile (*Chlorophyceae*): *an ecological and structural perspective.* MSc thesis, University of Rhode Island.

Wood, R. D. 1962. *Codium* is carried to Cape Cod. *Bulletin of the Torrey Botanical Club* **89**, 178–80.

Yang, M.-H., Blunden, G., Huang, F.-L. & Fletcher, R. L. 1997. Growth of a dissociated, filamentous stage of *Codium* species in laboratory culture. *Journal of Applied Phycology* **9**, 1–3.

Yokohama, Y., Kageyama, A., Ikawa, T. & Shimura, S. 1977. A carotenoid characteristic of chlorophycean seaweeds living in deep coastal waters. *Botanica Marina* **20**, 433–6.

Yotsui, T. & Migita, S. 1989. Cultivation of a green alga *Codium fragile* by regeneration of medullary threads. *Nippon Suisan Gakkaishi* **55**, 41–4.

Oceanography and Marine Biology: an Annual Review 1998, **36**, 65–96

UCL Press

CORAL/SEAWEED COMPETITION AND THE CONTROL OF REEF COMMUNITY STRUCTURE WITHIN AND BETWEEN LATITUDES

MARGARET WOHLENBERG MILLER
*Division of Marine Biology and Fisheries, Rosenstiel School of Marine and Atmospheric Science, University of Miami, 4600 Rickenbacker Causeway, Miami, FL 33149, USA**
** Present address: Southeast Fisheries Science Center, 75 Virginia Beach Drive, Miami, FL 33149, USA*

Abstract Many tropical coral reefs have recently manifested increasing dominance by seaweeds, in many cases a symptom of anthropogenic stresses. A pattern of increasing plant dominance also occurs over a gradient of increasing latitude, both at the present time and over evolutionary history. Competition from seaweeds has been hypothesized to be the direct cause of the restriction of scleractinian coral-built reef structures to tropical waters. Intense grazing and, to a lesser extent, low nutrient levels characteristic of low latitude habitats appear to be key determinants in maintaining the dominance of corals on tropical reefs. These ecological factors are less typical of temperate waters and so may provide mechanisms for the observed latitudinal shift in dominance of reef benthic communities. Also, anthropogenic alterations of these factors via overfishing and eutrophication are widespread in coastal areas, and so may account for recent shifts towards seaweed dominance observed in many tropical coral reefs. Although effects of altered grazing regimes on tropical reefs have been well demonstrated experimentally, controlled experiments on nutrient perturbations are few. Even fewer are studies designed to detect the interactive effects of perturbations in grazing and nutrient regimes on benthic community structure. Because overfishing and eutrophication often co-occur in the real world within a context of other natural disturbances such as storms, El Niño Southern Oscillation events, or epizootic die-offs, a rigorous experimental approach is necessary, although often lacking, for the discernment of the relative importance of each. Better understanding of such interactions in both plant- and animal-dominated communities, in temperate (rocky) and tropical (coral) reefs, is vital to successful management and conservation of reef ecosystems.

Coral reefs and algal reefs

Terrestrial communities display latitudinal gradients in quantitative characteristics such as species diversity while remaining qualitatively similar in structure. For example, both temperate and tropical regions have forests that are composed of canopy trees and understory herbs. Marine hard-bottom communities, in contrast, show qualitative variation in community structure over the latitudinal gradient. In tropical latitudes, scleractinian corals construct coral reefs, one of the most diverse communities on earth. However, scleractinian corals are virtually absent from temperate shallow marine hard-bottom communities. Those corals that do occur in temperate habitats are usually solitary, colonial but diminutive in size, or restricted to very deep water (e.g. Squires & Keyes 1967, Schuhmacher & Zibrowius 1985, Cairns 1994). Instead, kelps or other large brown seaweeds are commonly the primary occupiers of space and purveyors of physical structure on shallow temperate reefs (Schiel & Foster 1986). In addition to low temperature, competition from these large seaweeds at high latitudes has been suggested to limit corals and coral reef structures to tropical latitudes (Johannes et al. 1983). While the lack of coral reef development in many tropical habitats indicates that natural local factors within the tropics can also restrict corals (e.g. areas of upwelling or severe river runoff), a general latitudinal pattern of largely disjointed distribution of coral-dominated and plant-dominated community types is evident on a global scale (Fig. 1).

This global pattern of coral dominance at low latitudes and plant dominance at higher latitudes appears to be persistent over evolutionary time. Biogenic reefs have undergone repeated periods of expansion and regression (Copper 1994) and variation in latitudinal extent. However, much of this fluctuation occurred in the absence of scleractinian corals, which appeared relatively late, in the Middle Triassic (Coates & Jackson 1985). In other Periods, algae, sponges, bryozoans, brachiopods, and rudist bivalves were the predominant reef-formers (Coates & Jackson 1985). Notably, Copper (1994) reports that when biogenic reef formation extended well into temperate areas (40–50° latitude in the Siluro-Devonian) the reefs were formed by calcareous red and green algae, not corals.

Tropical coral reefs are threatened by human population pressures and land use changes in coastal areas (Hatcher et al. 1989) and, to a lesser extent, by aspects of global change (Smith & Buddemeier 1992, Gleason & Wellington 1993). One of the

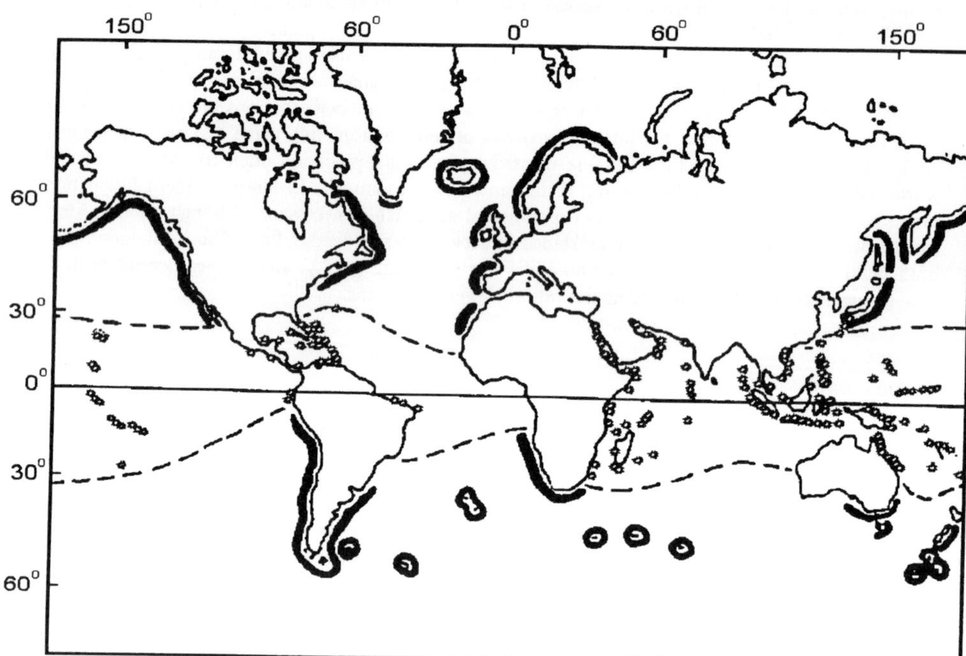

Figure 1 Map showing the global distribution of kelp beds (dark bars), communities dominated by large brown seaweeds, as more or less disjoint with global distribution of coral reefs (stars) that are generally restricted to within the depicted 20 °C winter isotherms (dashed line). The noticeable overlap occurs in southern Japan. Adapted from Ebeling & Hixon (1991) and Dubinski (1990).

most obvious manifestations of this reef degradation is the replacement of live coral cover with fleshy seaweeds (Lapointe 1989, Done 1992, Ogden & Ogden 1993, Hughes 1994). Overfishing, nutrient enrichment, and sedimentation are some of the modes of human impact that are believed to contribute to this alteration in reef community structure (Hatcher et al. 1989). Bell (1991) has suggested high macroalgal cover to be the best criterion for indicating eutrophication of coral reef communities. With the widespread recognition of this community shift in degraded tropical reefs, ecologists have been devoting increasing attention and effort to the understanding of the biotic and abiotic factors that maintain the dominance of scleractinian corals, as opposed to seaweeds, on tropical reefs. These investigations may provide insights on the poorly-understood mechanisms that maintain geographic separation of coral-dominated and seaweed-dominated habitat types on the latitudinal scale (see Fig. 1).

Thus, similar patterns of coral versus seaweed dominance are observed along gradients of increasing latitude (both currently and over evolutionary time) and along gradients of reef degradation within the tropics. Similar processes or mechanisms may or may not be driving these similar patterns. If the processes are similar, much of what we have learned about factors maintaining coral dominance in tropical reefs may be applicable in trying to understand latitudinal patterns in marine community structure. Similarly, whatever we may learn about mechanisms of latitudinal variation in reef structure may be relevant to the understanding of processes of reef degradation in the tropics.

In this review I use the mechanism of coral/seaweed competition as an organizing theme for integrating what we know about the factors structuring reef benthic communities (especially determination of coral- versus seaweed-dominance) both within and between latitudes. For the purposes of this review, the terms "seaweeds" and "macroalgae" will be used interchangeably to refer to fleshy, upright functional forms including calcareous types such as *Halimeda*; "temperate reef" will refer to rocky hard-bottom communities in temperate latitudes that are generally devoid of zooxanthellate coral structures; "high-latitude coral reef" will be used to refer to coral-dominated communities on structures accreted by zooxanthellate corals in marginal latitudes (for coral-built reefs, i.e. 24–34°) which are often distinct in structure and function from more equatorial reefs (Crossland 1988). I begin with a discussion of both physical and biotic factors that have been suggested to account for the restriction of zooxanthellate corals to low latitudes on ecological scales. Then, I shall examine the results from investigations of factors controlling the structure of benthic communities within (mainly tropical) latitudes and evaluate whether these factors may vary by latitude. Lastly, I shall explore how or whether these data might be extrapolated to explain between-latitude differences in community structure and composition, and how such an understanding might aid in predicting and managing the effects of anthropogenic alterations in these structuring factors in both temperate and tropical habitats.

Factors affecting between-latitude differences in coral dominance

Temperature

The classical explanation for the restricted distribution of coral reef communities to low latitudes (usually to less than 29° latitude) is temperature (Dana 1843). There is fairly consistent laboratory evidence that individual hermatypic (reef-building) corals do not persist at temperatures below 18 °C (Jokiel & Coles 1977, Coles & Jokiel 1978, others summarized in Veron 1995).

Numerous field correlative studies also support the contention that low temperature limits the rate at which tropical corals grow both between sites and within a site over time. Logan & Tomascik (1991) present evidence that coral growth rates (linear extension) in Bermuda are significantly lower than at other sites further south in the Caribbean (including Jamaica, Barbados, St Croix, Belize, and Florida). They suggest that temperature may provide the best correlation for this contrast in coral growth rates within the Caribbean. Other studies that have compared coral growth or calcification rates between sites and found positive correlation with sea surface temperature within normal ranges include those of Glynn & Stewart (1973) between different upwelling regimes in the Pearl Islands, Panama and Grigg (1981) for islands across Hawaii. Veron & Minchin (1992) show that the attenuation of coral community diversity at sites along the latitudinal gradient in Japan is explained by attenuation of minimum sea surface temperature.

Temperature effects on coral growth have also been examined over several temporal scales. Growth and calcification of *Acropora formosa* and *Pocillopora damicornis* at a high-latitude coral reef site in western Australia showed seasonal variation that, of the several physical parameters measured, correlated best with low winter temperature (Crossland 1981, 1984). This effect may have been confounded by competition from winter macroalgal blooms. Dodge & Lang (1983) studied density growth bands in colonies of *Montastrea annularis* from the northern Gulf of Mexico and performed correlations with historical physical and climatic data. Sea surface temperature provided the best correlate (positive correlation), although annual discharge from the adjacent Atchafalaya River was also significant (negative correlation). On an evolutionary scale, periods of reef collapse have been associated with both warming and cooling cycles of global climate change (Copper 1994).

Other types of data, however, suggest that temperature may not be an absolute, direct limitation to corals. First, there are many coral reefs in extreme environments where temperatures regularly reach down to 13 °C, at least for short periods, including the Arabian Gulf (Downing 1985, Coles 1988), the Red Sea (Sheppard 1988) and Japan (Tribble & Randall 1986). Indeed, Coles & Fadlallah (1991) report on an extreme cold event in the Arabian Gulf when sea temperature was 11 °C for four days and averaged 13 °C for a period of thirty days. They report that *Acropora pharaonis* and *Platygyra daedalea* displayed extensive but not complete mortality. However, *Porites compressa*, which they refer to as the "principal reef former", suffered only sublethal effects and recovery was complete within six months. Coles et al. (1976) showed that the upper thermal tolerances for individuals of a single species of coral from different sites varied. Individuals from Hawaii (subtropical) displayed an upper temperature tolerance two degrees less than corals from Enewetak Atoll (tropical). At least some hermatypic species, then, are capable of considerable selective adaptation or physiological acclimation (the two cannot be distinguished in this study) to local temperature extremes.

Secondly, while all hermatypic coral species are restricted to the tropics or subtropics, there are scleractinian species that live in temperate habitats. While most temperate species are asymbiotic and live in deep-water (Squires & Keyes 1967, Cairns 1994), several shallow water symbiotic forms do exist. In contrast to the massive hermatypic growth forms of some tropical species, these temperate species tend to be diminutive and do not contribute to the accretion of reef structure. Also, whereas tropical coral species are obligate associates of endosymbiotic plants (zooxanthellae), from which they receive most or all of their energy requirements (Muscatine & Porter 1977, Davies 1991), temperate species seem facultative in this association, occurring in both symbiotic and asymbiotic forms, depending on the environment in which they occur (Reed 1982, Jacques et al. 1983, Szmant-Froelich & Pilson 1984, Schuhmacher & Zibrowius 1985, Miller 1995). Examples of facultative symbioses include the corals *Astrangia danae* that ranges up into Maine in the northwest Atlantic, and several Mediterranean species such as *Madracis asperula*, *M. pharensis*, and *Oculina patagonica* (Schuhmacher & Zibrowius, 1985). *O. arbuscula* is endemic to nearshore and inshore hard substrata in North and South Carolina, off the east coast of the USA. Although average growth rates of *O. arbuscula* are diminished in the winter when inshore waters average 8–10 °C, growth remains positive during the coldest months (Miller, 1995) but no reef accretion occurs. Symbiotic colonies of another congener, *O. varicosa*, form reef-like thickets in deep water (80 m) off the subtropical east coast of Florida, USA (Reed, 1982). Because temperate zooxanthellate corals exist and grow at low temperatures but do not form reefs, it appears that other ecological factors are involved.

Thus, while low temperature clearly limits the distribution of tropical reef-building coral species, it is an incomplete explanation for the restriction of coral reefs to tropical habitats. A few zooxanthellate corals do occur in shallow temperate habitats, and some tropical corals show greater adaptive capabilities to temperature extremes than is often recognized. The deterrence of low temperatures does not explain why scleractinian corals have not more readily adapted and radiated into temperate regions over evolutionary time. The existence of a few shallow symbiotic temperate species suggests that there is no overarching adaptive constraint. However, Veron (1995) wisely distinguishes between a temperature minimum for corals versus that for reefs (i.e. coral productivity of sufficient vigour to accrete reef structure). Indeed, the direct effects of low temperature on coral growth rate may at least partially explain the lack of accreting reef structure in the few temperate habitats where symbiotic scleractinians are abundant.

Seaweed competition

The observation of increasing seaweed standing stocks with increasing latitude is the basis for an alternative hypothesis for the restriction of corals to the tropics. Indeed, most latitudinally marginal coral reefs (28–34 °N or S) described in the literature experience seasonal algal blooms or chronically high algal standing stocks, in contrast to pristine lower latitude reefs (i.e. less than 20° latitude) that generally have extremely low standing stocks. These sites include Japan at 34 °N (Tribble & Randall 1986), Taiwan at 25 °N (Birkeland & Randall 1981), Oman at 24 °N (Glynn 1993), Abaco, Bahamas at 26 °N (Lighty 1981), Lord Howe Island, South Pacific at 32 °S (Veron & Done 1979), the Arabian Gulf at 28–29 °N (Sheppard 1988), and the Red Sea at 28–29 °N (Coles 1988). The best studied high latitude coral reefs are off western Australia at the Houtman-Abrolos Islands (28–29 °S) (Johannes et al. 1983, Crossland et al. 1984, Hatcher 1985). This site has extensive coral reef structure and substantial coral diversity (37 genera) but, unlike more tropical reefs, it also contains areas dominated by the temperate kelp, *Ecklonia radiata*.

Johannes et al. (1983), working in the Houtman-Abrolos Islands, suggested that abiotic conditions (including, but not limited to, temperature) act indirectly by altering biotic interactions along the latitudinal gradient. The predominant biotic interaction that they saw as important was competition between corals and seaweeds (for space and/or light). They postulated a shift in competitive dominance from corals (at low latitude) to seaweeds (at higher latitudes). Thus, corals may be restricted to low latitudes by macroalgal competition. Because macroalgae have much higher potential growth rates than corals, they might be expected to outcompete corals in all cases. The fact that corals maintain dominance in low latitude sites is perhaps more remarkable, and ecological factors contributing to coral dominance will be discussed in later sections.

In contrast to corals, many seaweeds display suboptimal growth rates and productivity at tropical temperatures. Many tropical and subtropical seaweeds have higher growth rates in winter (De Wreede, 1976, Josselyn 1977, Ngan & Price 1980, Thorhaug & Marcus 1981). Growth rates of temperate kelps are often negatively correlated with temperature but this is generally attributed to increased nitrate availability in cooler waters (Jackson 1977, Zimmerman & Kremer 1984). Nonetheless, it appears that seaweeds may be advantaged by lower temperatures that are stressful to reef-building corals.

Several observational studies suggest that seaweed competition is an important detriment to corals. Sheppard (1988) surveyed coral reef communities in the Red Sea and Arabian Gulf along gradients of temperature, salinity, and sedimentation. He found a common pattern whereby communities in less stressful environments (i.e. moderate temperature, normal sea salinity, or low sedimentation) were dominated by corals in the genus *Acropora*. Community structure shifted to dominance by the smaller coral, *Porites*, at intermediate levels along the gradients and to dominance by large brown seaweeds (especially *Sargassum*) in stressful environments of all three types (i.e. high temperature, high salinity, or high sedimentation). Sheppard terms this phenomenon "algal replacement". His study suggests that hermatypic corals lose out in stressful environments but, because the study is observational, it cannot distinguish whether the loss of corals is due to direct physical stress or to algal competition.

In a study measuring *in situ* coral growth rates at a high-latitude coral reef, Crossland (1984) reported reduced growth in coral colonies that inadvertently became overgrown by seaweeds. Potts (1977) found reduced growth and inhibited recruitment of corals in the small algal patches of damselfish territories on tropical coral reefs. In a temperate setting, Coyer et al. (1993) documented greater occurrence of algal overgrowth and reduced activity of a solitary temperate coral when transplanted into a kelp bed zone from an "urchin barren", zone in California, USA. The extent of these zones was not controlled, however, and varied over the long timespan of this experiment. Survival was not different between the treatments, but this lack of significant effect may have been due to either inconsistency of the treatments or to unimportance of competition. Temperate kelps have also been shown to outcompete other invertebrates (mussels) in shallow habitats (Witman 1987).

The strongest evidence for seaweed competition severely limiting coral fitness comes from two controlled experimental field studies, one at a tropical, one at a temperate site. Tanner (1995) utilized coral outcrops in the Heron Island lagoon, Australia. This is a tropical site, but background seaweed cover in the lagoon is approximately 80%. Tanner (1995) experimentally removed seaweeds from part of these outcrops and followed the growth, mortality and reproductive output of several species of coral. He found strong increases in coral growth rate of two *Acropora* species on outcrops that were weeded, but no significant effect on growth of *Pocillopora damicornis*. Also, *Acropora palifera* on the outcrops where seaweed was removed had significantly greater production of planula larvae. Miller & Hay (1996), using controlled field experiments in a temperate habitat, demonstrated significant inhibition of growth and recruitment of the temperate coral *Oculina arbuscula* when subject to competition from ambient densities of seaweeds (primarily *Sargassum* sp.) at two reef sites in North Carolina, USA. This study also found disjoint distributions of the coral and seaweed populations both among depths within sites (scale of m) and between sites (scale of 10–100 km). Interestingly, in both these experimental studies, while significant effects on coral growth and reproduction or recruitment were demonstrated, no significant effect on coral mortality was detected, in one case even over a time frame of two years (Tanner 1995).

Controlled experimental evidence so far comes from only a few sites but shows that seaweed competition can limit coral growth, reproduction, and recruitment at both low and high latitudes. Thus, along with temperature, seaweed competition is a plausible proximate mechanism limiting coral distribution on both local and latitudinal scales. If macroalgae are outcompeting corals at high latitudes and/or in stressful or degraded tropical habitats, the mechanisms affecting this shift in dominance remain to be determined. That is, what variation in physical and ecological factors between coral-dominated and algal-dominated areas causes this competitive reversal? Below is an examination of shifts in community dominance within tropical reefs and studies concerning ecological factors regulating coral/seaweed competition in the tropics and subtropics. If these factors vary with latitude, they may be candidates to provide mechanisms for the latitudinal competitive shift in coral versus seaweed dominance postulated by Johannes et al. (1983).

Coral versus seaweed dominance within tropical latitudes

Several authors (e.g. Hatcher 1984, Done 1992, Knowlton 1992, Hughes 1994), have described coral and seaweed-dominated communities in tropical reefs as alternate stable states (*sensu* Sutherland 1974) determined by natural perturbations, anthropogenic influences or some combination of the two. Lighty (1981) described a macroalgal-dominated reef in the Bahamas as an alternate stable state to the normal coral reef, and suggested that the demise of the coral community began with lagoonal flooding during the Holocene Transgression and was maintained by a high-energy environment that excluded grazers. In Jamaica, the combined effects of Hurricane Allen, the epizootic mortality of dominant grazers (the herbivorous sea urchin *Diadema antillarum*), and natural predation on corals have effectively converted Jamaica's acroporid reefs to fields of macro- and turf algae (Hughes 1989, 1994, Knowlton et al. 1990). Done (1992) describes tropical reefs as undergoing anthropogenic "phase shifts" from coral-dominated to algal-dominated communities as a general scenario of reef degradation

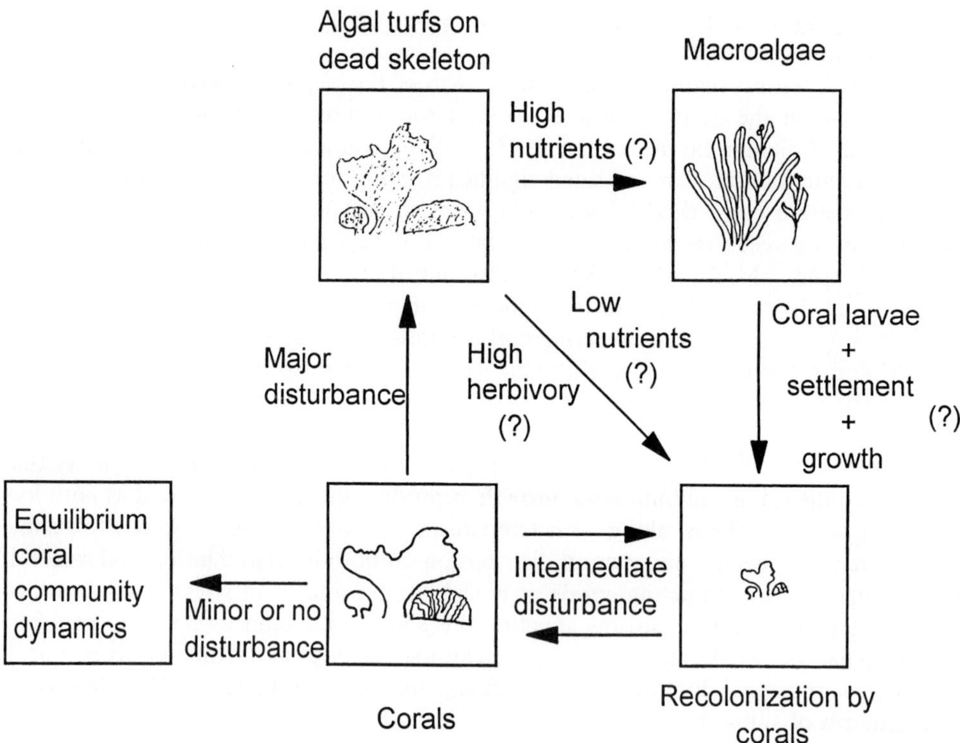

Figure 2 Schematic representation of "phase shift" (coral to algal dominance) scenario of tropical coral-reef degradation and recovery based on Done (1992). Note that temperate reefs characteristically exist in a macroalgal dominated phase, and do not display a coral-dominated phase.

(see Fig. 2). He also speculates that conditions such as reduced nutrient inputs, and increased coral recruitment favour the reverse phase shift, termed "recovery".

Hatcher (1984) describes what may be considered an example of this phase shift or macroalgal-dominated alternate stable state due to an anthropogenic disturbance. A ship grounding on a reef in Australia caused a temporary decrease in grazing pressure as fish were dispersed from the scene. During these several weeks, the unpalatable red alga *Asparagopsis taxiformis* overgrew much of the coral on the reef. When the fish returned, they would not eat the *A. taxiformis*, whereas before, small germlings of this alga had been consumed by the grazers incidentally. Hatcher (1984) suggests that this temporary reduction in grazing due to historical accident allowed the unpalatable alga to reach a refuge in size such that it could be recognized and avoided by the grazers. Thus, *A. taxiformis* dominance appears to be an alternate stable state for this community.

Incidental, alternate stable states have been described for temperate seaweed-dominated communities, usually induced by overgrazing by sea urchins. However, the described alternative to kelp-domination is a "barrens" or "crust" community, generally not involving corals (e.g. Fletcher 1987, Johnson & Mann 1988).

Ecologists investigating the basic ecology of tropical reefs have identified several characteristics of tropical coral reefs, including the abundance of herbivores and/or clear waters with low nutrient levels, as requisite factors for maintaining the dominance of corals over seaweeds and thus, the diversity and structure of coral reef benthic communities (Littler & Littler 1984, Lewis 1986, Birkeland 1988).

Grazing

As Hairston et al. (1960) so succinctly pointed out, the terrestrial world is green. That is, terrestrial herbivores rarely limit plant populations yielding communities with high standing plant biomass. Tropical coral reef ecosystems provide a clear contrast to this condition whereby high rates of herbivory restrict macrophyte biomass to virtually nil. This possibly unique characteristic of coral reefs may be the single most important factor in allowing corals to dominate. Indeed, corals are probably unable to "compete" with macroalgae even in tropical ecosystems. Rather, they probably require this virtual absence of macroalgae imposed by high herbivory levels.

Fishes and sea urchins

Levels of herbivory on tropical coral reefs are intense, with estimates of grazing intensity ranging up to 156000 bites m^{-2} day^{-1} (Carpenter 1986). Herbivorous fishes (primarily scarids and acanthurids) are recognized as profoundly affecting abundance and distribution of reef algae, often restricting them to refuge habitats (Hay 1991). Simultaneously, sea urchins are, or have been, capable of removing virtually all macroalgal standing stock on some Caribbean reefs (Carpenter 1986, Lessios 1988). Hay (1991) argues that this tight coupling of high reef algal production and consumption makes tropical reef plant/herbivore interactions among the most important structuring factors for these communities.

The importance of grazers in structuring tropical coral reef communities has been clearly demonstrated, although most studies are from the Caribbean. Multiple small-scale exclusions (or reductions) of herbivores have consistently produced radical changes in benthic community structure, namely a great increase in upright algae often at the expense of corals (Sammarco et al. 1974, Sammarco 1982, Hatcher & Larkum 1983, Hay & Taylor 1985, Carpenter 1986, Lewis 1986, Morrison 1988, Miller & Hay 1998). On a larger scale, the mass mortality of the herbivorous sea urchin, *Diadema antillarum*, throughout the Caribbean in 1983 resulted in extreme alteration in both benthic and fish community structures (reviewed by Lessios 1988). In Jamaica, Hughes et al. (1987) and Liddell & Ohlhorst (1986) reported a large and persistent increase in macroalgal cover and consequent decrease in coral cover (by 60%) following an 80–100% reduction in *Diadema* density. Similar effects were reported in Curaçao (Ruyter van Steveninck & Bak 1986) and St Croix (Carpenter 1990a). Population densities of herbivorous fishes did increase following this release from sea urchin competition in Panama (Robertson 1991) and in St Croix (Carpenter 1990b) but high standing macroalgal stocks have persisted despite increased fish grazing. Indeed, at least in Jamaica, the algal blooms following the *Diadema* die-off have persisted through the mid 1990s (Hughes 1996) and, in concert with other sources of coral death such as hurricane damage and intense predation (Hughes 1989, Knowlton et al 1990), have resulted in a macroalgal-dominated alternate stable state (Hughes 1994).

Prior to 1983, *Diadema* had been a dominant grazer on many Caribbean reefs, but this may have been an "unnatural" state resulting from historical overfishing of certain areas since the 1960s (Hughes 1994), and decimation of larger marine vertebrate grazers such as turtles and manatees possibly as far back as the 1800s (Jackson 1996). Hay (1984) showed that rates of fish grazing were higher and densities of sea urchins lower at several remote, less-populated sites he studied than at sites that appeared to be more heavily fished. McClanahan & Shafir (1990) also document increased sea urchin dominance in overfished areas in Kenya as compared with reserve areas where fishing is prohibited.

Several traits of sea urchins may make them especially important in affecting reef community structure in areas where they are abundant. Aspects of sea urchin physiology and population dynamics seem to make their densities relatively unresponsive to resource levels. Levitan (1988, 1989) showed that *D. antillarum* responds to starvation with negative individual growth but without significant levels of mortality. In fact, individual fitness is actually higher (due to increased external fertilization success) when population densities are high enough to display food-limitation, than when population densities are lower (Levitan 1989). Andrew (1989) postulated that great metabolic flexibility and resistance to starvation was a general trait of echinoids due to their high proportion of non-metabolic mass (skeleton and water) and low proportion of muscle and other metabolizing tissues. Thus, sea urchins can persist at extremely high population densities while overexploiting their algal food resources, especially in times and areas where heavy human fishing activity reduces predation pressure upon them (Hay 1984, Levitan 1992).

The extraordinarily high and persistent population densities that sea urchin populations can maintain may be detrimental to corals. Sammarco (1980) found high *Diadema* density severely inhibited coral recruitment because of incidental consumption of recent settlers. Intermediate sea urchin densities have been found to aid settlement, however, by controlling algal abundance, providing space, and excavating grooves that served as refuges from predation and benthic competitors (Birkeland & Randall 1981). The sea urchin, *Eucidaris thouarsii*, also attains extremely high densities in certain sites, and appears to limit coral reef growth in the Galápagos Islands (Glynn et al. 1978) by grazing directly on live corals while in other parts of their range, these sea urchins only eat algae. In Kenya, sea urchins released from predation pressure by human fishing also overgraze and cause reduction in coral cover (McClanahan & Shafir 1990). More recent work on reefs in the Florida Keys (a relatively high-latitude area, 25 °N), however, indicates that parrotfish predation on corals (especially *Porites* spp.) can also be substantial, and may help to explain low coral cover on some offshore bank reefs (Miller & Hay 1998, M. W. Miller et al. unpubl. data).

Overall, the intense herbivory by fishes and sea urchins typical of tropical coral reefs seems to be crucial in maintaining algal standing stocks at low levels and, thereby, preventing competitive pressure on corals. Extremely high densities of sea urchins or in some places, predatory impacts by *Sparisoma* parrotfishes, however, may obviate any advantage to corals of reduced seaweed abundance by hindering coral growth and recruitment via direct predation. The relative importance of fish versus sea urchin grazing varies geographically and historically according to natural and human influences (e.g. epizootic disease outbreaks and overfishing) in both temperate and tropical latitudes (Hay 1984, Aronson 1990, Levitan 1992), and these differences are often related to shifts in benthic community structure.

Complex interactions of grazers that influence coral/seaweed competition

There are several examples in the literature that describe grazers as mediating interactions between corals and algae or where algal competition in conjunction with grazers mediates interspecific competition between hard corals.

Coen (1988) described a relationship whereby the herbivorous crab *Mithrax sculptus* resides facultatively in the shelter provided by live *Porites porites*. When crabs were experimentally excluded from coral colonies, the colonies became overgrown in a matter of weeks by fleshy algae, especially *Dictyota* spp. Thus, it appears that by providing shelter to a symbiotic herbivore, the coral can overcome seaweed competition. *Mithrax sculptus* is also associated with the branching coralline alga *Neogoniolithon strictum* in seagrass beds in the Florida Keys and similarly aids this host against fleshy seaweed competitors (Stachowicz & Hay 1996).

Littler et al. (1989) determined that the mechanism of *Porites porites* providing shelter for small herbivores further mediated competitive interactions with the massive *P. astreoides*. Because *P. astreoides* cannot shelter small herbivores, this species of coral is restricted to areas of the reef with high fish density where the herbivorous fishes keep them clear of algal overgrowth. In these areas, they were also subjected to potentially higher levels of predation by corallivorous fishes. When transplanted into areas of the reef with high fish density, the *P. porites* showed much higher susceptibility to predation than *P. astreoides*. *P. porites*, by maintaining a resident herbivore, is able to utilize a spatial escape from predation. Without its crustacean symbiont, macroalgal competition appears to restrict corals to areas of high predation.

Wellington (1982) determined a similar zonation between branching and massive corals on a reef slope in Panamá. In this case the branching pocilloporid corals provide shelter for the territorial damselfish *Eupomacentrus acapulcoensis*. The damselfish most effectively kills the massive coral *Pavona* sp. in shallow areas, but benefits from shelter provided by the pocilloporids. Damselfish also serve to defend the pocilloporids from corallivorous fish to which they would otherwise be susceptible in shallow zones. In this case, damselfish allow branching corals (pocilloporids) to persist in the shallow parts of the reef where predatory pressure on corals is high while restricting the massive *Pavona* to deeper zones where fish are less common.

Territorial and algal farming behaviours of damselfish provide natural small-scale grazer exclusions on tropical reefs and are responsible for numerous other demonstrated indirect effects (Branch et al. 1992). Coral growth (Potts 1977) and recruitment (Vine 1974) in damselfish territories have been shown to be reduced, as a result of the encroachment of the filamentous algae/sediment mat. These territories can provide a refuge from predation, however, and enhance recruitment of some rarer coral species (Sammarco & Williams 1982). Reduced grazing within territories also appears to increase bioerosion rates (Risk & Sammarco 1982).

While there is evidence that mezograzers that eat epiphytes can improve the performance of certain temperate macrophytes (Steneck 1982, Norton & Benson 1983, Orth & Montfrans 1984, Duffy 1990), complex interactions involving grazers, corals, seaweeds, and predators have not been reported in temperate reefs. This may be because they have not been sought, or perhaps because they do not exist. The development of these complex associations including fishes and crustaceans in tropical reefs, however, seems to indicate that coral/seaweed competition may have been a significant selective force in these communities. These interactions seem important in mediating coral/seaweed competition and in maintaining the diversity of tropical benthic communities.

Nutrients

The classical conception of coral reefs is as oligotrophic systems because ambient concentrations of inorganic nitrogen and phosphorus in tropical reef waters are extremely low, often below measurable limits. Anthropogenic nutrification is often cited as a major cause of tropical reef degradation (Hatcher et al. 1989, Bell 1991). Excess nutrient input from natural sources has also been associated with coral reef demise and low carbonate production on geological timescales (Hallock & Schlager 1986). There are at least three mechanisms whereby increased nutrients on tropical coral reefs may impact corals and curtail reef development including (a) direct physiological disruption of coral/zooxanthellae symbiosis, (b) competition from increased production of seaweeds, and (c) increased bioerosion of coral skeletons. Below is an examination of the conceptual background and the evidence regarding the importance of these mechanisms in nutrient degradation of tropical coral reefs.

Physiological effects

Coral-dominated ecosystems are able to maintain high productivity in the "nutrient deserts" of tropical seas by efficient use and recycling of nutrients. The classical example of adaptation for low nutrient conditions is the symbiosis whereby many sessile reef invertebrates, including scleractinian corals, receive energy from endosymbiotic zooxanthellae (specialized dinoflagellates) that in turn are believed to benefit from the nutrients in the cellular waste of the animal host (Muscatine & Porter 1977).

This association with zooxanthellae, seemingly so adaptive in oligotrophic environments (Muscatine & Porter 1977), may be the source of physiological problems when nutrient levels are increased. Suboptimal growth of the symbiotic zooxanthellae populations is required from the coral host's perspective to ensure an excess of fixed carbon for translocation. It is believed that host regulation of zooxanthellae population growth is accomplished by restricting nutrient acquisition by the zooxanthellae. Experimental studies have shown that enriched nitrogen concentrations in the water column cause increased zooxanthellae population density within a tropical coral host (Hoegh-Guldberg & Smith 1989, Stambler et al. 1991, Stimson & Kinzie 1991). Indeed, Falkowski et al. (1993) suggested that under chronically high nutrient levels, the host restrictive mechanisms are ineffective, and zooxanthellae populations escape host control (i.e. fixed carbon beyond the maintenance needs of the zooxanthellae is converted to more zooxanthellae, not translocated to the animal host) possibly resulting in total expulsion of the symbionts and colony death. This mechanism of high nutrient stress might be expected to be more important to tropical than to shallow temperate corals since temperate corals seem to be much more plastic in their associations with zooxanthellae (able to exist in asymbiotic states, Schuhmacher & Zibrowius 1985, Miller 1995).

It should be noted that physiological studies of responses of tropical corals and zooxanthellae to nutrient enrichment have only shown significant effects at concentrations much higher than are likely to exist in even very polluted waters (often $>10\,\mu M$ inorganic N). Also, recent studies measuring actual translocation rates have shown that translocation per zooxanthella cell is reduced under conditions of ammonium enrichment in *Porites astreoides* but because zooxanthellae density increases with enrichment, total translocation per surface area of coral tissue does not change (McGuire 1997). Indeed, moderate (Meyer et al. 1983) to not so moderate (Atkinson et al. 1995) levels of nutrient enrichment have been found to enhance coral growth rates under certain circumstances. Overall, it is not clear if nitrogen enrichment of an ecologically relevant scale ($<10\,\mu M$) has a significant detrimental effect on coral/zooxanthellae physiology or not.

It has also been suggested that high phosphorus levels may cause a simple chemical inhibition of coral calcification. In a field experiment where nitrogen (40x ambient) and phosphorus (10x ambient) were applied directly to One Tree Reef of Australia's Great Barrier Reef, Kinsey & Davies (1979) observed a 50% increase in photosynthesis and a significant decrease in net calcification, but no algal overgrowth of corals. They used this observation to suggest direct chemical inhibition based on previous chemical studies (Simkiss 1964).

Seaweed blooms

While animal-plant symbionts, especially corals, dominate in oligotrophic conditions because of efficient (internal) nutrient recycling, macroalgae, with their higher potential growth rates, should make better use of slight to moderate increases in nutrient availability and dominate benthic communities in moderately high nutrient conditions (Birkeland 1988). At very high nutrient levels, benthic heterotrophs (especially filter feeders) are expected to become dominant as the food web shifts towards a planktonic base, a condition that Birkeland (1988) identified and attributed to high terrestrial runoff in the northern Gulf of Thailand. Tomascik et al. (1993) report this sort of replacement of live coral with other heterotrophic invertebrates on various Indonesian reefs. Birkeland (1977) documents complete suppression of coral recruitment at an Eastern Pacific coral reef subject to upwelling by rapidly accumulating spatial competitors (both plants and filter-feeding invertebrates).

The general importance to corals of seaweed competition has been discussed above. Here, the question is the extent to which increased nutrient loads to tropical reefs cause increased seaweed standing stocks. A necessary (but not sufficient) condition for nutrient enrichment to result in increased seaweed abundance is for seaweeds in pristine coral reefs to exist in a nutrient-limited state. Many studies have examined the degree of nutrient limitation of seaweeds in different geographical areas along natural or anthropogenic nutrient gradients. Nutrient limitation is assessed by assaying seaweed photosynthetic rates of plants after incubation in nutrient-enriched conditions (usually including +N, +P, and +N + P treatments) and comparing to controls incubated without nutrient enrichment. Also, activity of the plant enzyme alkaline phosphatase (which allows the plant to utilize organic phosphate compounds) is assayed as an indication of the P-limitation of the plant. These studies show fairly consistently that productivity of seaweeds from pristine, low-nutrient reefs in the Seychelles, Barbados, and Belize is enhanced by nutrient enrichment (especially by phosphorus) while that of seaweeds from naturally or anthropogenically enriched sites is not (Littler et al. 1991,

1992, Lapointe et al. 1992). Also, alkaline phosphatase activity is enhanced in seaweeds from low-nutrient areas. Benthic community descriptions also suggest that low-nutrient habitats that contained nutrient-limited seaweeds tended to have a coral-dominated benthic community while higher nutrient sites tended to house a benthic community dominated by macroalgae that were not nutrient-limited (Littler et al. 1991, 1992, Lapointe et al. 1992).

As noted above, nutrient-limited seaweed growth is a necessary but not sufficient condition for enrichment to result in seaweeds outcompeting corals. Hatcher & Larkum (1983) experimentally determined that addition of nitrogen to a tropical reef significantly increased the production of the epilithic algal community. However, increased production did not necessarily translate into increased standing stock because herbivores often removed this excess production. It is standing stock, not production, of algae that competes with (shades or abrades) coral.

The well-known case history of Kaneohe Bay, Hawaii (Smith et al. 1981, Maragos et al. 1985) is the primary case study indicating that enrichment can result in seaweed blooms and loss of corals. The reef there experienced an outbreak of the green seaweed *Dictyosphaeria cavernosa* shortly after a sewage outfall was installed in 1963. After the outfall was moved outside the bay in 1978 several hermatypic coral species have rebounded remarkably, and *D. cavernosa* has declined. In contrast to this fairly clear case of sewage causing severe seaweed bloom and loss of coral, recent studies on tropical reefs in Hawaii by Grigg (1994, 1995) report no impact by sewage outfalls on benthic community structure. Also, more recent studies of the ecology of *D. cavernosa* in Kaneohe Bay indicate that grazing intensity has a strong influence on *D. cavernosa* standing stocks (Stimson et al. 1996). Like Hatcher & Larkum (1983), Stimson et al. 1996 indicate that nutrient effects on seaweed growth are expressed as increased biomass primarily in the absence of grazing.

A weakness of most of the existing evidence for nutrient addition being the cause of seaweed blooms and/or declines in coral vigour and abundance is that most involve site comparisons between enriched or "eutrophic" and pristine sites (Littler et al. 1991, 1992, Lapointe et al. 1992, Wittenberg & Hunte 1992, Tomascik & Sander 1985, 1987, 1991) and thus the effects of nutrients are confounded by whatever other environmental factors vary between compared sites. Notably, several of these studies suggest that high-nutrient or degraded sites also have reduced grazer densities (Littler et al. 1991, Tomascik & Sander 1987) so it is unclear whether variation in seaweed and coral abundance result directly from nutrient enrichment or from indirect effects via grazer populations.

The much-touted ENCORE project in Australia (Larkum & Steven 1994) may go far towards filling the gap by providing controlled experimental evidence concerning nutrient effects on corals, macroalgae and other reef organisms at both the organismal and the community levels. This large-scale field experimental study is using automated nutrient delivery devices to enrich micro-atolls in the One Tree Island lagoon. To date, however, no results have been published, and unpublished reports indicate a general lack of significant effects of enrichment on many of the measured parameters including productivity of the epilithic algal community, growth of rodolithic algae, and coral growth (Larkum & Koop 1996, Larkum & Steven 1996, Steven & Broadbent 1996) although coral reproduction and larval settlement were inhibited by enrichment (Ward & Harrison 1996). Ongoing work is being conducted at a boosted level of nutrient addition in order to obtain measurable responses.

Increased bioerosion

The productivity hypothesis of Highsmith (1980) suggests that high levels of ocean productivity reduce coral reef accumulation on a global scale by enhancing bioerosion. Most bioeroding organisms (primarily sponges and molluscs) are filter feeders and, thus, may be more prevalent in areas of high productivity. This study quantified the degree of bioerosion in museum coral specimens from all over the world and determined that the degree of bioerosion was positively correlated with global patterns of ocean productivity or nutrient concentration. In other words, because corals in areas of high productivity experience faster bioerosion, they would have to grow even faster than those in oligotrophic areas to accomplish equal reef accumulation. In fact, we would expect not faster, but slower coral growth under these conditions due to competition from seaweeds or benthic heterotrophs. Similarly, Sammarco & Risk (1990) show higher rates of bioerosion in more productive nearshore than more oligotrophic offshore areas of Australia's Great Barrier Reef. Scott & Risk (1988) offer quantitative documentation that bioeroded corals suffer reduced structural integrity and are probably more susceptible to breakage from physical disturbances such as storms.

Thus, extant studies of nutrient effects on tropical reef seaweeds and bioeroders are primarily observational, along natural or anthropogenic nutrient gradients and, thus, confounded by variation in other environmental factors. Studies of nutrient effects on coral/zooxanthellae physiology are largely experimental but show significant impacts only at unrealistically high levels of enrichment. Most studies indicate some degree of nutrient limitation of reef algae in oligotrophic habitats, but the impacts of nutrient addition, on corals directly or on coral/seaweed competition, are not consistent. It is becoming increasingly recognized that many tropical coral reefs have developed under relatively high nutrient regimes. Szmant (1997) suggests the capacity of a reef to process high nutrient loads without significant changes in community structure depends on its physical and trophic complexity. This implies that reefs whose complex trophic regimes are intact will be much more robust in handling high nutrient perturbations while reefs with trophic systems disrupted by human harvesting activities will have a greater susceptibility to nutrient-induced benthic community shifts. Clearly, much more experimental work is required to establish the relative importance of nutrient enrichment (compared with or interacting with grazing regimes) as a cause of reef degradation.

Sedimentation

Increasing sedimentation in both temperate and tropical nearshore environments is a widespread result of accelerating anthropogenic alterations in coastal zones (Hatcher et al. 1989). While sedimentation has always been recognized by geologists as a pervasive process in marine environments, ecologists have become interested in its effects on coral reef community structure primarily as human activities and land-use changes, including deforestation (Hodgson 1993), have intensified. Effects of sedimentation on tropical reef corals and communities are reviewed by Rogers (1990).

Several studies of tropical coral reef sites that have developed under chronically (naturally?) high levels of sedimentation all indicate relatively low cover and low diversity of live corals at highly sedimented sites when compared with otherwise comparable sites with lower sedimentation: Costa Rica (Cortés & Risk 1985), Fanning Lagoon in the South Pacific (Roy & Smith 1971), and Puerto Rico (Loya 1976). Thus, chronic sedimentation seems to suppress coral abundance and diversity.

There are several distinct mechanisms whereby sedimentation might be expected to influence benthic organisms and benthic community structure over and above the direct nutrient effects described in the previous section. These mechanisms include (a) an increased potential heterotrophic food source for corals or heterotrophic coral competitors (in the form of organic particles of terrestrial or planktonic origin), (b) decreased light transmittance or turbidity reducing photosynthetic potential, and (c) direct abrasion, smothering or burial by settling particles with associated physiological stress and increased susceptibility to infection.

Food

We might expect the food value of high particulate concentrations to vary according to the level of organic content in the particles. Increased organic particulate concentrations might aid corals to the extent that they are heterotrophic, but most current work on coral nutrition indicates a relatively small contribution of planktonic consumption (Davies 1991). Thus, increased concentration of organic particles may be expected to be of greater benefit to purely heterotrophic benthic competitors of corals (Birkeland 1988). However, reduced light transmittance or turbidity associated with high particulate concentrations can also significantly reduce light levels and could conceivably increase coral dependence on heterotrophic nutrition. Although heterotrophic feeding can maintain some temperate corals which are less dependent on symbiosis (Miller 1995), it is not clear whether obligately-symbiotic tropical corals are capable of such compensatory feeding.

Light

Light reduction from turbidity in high sedimentation regimes might be expected to be more detrimental to the free-living plants than the mixotrophic corals. Turbidity and low light levels have been shown to limit severely photosynthesis and growth of tropical sand-plain seaweeds (Hay 1981). Also, Calem & Pierce (1993) found that the distribution of the seagrass, *Thalassia*, was negatively correlated with turbidity levels in the Twin Cays area of Belize. However, corals may be similarly disadvantaged by loss of photosynthetic efficiency. Davies (1991) has determined carbon budgets for three species of shallow reef corals under differing light regimes. Each obtained from photosynthesis at least 100% of their carbon requirements under all but very low light conditions ("cloudy day" treatment). Under these low-light conditions, lipid reserves (not heterotrophic consumption) made up for the net carbon deficit. A chronic decrease in light levels would deplete these reserves rapidly. Thus, the net beneficiary of light reduction in the coral versus algal race is not clear and probably depends on other local conditions and/or the species involved.

More generally, the interaction of light levels with disturbance is argued by Huston (1985) to be the primary determinant of universal depth patterns in coral communities, that is, low diversity at shallow depths (< 1 m), highest diversity between 5 m and 30 m, and low diversity below 30 m. Light is a primary determinant of growth rate in photosynthetic organisms. Thus, if light and growth are low, progress towards competitive exclusion is slow, and more likely to be interrupted by disturbance before exclusion occurs. Although Huston (1985) only addresses competition between tropical coral species, it would seem that photosynthetic seaweeds may be subject to the same effects of reduced light on growth. If seaweeds are the competitive dominant in a certain habitat, it would seem that reducing light and seaweed growth rates would reduce the rate of competitive exclusion of corals. This appears to be the case in a turbid habitat in North Carolina where abundance of the temperate facultatively zooxanthellate coral, *Oculina arbuscula*, was significantly greater below the euphotic zone for seaweeds (Miller & Hay 1996). In turbid waters at this site, the euphotic zone for seaweeds was only 3–4 m deep. Even though the coral grew better under well lighted or shallow conditions, it occurred with much greater abundance in deeper, darker habitats that appear to provide the coral with a refuge from seaweed competition. In this temperate area, this refuge is fairly large because turbidity is high. If the water were clearer, seaweeds might be expected to have higher growth rates and to restrict corals over a much greater depth range. Glynn (1993) reported a somewhat similar pattern in a lower latitude area subject to upwelling (southern coast of Oman, 24 °N). Here, upwelled nutrients probably contributed to macroalgal domination in shallow zones. However, turbidity was relatively high at these sites (3–5 m visibility) and so macroalgal cover diminished at depths of 5–10 m where corals became abundant.

Rogers (1979) performed a field manipulation completely excluding light in what she termed a "partial simulation of extreme turbidity". She found, not surprisingly, significant reductions in productivity and complete mortality of corals which had dominated the benthos. Following removal of the shade, macroalgae colonized the dead or damaged corals.

Abrasion/particle effects

Sediment burial or smothering effects, although almost certainly detrimental, are often not lethal, except at extremely high doses (Rogers 1983). The mechanism of coral mortality from sedimentation is not well understood, but microbial infection may be involved since the antibiotic tetracycline significantly reduced tissue necrosis when corals were exposed to sedimentation in experimental tanks (Hodgson 1990a). Coral species vary widely in their abilities to cope with sediment rain stress (Stafford-Smith & Ormond 1992). Lasker (1980) determined that *Montastrea cavernosa* was able to remove the ambient level of settling particles that it experienced at two sites in Panamá (345 mg 25 $cm^{-2} d^{-1}$). He found that polyp size and colony morphology affected the ease of removal and therefore the metabolic cost of keeping clear of sediment.

Dallmeyer et al. (1982) showed that fine suspended peat particles significantly increased expulsion of zooxanthellae and increased respiration of corals in a field respirometer. This response created a carbon deficit that would seem difficult to make up. Corals under high sedimentation regimes experience combined stresses of reduced light (turbidity), which limits the photosynthetic potential of their symbionts, and reduced ability to feed (despite potentially elevated prey availability) since polyps are busy removing sediment. So, even if sedimentation rates are not severe enough to bury or smother organisms, active clearing of sediment particles combined with reduced light availability can represent a severe energetic cost to the organism possibly reducing growth, reproduction, and competitive ability.

A study by Peters & Pilson (1985) emphasizes the issue of sublethal metabolic costs of sedimentation. They used a temperate coral species, *Astrangia danae*, which can be either symbiotic (photosynthetic) or asymbiotic (purely heterotrophic). They subjected both types of colonies to sedimentation levels of 200 cm^{-2} day^{-1} but could measure no effects when corals had sufficient food and light resources. Detrimental effects were observed, however, when the corals were starved and symbiotic colonies did worse than asymbiotic colonies. They conclude that sedimentation is not detrimental if corals have sufficient metabolic resources to handle it.

Sediment particles have also been shown to have significant effects on coral recruitment. Several laboratory studies examining settlement behaviour of coral larvae in conditions of high sedimentation have shown that coral larvae settle less successfully in high-sedimentation treatments and most settlement takes place on vertical surfaces where sediment does not accumulate (Hodgson 1990b, Babcock & Davies 1991). Te (1992) found no inhibition of settlement, but his high-sediment treatments displayed significantly higher reverse metamorphosis responses, termed "polyp bail-out". Although undocumented for tropical algae, it is conceivable that sedimentation could hinder algal recruitment analogously. Devinny & Volse (1978) demonstrated such an effect in the temperate kelp, *Macrocystis pyrifera* whereby sediment coating of the substratum prevented sporling settlement and this detrimental effect was exacerbated by scouring in conditions of high-water movement.

Interactive effects of sedimentation

Aside from these direct effects, several authors have suggested that sediments are an important mechanism in algal suppression of corals. Walker & Ormond (1982) observed coral death to result from leakage of phosphate dust in ship loading operations at Aqaba, Red Sea. They also observed increased algal growth, but attribute coral death to sediment trapping by these filamentous algae, not direct algal overgrowth. Birkeland (1977) and Potts (1977) also suggest that sediment trapping by algae is the important mechanism in coral suppression.

Thus, sedimentation appears to have strong impacts on reef community structure at certain times and places. The net benefits and detriments of the multiple aspects of sedimentation on coral versus algal competitive interactions are not intuitively obvious. Observations of high-sedimentation and high-nutrient (either natural or eutrophied) reefs, however, suggest that seaweeds or filamentous algae often prevail. Physiological effects of suspended particles on corals are measurable and may be a contributing factor in coral susceptibility to other stresses such as competition or disease.

Combined or interactive influence of factors on tropical reefs

High nutrient levels often are associated with pollution or terrestrial runoff and can result in high levels of planktonic production (Birkeland 1988). Both of these conditions mean that high-nutrient environments are often subject to high levels of turbidity and sedimentation. Many studies of anthropogenic alterations often refer simply to "eutrophication" effects, which generally involve a combination of nutrient and sedimentation increase resulting from terrestrial runoff or discharge of waste water from urban areas. Kaneohe Bay, Hawaii (described earlier) is the classical case study of eutrophication of a tropical coral reef, but Pastorok & Bilyard (1985) review additional studies of sewage pollution.

Eutrophication gradients often emanate from coastal urban centres. Studies of coral reefs along one such gradient in Barbados have found, progressing from pristine to impacted reefs, increased algal abundance (Tomascik & Sander 1987), increased coral growth to a point (that is attributed to increased heterotrophic food source) and then decreased coral growth (Tomascik & Sander 1985), decreased coral settlement on artificial substrata (Tomascik & Sander 1991), and increased juvenile coral mortality (Wittenberg & Hunte 1992). Unfortunately, all of these effects are confounded by a reported decrease in abundance of *Diadema* with increasing eutrophication. Thus, it remains unclear if eutrophication affects corals (or seaweeds) directly or indirectly by affecting grazer abundance.

In contrast, surveys of benthic and fish communities surrounding primary and secondary sewage outfalls in Hawaii in open water (as opposed to the restricted circulation of Kaneohe Bay) showed no effect of discharge on benthic community structure, including coral diversity and cover, according to Grigg (1994). He found enhanced abundances of fishes, especially planktivorous species, at the outfalls. High topographic complexity and increased planktonic food resources are suggested to account for these increased fish abundances. However, in this case where open circulation ameliorated the discharge, increased nutrient and turbidity levels by the discharge of sewage did not cause shifts in coral/algal dominance.

Another pattern in coral reef community structure associated with simultaneous variation in several environmental factors involves variation along an inshore/offshore gradient in the Great Barrier Reef of Australia. Inshore, reefs are reported to display reduced influence of grazers on epilithic algae (Scott & Russ 1987), increased biomass of seaweeds (McCook 1996), reduced coral diversity (Done 1982), increased bioerosion of corals (Sammarco & Risk 1990), and reduced coral skeletal density indicating reduced calcification efficiency (Risk & Sammarco 1991) in comparison with mid- and outer-shelf reefs. This gradient in benthic community structure is associated with greater terrestrial runoff, sedimentation, turbidity and nutrient levels, relatively depauperate herbivorous fish abundance (Russ 1984), and low overall diversity of fishes (Williams & Hatcher 1983) on the inshore, as opposed to offshore reefs. There is also evidence that certain competitive interactions between benthic species are altered along this cross-shelf gradient. Aliño et al. (1992) showed that octocorals are able to outcompete the scleractinian coral *Acropora longicyanthus* at inshore reefs but not at a mid-shelf reef. However, as with all these cross-shelf studies, because so many factors vary over the gradient, it was impossible to isolate causation; whether this competitive reversal was due to the inshore characteristic of increased nutrients, reduced predation, reduced light, or some combination or interaction of factors.

Littler & Littler (1984) proposed a conceptual model suggesting that an interaction of both nutrient levels and grazing disturbance determines which group of primary producers attains dominance on tropical coral reefs (Fig. 3). This "relative dominance paradigm" predicts that under long-term conditions of high-grazing disturbance and low nutrients, corals dominate. When grazing is reduced, algae will dominate with large frondose forms (seaweeds) at high-nutrient levels and smaller turf forms (with greater surface to volume ratio) at low-nutrient levels. Coralline algae are the poorest

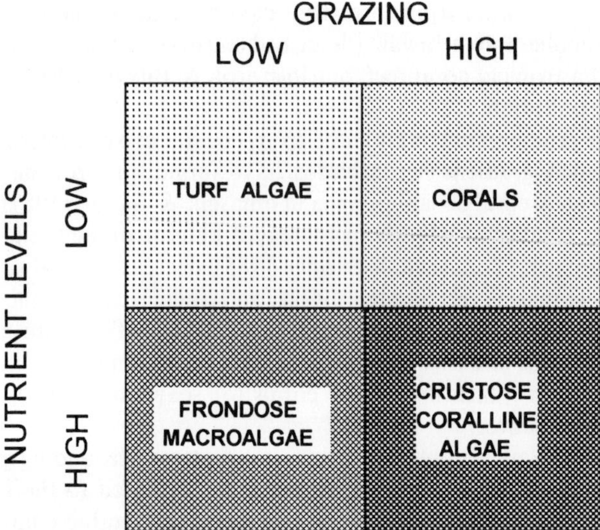

Figure 3 Schematic of the Relative Dominance Paradigm of Littler & Littler (1984) and Littler et al. (1991). This conceptual model emphasizes the interaction of grazing and nutrient regimes in determining dominant benthic producers in tropical reef communities.

competing primary producer guild but are extremely resistant to physical disturbance and so may dominate under the high-nutrient, high-disturbance regime.

Experimental manipulations to distinguish the relative importance of several structuring factors and their possible interactions are rare. Although Littler et al. described their 1991 study, discussed earlier as "a test of the relative-dominance paradigm", they did not control differences in grazing levels between their nutrient-limited and non-limited sites. Thus, the relative importance (or interaction) of grazing versus nutrients in controlling the increased seaweed abundance and decreased coral abundance at enriched sites was not addressed.

One study that has attempted rigorous testing of the relative importance of nutrients versus grazing in a tropical reef examined the epilithic algal community at One Tree Reef on Australia's Great Barrier Reef. Hatcher & Larkum (1983) performed controlled orthogonal field manipulations of herbivory (through caging) and nutrient levels (addition of nitrogen) and determined that algal productivity was greatly enhanced by nutrient addition. However, this increased production only translated into increased standing stock (that which may compete with corals) when grazers were excluded. Thus, production and standing stock of a tropical reef algal community were completely decoupled by the effects of herbivory.

McCook (1996) is the other study where relative roles of grazing versus water quality in determining the cross-shelf gradient in seaweed standing stock on Australia's Great Barrier Reef are addressed. Caged and uncaged transplants of the brown seaweed *Sargassum* were placed at inshore and offshore reefs to distinguish between the influence of water quality versus grazing regime in controlling the inshore/offshore gradient in standing stock of this seaweed. This study indicated that offshore water quality (including low nutrient levels) was less important than high-grazing pressure in limiting *Sargassum* abundance on offshore reefs since plants transplanted to offshore sites grew well and displayed no indication of nutrient limitation, but only if protected from grazers. In both these studies, only the algal community is directly addressed. Consequences for corals and benthic community structure overall can only be extrapolated, based on the assumption that corals are favoured by conditions that deter seaweeds.

Variation in structuring factors between latitudes and effects on temperate community structure

If algal competition restricts coral growth at high latitudes and if high grazing, low nutrients, and low sedimentation foster coral superiority in tropical areas, it is important to determine whether these factors vary between latitudes in order to determine their power to explain a latitudinal shift in competitive dominance.

Where documented, both abundance of herbivores and absolute consumption of algal biomass on high-latitude coral reefs are reduced from those on tropical reefs, in some cases by orders of magnitude. Birkeland & Randall (1981) report a mean of 178 scarids in a five-minute visual count in Guam (13 °N) while the mean was only 0.1 in northern Taiwan (25 °N). The corresponding measures of fleshy algal cover were zero and 43.9%, respectively. In this case, however, the latitudinal comparison may be confounded by historically heavy fishing in Taiwan. At the Houtman Abrolhos Islands off Western Australia, one of the best studied high-latitude reefs, Hatcher & Rimmer (1985) found that absolute grazing rates were 4 to 20 times less than in analogous habitats on tropical reefs.

It is a well recognized pattern that herbivory is reduced in marine benthic systems as latitude increases (Gaines & Lubchenco 1982, Horn 1989). This is due primarily to the lessened influence of herbivorous fishes (compared with invertebrates) in temperate areas. Reported accounts of temperate herbivorous fishes' impact on seaweed abundance and community structure indicate weak or very patchy effects (Andrew & Jones 1990, Jones 1992, Miller & Hay 1996). Invertebrates seem to be the dominant herbivores in temperate communities (Gaines & Lubchenco 1982, Hawkins & Hartnoll 1983) while Klumpp & Pulfrich (1989) found herbivorous invertebrates in the tropical central Great Barrier Reef (Australia) at their highest abundance consume only one-third the amount that fishes do. Diversity of herbivorous fishes is undeniably greater in the tropics although total biomass may not be (Horn 1989). It is suggested that fish are not able to go as long without food as invertebrates and so may not be able to cope as well with seasonality in algal production that might be expected to occur in temperate areas. Such seasonal variability in temperate algal productivity has not been well documented, however. Alternatively, fish digestion of algal material may be less efficient at low temperatures making herbivory an energetically precarious habit for fishes in temperate areas. Also, since fish are poikilothermic, those that are herbivorous in temperate areas should require less total consumption than the same biomass of fish in a tropical area where temperatures are higher (Hay & Steinberg 1992). In fact, Hatcher (1981) did find grazing rates (bites per unit time) of tropical herbivorous fish to be well correlated with temperature.

The dominance of invertebrates, as opposed to fishes, in the herbivore guild of temperate marine habitats has great influence on algal community structure. The invertebrate herbivores such as sea urchins and snails have rather limited perceptual ability and mobility (compared with fish) so their ability to seek out food plants is diminished. Many temperate seaweeds may escape being eaten simply because they are not found by herbivores (Hawkins & Hartnoll 1983). Spatial refuges from grazing are also important to tropical seaweeds (Hay 1981, Steneck 1988) but may be more available and effective in temperate communities due to the relatively limited mobility and perceptual capacity of invertebrate (compared with fish) herbivores and may contribute to greater macroalgal standing stocks in temperate areas (Gaines & Lubchenco 1982).

Temperate invertebrate herbivores, especially sea urchins (Andrew 1989), certainly are capable of reducing algal biomass, such as the creation of "urchin barrens" (e.g. Lawrence 1975, Andrew 1993). However, these effects on temperate reefs seem to be patchy in time and space (e.g. Lubchenco & Cubit 1980, Andrew & Jones 1990, Watanabe & Harrold 1991, Foster 1992, Coyer et al. 1993, Hagen 1995) in contrast to the relentless assault of tropical reef herbivores. The mechanism controlling the shift between kelp versus barren communities is unclear (Johnson & Mann 1988, Elner & Vadas 1990, Scheibling 1996), but in some cases has been attributed to variation in recruitment of kelp (Harrold & Reed 1985), not control by grazers.

As an example of differential herbivory between a temperate and tropical community, Lewis (1985) demonstrated that algae of many species including *Sargassum polyceratium*, *Padina jamaicensis*, and *Lobophora variegata* were completely consumed in a matter of hours on a Belizean reef crest by fishes. Paul & Hay (1986), at an intermediate latitude in Florida, assayed the susceptibility of various seaweeds to herbivory. They found *Sargassum* sp. and *Padina vickersiae* to be almost completely consumed (70–100%) within 3–18 h at three different reef sites. *Lobophora variegata*, however, was not consumed in Florida (Paul & Hay 1986). In temperate North Carolina, closely related species of *Sargassum* and *Padina* dominate shallow subtidal habitats where co-occurring fishes and sea urchins find them unpalatable (Hay et al. 1987). Also, the same species of *Lobophora* is abundant on offshore rock outcrops in North Carolina (pers. obs.). Apparently, the omnivorous consumers present in the temperate community do not overcome the modest morphological and/or chemical defenses in these seaweeds while the tropical herbivorous fishes do. This difference is compounded by results of recent work by Bolser & Hay (1996) showing that both temperate and tropical herbivorous sea urchins find temperate algae more palatable than tropical algae of the same genus. That is, tropical algae are less palatable to herbivores than temperate congeneric or conspecific algae (in most cases related to increased chemical defenses (Bolser & Hay 1996)), but herbivore pressures are so great in the tropics that many seaweeds are still eaten to local extinction on shallow tropical reefs except in refuge habitats. Thus, herbivore tolerance of unpalatable foods as well as herbivore abundance may vary with latitude and contribute to greater seaweed abundance and greater seaweed competitive effects on corals in temperate reefs.

There is also some empirical evidence concerning the alternative (or complementary) explanation that nutrients are limiting to algae in tropical habitats and that algae can gain competitive advantage over corals when ambient nutrient levels are higher (including temperate habitats). Ambient inorganic nutrient levels on tropical coral reefs reported in the literature are wide ranging and difficult to compare (Pilson & Betzer 1973, Kinsey & Davies 1979, Andrews & Muller 1983, Adey & Goertemiller 1987).

"Normal" levels ranged from about 0.2–0.5 μM inorganic N and less than 0.1 μM soluble reactive phosphorous. Indeed, Bell (1991) proposed eutrophication "threshold" nutrient levels for tropical reef waters of 1 μM inorganic N and 0.1–0.2 μM phosphate. However, given the variation in natural nutrient regimes under which tropical reefs have developed, such threshold values are probably meaningless (Szmant 1997).

Crossland et al. (1984) describe nutrient levels for high-latitude coral reefs in the Houtman Abrolhos Islands. They reported nitrate levels ranging from 0.83–1.5 μM and reactive phosphorus concentrations of 0.26–0.5 μM at sites around the island group, substantially higher than the "normal" levels just quoted for low-latitude coral reefs. These concentrations are also consistently higher than those measured in adjacent oceanic waters (Crossland et al. 1984), indicating enrichment by the reef system. The authors suggest that the abundant standing stock of macroalgae providing for a detrital food chain and decomposition may be the source of this enrichment. Unfortunately, this provides a circular argument for the explanation that higher nutrient levels cause more algae and less coral. Hopkinson et al. (1991) provide a comprehensive account of nutrients and metabolism for a temperate reef in Georgia, but their data for ambient nutrient concentrations in a temperate hardbottom community are in units ($\mu g\, m^{-2}$) that are not comparable with the above figures for coral reefs ($\mu g\, m^{-3}$). On deep hard bottoms off the North Carolina coast, nutrient levels vary seasonally in the range of nil to 1.4 μM for nitrate and 0.1–0.25 μM for phosphate (Peckol 1980).

There does seem to be potential for the mechanisms of grazing intensity and nutrient levels to explain the latitudinal pattern of high algal abundance at high latitudes and, in turn, competitive suppression of corals. The only extant study of grazing and nutrient impacts on competition between a temperate coral and dominant co-occurring seaweeds and relative control by grazing and nutrients was conducted at an inshore artificial reef and at offshore continental slope reefs in North Carolina, USA (Miller & Hay 1996). This study found that cage exclusion of larger herbivores at three sites significantly altered macroalgal species composition with palatable red algal species replacing the less palatable brown, *Sargassum*. However, the caging treatment had no significant effect on total macroalgal cover or biomass and no significant effect on growth of the facultatively zooxanthellate coral, *Oculina arbuscula*. In contrast, a factorial experiment manipulating both grazing and nutrient levels did show a significant decrease of coral growth in grazer exclusion cages and a significant interaction of grazing and nutrient enrichment in affecting red algal cover. Palatable red algae (abundant in grazer exclusion cages) appear to be nutrient-limited and exert greater competitive effects on the coral in contrast to the unpalatable brown algae (which dominate under the ambient grazing regime).

Overall, evidence suggests that the intensity of herbivory on temperate reefs is markedly less than on tropical coral reefs and this may account for higher seaweed abundance at high latitudes. While effects of temperate grazers may be more subtle than those on tropical reefs, that is, altering species composition instead of biomass of seaweeds, these differences are real and can produce variation in coral growth rates. Published descriptions of nutrient levels on reefs at different latitudes indicate that nutrient levels are chronically higher in temperate than in tropical coastal waters. In North Carolina, nutrient addition has an interactive effect with grazing regime in determining seaweed abundance and coral growth rates, but the broader implications of this geographical difference in nutrient regimes for processes structuring benthic communities have not been investigated.

Conclusion

Figure 4 summarizes the physical and ecological interactions important in structuring reef benthic communities. Temperature is clearly important in limiting tropical hermatypic coral species. Although experimental evidence comes from only a few sites, it does appear that seaweeds compete with corals and observational evidence on various scales in both temperate and tropical regions suggests that exclusion of corals can result. Controlled experimental studies are needed from more sites. Similar factors, namely variation in grazing and nutrient regimes appear to be influential in determining benthic community structure along gradients both within and between latitudes. Controlled experiments on nutrient and sedimentation effects at both high and low latitudes are needed, especially to distinguish the relative importance of their many plausible mechanisms of impact (e.g. by alteration of light or turbidity, bioerosion, recruitment inhibition, etc). Most studies to date have examined factors regulating coral or algal components of the community; rarely have both coral and algal components of the benthos been examined together, nor have multiple factors been manipulated simultaneously to begin to discern the complicated interactions of these factors (with each other and with natural disturbances such as storms or El Niño events) that most likely operate in natural coral reefs.

From a management perspective, these issues of relative importance and interactive effects are of vital importance. Coral reef degradation, as manifested by replacement of corals by seaweeds, probably results from an interaction of causal factors (Tomascik & Sander 1987, Stimson et al 1996, Szmant 1997), although simple, single-factor explanations are often sought, and accepted (Smith et al. 1981, Bell 1991). Anthropogenic

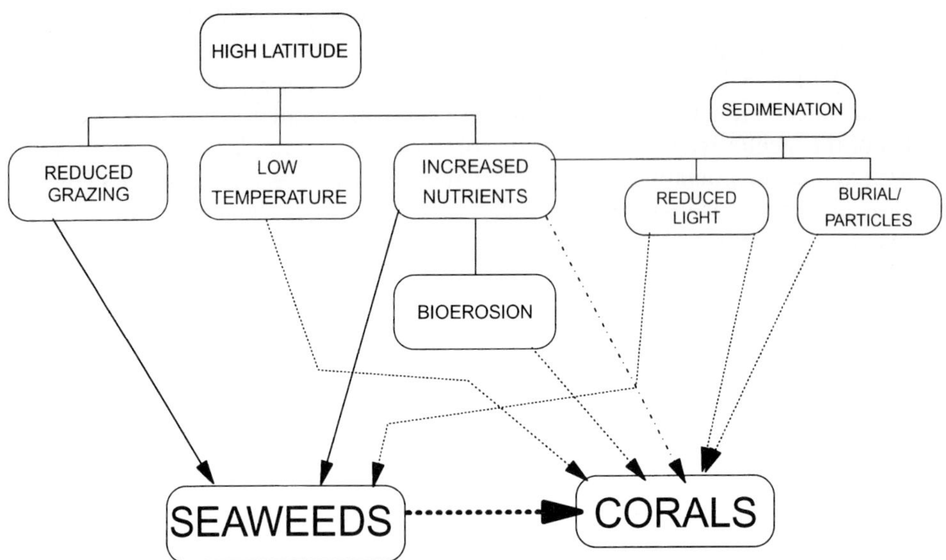

Figure 4 Conceptual model illustrating physical and ecological factors influencing coral versus seaweed dominance of reef communities within and between latitudes. Solid arrows indicate positive influences, dotted arrows indicate negative influences, and the dashed arrow indicates a positive influence at moderate levels and negative influence at high levels.

perturbations in nutrient, sedimentation, and fishing or grazing regimes are proceeding unabated in both temperate and tropical areas. We need to understand which of these perturbations, in what combinations, cause the most harm and curtail them if we are to conserve reef communities, the biodiversity they harbour, and the aesthetic and economic values they provide.

Acknowledgements

Financial support for this paper was provided by funding from the NOAA Coastal Ocean Program and the US Army Corps of Engineers through the Center for Marine and Environmental Analysis at the University of Miami and from the Graduate School of the University of North Carolina at Chapel Hill. Comments and discussion from M. E. Hay, A. M. Szmant, P. W. Glynn, J. J. Stachowicz, S. R. Reice, and C. H. Peterson are greatly appreciated.

References

Adey, W. H. & Goertemiller, T. 1987. Coral reef algal turfs: master producers in nutrient-poor seas. *Phycologia* **26**, 374–86.

Aliño, P. M., Sammarco, P. W. & Coll, J. C. 1992. Competitive strategies in soft corals (Coelenterata, Octocoralia). IV. Environmentally induced reversals in competitive superiority. *Marine Ecology Progress Series* **81**, 129–45.

Andrew, N. L. 1989. Contrasting ecological implications of food limitation in sea urchins and herbivorous gastropods. *Marine Ecology Progress Series* **51**, 189–93.

Andrew, N. L. 1993. Spatial heterogeneity, sea urchin grazing, and habitat structure on reefs in temperate Australia. *Ecology* **74**, 292–302.

Andrew, N. L. & Jones, G. P. 1990. Patch formation by herbivorous fish in a temperate Australian kelp forest. *Oecologia* **85**, 57–68.

Andrews, J. C. & Muller, H. 1983. Space-time variability of nutrients in a lagoonal patch reef. *Limnology and Oceanography* **28**, 215–27.

Aronson, R. B. 1990. Onshore-offshore patterns of human fishing activity. *Palaois* **5**, 88–93.

Atkinson, M., Carlson, B. & Crow, G. I. 1995. Coral growth in high-nutrient, low pH seawater: a case study of corals cultured at the Waikiki Aquarium, Honolulu, Hawaii. *Coral Reefs* **14**, 215–23.

Babcock, R. & Davies, P. 1991. Effects of sedimentation on settlement of *Acropora millepora. Coral Reefs* **9**, 205–8.

Bell, P. R. F. 1991. Status of eutrophication in the Great Barrier Reef Lagoon. *Marine Pollution Bulletin* **23**, 89–93.

Birkeland, C. 1977. The importance of rate of biomass accumulation in early successional stages of benthic communities to the survival of coral recruits. *Proceedings of the 3rd International Coral Reef Symposium* **1**, 15–21.

Birkeland, C. 1988. Geographic comparisons of coral-reef community processes. *Proceedings of the 6th International Coral Reef Symposium.* **1**, 211–20.

Birkeland, C. & Randall, R.H. 1981. Facilitation of coral recruitment by echinoid excavations. *Proceedings of the 4th International Coral Reef Symposium* **1**, 695–8.

Bolser, R. C. & Hay, M. E. 1996. Are tropical plants better defended? Palatability and defenses of temperate vs. tropical seaweeds. *Ecology* **77**, 2269–86.

Branch, G. M., Harris, J. M., Parkins, C., Bustamante, R. H. & Eekhout, W. 1992. Algal "gardening" by grazers: a comparison of the ecological effects of territorial fish and limpets. In *Plant-animal interactions in the marine benthos*, D. M. John et al. (eds). Oxford: Clarendon Press, 405–24.

Cairns, S. D. 1994. *Scleractinia* of the temperate North Pacific. *Smithsonian Contributions to Zoology* **557**, 1–108.

Calem, J. A. & Pierce, J. W. 1993. Distribution and control of seagrasses by light availability, Twin Cays, Belize, C. A. *Atoll Research Bulletin* **387**, 1–13.

Carpenter, R. C. 1986. Partitioning herbivory and its effects on coral reef algal communities. *Ecological Monographs* **56**, 345–65.

Carpenter, R. C. 1990a. Mass mortality of *Diadema antillarum*. I. Long term effects on sea urchin population dynamics and coral reef algal communities. *Marine Biology* **104**, 67–77.

Carpenter, R. C. 1990b. Mass mortality of *Diadema antillarum*, II. Effects on population densities and grazing intensities of parrotfishes and surgeonfishes. *Marine Biology* **104**, 79–86.

Coates, A. G. & Jackson, J. B. C. 1985. Morphological themes in the evolution of clonal and aclonal marine invertebrates. In *Population biology and evolution of clonal organisms*, J. B. C. Jackson et al. (eds). New Haven: Yale University Press, 65–106.

Coen, L. D. 1988. Herbivory by crabs and the control of algal epibionts on Caribbean host corals. *Oecologia* **75**, 198–203.

Coles, S. L. 1988. Limitations on reef coral development in the Arabian Gulf: temperature or algal competition? *Proceedings of the 6th International Coral Reef Symposium* **3**, 211–16.

Coles, S. L. & Fadlallah, Y. H. 1991. Reef coral survival and mortality at low temperatures in the Arabian Gulf: new species-specific lower temperature limits. *Coral Reefs* **9**, 231–7.

Coles, S. L. & Jokiel, P. L. 1978. Synergistic effects of temperature, light, and salinity on the hermatypic coral *Montipora verrucosa*. *Marine Biology* **49**, 188–95.

Coles, S. L., Jokiel, P. L. & Lewis, C. R. 1976. Thermal tolerance in tropical versus subtropical Pacific reef corals. *Pacific Science* **30**, 159–66.

Copper, P. 1994. Ancient reef ecosystem expansion and collapse. *Coral Reefs* **13**, 3–11.

Cortés, J. N. & Risk, M. J. 1985. A reef under siltation stress: Cahuita, Cost Rica. *Bulletin of Marine Science* **36**, 339–56.

Coyer, J. A., Ambrose, R. F., Engle, J. M. & Carroll, J. C. 1993. Interactions between corals and algae on a temperate zone rocky reef: mediation by sea urchins. *Journal of Experimental Marine Biology and Ecology* **167**, 21–37.

Crossland, C. J. 1981. Seasonal growth of *Acropora formosa* and *Pocillopora damicornis* on a high latitude reef. *Proceedings of the 4th International Coral Reef Symposium* **1**, 663–7.

Crossland, C. J. 1984. Seasonal variations in the rates of calcification and productivity in the coral *Acropora formosa* on a high latitude reef. *Marine Ecology Progress Series* **15**, 135–40.

Crossland, C. J. 1988. Latitudinal comparisons of coral reef structure and function. *Proceedings of the 6th International Coral Reef Symposium* **1**, 221–6.

Crossland, C. J., Hatcher, B. G., Atkinson, M. J. & Smith, S. V. 1984. Dissolved nutrients of a high-latitude coral reef, Houtman Abrolos Islands, Western Australia. *Marine Ecology Progress Series* **14**, 159–63.

Dallmeyer, D. G., Porter, J. W. & Smith, G. J. 1982. Effects of particulate peat on the behavior and physiology of the Jamaican reef-building coral *Montastrea annularis*. *Marine Biology* **68**, 229–33.

Dana, J. D. 1843. On the temperature limiting the distribution of corals. *American Journal of Science* **45**, 130–31.

Davies, P. S. 1991. Effects of daylight variations on the energy budgets of shallow water corals. *Marine Biology* **108**, 137–44.

Devinny, J. S. & Volse, L. A. 1978. Effects of sediments on development of *Macrocystis pyrifera* gametophytes. *Marine Biology* **48**, 343–8.

De Wreede, R. E. 1976. The phenology of three species of *Sargassum* (Sargassaceae, Phaeophyta) in Hawaii. *Phycologia* **15**, 175–83.

Dodge, R. E. & Lang, J. C. 1983. Environmental correlates of hermatypic coral (*Montastrea annularis*) growth on the East Flower Gardens Bank, northwest Gulf of Mexico. *Limnology and Oceanography* **28**, 228–40.

Done, T. J. 1982. Patterns in the distribution of coral communities across the Great Barrier Reef. *Coral Reefs* **1**, 95–108.

Done, T. J. 1992. Phase shifts in coral reef communities and their ecological significance. *Hydrobiologia* **247**, 121–32.

Downing, N. 1985. Coral reefs in an extreme environment: the northwest Arabian Gulf. *Proceedings of the 5th International Coral Reef Symposium* **6**, 343–8.

Dubinski, Z. (ed.) 1990. *Coral reefs, Ecosystems of the World, Vol. 25*. Amsterdam: Elsevier.

Duffy, J. E. 1990. Amphipods on seaweeds: partners or pests? *Oecologia* **83**, 267–76.

Ebeling, A. W. & Hixon, M. A. 1991. Tropical and temperate reef fishes: comparison of community structures. In *The ecology of fishes on coral reefs*, P. Sale (ed.). San Diego: Academic Press, 509–63.

Elner, R. W. & Vadas, R. L. 1990. Inference in ecology: the sea urchin phenomenon in the North-west Atlantic. *American Naturalist* **136**, 108–25.

Falkowski, P. G., Dubinsky, Z., Muscatine, L. & McCloskey, L. 1993. Population control in symbiotic corals. *BioScience* **43**, 606–11.

Fletcher, W. J. 1987. Interactions among subtidal Australian sea urchins, gastropods and algae: effects of experimental removals. *Ecological Monographs* **57**, 89–109.

Foster, M. S. 1992. How important is grazing to seaweed evolution and assemblage structure in the north-east Pacific? *In Plant-animal interactions in the marine benthos*, D. M. John, et al. (eds). Oxford: Clarendon Press, 61–85.

Gaines, S. D. & Lubchenco, J. 1982. A unified approach to marine plant-herbivore interactions. II. Biogeography. *Annual Review of Ecology and Systematics* **13**, 111–38.

Gleason, D. F. & Wellington, G. M. 1993. Ultraviolet radiation and coral bleaching. *Nature* **365**, 836–8.

Glynn, P. W. 1993. Monsoonal upwelling and episodic *Acanthaster* predation as probable controls of coral reef distribution and community structure in Oman, Indian Ocean. *Atoll Research Bulletin* **379**, 1–66.

Glynn, P. W. & Stewart, R. H. 1973. Distribution of coral reefs in the Pearl Islands in relation to thermal conditions. *Limnology and Oceanography* **18**, 367–79.

Glynn, P. W., Wellington, G. M. & Birkeland, C. 1978. Coral reef growth in the Galapagos: limitation by sea urchins. *Science* **203**, 47–9.

Grigg, R. W. 1981. Coral reef development at high latitudes in Hawaii. *Proceedings of the 4th International Coral Reef Symposium* **1**, 687–93.

Grigg, R. W. 1994. Effects of sewage discharge, fishing pressure, and habitat complexity on coral ecosystems and reef fishes in Hawaii. *Marine Ecology Progress Series* **103**, 25–34.

Grigg, R. W. 1995. Coral reefs in an urban embayment in Hawaii: a complex case history controlled by natural and anthropogenic stress. *Coral Reefs* **14**, 253–66.

Hagen, N. T. 1995. Recurrent destructive grazing of successionally immature kelp forests by green sea urchins in Vestfjorden, Northern Norway. *Marine Ecology Progress Series* **123**, 95–106.

Hairston, N. G., Smith, F. E. & Slobodkin, L. B. 1960. Community structure, population control, and competition. *American Naturalist* **94**, 421–25.

Hallock, P. & Schlager, W. 1986. Nutrient excess and the demise of coral reefs and carbonate platforms. *Palaios* **1**, 389–98.

Harrold, C. & Reed, D. C. 1985. Food availability, sea urchin grazing, and kelp forest community structure. *Ecology* **66**, 1160–69.

Hatcher, B. G. 1981. The interaction between grazing organisms and the epilithic algal community of a coral reef. *Proceedings of the 4th International Coral Reef Symposium* **2**, 515–24.

Hatcher, B. G. 1984. A maritime accident provides evidence for alternate stable states in benthic communities on coral reefs. *Coral Reefs* **3**, 199–204.

Hatcher, B. G. 1985. Ecological research at the Houtman Abrolhos: high latitude reefs of Western Australia. *Proceedings of the 5th International Coral Reef Symposium* **6**, 291–7.

Hatcher, B. G., Johannes, R. E. & Robertson, A. I. 1989. Review of research relevant to the conservation of shallow tropical marine ecosystems. *Oceanography and Marine Biology: an Annual Review* **27**, 337–414.

Hatcher, B. G. & Larkum, A. W. D. 1983. An experimental analysis of factors controlling the standing crop of the epilithic algal community on a coral reef. *Journal of Experimental Marine Biology and Ecology* **69**, 61–84.

Hatcher, B. G. & Rimmer, D. W. 1985. The role of grazing in controlling benthic community structure on a high latitude coral reef: measurements of grazing intensity. *Proceedings of the 5th International Coral Reef Symposium* **6**, 229–36.

Hawkins, S. J. & Hartnoll, R. G. 1983. Grazing of intertidal algae by marine invertebrates. *Oceanography and Marine Biology: an Annual Review* **21**, 195–282.

Hay, M. E. 1981. Herbivory, algal distribution, and the maintenance of between-habitat diversity on a tropical fringing reef. *American Naturalist* **118**, 520–40.

Hay, M. E. 1984. Patterns of fish and urchin grazing on Caribbean coral reefs: are previous results typical? *Ecology* **65**, 446–54.

Hay, M. E. 1991. Fish-seaweed interactions on coral reefs: effects of herbivorous fishes and adaptations of their prey. In *The ecology of fishes on coral reefs*, P. Sale (ed.). San Diego: Academic Press, 96–119.

Hay, M. E., Duffy, J. E., Pfister, C. A. & Fenical, W. 1987. Chemical defense against different marine herbivores: are amphipods insect equivalents? *Ecology* **68**, 1567–80.

Hay, M. E. & Steinberg, P. D. 1992. The chemical ecology of plant-herbivore interactions in marine versus terrestrial communities. In *Herbivores: their interactions with secondary plant metabolites, 2nd edn. Vol. II. Evolutionary and ecological processes*, G. A. Rosenthal & M. R. Berenbaum (eds). San Diego: Academic Press, 371–413.

Hay, M. E. & Taylor, P. R. 1985. Competition between herbivorous fishes and urchins on Caribbean reefs. *Oecologia* **65**, 591–8.

Highsmith, R. C. 1980. Geographical patterns of coral bioerosion: a productivity hypothesis. *Journal of Experimental Marine Biology and Ecology* **46**, 177–96.

Hodgson, G. 1990a. Tetracycline reduces sedimentation damage to corals. *Marine Biology* **104**, 493–6.

Hodgson, G. 1990b. Sediment and the settlement of larvae of the reef coral *Pocillopora damicornis. Coral Reefs* **9**, 41–43.

Hodgson, G. 1993. Sedimentation damage to reef corals. In *Proceedings of the Colloquium on Global Aspects of Coral Reefs: Health, Hazards, and History*. R. N. Ginsburg (Compiler), Rosenstiel School of Marine and Atmospheric Science, University of Miami. 298–303.

Hoegh-Guldberg, O. & Smith, G. J. 1989. Influence of the population density of zooxanthellae and supply of ammonium on the biomass and metabolic characteristics of the reef corals *Seriatopora hystix* and *Stylophora pistillata. Marine Ecology Progress Series* **57**, 173–86.

Hopkinson, C. S., Fallon, R. D., Jansson, B. O. & Schubauer, J. P. 1991. Community metabolism and nutrient cycling at Gray's Reef, a hardbottom habitat in the Georgia Bight. *Marine Ecology Progress Series* **73**, 105–20.

Horn, M. H. 1989. Biology of marine herbivorous fishes. *Oceanography and Marine Biology: an Annual Review* **27**, 167–272.

Hughes, T. P. 1989. Community structure and diversity of coral reefs: the role of history. *Ecology* **70**, 275–9.

Hughes, T. P. 1994. Catastrophes, phase shifts, and large-scale degradation of a Caribbean coral reef. *Science* **265**, 1547–51.

Hughes, T. P. 1996. Demographic approaches to community dynamics: a coral reef example. *Ecology* **77**, 2256–60.

Hughes, T. P., Reed, D. C. & Boyle, M. J. 1987. Herbivory on coral reefs: community structure following mass mortality of sea urchins. *Journal of Experimental Marine Biology and Ecology* **113**, 39–59.

Huston, M. A. 1985. Patterns of species diversity on coral reefs. *Annual Review of Ecology and Systematics* **16**, 149–77.

Jackson, G. A. 1977. Nutrients and production of giant kelp, *Macrocystis pyrifera*, off southern California. *Limnology and Oceanography* **22**, 976–96.

Jackson, J. B. C. 1996. Reefs since Columbus. *Abstracts: 8th International Coral Reef Symposium*, 98 only.

Jacques, T. G., Marshall, N. & Pilson, M. E. Q. 1983. Experimental ecology of the temperate scleractinian coral *Astrangia danae*. II Effect of temperture, light intensity and symbiosis with zooanthellae on metabolic rate and calcification. *Marine Biology* **76**, 135–48.

Johannes, R. E., Weibe, W. J., Crossland, C. J., Rimmer, D. W. & Smith, S. V. 1983. Latitudinal limits to coral reef growth. *Marine Ecology Progress Series* **11**, 105–11.

Johnson, C. R. & Mann, K. H. 1988. Diversity, patterns of adaptation, and stability of Nova Scotian kelp beds. *Ecological Monographs* **58**, 129–54.

Jokiel, P. W. & Coles, S. L. 1977. Effects of temperature on the mortality and growth of Hawaiian reef corals. *Marine Biology* **43**, 201–8.

Jones, G. P. 1992. Interactions between herbivorous fishes and macroalgae on a temperate rocky reef. *Journal of Experimental Marine Biology and Ecology* **159**, 217–35.

Josselyn, M. N. 1977. Seasonal changes in the distribution and growth of *Laurencia poitei* (Rhodophyceae, Ceramiales) in a subtropical lagoon. *Aquatic Botany* **3**, 217–29.

Kinsey, D. W. & Davies, P. J. 1979. Effects of elevated nitrogen and phosphorus on coral reef growth. *Limnology and Oceanography* **24**, 935–9.

Klumpp, D. W. & Pulfrich, A. 1989. Trophic significance of herbivorous macroinvertebrates on the central Great Barrier Reef. *Coral Reefs* **8**, 135–44.

Knowlton, N. 1992. Thresholds and multiple stable states in coral reef community dynamics. *American Zoologist* **32**, 674–9.

Knowlton, N., Lang, J. C. & Keller, B. D. 1990. Case study of a natural population collapse: post-hurricane predation on Jamaican staghorn corals. *Smithsonian Contributions to the Marine Sciences* **31**, 1–25.

Lapointe, B. E. 1989. Caribbean coral reefs: are they becoming algal reefs? *Sea Frontiers* **35**, 82–91.

Lapointe, B. E., Littler, M. M. & Littler, D. S. 1992. Modification of benthic community structure by natural eutrophication: the Belize Barrier Reef. *Proceedings of the 7th International Coral Reef Symposium* **1**, 323–34.

Larkum, A. W. D. & Koop, K. 1996. ENCORE, algal productivity and possible paradigm shifts. *Abstracts: 8th International Coral Reef Symposium, Panama*, 113 only.

Larkum, A. W. D. & Steven, A. D. L. 1994. ENCORE: The effect of nutrient enrichment on coral reefs. 1. Experimental design and research programme. *Marine Pollution Bulletin* **29**, 112–20.

Larkum, A. W. D. & Steven, A. D. L. 1996. ENCORE and the growth of crustose coralline algae. *Abstracts: 8th International Coral Reef Symposium, Panama*, 114 only.

Lasker, H. R. 1980. Sediment rejection by reef corals: the roles of behavior and morphology in *Montastrea cavernosa* (Linnaeus). *Journal of Experimental Marine Biology and Ecology* **47**, 77–87.

Lawrence, J. M. 1975. On the relationship between marine plants and sea urchins. *Oceanography and Marine Biology: an Annual Review* **13**, 213–86.

Lessios, H. A. 1988. Mass mortality of *Diadema antillarum* in the Caribbean: what have we learned? *Annual Review of Ecology and Systematics* **19**, 371–93.

Levitan, D. R. 1988. Density-dependent size regulation and negative growth in the sea urchin *Diadema antillarum* Philippi. *Oecologia* **76**, 627–9.

Levitan, D. R. 1989. Density-dependent size regulation in *Diadema antillarum*: effects on fecundity and survivorship. *Ecology* **70**, 1414–24.

Levitan, D. R. 1992. Community structure in times past: influence of human fishing pressure on algal-urchin interactions. *Ecology* **73**, 1597–605.

Lewis, S. M. 1985. Herbivory on coral reefs: algal susceptibility to herbivorous fishes. *Oecologia* **65**, 370–75.

Lewis, S. M. 1986. The role of herbivorous fishes in the organization of a Caribbean reef community. *Ecological Monographs* **56**, 183–200.

Liddell, W. D. & Ohlhorst, S. L. 1986. Changes in benthic community composition following mass mortality of *Diadema* at Jamaica. *Journal of Experimental Marine Biology and Ecology* **95**, 271–8.

Lighty, R. G. 1981. Fleshy algal domination of a modern Bahamian Barrier reef: example of an alternate climax reef community. *Proceedings of the 4th International Coral Reef Symposium* **1**, 722 only.

Littler, M. M. & Littler, D. S. 1984. Models of tropical reef biogenesis: the contribution of algae. *Progress in Phycological Research* **3**, 323–64.

Littler, M. M., Littler, D. S. & Lapointe, B. E. 1992. Modification of tropical reef community structure due to cultural eutrophication: the southwest coast of Martinique. *Proceedings of the 7th International Coral Reef Symposium* **1**, 335–43.

Littler, M. M., Littler, D. S. & Titlyanov, E. A. 1991. Comparisons of N- and P-limited productivity between high granitic islands versus low carbonate atolls in the Seychelles archipelago: a test of the relative-dominance paradigm. *Coral Reefs* **10**, 199–209.

Littler, M. M., Taylor, P. R. & Littler, D. S. 1989. Complex interactions in the control of coral zonation on a Caribbean reef flat. *Oecologia* **8**, 331–40.

Logan, A. & Tomascik, T. 1991. Extension growth rates in two coral species from high latitude reefs of Bermuda. *Coral Reefs* **10**, 155–60.

Loya, Y. 1976. Effects of water turbidity and sedimentation on the community structure of Puerto Rican corals. *Bulletin of Marine Science* **26**, 450–66.

Lubchenco, J. & Cubit, J. 1980. Heteromorphic life histories of certain marine algae as adaptations to variations in herbivory. *Ecology* **61**, 676–87.

Maragos, J. E., Evans, C. & Holthus, P. 1985. Reef corals in Kaneohe Bay six years before and after termination of sewage discharges (Oahu, Hawaiian Archipelago). *Proceedings of the 5th International Coral Reef Symposium* **4**, 189–94.

McClanahan, T. R. & Shafir, S. H. 1990. Causes and consequences of sea-urchin abundance and diversity in Kenyan coral reef lagoons. *Oecologia* **83**, 362–70.

McCook, L. J. 1996. Effects of herbivores and water quality on *Sargassum* distribution on the central Great Barrier Reef: cross shelf transplants. *Marine Ecology Progress Series* **139**, 179–92.

McGuire, M. P. 1997. *The biology of the coral* Porites astreoides: *reproduction, larval settlement behavior, and responses to ammonium enrichment*. PhD dissertation, University of Miami.

Meyer, J. L., Schultz, E. T. & Helfman, G. S. 1983. Fish schools: an asset to corals. *Science* **220**, 1047–49.

Miller, M. W. 1995. Growth of a temperate coral: effects of temperature, light, depth, and heterotrophy. *Marine Ecology Progress Series* **122**, 217–25.

Miller, M. W. & Hay, M. E. 1996. Coral/seaweed/grazer/nutrient interactions on temperate reefs. *Ecological Monographs* **66**, 323–44.

Miller, M. W. & Hay, M. E. 1998. Effects of fish predation and seawood competition on the survival and growth of corals. *Oecologia* **113**, 231–8.

Morrison, D. 1988. Comparing fish and urchin grazing in deep and shallow coral reef algal communities. *Ecology* **69**, 1367–82.

Muscatine, L. & Porter, J. W. 1977. Reef corals: mutualistic symbioses adapted to nutrient-poor environments. *BioScience* **27**, 454–60.

Ngan, Y. & Price, I. R. 1980. Seasonal growth and reproduction of intertidal algae in the Townsville region (Queensland, Australia). *Aquatic Botany* **9**, 117–34.

Norton, T. A. & Benson, M. R. 1983. Ecological interactions between the brown seaweed *Sargassum muticum* and its associated fauna. *Marine Biology* **75**, 169–77.

Ogden, J. C. & Ogden, N. B. 1993. The coral reefs of the San Blas Islands: revisited after 20 years. In *Proceedings of the Colloquium on Global Aspects of Coral Reefs: Health, Hazards, and History*. R. N. Ginsburg (Compiler), Rosenstiel School of Marine and Atmospheric Science, University of Miami, 267–70.

Orth, R. J. & Montfrans, J. van. 1984. Epiphyte-seagrass relationships with an emphasis on the role of micrograzing: a review. *Aquatic Botany* **18**, 43–69.

Pastorok, R. A. & Bilyard, G. R. 1985. Effects of sewage pollution on coral reef communities. *Marine Ecology Progress Series* **21**, 175–89.

Paul, V. J. & Hay, M. E. 1986. Seaweed susceptibility to herbivory: chemical and morphological correlates. *Marine Ecology Progress Series* **33**, 255–64.

Peckol, P. M. 1980. *Seasonal and spatial organization and development of a Carolina continental shelf community*. PhD dissertation, Duke University.

Peters, E. C. & Pilson, M. E. Q. 1985. A comparative study of the effects of sedimentation on symbiotic and asymbiotic colonies of the coral *Astrangia danae* Milne Edwards and Haime 1849. *Journal of Experimental Marine Biology and Ecology* **92**, 215–30.

Pilson, M. E. & Betzer, S. 1973. Phosphorus flux across a coral reef. *Ecology* **54**, 581–8.

Potts, D. C. 1977. Suppression of coral populations by filamentous algae within damselfish territories. *Journal of Experimental Marine Biology and Ecology* **28**, 207–16.

Reed, J. K. 1982. *In situ* growth rates of the scleractinian coral *Oculina varicosa* occurring with zooxanthellae on 6 m reefs and without on 80 m banks. *Proceedings of the 4th International Coral Reef Symposium* **2**, 201–6.

Risk, M. J. & Sammarco, P. W. 1982. Bioerosion of corals and the influence of damselfish territoriality: a preliminary study. *Oecologia* **52**, 376–80.

Risk, M. J. & Sammarco, P. W. 1991. Cross shelf trends in skeletal density of the massive coral *Porites lobata* from the Great Barrier Reef. *Marine Ecology Progress Series* **69**, 195–200.

Robertson, D. R. 1991. Increases in surgeonfish populations after mass mortality of the sea urchin *Diadema antillarum* in Panama indicate food limitation. *Marine Biology* **111**, 437–44.

Rogers, C. S. 1979. The effect of shading on coral reef structure and function. *Journal of Experimental Marine Biology and Ecology* **41**, 269–88.

Rogers, C. S. 1983. Sublethal and lethal effects of sediments applied to common Caribbean reef corals in the field. *Marine Pollution Bulletin* **14**, 378–82.

Rogers, C. S. 1990. Responses of coral reefs and reef organisms to sedimentation. *Marine Ecology Progress Series* **62**, 185–202.

Roy, K. J. & Smith, S. V. 1971. Sedimentation and coral reef development in turbid waters: Fanning Lagoon. *Pacific Science* **25**, 234–48.

Russ, G. 1984. Distribution and abundance of herbivorous grazing fishes in the central Great Barrier Reef. I. Levels of variability across the entire continental shelf. *Marine Ecology Progress Series* **20**, 23–34.

Ruyter van Steveninck, E. D. de & Bak, R. P. M. 1986. Changes in abundance of coral-reef bottom components related to mass mortality of the sea urchin *Diadema antillarum*. *Marine Ecology Progress Series* **34**, 87–94.

Sammarco, P. W. 1980. *Diadema* and its relationship to coral spat mortality: grazing, competition, and biological disturbance. *Journal of Experimental Marine Biology and Ecology* **45**, 245–72.

Sammarco, P. W. 1982. Echinoid grazing as a structuring force in coral communities: whole reef manipulations. *Journal of Experimental Marine Biology and Ecology* **61**, 31–55.

Sammarco, P. W., Levinton, J. S. & Ogden, J. C. 1974. Grazing and control of coral reef community structure by *Diadema antillarum* Phillipi (Echinodermata: Echinoidea): a preliminary study. *Journal of Marine*

Research **32**, 47–53.

Sammarco, P. W. & Risk, M. J. 1990. Large-scale patterns in internal bioerosion of *Porites*: cross continental shelf trends in the Great Barrier Reef. *Marine Ecology Progress Series* **59**, 145–56.

Sammarco, P. W. & Williams, A. H. 1982. Damselfish territoriality: influence on *Diadema* distribution and implications for coral community structure. *Marine Ecology Progress Series* **8**, 53–9.

Scheibling, R. E. 1996. The role of predation in regulating sea urchin populations in eastern Canada. *Oceanologica Acta* **19**, 421–30.

Schiel, D. R. & Foster, M. S. 1986. The structure of subtidal algal stands in temperate waters. *Oceanography and Marine Biology: an Annual Review* **24**, 265–307.

Schuhmacher, H. & Zibrowius, H. 1985. What is hermatypic? A redefinition of ecological groups of corals and other organisms. *Coral Reefs* **4**, 1–9.

Scott, F. J. & Russ, G. R. 1987. Effects of grazing on species composition of the epilithic algal community on coral reefs of the central Great Barrier Reef. *Marine Ecology Progress Series* **39**, 293–304.

Scott, P. J. B. & Risk, M. J. 1988. The effect of *Lithophaga* (Bivalvia: Mytilidae) boreholes on the strength of the coral *Porites lobata*. *Coral Reefs* **7**, 145–51.

Sheppard, C. R. C. 1988. Similar trends, different causes: responses of corals to stressed environments in Arabian Seas. *Proceedings of the 6th International Coral Reef Symposium* **3**, 297–302.

Simkiss, K. 1964. Phosphates as crystal poisons of calcification. *Biological Reviews* **39**, 487–505.

Smith, S. V. & Buddemeier, R. W. 1992. Global change and coral reef ecosystems. *Annual Review of Ecology and Systematics* **23**, 89–118.

Smith, S. V., Kimmerer, W. J., Laws, E. A., Brock, R. E. & Walsh, T. W. 1981. Kaneohe Bay sewage diversion experiment: perspectives on ecosystem responses to nutritional perturbation. *Pacific Science* **35**, 279–395.

Squires, D. F. & Keyes, I. W. 1967. Marine fauna of New Zealand: scleractinian corals. *Bulletin of New Zealand Department of Scientific and Industrial Research* **185**, 9–46.

Stachowicz, J. J. & Hay, M. E. 1996. Facultative mutualism between an herbivorous crab and a coralline alga: advantages of eating noxious seaweeds. *Oecologia* **105**, 377–87.

Stafford-Smith, M. & Ormond, R. F. G. 1992. Sediment rejection mechanisms of 42 species of Australian scleractinian corals. *Australian Journal of Marine and Freshwater Research* **43**, 683–706.

Stambler, N., Popper, N., Dubinsky, Z. & Stimson, J. 1991. Effects of nutrient enrichment and water motion on the coral *Pocillopora damicornis*. *Pacific Science* **45**, 299–307.

Steneck, R. S. 1982. A limpet-coralline alga association: adaptations and defenses between a selective herbivore and its prey. *Ecology* **63**, 507–22.

Steneck, R. S. 1988. Herbivory on coral reefs: a synthesis. *Proceedings of the 6th International Coral Reef Symposium* **1**, 37–49.

Steven, A. D. L. & Broadbent, A. D. 1996. Effects of nutrient enrichment on the growth and bioenergetics of the coal *Acropora palifera*. *Astracts: 8th International Coral Reef Symposium, Panama*, 188 only.

Stimson, J. & Kinzie III, R. A. 1991. The temporal pattern and rate of release of zooxanthellae from the reef coral *Pocillopora damicornis* (Linnaeus) under N-enrichment and control conditions. *Journal of Experimental Marine Biology and Ecology* **153**, 63–74.

Stimson, J., Learned, S. & McDermid, K. 1996. Seasonal growth of the coral reef macroalga *Dictyosphaeria cavernosa* (Forskål) Børgesen and the effects of nutrient availability, temperature and herbivory on growth rate. *Journal of Experimental Marine Biology and Ecology* **196**, 53–77.

Sutherland, J. P. 1974. Multiple stable points in natural communities. *American Naturalist* **108**, 859–73.

Szmant, A. M. 1997. Nutrient effects on coral reefs: a hypothesis on the importance of topographic and trophic complexity to reef nutrient dynamics. *Proceedings of the 8th International Coral Reef Symposium* **2**, 1527–32.

Szmant-Froelich, A. & Pilson, M. E. Q. 1984. Effects of feeding frequency and symbiosis with zooxanthellae on nitrogen metabolism and respiration of the coral *Astrangia danae*. *Marine Biology* **81**, 153–62.

Tanner, J. E. 1995. Competition between hard corals and macroalgae: an experimental analysis of growth survival and reproduction. *Journal of Experimental Marine Biology and Ecology* **190**, 151–68.

Te, F. T. 1992. Response to higher sediment loads by *Pocillopora damicornis* planulae. *Coral Reefs* **11**, 131–4.

Thorhaug, A. & Marcus, J. H. 1981. Effects of temperature and light on attached forms of tropical and semi-tropical macro-algae potentially associated with ocean thermal energy conversion machine operation. *Botanica Marina* **24**, 393–8.

Tomascik, T. & Sander, F. 1985. Effects of eutrophication on reef-building corals, growth rate of the reef-building coral *Montastrea annularis*. *Marine Biology* **87**, 143–55.

Tomascik, T. & Sander, F. 1987. Effects of eutrophication on reef building corals. II. structure of scleractinian coral communities on fringing reefs, Barbados, W.I. *Marine Biololgy* **94**, 53–75.

Tomascik, T. & Sander, F. 1991. Settlement patterns of Caribbean scleractinian corals on artificial substrata along a eutrophication gradient, Barbados, W.I. *Marine Ecology Progress Series* **77**, 261–9.

Tomascik, T., Suharsono & Mah, A. J. 1993. Case histories: a historical perspective of the natural and anthropogenic impacts in the Indonesian Archipelago. In *Proceedings of the Colloquium on Global Aspects of Coral Reefs: Health, Hazards, and History*. R. N. Ginsburg (Compiler), Rosenstiel School of Marine and Atmospheric Science, University of Miami, 304–10.

Tribble, G. W. & Randall, R. H. 1986. A description of the high-latitude shallow water coral communities of Miyake-jima, Japan. *Coral Reefs* **4**, 151–9.

Veron, J. E. N. 1995. *Corals in space and time, the biogeography and evolution of the* Scleractinia. Ithaca: Comstock/Cornell.

Veron, J. E. N. & Done, T. J. 1979. Corals and coral communities of Lord Howe Island. *Australian Journal of Marine and Freshwater Research* **30**, 203–36.

Veron, J. E. N. & Minchin, P. R. 1992. Correlations between sea surface temperature, circulation patterns and the distribution of hermatypic corals of Japan. *Continental Shelf Research* **12**, 835–57

Vine, P. J. 1974. Effects of algal grazing and aggressive behaviour of the fishes *Pomacentrus lividus* and *Acanthurus sohal* on coral-reef ecology. *Marine Biology* **24**, 131–6.

Walker, D. I. & Ormond, R. F. G. 1982. Coral death from sewage and phosphate pollution at Aqaba, Red Sea. *Marine Pollution Bulletin* **13**, 21–25.

Ward, S. & Harrison, P. L. 1996. The effects of elevated nutrients on settlement of larvae in scleractinian corals. *Abstracts: 8th International Coral Reef Symposium, Panama*, 205 only.

Watanabe, J. M. & Harrold, C. 1991. Destructive grazing by sea urchins, *Strongylocentrotus* spp., in a central California kelp forest: potential roles of recruitment, depth, and predation. *Marine Ecology Progress Series* **71**, 125–44.

Wellington, G. M. 1982. Depth zonation of corals in the Gulf of Panama: control and facilitation by resident reef fishes. *Ecological Monographs* **52**, 223–41.

Williams, D. McB. & Hatcher, A. I. 1983. Structure of the fish communities on outer slopes of inshore, mid-shelf, and outer shelf reefs of the Great Barrier Reef. *Marine Ecology Progress Series* **10**, 239–50.

Witman, J. D. 1987. Subtidal coexistence: storms, grazing, mutualism, and the zonation of kelps and mussels. *Ecological Monographs* **57**, 167–87.

Wittenberg, M. & Hunte, W. 1992. Effects of eutrophication and sedimentation on juvenile corals. I. Abundance, mortality and community structure. *Marine Biology* **112**, 131–8.

Zimmerman, R. C. & Kremer, J. N. 1984. Episodic nutrient supply to a kelp forest ecosystem in Southern California. *Journal of Marine Research* **42**, 591–604.

Oceanography and Marine Biology: an Annual Review 1998, **36**, 97–125
© A. D. Ansell, R. N. Gibson and Margaret Barnes, Editors
UCL Press

ECOLOGICAL IMPACT OF GREEN MACROALGAL BLOOMS

DAVID G. RAFFAELLI,[1] JOHN A. RAVEN[2] & LYNDA J. POOLE[2]
[1] *Culterty Field Station, University of Aberdeen, Newburgh, Ellon, Aberdeen AB41 0AA, Scotland, UK*
[2] *Biological Sciences, University of Dundee, Dundee, DD1 4HN, Scotland, UK*

Abstract Blooms of opportunistic macroalgae are an increasing feature of shallow-water marine areas. By virtue of their simple morphology and broad physiological tolerances, species within the genera *Enteromorpha, Ulva, Chaetomorpha* and *Cladophora* are able to utilize enhanced nutrients (nitrogen and phosphorus) and outcompete other seaweeds as well as seagrasses, and sometimes phytoplankton. High biomasses of bloom-forming macroalgae provide a refuge for small fishes, crustaceans and gastropods but generate a hostile physico-chemical environment in the underlying sediment. The associated sediment infauna responds dramatically: burrowing bivalves are forced to the surface and surface deposit feeders may be excluded. Blooms directly reduce the abundance of the invertebrate prey of fishes and shorebirds and also physically interfere with the feeding behaviour of some predators. Only a small proportion of the organic matter fixed by blooms is consumed by grazers and most of this material enters the local decomposer cycle or subsidizes more distant areas.

Introduction

Mass blooms of green macroalgae, mainly of the genera *Enteromorpha, Chaetomorpha, Cladophora* and *Ulva*, are now widespread on intertidal flats and shallow sublittoral areas along much of the world's coastlines. Fletcher (1996a) has provided an excellent review of the reported occurrences of such blooms. They present a significant nuisance problem, especially when loose mats accumulate on shorelines and decompose. Blooms (not all of which are due entirely to green algae; see p. 101, 103) can also have major ecological impacts on coastal systems, especially in sheltered bays, lagoons and estuaries that are often important as nursery grounds for commercial fish, as overwintering areas for migratory shorebirds, and which may support local fisheries, shellfisheries and mariculture.

The causes of such blooms have attracted considerable speculation but relatively few of the studies listed by Fletcher (1996a) are supported by rigorous investigations of the possible factors (Table 1). A wealth of information exists on the ecological requirements and physico-chemical tolerances of bloom-forming species but this information is often derived from laboratory investigations and is difficult to transfer to the complex conditions and interactions that occur in the field. Such studies have been reviewed in detail elsewhere (e.g. Dodds & Gudder 1992, Poole & Raven 1997) and here we provide only a brief account.

Factors limiting macroalgal blooms

The factors to which blooms are commonly attributed are increased availability of combined nitrogen (nitrate or ammonia) or phosphorus, or changes in the hydrography of the location (Table 1). Nitrogen has been identified as the most limiting nutrient in many of the studies shown in Table 1, but not always unequivocally, because phosphorus is also implicated in many blooms. The case for the involvement of increased nitrogen availability in bloom development has somctimes been made on the basis of focused experimentation, but there is often an assumption that nitrogen is involved because of the known low availability of nitrogen relative to other resources needed for macroalgal growth in coastal waters. Increases in phosphorus alone may cause some blooms, and changes in hydrography, induced by road construction or nearshore geological events, can also promote bloom conditions (Table 1). It seems likely that many blooms are caused by various combinations of changes in all three of these factors (nitrogen, phosphorus and hydrography) and perhaps others (Lowthian et al. 1985).

Table 1 Examples of locations experiencing blooms of green opportunistic macroalgae for which causes have been proposed, specifically changes in the availability of nitrogen (N) and phosphorus (P) and in the hydrography (H) of the location.

Location	Algae	N	P	H	Reference
Peel Inlet, Australia	*Chaetomorpha, Cladophora*		•		McComb & Humphries 1992
Wadden Sea, Germany	*Ulva, Enteromorpha, Chaetomorpha*	•			Reise & Siebert 1994
Funen Island and other areas of Denmark	*Ulva, Enteromorpha, Chaetomorpha*	•	•		Lyngby & Mortensen 1993, Funen Island Council 1991, Nienhuis 1996
Langstone Harbour, UK	*Enteromorpha, Ulva*	•	•		Nicholls et al. 1981, Soulsby et al. 1982
Venice Lagoon	*Ulva*	•			Sfriso et al. 1992, Sfriso & Marcomini 1996, De Vries er al. 1996
Rogerstown Island	*Enteromorpha*			•	Fahy et al. 1975
Avon-Heathcote, New Zealand	*Ulva*	•	•	?	Knox & Kilner 1973
Dublin Bay, Ireland	*Enteromorpha, Ulva*	•	•		Jeffrey 1993, Jeffrey et al. 1992
Manawatu, New Zealand	*Enteromorpha*			•	McBride et al. 1992
Bermuda	*Cladophora*	•	•		Schramm & Booth 1981, Lapointe & O'Connell 1989
Ythan, UK	*Enteromorpha, Ulva, Chaetomorpha*	•	?		Hull 1987, Raffaelli et al. 1989, in prep.
Ebro Delta, Spain	*Ulva*	•			Romero et al. 1996
Brittany, France	*Ulva*	•			Menesguen 1992, Dion & Bosec 1996

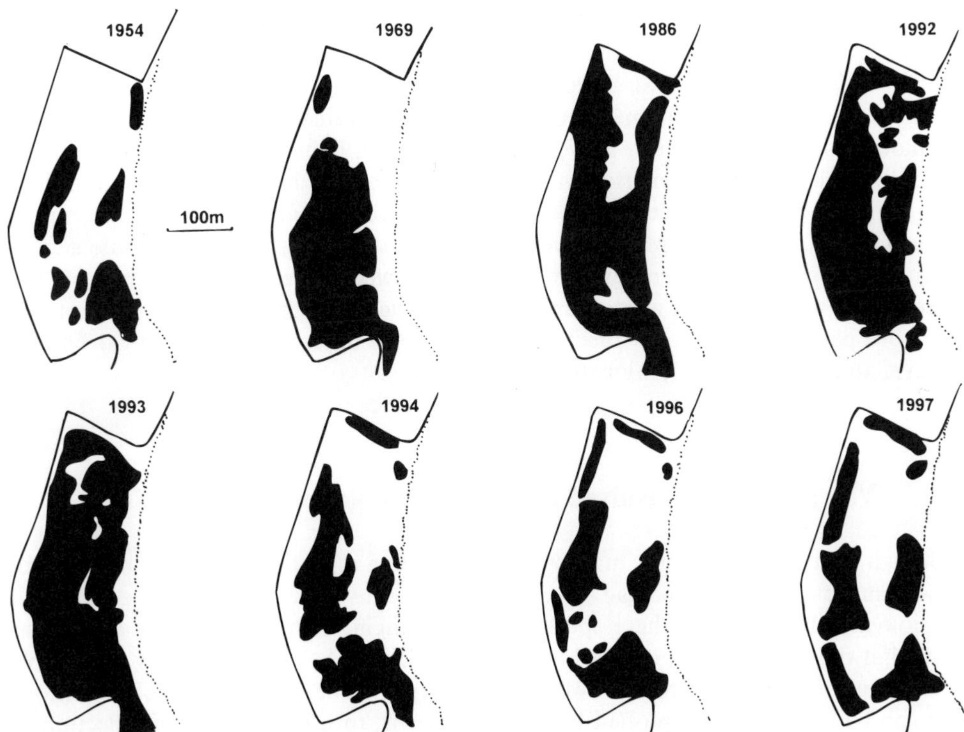

Figure 1 Between-year variation in the size and location of patches ($>1\,kg\,m^{-2}$) of *Enteromorpha* at Newburgh Quay, Ythan estuary, Aberdeenshire.

In many of the studies listed by Fletcher (1996a) it is likely that nutrients delivered to the bloom location are in excess of those required by the bloom (Poole & Raven 1997). Under otherwise favourable conditions for the accumulation of algal biomass (hydrography, temperature, incident light, grazer abundance), the rate at which the algae can grow may be limited to less than the rate of supply of nitrogen and phosphorus by self-shading or by the availability of inorganic carbon relative to nitrogen and phosphorus (Poole & Raven 1997). Further restriction is imposed by sub-optimal light, temperature and grazer conditions even when nutrient supply and hydrography

Table 2 Locations where data are available on year-to-year variation in macroalgal blooms.

Location	Reference
Cape Schank, Southeast Australia	Brown et al. 1990
Langstone Harbour, UK	reviewed in Fletcher 1996b
Ythan estuary, UK	Raffaelli et al. 1989, in prep.
Lakes Grevelingen and Veere, Netherlands	reviewed in De Vries et al. 1996
Venice lagoon	Sfriso et al. 1992, Sfriso & Marcomini 1996
Southern Baltic	reviewed in Schramm 1996
Funen Island, Denmark	Funen Island Council 1991, Nienhuis 1996
Laholm Bay, Sweden	Wennberg 1987, 1992, Rosenberg et al. 1990
Konigshaven, Germany	Reise et al. 1989, Kolbe et al. 1995
Avon-Heathcote estuary, New Zealand	Knox & Kilner 1973.

favour bloom development. In other words, even with large-scale increases in nutrients or changes in hydrography, blooms will only occur when other factors are not limiting (Aubert 1990, De Vries et al. 1996, Fletcher 1996b). Cold temperatures during the spring and excessively high temperatures during the summer may reduce the extent of the bloom considerably (Green 1977, Jeffrey et al. 1992, Fletcher 1996b, Nienhuis 1996, Raffaelli et al. in prep). Similarly, unusual combinations of limiting factors could lead to individual blooms. Once bloom-forming algae are established at a site, future blooms often build on the previous year's biomass, which may also contribute significantly to the store of nutrients in the sediment. Between-year variation in blooms can therefore be marked, often accompanied by great variation in the spatial distribution of bloom patches (Fig. 1), but medium- to long-term datasets on macroalgal blooms are available for relatively few locations (Table 2).

Why are species involved in macroalgal blooms so successful?

Although nutrient availability may be a major factor in the increasing dominance of opportunistic green macroalgae in nearshore coastal environments, the reported physiological responses of these species to wide-ranging light and salinity conditions show that the area of shore covered is often limited only by the availability of suitable substratum on which to grow (Poole & Raven 1997). The broad range of conditions that can be tolerated or exploited by these species is, to a large extent, a reflection of their morphology. *Enteromorpha*, *Ulva*, *Chaetomorpha* and *Cladophora* can take advantage of their high surface area per unit volume for intercepting incident light as well as nutrients (Littler & Littler 1980, Carpenter 1990). Tubular forms of *Enteromorpha* with entrapped gas can float at the surface of shallow water bodies, thus maximizing light supply for photosynthesis (but also maximizing the possibility of photoinhibition) relative to *Enteromorpha linza* and to *Ulva*, *Chaetomorpha* and *Cladophora* species. The extent to which chloroplast phototaxis, which occurs in *Ulva* and *Enteromorpha* but not in *Chaetomorpha* and *Cladophora* (Burrows 1959), can help to maximize photon absorption at low photon flux densities and minimize damage from photoinhibition, depends on the occurrence of essentially unidirectional photon supply and on stability of the thallus relative to this light supply (Raven 1984, Carpenter 1990).

The simple morphology of these species also enables them to tolerate fluctuating salinities, in that each cell is exposed to the bathing medium so that the cells can osmoregulate individually. The smaller cell size of *Enteromorpha* and *Ulva* relative to those of *Chaetomorpha* and *Cladophora* permits internal ion content (especially K^+; Young et al. 1987a) to be adjusted rapidly in response to tidal changes in osmolarity (Black & Weeks 1972). Long-term acclimation to high salinity seems to be a result of the accumulation of the organic solute dimethylsulphoniopropionate (DMSP) (Young et al. 1987a,b, Edwards et al. 1988). Examples of the salinities at which optimum growth of *Enteromorpha* occurs are given in Table 3.

These species are also able to tolerate and exploit the fluctuating and high temperatures often encountered in shallow water environments and tidal flats (Table 4). Upper temperature limits of *Enteromorpha* range from 31 °C for *E. clathrata* and 28 °C for *E. intestinalis* from Disko Island, Greenland (Bischoff & Wiencke 1993) and up to 30 °C in the brackish water of Chesapeake Bay (Ott 1973). In contrast, *E. bulbosa*

Table 3 The salinities giving maximal growth rate of *Enteromorpha* spp.

Species	Salinity	Reference
E. intestinalis	24‰	Kim & Lee 1996
E. intestinalis	17–34‰	Koeman & van den Hoek 1982
E. intestinalis (germlings): laboratory	16–32‰	Kim & Lee 1996
: field	16–28‰	
E. clathrata	10‰	Shellem & Josselyn 1982

grows subtidally at temperatures that are uniformly as low as −1.8 °C in Antarctica (Post & Larkum 1993). Reproduction is significantly affected by temperature: the rate of germination (increase in the percentage of germinated zoospores with time immediately after settlement) is twice as high at 20 °C as at 10 °C in *E. intestinalis* (Woodhead & Moss 1975).

It should be noted that the structural, physiological and life-history features of *Enteromorpha*, *Ulva*, *Chaetomorpha* and *Cladophora* described above are shared by many other species that are generally absent from estuarine and coastal blooms. Genera such as *Porphyra* (Rhodophyta) and *Ectocarpus* (Phaeophyta) occasionally participate in blooms (Fletcher 1996a; see also many of the Scandinavian studies discussed below), while the red filamentous alga *Gracilaria tickvahiae* can compete well with *Cladophora vagabunda* under eutrophic conditions (Peckol & Rivers 1995). Our inability to predict which species are likely to predominate in blooms is, of course, part of the more widespread problem of mechanistic prediction of the occurrence of plant species as a function of environmental conditions (Peckol & Rivers 1995).

General ecological effects of macroalgal blooms

Blooms of opportunistic macroalgae may have effects on the microbiology and chemistry of the underlying sediments, on the physical environment at the sediment–water interface, on other plants, both planktonic and pelagic, and on animals that may utilize the resource provided by the plants or are directly negatively affected by the plants. These effects are complex, vary with macroalgal species, can be both direct and indirect, are certainly interactive and many are probably scale-dependent and non-monotonic (Fig. 2). For instance, small amounts of macroalgal material may provide a refuge for invertebrates from epibenthic predators and promote localized deposition of

Table 4 Temperature giving maximum rates of growth of *Enteromorpha* and *Ulva*.

Species	Location	Temperature	Reference
Enteromorpha sp.	Helgoland, North Sea	10–15 °C	Fortes & Lüning 1980
Ulva sp.	Helgoland, North Sea	10–15 °C	Fortes & Lüning 1980
E. intestinalis	Eastern Coast Korea	15 °C	Kim & Lee 1996
E. clathrata	Disko Island, Greenland	10–20 °C	Bischoff & Wiencke 1993
E. clathrata	Southern Chile	5–15 °C	Wiencke & tom Dieck 1990
E. clathrata	San Francisco Bay	30 °C	Shellem & Josselyn 1982
E. compressa	Israeli Mediterranean Coast	25 °C	Einav et al. 1995

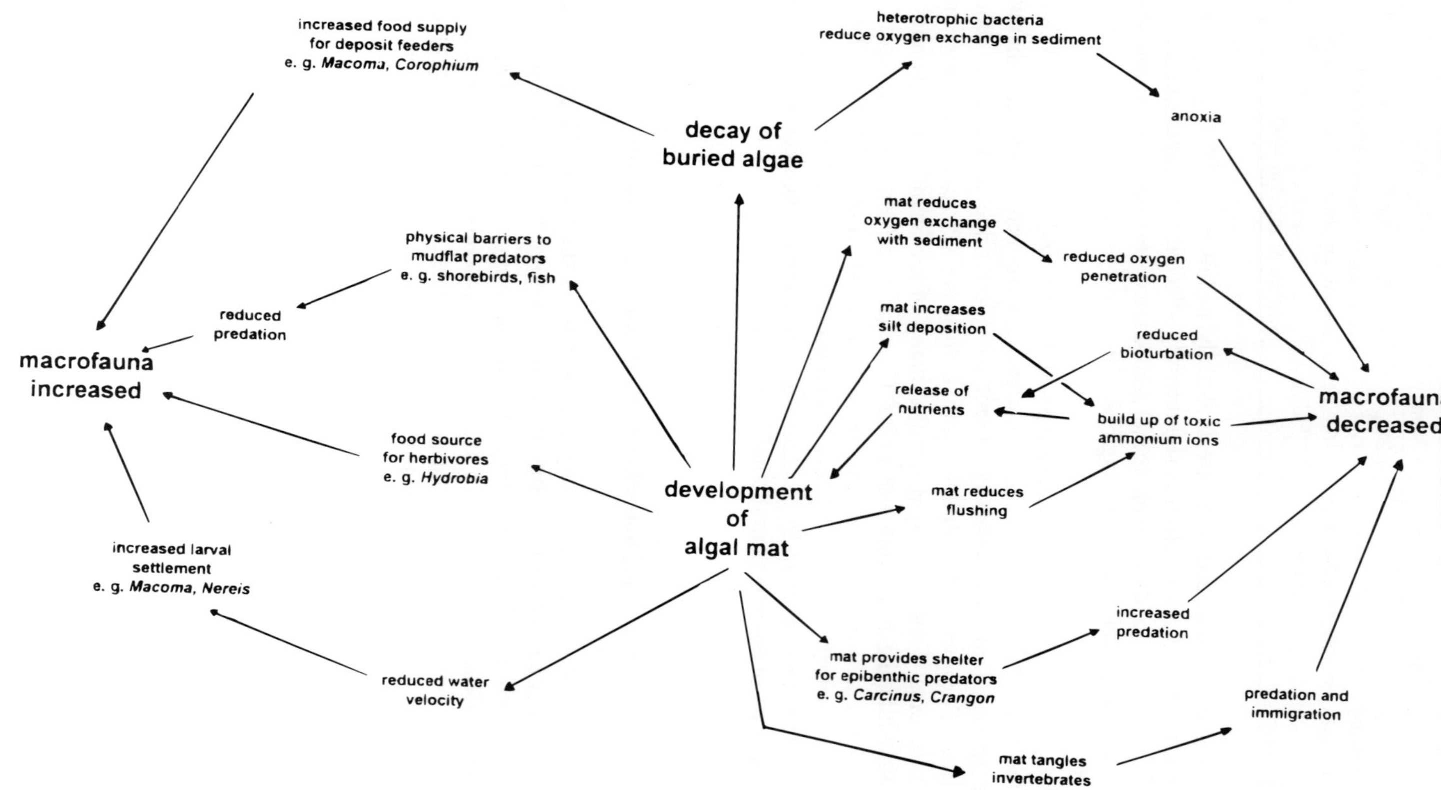

Figure 2 Effects of macroalgal mats on sediment physico-chemistry, invertebrates and predators (modified from Hull 1987).

fine particulate material that provides food resources for the infauna. Denser blooms may interfere with the foraging behaviour of invertebrates by generating a hostile sediment environment for the infauna, thereby forcing them to the sediment – water interface and increasing their availability to epibenthic predators. If the bloom is spatially and temporally heterogeneous (Fig. 1), then those invertebrates that are negatively affected can persist at high densities in the patches between macroalgal-affected areas, perhaps utilizing the organically enriched patches when the bloom subsides. However, if the affected area is extensive a mosaic of affected and non-affected areas can never develop and the system as a whole may change dramatically. This scale-dependency and the non-monotonicity of the response of the food web to macroalgal blooms makes comparisons between studies difficult, unless the extent and intensity of bloom is rigorously defined. We believe that some of the seemingly contrary responses of organisms to macroalgal blooms (see pp. 103–18) are due mainly to these problems, although the species of macroalgae involved and whether the plants are attached or free-floating may also modify the impact of the bloom.

Interactions with phytoplankton and benthic microalgae

By virtue of their thallus architecture and physiology, opportunistic green macroalgae such as *Ulva* and *Enteromorpha* (thin-walled, sheet-like and tubular forms) and *Chaetomorpha* and *Cladophora* (finely filamentous forms) are capable of rapid uptake of nutrients. If the nutrient concentration is sufficiently high, not only is growth saturated but nutrients can be stored. Indeed, *Ulva* species are capable of outcompeting phytoplankton for nutrients in mesocosm studies (Smith & Horne 1988) and Thybo-Christesen et al. (1993) have shown that *Cladophora sericea* outcompetes phytoplankton in Danish waters. The demands of these macroalgae for nutrients and the rates of nutrient recycling in systems like the Lagoon of Venice (Sfriso et al. 1992, Sfriso & Marcomini 1996) suggest that this greater competitive ability could be a general phenomenon. The consequences for planktonic food chains are unknown but they are likely to be significant. Whereas attached benthic macroalgae can outcompete phytoplankton, especially when nutrient concentrations are low, they may be outcompeted in low-light situations by phytoplankton or floating macroalgae that are in a better position in the water column for light absorption (Funen Island Council 1991, Duarte 1995, Schramm 1996; Fig. 3).

Smaller, and less-competitive benthic micro- and macroalgae might be expected to be excluded by macroalgal blooms but there is some evidence to the contrary. An increasing feature of Scandinavian (Kattegat and Skagerak) and Baltic shallow coastal waters over the last 20 years is the presence of drifting mats of macroalgae comprising a mixture of green, red and brown species (*Enteromorpha, Ulva, Cladophora, Chaetomorpha, Ectocarpus and Gracilaria* spp.). This change is attributed to increased nutrient discharges into these seas (Kautsky et al. 1986, Lundalv et al. 1986, Wennberg 1987, Breuer & Schramm 1988, Svane & Grondahl 1989, Rosenberg et al. 1990, Funen Island Council 1991, Bonsdorff 1992, Pihl et al. 1995, 1996, Bonsdorff et al. 1997). The effects of these mats on underlying phytobenthos can be severe (Fig. 3) and have been investigated experimentally for microbial mats comprising large motile diatoms, mainly *Gyrosigma balticum* and *Pleurosigma formosum*, on the west coast of Sweden (Sundback 1994, Sundback et al. 1990). The results of these experiments suggest that

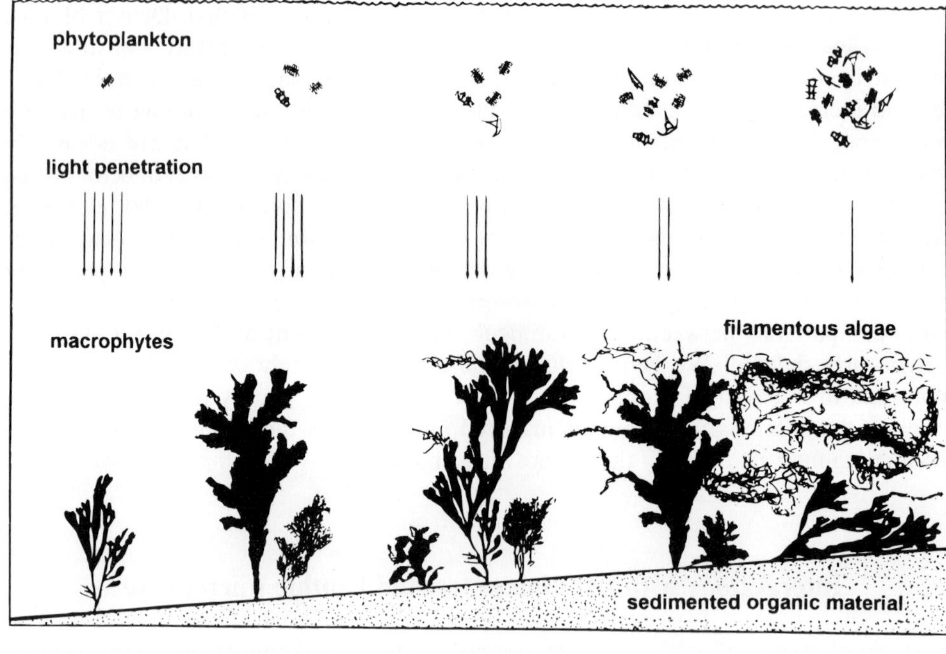

Figure 3 Changes in phytobenthos in coastal waters with increased nutrient loading (modified from Funen Island Council 1991).

despite the changes in light climate and oxygen environment induced by the drifting mats, there were no adverse effects on the underlying microbial assemblages. Indeed, an initial stimulus in biomass was observed. The authors stress that the amounts of macroalgal material added in their experiments was "average" (range $\sim$350–590 $g\,m^{-2}$ DW) and that the mats were floating. Presumably higher biomasses of macroalgal mats or mats in close contact with the sediment might have more negative impacts on the phytobenthos (Funen Island Council 1991, Norkko & Bonsdorff 1996).

Interactions with seagrasses and other seaweeds

In several regions there have been significant declines in seagrasses, long-lived seaweed populations and their associated assemblages concomitant with increases in opportunistic macroalgae. Examples include: the Orbetello lagoon, Tuscany, Italy, where the seagrasses *Posidonia oceanica* and *Cymodocea nodosa* have been replaced by *Chaetomorpha linum*, *Cladophora vagabunda* and the red alga *Gracilaria verrucosa* from 1976 to date (Bombelli & Lenzi 1996); the Black Sea where *Phyllophora nervosa* (Rhodophyta) beds have been reduced by 50% or more while there have been increasing blooms of *Enteromorpha linza*, *Cladophora laetevirens* and *Ulva rigida* over the last 20 years (Kalughina-Gutnik 1975, Kaminer 1978, Vasilu 1980, 1996); the Miro and Modego estuaries, Portugal, where there has been a progressive increase in *Enteromorpha* and a decline in the seagrasses *Zostera marina* and *Z. noltii* (Marques et al. in

Oliveira & Cabecadas 1996); the Wadden Sea, where the decline in *Zostera marina* in the 1930s seem to have been matched by an increase in macroalgae, especially *Enteromorpha, Cladophora* and *Ulva* (Reise et al. 1989, Den Hartog & Polderman 1975, Nienhuis 1996); the Danish Kattegat where *Zostera marina* has declined and *Ulva* and *Enteromorpha* and *Chaetomorpha* have increased (Rasmussen 1973, Rosenberg et al. 1990, Funen Island Council 1991, Nienhuis 1996); the Venice lagoon, where *Ulva rigida* has increased and *Cymodocea nodosa* and *Zostera noltii* have declined (Sfriso 1987); the Palmones estuary, Spain where *Ulva* blooms are linked to the disappearance of *Zostera noltii* beds (Perez-Llorens 1991, Niell et al. 1996); many parts of the southern Baltic, where *Z. marina* has declined and filamentous brown and green algae have increased (Plinski & Florszyk 1984, Kruk-Dowgiallo 1991, Messner & von Oertzen 1991, Zmudzinski & Osowiecki 1991, Ciszewski et al. 1992).

Given these widespread concomitant changes in opportunistic macroalgae and seagrass beds and the likely competition for light (Cowper 1978, Funen Island Council 1991), it is tempting to postulate causal interactions. However, such interactions can only be rigorously examined using controlled manipulative field experiments, which, to the best of our knowledge, have not been attempted for seagrass–macroalgae interactions. Whereas the evidence for direct interactions between macroalgae and seagrasses, as proposed by Niell et al. (1996) for the Palmones estuary, Spain, and Den Hartog (1994) for Langstone Harbour, England, is compelling, other factors may also be important. At some locations the observed seagrass declines and increases in opportunistic macroalgae may be, at least in part, manifestations of the same background process, namely, nutrient enrichment. For instance, around much of the coastline of the Island of Funen, Denmark, the depth limit for *Z. marina* had been reduced from 9–10 m in 1890 to 5–6 m in 1991, and for seaweeds from about 30 m to 10 m (Funen Islands Council 1991). This has been attributed in part to shading by an increased phytoplankton biomass and not solely to filamentous algae and *Ulva lactuca* (Fig. 3). Also, the burden of epiphytes on seagrass blades may be greater under eutrophic conditions (Short et al. 1995, Taylor et al. 1995). Re-establishment of seagrass populations in areas where they have been replaced by macroalgae may be a difficult and lengthy process, even after nutrients have been reduced and water transparency improved (Olesen & Sand-Jensen 1994).

Where *Enteromorpha* dominance occurs in a semi-closed habitat, such as a rock pool, a positive feedback mechanism generated by *Enteromorpha* may prevent other organisms from establishing. Oxygen concentrations produced by photosynthesis can reach as much as 4 times (i.e. about 0.8 mol m^{-3}) that found in air-saturated sea water and concentrations increase from levels below air saturation at night to $>150\%$ of air saturation by midday throughout the year (Ganning 1970, 1971, Larsson et al. 1997, L. J. Poole unpubl. data). In rockpools with a high oxygen concentration, the pH is higher than normal sea water ranging from 8.5 to 10.1 (Ganning 1970). The high oxygen levels may increase production of superoxide, hydroxyl radicals, hydrogen peroxide and singlet oxygen; these radicals can cause damage to the photosynthetic apparatus if they are not inactivated by internal antioxidant systems (for review see Lidon & Henriques 1993). *Ulva rigida* successfully scavenges H_2O_2, a product of photosynthesis (Collén et al. 1995) in high oxygen environments from the Mehler reaction (Polle 1996). *Enteromorpha* species survive exceptionally well under conditions of high oxygen concentration, suggesting that the genus has a mechanism allowing particular adaptation or acclimation to these conditions, a feature lacking in many other organisms, which, therefore, cannot successfully compete with *Enteromorpha* (Ganning 1970, 1971, L. J. Poole unpubl. data).

Interactions between macroalgal mats, water and the sediment environment

Macroalgal blooms can have both positive and negative effects on invertebrates, often on the same species depending on the scale of the bloom. In depositional environments *Enteromorpha, Chaetomorpha,* and *Cladophora* and sometimes *Ulva,* are intimately associated with the sediment, often with considerable biomass anchored just beneath the surface. In addition to changing the flow structure within the stand itself (Escartin & Aubrey 1995), the presence of this material at the sediment–water interface has significant effects on near-bed hydrographic conditions, promoting the settlement of fine particulate material (Joint 1978, Frostick & McCave 1979, Hull 1987, Worcester 1995), thereby further anchoring the filaments within the sediment and perhaps enhancing the levels of particulate-bound nutrients (Jeffrey et al. 1992) The interactions between the hydrodynamics of flowing water and the growth of freshwater benthic algae have been described by Reiter (1986, 1989) and for the seagrass *Zostera marina* by Worcester (1995).

In addition to the effects on sediments through changes to near-bed hydrography, there are major impacts of the macroalgal biomass itself on the underlying sediment. Growth of macroalgal mats involves the production of oxygen in at least equimolar quantities with the organic carbon in the algae. However, this additional oxygen is lost by evasion to the atmosphere and by hydrodynamic removal. Decomposing macroalgal mats not only do not have access to this additional oxygen, but also restrict the availability of the near-equilibrium oxygen levels by limiting oxygen exchange with the sediment. In unaffected marine sediments free oxygen is available only within the top few mm (Parks & Buckingham 1986), so it is no surprise that the redox potential of the sediment underneath such algal mats is markedly negative indicating an environment low in oxidized and high in reduced compounds.

Under these highly reduced conditions the sediment supports remineralization and recycling of macroalgal nutrients (McComb et al. 1981, Jossleyn & West 1985, Jeffrey et al. 1992). Owens & Stewart (1983) showed that the ammonium-N release from sediments of the Eden estuary, Scotland, was greatest at the time when macroalgal (*Enteromorpha*) biomass was declining, inputs of decomposing material were great and ammonifying microbial activity was highest. Similarly, Lavery & McComb (1991) showed that decomposing mats of *Chaetomorpha linum* release large amounts of ammonia and phosphate from sediments in a redox-dependent manner and that this may provide a source of nutrients to dense algal mats in the summer. These interactions between macroalgal mats and sediment chemistry not only have the potential to sustain macroalgal growth but they also generate a hostile physico-chemical environment for macro-invertebrates. Many of these invertebrate species are significant bioturbators of marine sediments and their activities generate less-reduced conditions by drawing water containing oxygen down to several cm depth. However, complex interactions between these species, the sediment environment and macroalgae may promote the formation of blooms (Fig. 1). For instance, the bioturbatory activities of the lugworm *Arenicola marina* draws macroalgal filaments into its burrow, providing a secure anchorage for the plant and facilitating locally-catastrophic blooms (Reise 1983). The burrowing amphipod *Corophium volutator* can attain densities of many thousands m^{-2} on tidal flats where its burrow irrigation activity may play a significant role in maintaining a favourable sedimentary environment for other species (Limia & Raffaelli 1997). However, *Corophium* is highly sensitive to the presence of macroalgal mats (Raffaelli et al. 1991) and its rapid departure from affected areas may lead to reduced conditions in the sediment and release of nutrients that promote further algal growth. The tube-worm *Lanice conchilega* may assimilate particles from the water column and in the process of metabolizing them, liberates mineralized nitrogen and phosphorus as excretory products that are absorbed locally by filamentous algae (*Ectocarpus*) growing on the tubes (Jeffrey et al. 1992).

There is a wealth of literature on the physical effects of microbial assemblages and the mats they form on mudflats (see Krumbein et al. 1994 for an excellent compendium). Much of this research has focused on the biostabilization effects of extracellular products generated by unicellular algae and microbes, but mats and biofilms often include filamentous algae such as the coenocytic *Vaucheria* (Tribophyceae) as well as settled propagules of macroalgae. The effects of extracellular products on the physical stability and associated physico-chemical characteristics of sediments are probably significant for the recruitment and initial development of blooms (Paterson et al. 1994, Riege & Villbrandt 1994, Stahl et al 1994, Sundback 1994, Yallop 1994, Yallop et al. 1994, Paterson 1997) and would repay further investigation.

Effects of macroalgal mats on benthic invertebrates

Blooms of macroalgae attain high biomasses *in situ*, often up to several kg ww m^{-2} over extensive areas, with major impacts on the underlying sediment biota. The invertebrate assemblages of sediments affected by blooms of green macroalgae have been recorded for many shallow water areas (Perkins & Abbot 1972, Wharfe 1977, Dauer & Conner 1980, Nicholls et al. 1981, Soulsby et al. 1982, Reise 1983, Hull 1987, Raffaelli et al. 1989, 1991, Everett 1991, 1994, Jeffrey et al. 1992, Norkko & Bonsdorff 1996). Generalities in invertebrate response are difficult to make for several reasons. The genus of macroalga causing the bloom often differs between studies and some authors have focused on particular taxa, so that the response of other species may have been overlooked. Other apparent differences in response may have been caused by differences in the biomass (sometimes unstated) of the bloom at the time of sampling, whether the algae are attached or floating, and whether the responses were determined by controlled field experiments or comparisons of macroalgal-affected and macroalgal-free areas of mudflats.

Effects of different macroalgal species on the benthos

Given the diversity of physical form among those species of macroalgae responsible for "green tides" (Fletcher 1996a), one would not be surprised if blooms of these species had different impacts on the fauna in the underlying sediments. Most species of *Enteromorpha* have a hollow tubular thallus, while *Ulva* species and *Enteromorpha linza* are laminar. *Chaetomorpha* spp. have unbranched filaments of cells that are large relative to those of the other genera, whereas *Cladophora* spp. have branched filaments of elongate cells. In situations like the Ythan estuary, Scotland, *Chaetomorpha* is much more intimately associated with the sediment than are *Enteromorpha*, *Ulva* or *Cladophora*, with substantial amounts of biomass anchored firmly beneath the surface, although this is not the case everywhere, for example, Denmark (Funen Island Council 1991). Macroalgal blooms are rarely monospecific and may comprise several species or genera that bloom at different times and at different locations (Fig. 4). These effects are compounded by uncertainties in algal identification (invertebrate ecologists may not always be the best algal taxonomists!) and the systematic status of even the most familiar species, such as those within the genus *Enteromorpha*, are more complex than originally supposed (Koeman & van den Hoek 1982, Burrows 1991, Poole & Raven 1997). Finally, there may be between-species differences in the production of extracellu-

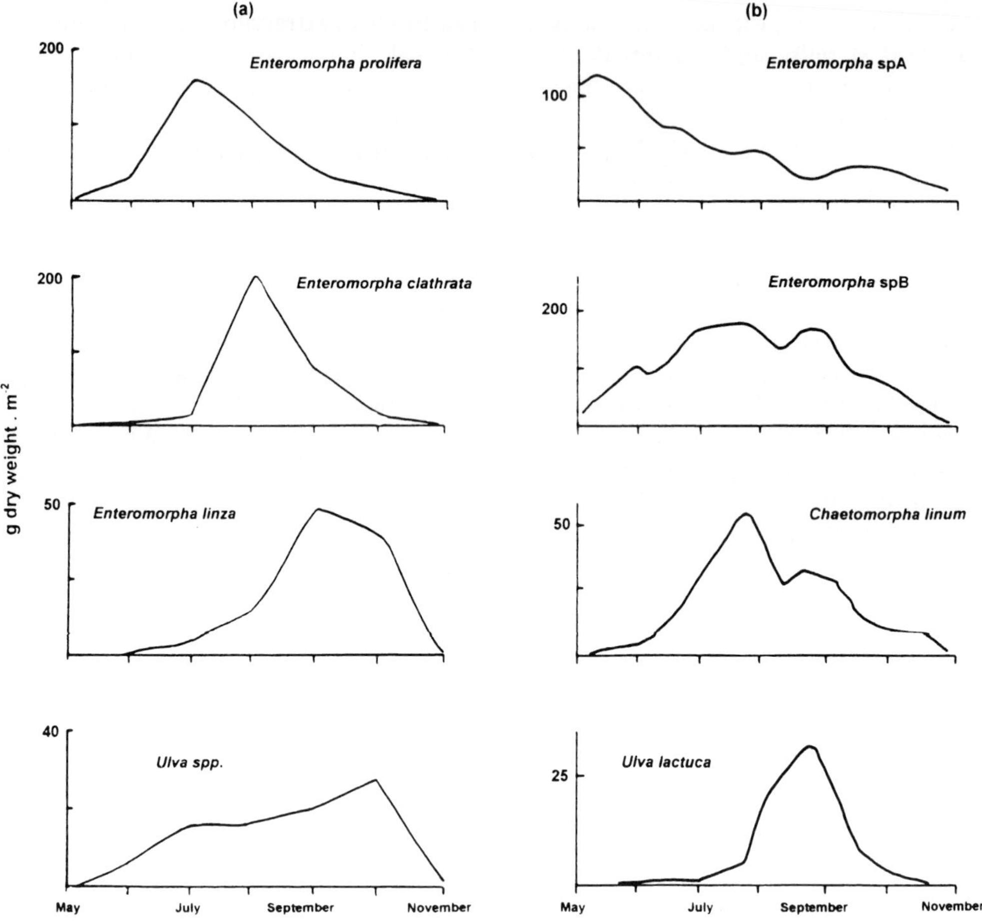

Figure 4 Seasonal succession of bloom macroalgae (a) Coos Bay, USA (modified from Pregnal & Rudy 1985); (b) Ythan estuary, UK (Raffaelli, unpubl.).

lar products some of which have been shown to affect invertebrates (Johnson & Welsh 1985, Poole & Raven 1997).

Despite this diversity of morphologies there is little evidence that different green macroalgae have different impacts on invertebrates. Everett (1994) suggested that the differences in response of the bivalve *Macoma* to macroalgal mats in his experiments and in experiments carried out by Hull (1987) might be because the mats consisted of *Ulva expansa* and *Enteromorpha intestinalis*, respectively. However, a more likely explanation is that the *Macoma* in Hull's study were settling spat, whereas those in Everett's experiments were adults and their responses would have been quite different (see pp. 110–14).

Effects of drifting macroalgal mats on the benthos

The effects of macroalgae on invertebrates may be direct through physical interference with invertebrate feeding behaviours, for example, the amphipod *Corophium volutator* (Raffaelli et al. 1991) and spionid polychaetes (Soulsby et al. 1982, Everett 1994) or indirect, via changes in the physico-chemical environment of the sediment (see p. 106). These changes are likely to be more intensive where macroalgal material is in intimate contact with the sediment. In tidal, low energy environments, and in the case of the drifting algal mats reported from many microtidal Scandinavian locations (Kautsky et al. 1986, Lundalv et al. 1986, Wennberg 1987, Breuer & Schramm 1988, Svane & Grondahl 1989, Funen Island Council 1991, Bonsdorff 1992, Thybo-Christesen et al. 1993, Bonsdorff et al. 1996a,b, 1997, Pihl et al. 1995, 1996), algal material is not always in close contact with the sediment and the effects on invertebrates might be expected to differ from those of attached algae, as described by Sundback et al. (1990) for underlying microbial assemblages. This proposition has been examined by Norkko & Bonsdorff (1996) through a manipulative and controlled field experiment carried out at 7.5 m depth at Åland (Finland) in the central northern Baltic Sea. Drifting algal material (a mixture of green, red and brown filamentous species) was anchored to the seabed in net bags so that they simulated the behaviour of natural mats. The underlying sediment became anaerobic, species diversity, total abundance and total biomass declined and adults of the bivalve *Macoma balthica* moved close the sediment surface. In addition, *Corophium volutator* was absent from the algal treatments, opportunists (oligochaetes) became the dominant faunal component and epibenthic species (*Hydrobia*, *Gammarus* and isopods) colonized the algae. These responses are broadly similar to those recorded elsewhere for attached algae and discussed below (Perkins & Abbot 1972, Wharfe 1977, Dauer & Conner 1980, Nicholls et al. 1981, Wiltse et al. 1981, Soulsby et al. 1982, Reise 1983, Hull 1987, Raffaelli et al. 1989, 1991, Everett 1991, 1994).

Effects of macroalgal mats on benthic larvae

The response of invertebrate assemblages to macroalgal blooms are likely to be non-monotonic. At relatively low biomasses, stands of macroalgae may facilitate the recruitment of sediment infauna through the entrainment of larvae into the low-flow environment. This effect has been reported for *Macoma balthica*, *Nereis diversicolor* and *Mytilus edulis* (Hull 1987, 1988, Olafsson 1988, Bonsdorf 1992, Bonsdorf et al. 1996a,b, Ragnarsson 1996), However, at high macroalgal biomasses there may subsequently be heavy mortality of these recruits as a result of the high sediment silt content and a hostile chemical environment.

Dispersing post-larvae may actively recruit into areas where blooms are developing because of the altered physical environment. For instance, the crustacean amphipod *Corophium volutator* is not usually found in coarse sand sediments because it prefers finer grades. Experimental addition of *Enteromorpha* to coarse sand in the Ythan estuary, Aberdeenshire, resulted in a fine layer of silt at the sediment surface and this was rapidly colonized by juvenile *Corophium* that were dispersing in the water column (Ragnarsson 1996). In contrast, *Corophium* is highly sensitive to the presence of even moderate biomasses of macroalgae in silty sediments (Hull 1987, Raffaelli et al. 1991; see below).

Effects of attached macroalgal mats on benthic invertebrates

Field experiments

Field experiments like those carried out by Norkko & Bonsdorff (1996) are the least equivocal way of unravelling interactions between species and for testing hypotheses concerning cause-effect relationships (Hall et al. 1991). Only a few studies on the effects of macroalgal mats have adopted this experimental approach (Hull 1987, Raffaelli et al. 1991, Everett 1994, Dayawansa 1995, Ragnarsson 1996, Cha, in prep). Hull (1987) carried out two complementary field experiments in the Ythan estuary Aberdeenshire, Scotland. In the first, a range of biomasses (0, 300 g, 1 kg and 3 kg ww m^{-2}) of live *Enteromorpha* spp. were anchored to the surface of a mudflat without any previous history of macroalgal blooms in a fully replicated randomized block design with treatment plots of 5 m^2. Densities of invertebrates were estimated after 10 weeks and at 22 weeks when the mats had been largely removed or buried by wave action. There were clear dose-dependent responses at 10 weeks in both physical and biological variables (Hull 1987; Fig. 5). The amphipod *Corophium volutator* was the species most dramatically reduced by the presence of the algal mats, but there were also declines in the spionid polychaete *Pygospio elegans* under the highest algal biomasses. The polychaetes *Capitella capitata* and *Nereis diversicolor* were more abundant in algal-covered plots, as were the bivalve *Macoma balthica* and the gastropod *Hydrobia ulvae* (Fig. 5). Following the loss of algal cover across all treatments in Week 22, *Corophium* densities recovered completely, mainly through immigration of juveniles, *Hydrobia* densities were now not significantly different between treatments, but *Capitella* densities remained high in the previously high algal biomass plots (Fig. 5). The rapid recovery by *Corophium* suggests that its absence from algal-affected areas is due to the physical presence of macroalgae, perhaps through interference with the amphipod's feeding behaviour, rather than to changes in sediment physico-chemistry, since redox potentials remained very low at Week 22, consistent with the high densities of *Capitella* (Fig 5).

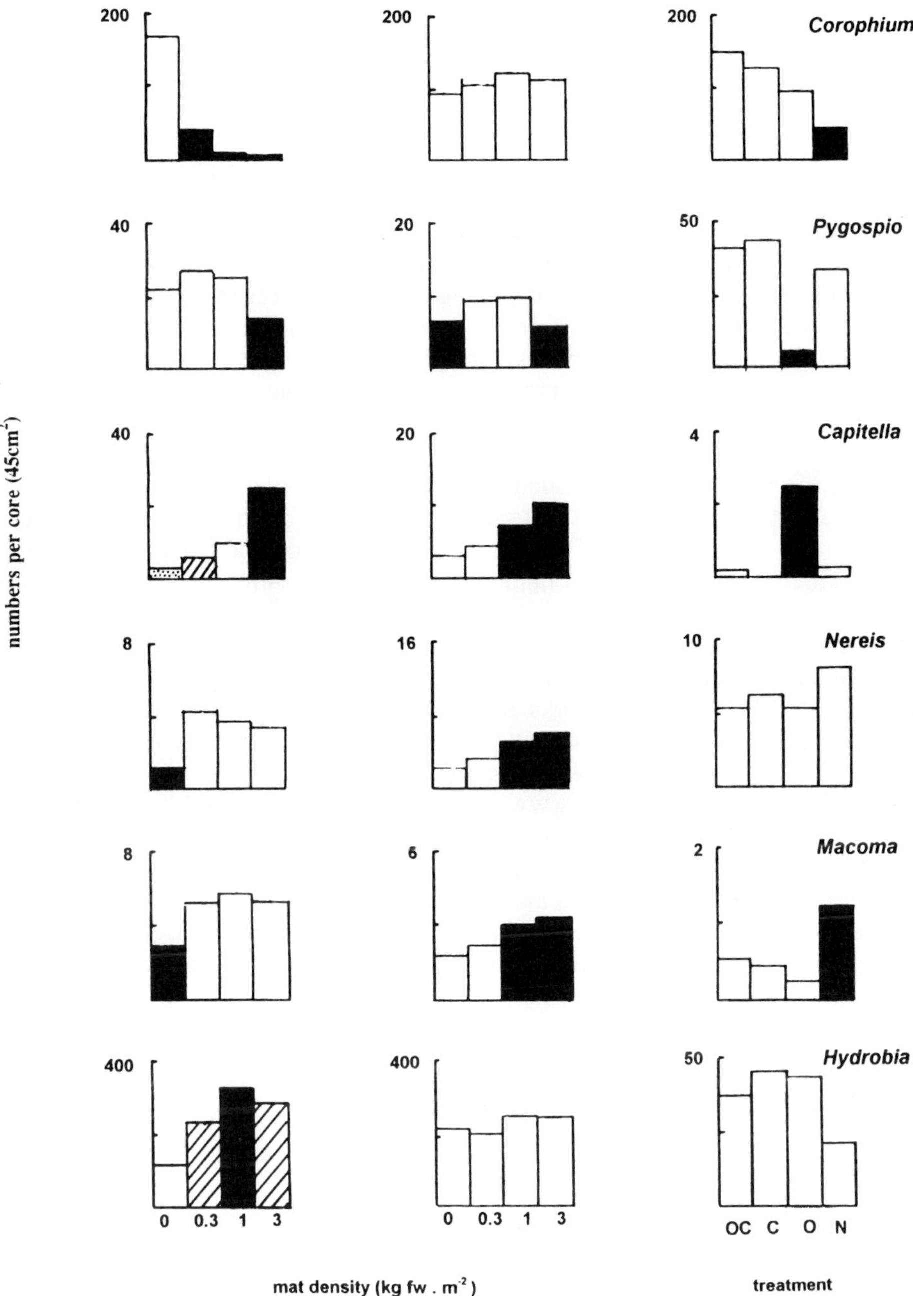

Figure 5 Abundance of macrofauna under different amounts of *Enteromorpha* after 10 weeks and 22 weeks, and the effects of changes in sediment chemistry (organic treatment, O) and physical interference (netting treatment, N), Ythan estuary, UK. C is an untreated control, OC is a control for the organic treatment where the sediment was physically disturbed in the same way as for the addition of organic material (see Raffaelli et al. 1991). Shaded bars represent different homogeneous subsets (ANOVA followed by Tukey test). Modified from Hull (1988).

This proposition was further examined through a second experiment where the treatments consisted of the addition of either organic material or nylon filaments, designed to mimic the enrichment and physical effects of filamentous macroalgae respectively (Hull 1988, Raffaelli et al. 1991) (Fig. 5). These results suggested that the main effect of macroalgae on *Corophium* is physical interference with the amphipod's behaviour, since there were significantly fewer *Corophium* in the nylon filament treatment. In contrast, *Capitella* densities were only high in the organic enrichment treatment. *Pygospio* densities were reduced in the organic treatment only, suggesting that its decline in the first experiment was due to an enrichment effect, whereas *Macoma* densities were much higher in the nylon filament plots, suggesting a physical effect in the first experiment. Most *Macoma* individuals were juveniles, consistent with the entrainment effects of filamentous macroalgae on this species noted by Olafsson (1988), Bonsdorff (1992) and Bonsdorff et al. (1996a,b). Interestingly, the response of the small polychaete *Manayunkia aestuarina* was different in the two experiments (Fig. 5).

A similar experiment was carried out by Ragnarsson (1996) on a sandy beach of the Ythan estuary, rather than in the muddy sites studied by Hull (1987), Cha (in prep) and Dayawansa (1995). Ragnarsson (1996) focused specifically on the effects of macroalgal mats on the colonization of azoic sediments. Mats of *Enteromorpha* spp. ($1.2\,\mathrm{kg\,ww\,m^{-2}}$) were anchored to randomised plots of sterilized medium-coarse sand and the invertebrate colonization of this initially azoic sediment was compared between treatments after 4 and 31 days. *Capitella* was more abundant and *Pygospio* less abundant after 1 month within the macroalgal plots. Epibenthic species (*Hydrobia*, *Gammarus* spp, *Jaera albifrons* and harpacticoid copepods) were all more abundant within the algal mats. Dayawansa's (1995) experiments were also carried out on the Ythan estuary, but designed to assess the foraging behaviour of shorebirds feeding in large ($16\,\mathrm{m^2}$) areas dosed with $3\,\mathrm{kg\,ww\,m^{-2}}$ *Enteromorpha* spp. and in natural, algal-free areas. Like Hull (1987), he recorded significantly lower densities of *Corophium* in plots to which macroalgae had been added.

Whereas Hull (1987, 1988) added macroalgae to algal-free areas of sediment, Everett (1994) removed *Ulva expansa* biomass from plots on a tidal flat in Bodega Harbour, California and compared invertebrate densities in these plots and in macroalgal-covered areas. Biogeographical differences between California and Scotland preclude detailed comparisons of species responses but Everett's findings are broadly similar to those of Hull (1987, 1988). Epifaunal species increased and those species that feed at the sediment–water interface declined under macroalgal mats. Everett (1994) recorded declines in large bivalves (*Macoma* spp.) under the mats, an effect also seen in *Tapes decussatus* (Breber 1985), *Macoma balthica* (Perkins & Abbott 1972, Norkko & Bonsdorff 1996), *Mya arenaria* (Vadas & Beal 1987), *Cerastoderma edule* (Perkins & Abbott 1972, Nicholls et al. 1981, Den Hartog 1994) and *Katelysia scalarina* (Peterson et al. 1994). In contrast, Hull (1987, 1988) did not record negative effects of macroalgal mats on large bivalves. *Cerastoderma edule* densities are very low at Hull's study site (pers. obs.) and the positive effects he noted for *Macoma balthica* seem to be associated with the enhancement of recruitment rather than effects on adults (cf. Everett 1994). Algal (*Enteromorpha prolifera* and *E. clathrata*) removal experiments carried out by Cha (in prep.) in the Ythan estuary confirmed that *Corophium* is negatively affected by the presence of the algal mats and that *Capitella*, *Macoma* and *Hydrobia* abundances are enhanced but, in contrast to Hull's findings, numbers of *Pygospio* were greater in macroalgal-affected areas.

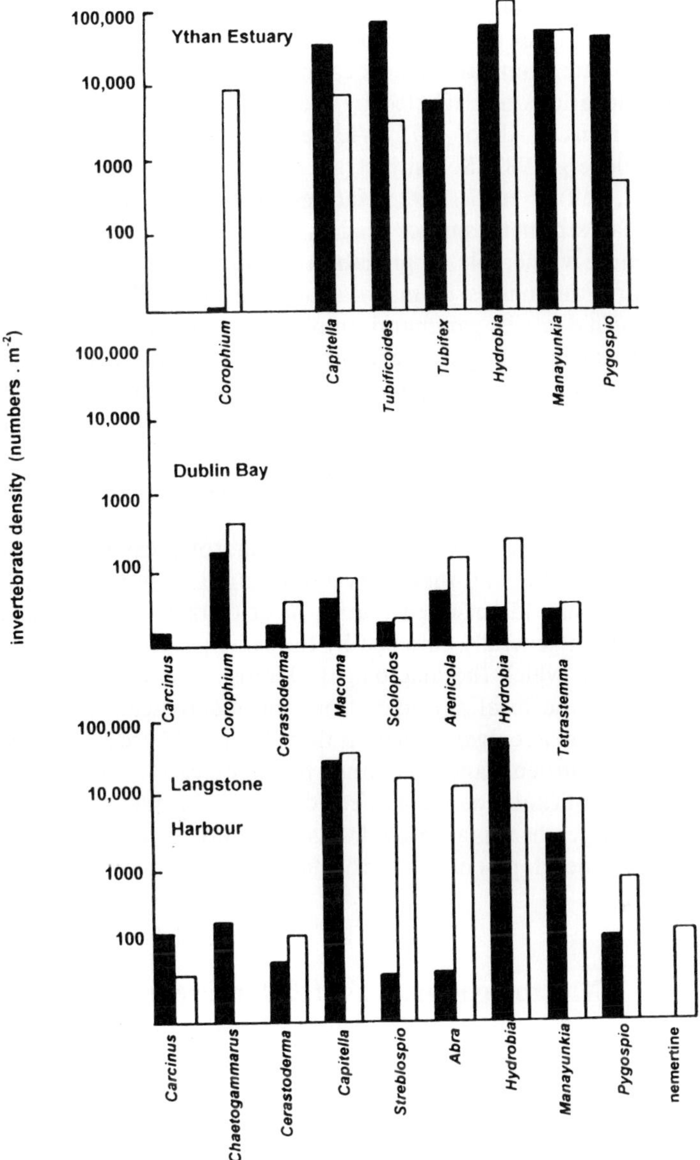

Figure 6 Abundance of invertebrates under macroalgal mats (shaded bars) and in unaffected areas (open bars) in the Ythan estuary, UK (Limia 1989), Dublin Bay, Eire (Jeffrey et al. 1992) and Langstone Harbour, UK (Soulsby et al. 1982).

Broad-scale surveys

Manipulative field experiments (Hull 1987, 1988, Raffaelli et al. 1991, Everett 1994, Norkko & Bonsdorff 1996) provide the best approach to investigate suspected cause–effect relationships but they tend to be small-scale, typically less than a few m^2 in area

(Hall et al. 1993, Moller & Raffaelli 1998). Information on interactions between macroalgal mats and invertebrates at larger spatial scales has been obtained through comparisons of invertebrate abundances in locations supporting an obvious macroalgal bloom with other locations not affected by blooms, or in the same area over a long time period (e.g. Perkins & Abbot 1972, Wharfe 1977, Nicholls et al. 1981, Wiltse et al. 1981, Price & Hylleberg 1982, Soulsby et al. 1982, Reise 1983, Raffaelli et al. 1989, 1991, Everett 1991, Jeffrey et al. 1992, Mathieson & Atkins 1995). In general, these studies indicate that sediment surface feeders (such as spionids) decline, and large bivalves become more vulnerable to predation by moving to the sediment surface in the presence of large biomasses of macroalgae (Fig. 6). Epibenthic species such as hydrobiid and littorinid snails, gammarid amphipods (except *Corophium*) and juvenile shore crabs (*Carcinus maenas*) reach higher densities, often within the macroalgal mats themselves. Juveniles (spat and post-larvae) of bivalves can attain very high densities within the macroalgal mats. Opportunistic infauna, such as the polychaetes *Capitella* and *Scololepis* and some oligochaetes such as *Tubificoides* (*Peloscolex*) *benedini*, became abundant in the organically enriched and reduced sediments under macroalgal mats (Fig. 6).

Although broadly consistent with the more rigorous experimental approaches described above, the results of all these survey-type studies are potentially confounded by location effects. It should be noted that in most of these surveys, algae-free areas were likely to have been generally enriched by the same processes (sewage discharge or agricultural run-off) to which the macroalgal blooms were attributed in the affected areas, or enriched by the local subsidies from the macroalgal mats themselves (see p. 103). The absence of macroalgal growth in these "control" areas is therefore unlikely to be due to a lack of nutrients and there must be biological, sediment or hydrographic features that differ significantly between affected and algal-free sites (Jeffrey et al. 1992). Resolution of such differences may provide insights into the factors other than nutrients that promote the development of macroalgal blooms.

Macroalgal mats as structural refuges

There is some evidence that stands of opportunistic macroalgae provide a structural refuge from predation for small epimobile species, as has been argued for seagrass beds and other seaweeds (e.g. Orth & Heck 1980, Orth et al. 1984, Lubbers et al. 1990, Sogard & Able 1991). Testing this hypothesis in the field is not straightforward and most rigorous investigations have been carried out in the laboratory or in outdoor mesocosm facilities. Isaksson et al. (1994) and Pihl et al. (1995) varied the degree of cover by *Enteromorpha prolifera*, *E. intestinalis* or *Cladophora* spp. to which three crustacean prey (*Crangon crangon*, *Carcinus maenus* and *Palaemon adspersus*) had access as refuges from predatory cod *Gadus morhua*. They found significant effects for *Crangon* and *Carcinus* and suggested that the general increase in *Enteromorpha* and *Cladophora* in the Skagerak and Kattegat over the last 20 years may have altered community organization in this region. On the east coast of the USA, *Ulva lactuca* provides an effective predation refuge for juvenile blue crabs *Callinectes sapidus* (Wilson et al. 1990) and other small fishes and crustaceans (Sogard & Able 1991), whereas growth rates of juveniles of the fishes *Pseudopleuronectes americanus*, *Tautoga onitis* and *Gobiosoma bosci* were highest at locations with patchy accumulations of *Ulva lactuca* (Sogard 1992).

Can grazers modify macroalgal blooms?

Consumers of living green macroalgal material include the gastropods *Hydrobia ulvae* and *Littorina* spp. (Green 1977, Hodder 1987, Schories & Reise 1993), the amphipods *Hyale nilsoni, Eogammarus confervicolus, Chaetogammarus marinus* (Pomeroy & Levings 1980, Price & Hylleberg 1982, Warwick et al. 1982, Hodder 1987, Hall & Raffaelli 1991), the shrimps *Crangon crangon, Palaemon serratus* (Warwick et al. 1982), shore crabs *Carcinus maenas* (Warwick et al. 1982), the polychaetes *Nereis diversicolor, N. vexillosa* and *Platynereis bicaniculata* (Woodin 1977, Hughes 1997) and shorebirds such as the widgeon *Anas penelope*, mute swan *Cygnus olor* and brent geese *Branta bernicla bernicla* (Green 1977, Tubbs & Tubbs 1980, Hodder 1987, Raffaelli et al. 1989, Summers 1990, Jeffrey et al. 1992).

Many of these consumers reach high densities when macroalgal mats are present and may have the potential to prevent the bloom, reduce its intensity or accelerate its decline. Brent geese and widgeon generally arrive to feed on tidal flats during the autumn, usually after the peak of macroalgal production, and may well accelerate the decline of the macroalgal biomass. This might also be the case for the many other invertebrates that generally reach maximum densities in mid to late summer after the peak in algal growth. Enclosure experiments by Warwick et al. (1982) in the Lynher estuary, England, suggest that small crustaceans (*Crangon, Carcinus* and *Chaetogammarus*) can accelerate the decline of *Enteromorpha* during the autumn and that variations in the interactions between these grazers and their fish predators may account for some of the year-to-year variation in macroalgal growth at this site. Everett (1994) suggests that similar interactions between macroalgae, amphipods and their fish and gull predators may occur in Bodega Harbour, California.

Removal of macroalgal biomass during the autumn could reduce the overwintering biomass that partly fuels the next year's bloom. However, there is also evidence that consumers can prevent a bloom developing in the same year (Geertz-Hansen et al. 1993). In Denmark, grazing pressure on *Ulva lactuca* matched or exceeded the algal growth rate in the outer reaches of an estuary where *Ulva* was scarce, but was negligible in the inner parts of the estuary where *Ulva* biomass accumulated (Geertz-Hansen et al. 1993). In combination with other limiting factors, such as removal of the overwintering biomass or mortality of settling propagules by exceptional hydraulic events (Talbot et al. 1990; Fig. 7), ice formation and scour, it might be possible for grazer populations to develop sufficiently rapidly to control the bloom or even prevent it altogether.

Effects on higher trophic levels

In the preceding section, we have presented overwhelming evidence for significant and largely consistent effects of macroalgal mats on the relative abundance of mud- and sandflat invertebrates, many of which are prey for other organisms. On a local scale, macroalgal mats will affect predation rates through reductions in prey abundance (especially of surface feeders), increases in prey abundance (especially of epibenthic species) and by forcing deeper-living species such as large bivalves and polychaetes to the sediment surface thereby increasing their predation risk (Perkins & Abbott 1972, Jeffrey et al. 1992, Bonsdorff et al. 1996a). Mats may also provide a prey refuge by physically interfering with the predators' foraging behaviour, or by providing a refuge for small predators, themselves prey of larger species.

The refuge effects for small fishes and invertebrate epibenthic predators have already been described (Sogard & Able 1991, Isaksson et al. 1994, Pihl et al. 1995). There is evidence that flatfishes feed more intensely in patches not affected by macroalgal mats (Bonsdorff et al. 1996a) and this probably also applies to shorebirds. Detailed studies in Langstone Harbour, England, provide compelling evidence that changes in the numbers of several species of shorebirds (Tubbs 1977, Tubbs & Tubbs 1980, 1983) are associated with changes in the distribution and abundance of mudflat invertebrates caused by extensive and thick macroalgal mats (Nicholls et al. 1981). Soulsby et al. (1982) argued that shorebirds in Langstone Harbour were unlikely to be affected because alternative prey (*Hydrobia*) were in high abundance within the macroalgal mats and that herbivorous wildfowl, brent geese and widgeon, would benefit from the increased algal food supply (see also Jeffrey et al. 1992). These suggestions were countered by Tubbs & Tubbs (1983) who pointed out that any increases in brent geese and widgeon were probably due to an increase in the population of their preferred food *Zostera*, not an increase in macroalgae. Furthermore, two of the species in decline at Langstone, redshank *Tringa totanus* and shelduck *Tadorna tadorna*, do not normally feed on algal mats and therefore cannot utilize the large numbers of *Hydrobia* present. Indeed, oystercatcher *Haematopus ostralegus*, grey plover *Pluvialis squatarola*, knot *Calidris canutus*, redshank, black-tailed and bar-tailed godwit *Limosa limosa* and *L. lapponica*, curlew *Numenius arquata* and shelduck had feeding distributions that were the inverse of the distribution of macroalgal mats; only dunlin *Calidris alpina* appeared to feed everywhere in Langstone Harbour (Tubbs & Tubbs 1980). Those individuals that were recorded feeding within areas of the harbour characterized by macroalgal mats were found on closer inspection to be using small algal-free patches of tidal flat. They concluded that macroalgal mats inhibit feeding by most species but *Zostera* has little effect (Tubbs & Tubbs 1980). The only wildfowl species that seem to have increased in line with macroalgal blooms are the mute swan *Cygnus olor* (Raffaelli et al. 1989, Pehrsson 1990) and possibly brent geese (Jeffrey et al. 1992).

Murias et al. (1997) found that most species of wader on the Mondego estuary, Portugal, showed a tendency to avoid areas with dense macroalgal mats ($\sim 50\,\mathrm{g\,DW\,m^{-2}}$), but that distributions were not as strongly influenced by the mats as, for example, at Langstone Harbour (Tubbs & Tubbs 1980). There were no striking differences in the foraging of the "representative" species dunlin and grey plover in algal-free and algal-covered areas and they suggested that most waders can adapt through subtle behavioural changes that are difficult to detect. Dayawansa (1995) also found little effect of macroalgal mats $< 1\,\mathrm{kg\,ww\,m^{-2}}$ on the foraging behaviour of redshank, curlew, oystercatcher and bar-tailed godwit, although at higher macroalgal biomasses ($> 3\,\mathrm{kg\,ww\,m^{-2}}$) there were negative effects on the food intake of redshank, curlew and bar-tailed godwit, probably because of the screening of visual cues. Shelduck appear particularly sensitive to the presence of macroalgae. This species feeds by pushing its bill along the sediment surface to filter out small invertebrates, such as *Hydrobia* and *Corophium*, and this behaviour is probably impossible in areas with macroalgal mats (Atkinson-Willes 1976).

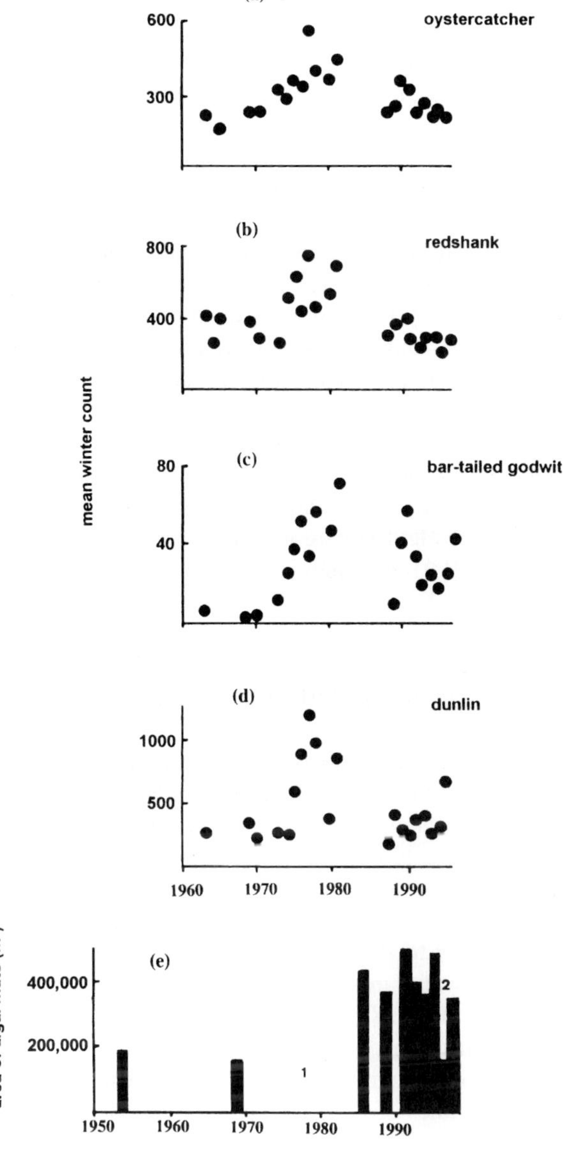

Figure 7 Long-term changes in the number of shorebirds (a–d) and cover by macroalgal mats, as estimated from aerial photographs, (e), Ythan estuary, Aberdeenshire. Winter shorebird counts are the mean of the mean counts of the three months December, January and February. 1, Aerial photographs are not available for the period 1970–84, but other data show that macroalgal biomass increased generally over this period (Raffaelli et al. 1989). 2, the low value recorded in 1996 was due to a catastrophic flood the previous autumn that removed most of the 1995–96 overwintering macroalgal biomass. From Raffaelli et al. (in prep).

Given these effects of dense macroalgal mats on shorebirds, it is reasonable to ask whether there are any effects detectable at the population level. Such analyses are not straightforward. Many factors apart from food availability will affect the numbers of shorebirds present at a site, including success on breeding grounds and protection from persecution, so that local changes in abundance need to be seen in the context of regional trends. Declines in shelduck seen at Langstone Harbour (Tubbs 1977) and on the Ythan (Raffaelli et al. 1989) over the time that macroalgal mats have developed are against the national trends for this species and therefore probably directly related to the spread of algal mats. Analyses are also complicated by the enhanced biomass and production of invertebrates in algal-free areas due to general eutrophication of the estuary. Numbers may actually increase overall in estuaries with a macroalgal bloom problem as shorebirds redistribute into the more productive algal-free areas, similar to the effects reported by van Impe (1985) and Desperez et al. (1992) for eutrophic estuaries and as proposed by Bonsdorf et al. (1996a) for benthic feeding flatfishes. However, if macroalgae continue to spread to previously unaffected areas then shorebird numbers are expected to decline. This seems to be the case for the Ythan estuary, Scotland where numbers of several species increased against national trends for many years but then decreased as algal mats continued to spread into previously unaffected areas (Raffaelli et al. 1989, 1991, Raffaelli in prep.) (Fig. 7).

Effects at larger scales

Many of the studies described above deal with local-scale interactions between macroalgal blooms and other flora and fauna. In many studies there were also likely to have been effects on benthic systems outwith these areas caused by the same eutrophication processes responsible for the blooms. In addition, substantial quantities of organic matter and nutrients are produced by the blooms and much of this material is exported in particulate and dissolved forms. The simple morphology that allows these macroalgae opportunistically to exploit eutrophic environments makes them especially vulnerable to attrition by water movement and to loss of dissolved organics. Pregnal (1981) and Pregnal and Rudy (1985) suggested that *Enteromorpha* contributes to estuarine production chiefly through dissolved organic carbon (DOC), burial of material in the sediment and herbivore-enhanced fragmentation, although no quantitative estimates were provided. In studies of *Enteromorpha* in the Eden estuary, Scotland, Owens & Stewart (1983) concluded that a considerable quantity of material was buried annually ($\sim 320\,g\,C\,m^{-2}$). For *Enteromorpha* blooms in the Ythan estuary, Aberdeenshire, Green (1977) estimated the dissolved organic matter DOM losses as 26%, with 60% to 70% (depending on location) of the monthly production lost to grazers (widgeon ($<2\%$), mute swan ($<2\%$) and *Hydrobia ulvae* ($\sim 10\%$) and attrition or burial. Substantial amounts of material were transported to the strandline or out of the estuary. Detached material will be transported to low energy environments or stranded above the high tide mark, often with serious ecological and aesthetic impacts (Bonsdorff 1992, Jeffrey et al. 1992, Dion & Le Bosec 1996). There is compelling evidence that this material can subsidize food webs in other locations not directly affected by the blooms (Raffaelli et al. 1991, in prep), but quantitative estimates of the role of this material in organizing shallow-water systems are lacking (Duarte 1995).

Acknowledgements

We wish to thank SOAFD and NERC for providing support for LP and to Eric Bonsdorff for providing constructive comments.

References

Atkinson-Willes, G. L. 1976. The numerical distribution of ducks, swans and geese as a guide to assessing the importance of wetlands. *Proceedings of the International Conference of Wetlands and Waterfowl, Heilingenhafen, 1974*, 199–254.

Aubert, M. 1990. Mediators of microbiological origin and eutrophication phenomena. *Marine Pollution Bulletin* **21**, 24–29.

Bischoff, B. & Wiencke, C. 1993. Temperature requirements for growth and survival of macroalgae from Disko Island (Greenland). *Helgoländer Meeresuntersuchungen* **47**, 167–91.

Black, D. R. & Weeks, D. C. 1972. Ionic relationships of *Enteromorpha intestinalis*. *New Phytologist* **71**, 119–27.

Bombelli, V. & Lenzi, M. 1996. Italy – the Orbetello lagoon and the Tuscan coast. In *Marine benthic vegetation: recent changes and the effects of eutrophication*, W. Schramm & P. H. Nienhuis (eds). Berlin: Springer 331–8.

Bonsdorff, E. 1992. Drifting algae and zoobenthos – effects on settling and community structure. *Netherlands Journal of Sea Research* **30**, 57–62.

Bonsdorff, E., Blomqvist, E. M., Mattila, J. & Norkko, A. 1996b. Long-term changes and coastal eutrophication. Examples from the Aland Islands and the Archipelago Sea, northern Baltic Sea. *Oceanologica Acta* **20**, 319–29.

Bonsdorff, E., Blomqvist, E. M., Mattila, J. & Norkko, A. 1997. Coastal eutrophication: causes, consequences and perspectives in the archipelago areas of the northern Baltic Sea. *Estuarine, Coastal and Shelf Science* **44** (suppl. A), 63–72.

Bonsdorff, E., Norkko, A. & Bostrom, C. 1996a Recruitment and population maintenance of the bivalve *Macoma balthica* (L.) – factors affecting settling success and early survival on shallow sandy bottoms. In *Biology and ecology of shallow coastal waters*. A. Eleftheriou et al. (eds). Denmark: Olsen & Olsen.

Breber, P. 1985. On-growing of the carpet-shell clam (*Tapes decussatus* (L.)): two years experience in Venice Lagoon. *Aquaculture* **44**, 51-6.

Breuer, G. & Schramm, W. 1988. Changes in macroalgal vegetation of Kiel Bight (Western Baltic Sea) during the past 20 years. *Kieler Meeresforchungen Sonderheft* **6**, 241–55.

Brown, V. B., Davies, S. A. & Synnot, R. N. 1990. Long-term monitoring of the effects of treated sewage effluent on the intertidal community near Cape Schank, Victoria, Australia. *Botanica Marina* **33**, 85–98.

Burrows, E. M. 1959. Growth form and environment in *Enteromorpha*. *Botanical Journal of the Linnean Society* **56**, 204–6.

Burrows, E. M. 1991. *Seaweeds of the British Isles*. Vol. 2. *Chlorophyta*. London: Natural History Museum Publications.

Carpenter, R. C. 1990. Competition among marine macroalgae: a physiological perspective. *Journal of Phycology* **26**, 6–12.

Ciszewski, P., Kruk-Dowgiallo, L. & Zmudzinski, L. 1992. Deterioration of the Puck Bay and biotechnical approaches to its reclamation. *Proceedings of the 12th Baltic Marine Biology Symposium 1991, Helsingor*, 43–6.

Collén, J., Delrio, M. J., Garciareina, G. & Pedersen, M. 1995. Photosynthetic production of hydrogen-peroxide by *Ulva rigida* C. Ag. (Chlorophyta). *Planta* **196**, 225–30.

Cowper, S. W. 1978. The drift algae community of seagrass beds in Redfish Bay, Texas. *Contributions to Marine Science* **21**, 125–32.

Dauer, D. M. & Connor, W. G. 1980. Effect of moderate sewage input on benthic polychaete populations. *Estuarine and Coastal Marine Science* **10**, 335–46.

Dayawansa, P. N. 1995. *The distribution and foraging behaviour of wading birds on the Ythan estuary, Aberdeenshire, in relation to macroalgal mats*. PhD thesis, University of Aberdeen.

Den Hartog, C. 1994. Suffocation of a littoral *Zostera* bed by *Enteromorpha radiata*. *Aquatic Botany* **47**, 21–28.

Den Hartog, C. & Polderman, P. J. G. 1975. Changes in the seagrass populations of the Dutch Wadden Sea. *Aquatic Botany* **1**, 141–7.

Desperez, M., Rybarczyk, H., Wilson, J. G., Ducrotoy, J. P., Sueur, F. & Olives, R. 1992. Biological impact of eutrophication in the Bay of Somme and the induction and impact of anoxia. *Netherlands Journal of Sea Research* **30**, 149–59.

De Vries, I., Philippart, C. J. M., De Groot, E. G. & van der Toll, M. W. M. 1996. Coastal eutrophication and marine benthic vegetation: a model analysis. In *Marine benthic vegetation: recent changes and the effects of eutrophication*, W. Schramm & P. H. Nienhuis (eds). Berlin: Springer, 79–114.

Dion, P. & Le Bosec, S. 1996. The French Atlantic coasts. In *Marine benthic vegetation: recent changes and the effects of eutrophication*, W. Schramm & P. H. Nienhuis (eds). Berlin: Springer, 251–64.

Dodds, W. K. & Gudder, D. A. 1992. The ecology of *Cladophora*. *Journal of Phycology* **28**, 415–27.

Duarte, C. M. 1995. Submerged aquatic vegetation in relation to different nutrient regimes. *Ophelia* **41**, 87–112.

Edwards, D. M., Reed, R. H. & Stewart, W. D. P. 1988. Osmoacclimation in *Enteromorpha intestinalis*: long-term effects of osmotic stress on organic solute accumulation. *Marine Biology* **98**, 467–76.

Einav, R., Breckle, S. & Beer, S. 1995. Ecophysiological adaptation strategies of some intertidal marine macroalgae of the Israeli Mediterranean Coast. *Marine Ecology Progress Series* **125**, 219–28.

Escartin, J. & Aubrey, D. G. 1995. Flow structure and dispersion within algal mats. *Estuarine Coastal Shelf Science* **40**, 451–72.

Everett, R. A. 1991. Intertidal distribution of infauna in a central California lagoon: the role of seasonal blooms of macroalgae. *Journal of Experimental Marine Biology and Ecology* **150**, 223–47.

Everett, R. A. 1994. Macroalgae in marine soft-sediment communities: effects on benthic faunal assemblages. *Journal of Experimental Marine Biology and Ecology* **175**, 253–74.

Fahy, E., Goodwillie, R., Rochford, J. & Kelly, D. 1975. Eutrophication of a partially enclosed estuarine mudflat. *Marine Pollution Bulletin* **6**, 29–31.

Fletcher, R. L. 1996a. The occurrence of "Green tides" – a review. In *Marine benthic vegetation: recent changes and the effects of eutrophication*, W. Schramm & P. H. Nienhuis (eds). Berlin: Springer, 7–43.

Fletcher, R. L. 1996b. The British Isles. In *Marine benthic vegetation: recent changes and the effects of eutrophication*, W. Schramm & P. H. Nienhuis (eds). Berlin: Springer, 223–50.

Fortes, M. D. & Lüning, K. 1980. Growth rates of North Sea macroalgae in relation to temperature, irradiance and photoperiod. *Helgoländer Meeresuntersuchungen* **34**, 15–29.

Frostick, L. E. & McCave, I. N. 1979. Seasonal shifts of sediment within an estuary mediated by algal growth. *Estuarine and Coastal Marine Science* **9**, 569–76.

Funen Island Council 1991. *Eutrophication of coastal waters. Coastal water quality management in the county of Funen, Denmark 1976–1990*. Denmark: Funen Island Council.

Ganning, B. 1970. Population dynamics and salinity tolerance of *Hyadesia fusca* (Lohman) (Acarina, Sarcoptiformes) from brackish water rockpools, with notes on the microenvironment inside *Enteromorpha* tubes. *Oecologia* **5**, 127–37.

Ganning, B. 1971. Studies on chemical, physical and biological conditions in Swedish rockpool ecosystems. *Ophelia* **9**, 51–105.

Geertz-Hansen, O., Sand-Jensen, K., Hansen, D. F. & Christiansen, A. 1993. Growth and grazing control of abundance of the marine macroalgae *Ulva lactuca* L. in a eutrophic Danish estuary. *Aquatic Botany* **46**, 101–9.

Green, A. J. 1977. *The production and utilisation of Enteromorpha spp. in the Ythan estuary, Aberdeenshire*. PhD thesis, University of Aberdeen.

Hall, S. & Raffaelli, D. 1991. Food web patterns: lessons from a species-rich web. *Journal of Animal Ecology* **60**, 823–4.

Hall, S., Raffaelli, D. G. & Thrush, S. 1993. Spatial scale, disturbance and patch dynamics in soft sediment systems. In *Aquatic ecology: scale, pattern and process*, P. Giller et al. (eds). Oxford: Blackwells.

Hall, S., Raffaelli, D. G. & Turrell, W. 1991. Predator caging experiments in marine systems: a re-examination of their value. *American Naturalist* **136**, 657–72.

Hodder, J. 1987. Production biology of an estuarine population of the green algae Ulva spp. in Coos bay, Oregon. PhD thesis, University of Oregon.

Hughes, R. G. 1997. Saltmarsh erosion and management of saltmarsh restoration; the effects of infaunal invertebrates. *Aquatic Conservation* (in press).

Hull, S. C. 1987. Macroalgal mats and species abundances: a field experiment. *Estuarine, Coastal and Shelf Science* **25**, 519–32.

Hull, S. C. 1988. *The growth of macroalgal mats on the Ythan Estuary, with respect to their effects on invertebrate abundance*. PhD thesis, University of Aberdeen.

Isaksson, I., Pihl, L. & van Montfrans, J. 1994. Eutrophication-related changes in macrovegetation and foraging of young cod (*Gadus morhua* L.): a mesocosm experiment. *Journal of Experimental Marine Biology and Ecology* **177**, 203–17.

Jeffrey, D. W. 1993. Sources of nitrogen for nuisance macroalgal growths in Dublin Bay, Republic of Ireland. *The Phycologist* **34**, 30 only.

Jeffrey, D. W., Madden, B., Rafferty, B., Dwyer, R., Wilson, J. & Arnott, N. 1992. *Dublin Bay water quality management plan. Technical report 7. Algal growths and foreshore quality*. Dublin: Environmental Research Unit.

Johnson, D. A. & Welsh, B. L. 1985. Detrimental effects of *Ulva lactuca* (L.) exudates and low oxygen on estuarine crab larvae. *Journal of Experimental Marine Biology and Ecology* **86**, 73–83.

Joint, I. R. 1978. Microbial production of an estuarine mudflat. *Estuarine and Coastal Marine Science* **7**, 185–95.

Jossleyn, M. N. & West, J. A. 1985. The distribution and temporal dynamics of the estuarine macroalgal community of San Francisco Bay. *Hydrobiology* **129**, 139–52.

Kalughina-Gutnik, A. A. 1975. *Fitobenthos Carnogo moria*. Kiev: Izd Naukova.

Kaminer, K. M. 1978. Cernomorskaia filofora v usloviah anthropoghennogo vozdeistva. *Tezisi docl II confer po biologhia selfa, Sevastopol* **1**, 50–61.

Kautsky, N., Kautsky, H., Kautsky, U. & Waern, M. 1986. Decreased depth penetration of *Fucus vesiculosus* (L.) since the 1940s indicates eutrophication of the Baltic Sea. *Marine Ecology Progress Series* **28**, 1–8.

Kim, K. Y. & Lee, I. K. 1996. The germling growth of *Enteromorpha intestinalis* (Chlorophyta) in laboratory culture under different combinations of irradiance and salinity and temperature and salinity. *Phycologia* **35**, 327–31.

Knox, G. A. & Kilner, A. R. 1973. *The ecology of the Avon-Heathcote estuary*. Report to the Christchurch Drainage Board, Christchurch, New Zealand.

Koeman, R. P. T. & van den Hoek, C. 1982. The taxonomy of *Enteromorpha* Link, 1820, (Chlorophyceae) in the Netherlands. *Archiv für Hydrobiologie Supplement* **63**, 279–330.

Kolbe, K., Kaminiski, H., Michaelis, H., Obert, B. & Rahmel, J. 1995. Macroalgal mass development in the Wadden Sea: first experiences with a monitoring system. *Helgoländer Meeresunterschungen* **49**, 519–28.

Kruk-Dowgiallo, L. 1991. Long-term changes in the structure of underwater meadows of the Puck Lagoon. *Acta Ichthyol Piscator* (Suppl.) **22**, 77–84.

Krumbein, W. E., Paterson, D. M. & Stahl, L. J. 1994. *Biostabillization of sediments*. Oldenburg: BIS.

Lapointe, B. E. & O'Connell, J. 1989. Nutrient-enhanced growth of *Cladophora prolifera* in Harrington Sound, Bermuda: eutrophication of a confined, phosphorus-limited marine system. *Estuarine, Coastal and Shelf Science* **28**, 347–60.

Larsson, C., Axelsson, L., Ryberg, H. & Beer, S. 1997. Photosynthetic carbon utilisation by *Enteromorpha intestinalis* (Chlorophyta). *European Journal of Phycology* **32**, 49–54.

Lavery, P. S. & McComb, A. J. 1991. Macroalgal-sediment nutrient interactions and their importance to macroalgal nutrition in a eutrophic estuary. *Estuarine, Coastal and Shelf Science* **32**, 281–95.

Lidon, F. C. & Henriques, F. S. 1993. Oxygen metabolism in higher plant chloroplasts. *Photosynthetica* **29**, 249–79.

Limia, J. 1989. *Bioturbation of intertidal sediments: an experimental approach involving the amphipod* Corophium volutator (*Pallas*). PhD thesis, University of Aberdeen.

Limia, J. & Raffaelli, D. 1997. Effects of burrowing by the amphipod *Corophium volutator* on the ecology of intertidal sediments. *Journal of the Marine Biological Association of the United Kingdom* **77**, 409–13.

Littler, M. M. & Littler, D. S. 1980. The evolution of thallus form and survival strategies in benthic marine macroalgae: field and laboratory tests of a functional form model. *American Naturalist* **116**, 25–44.

Lowthian, D., Soulsby, P. G. & Houston, M. C. M. 1985. Investigation of a eutrophic tidal basin. I. Factors affecting the distribution and biomass of macroalgae. *Marine Environmental Research* **15**, 262–84.

Lubbers, L., Boyton, W. R. & Kemp, W. M. 1990. Variations in the structure of estuarine fish communities in relation to abundance of submersed vascular plants. *Marine Ecology Progress Series* **65**, 1–14.

Lundalv, T., Larsson, C. & Axelsson, L. 1986. Long term trends in algal-dominated rocky subtidal communities on the Swedish west coast – a transitional system? *Hydrobiologia* **142**, 81–95.

Lyngby, J. E. & Mortensen, S. M. 1993. Biomonitoring of eutrophication levels in shallow water coastal ecosystems. In *Marine eutrophication and population dynamics*, G. Colombo et al. (eds). Fredensurg: Olsen & Olsen, 39–44.

Mathieson, S. & Atkins, S. M. 1995. A review of nutrient enrichment in the estuaries of Scotland: implications for the natural heritage. *Netherlands Journal of Aquatic Ecology* **29**, 437–48.

McBride, G. B., Quinn, J. M. & Smith, R. K. 1992. Manawatu river estuary: a review of water quality issues. DSIR Report, Water Quality Centre, Hamilton, New Zealand, 1–27.

McComb, A. J., Atkins, P. B., Birch, P. B., Gordon, D. M. & Lukatelich, R. J. 1981. Eutrophication in the Peel-Harvey estuarine system, Western Australia. In *Estuaries and nutrients*, B. Nielson & E. Cronin (eds). New Jersey: Humana, 323–42.

McComb, A. J. & Humphries, R. 1992. Loss of nutrients from catchments and their ecological impacts in the Peel-Harvey estuarine system, Western Australia. *Estuaries* **15**, 529–37.

Menesguen, A. 1992. Modelling coastal eutrophication: the case of French *Ulva* mass blooms. In *Marine Coastal Eutrophication*, R. A. Vollenweider et al. (eds). *Science of the Total Environment 1992* (Suppl.), 979–92.

Messner, U. & von Oertzen, J. A. 1991. Long-term changes in the vertical distribution of macrophytobenthic communities in the Greifswalder Bodden. *Acta Ichthyol Piscator* (Suppl.) **21**, 135–43.

Moller, H. & Raffaelli, D. G. 1998. Predicting risks from new organisms: the potential of community press experiments. In *Statistics in ecology and environmental monitoring: risk assessment and decision making in biology*, D. J. Fletcher et al. (eds). Dunedin: Otago University Press, 131–56.

Murias, T., Cabral, J. A., Marques, J. C. & Goss-Custard, J. D. 1997. Short-term effects of intertidal macroalgal blooms on the macrohabitat selection and feeding behaviour of waders in the Mondego estuary (West Portugal). *Estuarine, Coastal and Shelf Science* **43**, 677–88.

Nicholls, D. J., Tubbs, C. R. & Haynes, F. N. 1981. The effect of green algal mats on intertidal macrobenthic communities and their predators. *Kieler Meeresforschungen* **5**, 511–20.

Niell, F. X., Fernandez, C., Figueroa, F. L., Figueiras, F. G., Fuentes, J. M., Perez-Llorens, J. L., Garcia-Sanchez, M. J., Hernandez, I., Fernandez, J. A., Espejo, M., Buela, J., Garcia-Jimenez, M. C., Clavero, V. & Jimenez, C. 1996. Spanish Atlantic coasts. In *Marine benthic vegetation: recent changes and the effects of eutrophication*, W. Schramm & P. H. Nienhuis (eds). Berlin: Springer, 265–81.

Nienhuis, P. H. 1996. The north sea coasts of Denmark, Germany and the Netherlands. In *Marine benthic vegetation: recent changes and the effects of eutrophication*, W. Schramm & P. H. Nienhuis (eds). Berlin: Springer, 187–222.

Norkko, A. & Bonsdorff, E. 1996. Rapid zoobenthic community responses to accumulations of drifting algae. *Marine Ecology Progress Series* **131**, 143–57.

Olafsson, E. B. 1988. Inhibition of larval settlement to a soft bottom benthic community by drifting algal mats: an experimental test. *Marine Biology* **97**, 517–74.

Olesen, B. & Sand-Jensen, K. S. 1994. Patch dynamics of eelgrass *Zostera marina. Marine Ecology Progress Series* **106**, 147–56.

Oliveira, J. C. & Cabecadas, G. 1996. Portugal. In *Marine benthic vegetation: recent changes and the effects of eutrophication*, W. Schramm & P. H. Nienhuis (eds). Berlin: Springer, 283–94.

Orth, R. J. & Heck, J. K. L. 1980. Structural components of the eelgrass (*Zostera marina*) meadows in the lower Chesapeake Bay – fishes. *Estuaries* **3**, 278–88.

Orth, R. J., Heck, J. K. L. & van Montfrans, J. 1984. Faunal communities in seagrass beds: a review of the influence of plant structure and prey characteristics on predator-prey relationships. *Estuaries* **7**, 339–50.

Ott, F. D. 1973. The marine algae of Virginia and Maryland including the Chesapeake Bay area. *Rhodara* **75**, 258–96.

Owens, N. J. P. & Stewart, W. D. P. 1983. *Enteromorpha* and the cycling of nitrogen in a small estuary. *Estuarine, Coastal and Shelf Science* **17**, 287–96.

Parks, R. J. & Buckingham, W. J. 1986. The flow of organic carbon through aerobic respiration and sulphate reduction in inshore marine sediments. *Proceedings of the 4th International Symposium on Microbial Ecology*, 617–24.

Paterson, D. M. 1997. Biological mediation of sediment erodibility: ecology and physical dynamics. In *Cohesive sediments*, N. Burt et al. (eds). New York: Wiley & Sons, 215–29.

Paterson, D. M., Yallop, M. L. & George, C. 1994. Stabilization. In *Biostabilisation of sediments*, W. E. Krumbein et al. (eds). Oldenburg: BIS, 401–32.

Peckol, P. & Rivers, J. S. 1995. Competitive interactions between the opportunistic macroalgae *Cladophora vagabunda* (Chlorophyta) and *Gracilaria tickvahiae* (Rhodophyta). *Journal of Phycology* **31**, 229–32.

Pehrsson, O. 1990. Which environmental changes are indicated by waterfowl on the Swedish west coast? *Fauna and Flora Stockholm* **85**, 166–74.

Perez-Llorens, J. L. 1991. *Estimaciones de biomasa y contenido interno de nutrientes. ecofisiologia de incorpor-*

acion de carbono y fostato en Zostera noltii *Hornem.* PhD thesis, University of Malaga.

Perkins, E. J. & Abbott, O. J. 1972. Nutrient enrichment and sand-flat fauna. *Marine Pollution Bulletin* **3**, 70–2.

Peterson, C. H., Irlandi, E. A. & Black, R. 1994. The crash of suspension-feeding bivalve populations (*Katelysia* spp.) in Princess Royal Harbour: an unexpected consequence of eutrophication. *Journal of Experimental Marine Biology and Ecology* **176**, 39–53.

Pihl, L., Isaksson, I., Wennhage, H. & Moksnes, P.-O. 1995. Recent increases of filamentous algae in shallow Swedish Bays: effects on the community structure of epibenthic fauna and fish. *Netherlands Journal of Aquatic Ecology* **29**, 349–58.

Pihl, L., Magnusson, G., Isaksson, I & Wallentinus, I. 1996. Distribution and growth dynamics of ephemeral macroalgae in shallow bays on the Swedish west coast. *Journal of Sea Research* **35**, 169–80.

Plinski, M. & Florszyk, I. 1984. Changes in the phytobenthos resulting from the eutrophication of the Puck Bay. *Limnologica* **15**, 325–27.

Polle, A. 1996. Mehler reaction: friend or foe in photosynthesis. *Botanica Acta* **109**, 84–9.

Pomeroy, W. M. & Levings, C. D. 1980. Association and feeding relationships between *Eogammarus confevicolus* (Amphipoda, Gammaridae) and benthic algae on Sturgeon and Roberts Banks, Fraser river estuary. *Canadian Journal of Fisheries and Aquatic Sciences* **37**, 1–10.

Poole, L. J. & Raven, J. A. 1997. The biology of *Enteromorpha. Progress in Phycological Research* **12**, 1–147.

Post, A. & Larkum, A. W. D. 1993. UV-absorbing pigments, photosynthesis and UV exposure in Antarctica: comparison of terrestrial and marine algae. *Aquatic Botany* **45**, 231–43.

Pregnal, A. M. 1981. Release and potential for microbial utilisation of dissolved organic carbon from intertidal macroalgal mats. *Estuaries* **4**, 247 only.

Pregnal, A. M. & Rudy, P. 1985. Contribution of green macroalgal mats (*Enteromorpha* spp.) to seasonal production in an estuary. *Marine Ecology Progress Series* **24**, 167–76.

Price, L. H. & Hylleberg, J. 1982. Algal-faunal interactions in mat of *Ulva fenestrata* in False bay, Washington. *Ophelia* **21**, 75–88.

Raffaelli, D. G., Hull, S. & Milne, H. 1989. Long-term changes in nutrients, macroalgal mats and shorebirds in an estuarine system. *Cahiers de Biologie Marine* **30**, 259–70.

Raffaelli, D. G., Limia, J., Hull, S. & Pont, S. 1991. Interactions between invertebrates and macroalgal mats on estuarine mudflats. *Journal of the Marine Biological Association of the United Kingdom* **71**, 899–908.

Ragnarsson, S. A. 1996. *Successional patterns and biotic interactions in intertidal sediments.* PhD thesis, University of Aberdeen.

Rasmussen, E. 1973. Systematics and ecology of the Isefjord marine fauna (Denmark). With a survey of the seagrass (*Zostera*) vegetation and its communities. *Ophelia* **11**, 81–507.

Raven, J. A. 1984. *Energetics and transport in aquatic plants.* New York: A. R. Liss.

Reise, K. 1983. Indirect effects of sewage on a sandy tidal flat in the Wadden Sea. *Netherlands Journal for Sea Research* **10**, 159–64.

Reise, K., Heere, E. & Sturm, K. 1989. Historical changes in the benthos of the Wadden Sea around the island of Sylt in the North Sea. *Helogoländer Meeresunterschungen* **43**, 417–33.

Reise, K. & Siebert, I. 1994. Mass occurrence of green algae in the German Wadden Sea. *German Journal of Hydrography* (Suppl.) **1**, 171–80.

Reiter, M. A. 1986. Interactions between the hydrodynamics of flowing water and the development of a benthic algal community. *Journal of Freshwater Ecology* **3**, 511–17.

Reiter, M. A. 1989. The effect of a developing algal assemblage on the hydrodynamics of near bed flows over substrates of different sizes. *Archive für Hydrobiogie* **115**, 221–44.

Riege, H. & Villbrandt, M. 1994. Norderney survey. In *Biostabilisation of sediments*, W. E. Krumbein et al. (eds). Oldenburg: BIS, 339–60.

Romero, J., Niell, F. X., Martinez-Arroyo, A., Perez. M. & Camp, J. 1996. The Spanish Mediterranean coasts. In *Marine benthic vegetation: recent changes and the effects of eutrophication*, W. Schramm & P. H. Nienhuis (eds). Berlin: Springer, 295–306.

Rosenberg, R., Elmgren, R., Fleischer, S., Jonsson, P., Persson, G. & Dahlin, H. 1990. Marine eutrophication case studies in Sweden. *Ambio* **19**, 102–8.

Schramm, W. 1996. The Baltic sea and its transition zones. In *Marine benthic vegetation: recent changes and the effects of eutrophication*, W. Schramm & P. H. Nienhuis (eds). Berlin: Springer, 131–64.

Schramm, W. & Booth, W. 1981. Mass bloom of the alga *Cladophora prolifera* in Bermuda: productivity and phosphorus accumulation. *Botanica Marina* **24**, 419–26.

Schories, D. & Reise, K. 1993. Germination and anchorage of *Enteromorpha* spp. in sediments in the

Wadden Sea. *Helgoländer Meeresuntersuchungen* **47**, 275–85.

Sfriso, A. 1987. Flora and vertical distribution of macroalgae in the lagoon of Venice: a comparison with previous studies. *Giornale Botanica Italiano* **121**, 69–85.

Sfriso, A. & Marcomini, A. 1996. Italy – The lagoon of Venice. In *Marine benthic vegetation: recent changes and the effects of eutrophication*, W. Schramm & P. H. Nienhuis (eds). Berlin: Springer, 339–68.

Sfriso, A., Pavoni, B., Marcomini, A. & Orio, A. A. 1992. Macroalgae, nutrient cycles and pollutants in the lagoon of Venice. *Estuaries* **15**, 517–28.

Shellem, B. H. & Josselyn, M. N. 1982. Physiological ecology of *Enteromorpha clathrata* (Roth) Grev. on a salt marsh mudflat. *Botanica Marina* **25**, 541–49.

Short, F. T., Burdick, D. M. & Kaldy, J. E. 1995. Mesocosm experiments quantify the effects of eutrophication on eelgrass, *Zostera marina*. *Limnology and Oceanography* **40**, 740–49.

Smith, D. W. & Horne, A. J. 1988. Experimental measurement of resource competition between planktonic microalgae and macroalgae (seaweeds) in mesocosms simulating the San Francisco Bay estuary, California. *Hydrobiologia* **159**, 259–68.

Sogard, S. M. 1992. Variability in growth rates of juvenile fishes in different estuarine habitats. *Marine Ecology Progress Series* **85**, 35–53.

Sogard, S. M. & Able, K. W. 1991. A comparison of eelgrass, sea lettuce macroalgae and marsh creeks as habitats for epibenthic fishes and decapods. *Estuarine, Coastal and Shelf Science* **33**, 501–19.

Soulsby, P. G., Lowthian, D. & Houston, M. 1982. Effects of macroalgal mats on the ecology of intertidal mudflats. *Marine Pollution Bulletin* **13**, 162–6.

Stahl, L. J., Villbrandt, M. & de Winder, B. 1994. Ecophysiology. In *Biostabilisation of sediments*, W. E. Krumbein et al. (eds). Oldenburg: BIS, 377–400.

Summers, R. W. 1990. The exploitation of beds of green algae by brent geese. *Estuarine, Coastal and Shelf Science* **31**, 107–12.

Sundback, K. 1994. The response of shallow-water sediment communities to environmental changes. In *Biostabilisation of sediments*, W. E. Krumbein et al. (eds). Oldenburg: BIS, 17-40.

Sundback, K., Jonsson, B., Nilsson, P. & Lindstrom, I. 1990. Impact of accumulating drifting macroalgae on a shallow-water sediment system: an experimental study. *Marine Ecology Progress Series* **58**, 261–74.

Svane, I. & Grondahl, F. 1989. Epibioses of Gullmarsfjorden: an underwater stereophotographical transect analysis in comparison with Gilsen in 1926–29. *Ophelia* **28**, 95–110.

Talbot, M. M. B., Knoop, W. T. & Bate, G. C. 1990. The dynamics of estuarine macrophytes in relation to flood/siltation cycles. *Botanica Marina* **33**, 159–64.

Taylor, D., Nixon, S., Granger, S. & Buckley, B. 1995. Nutrient limitation and the eutrophication of coastal lagoons. *Marine Ecology Progress Series* **127**, 235 only.

Thybo-Christesen, M., Rasmussen, M. B. & Blackburn, T. H. 1993. Nutrient fluxes and growth of *Cladophora sericea* in a shallow Danish Bay. *Marine Ecology Progress Series* **100**, 273–81.

Tubbs, C. R. 1977. Wildfowl and waders in Langstone Harbour. *British Birds* **70**, 177–99.

Tubbs, C. R. & Tubbs, J. M. 1980. Waders and shelduck feeding distribution in Langstone Harbour, Hampshire. *Bird Study* **27**, 239–48.

Tubbs, C. R. & Tubbs, J. M. 1983. Macroalgal mats in Langstone Harbour, Hampshire, England. *Marine Pollution Bulletin* **14**, 148–9.

Vadas, R. L. & Beal, B. 1987. Green algal ropes: a novel estuarine phenomenon in the Gulf of Maine. *Estuaries* **10**, 171–6.

van Impe, J. 1985. Estuarine pollution as a probable cause of increase of estuarine birds. *Marine Pollution Bulletin* **16**, 271–6.

Vasilu, F. 1980. La production des especes d' *Enteromorpha* du littoral roumain de la mer Noire. *Institut Roman de Cercetari Marine* (*IRCM*) – *Cercetari Marina* **13**, 147–61.

Vasilu, F. 1996. The Black Sea. In *Marine benthic vegetation: recent changes and the effects of eutrophication*, W. Schramm & P. H. Nienhuis (eds). Berlin: Springer, 435–48.

Warwick, R. M., Davey, J. T., Gee, J. M. & George, C. L. 1982. Faunistic control of *Enteromorpha* blooms: a field experiment. *Journal of Experimental Marine Biology and Ecology* **56**, 23–31.

Wennberg, T. 1987. *Long term changes in the composition and distribution of the macroalgal vegetation in the southern part of Laholm bay, south-west Sweden, during the last thirty years*. Swedish Environmental Protection Agency Report 3290.

Wennberg, T. 1992. Colonization and succession of macroalgae on a breakwater in Laholm Bay, a eutrophicated brackish water area (SW Sweden). *Acta Phytogeographica Suesica* **78**, 65–77.

Wharfe, J. R. 1977. An ecological survey of the benthic invertebrate macrofauna of the lower Medway estuary, Kent. *Journal of Animal Ecology* **46**, 93–113.

Wiencke, C. & tom Dieck, I. 1990. Temperature requirements for growth and survival of macroalgae from Antarctica and Southern Chile. *Marine Ecology Progress Series* **59**, 157–70.

Wilson, K. A., Able, K. W. & Heck, K. L. 1990. Predation rates on juvenile blue crabs in estuarine nursery habitats: evidence for the importance of macroalgae (*Ulva lactuca*). *Marine Ecology Progress Series* **58**, 243–51.

Wiltse, W., Foreman, K., Valiela, I. & Teal, J. 1981. Regulation of benthic invertebrates in salt marsh creeks. *Estuaries* **4**, 254 only.

Woodhead, P. & Moss, B. 1975. The effects of light and temperature on settlement and germination of *Enteromorpha*. *British Phycological Journal* **10**, 269–72.

Woodin, S. A. 1977. Algal "gardening" behaviour by nereid polychaetes: effects on soft-bottom community structure. *Marine Biology* **44**, 39–42.

Worcester, S. E. 1995. Effects of eelgrass beds on advection and turbulent mixing in low current and low shoot density environments. *Marine Ecology Progress Series* **126**, 223–32.

Yallop, M. L. 1994. Survey of the Severn estuary. In *Biostabilisation of sediments*, W. E. Krumbein et al. (eds). Oldenburg: BIS, 279–326.

Yallop, M. L., de Winter, B. & Paterson, D. M. 1994. Texel survey. In *Biostabilisation of sediments*, W. E. Krumbein et al. (eds). Oldenburg: BIS, 327–38.

Young, A. J., Collins, J. C. & Russell, G. 1987a. Ecotypic variation in the osmotic responses of *Enteromorpha intestinalis* (L.) Link. *Journal of Experimental Botany* **38**, 1309–24.

Young, A. J., Collins, J. C. & Russell, G. 1987b. Solute regulation in the euryhaline marine alga *Enteromorpha prolifera* (O. F. Mull) J. Ag. *Journal of Experimental Botany* **38**, 1298–1308.

Zmudzinski, L. & Osowiecki, A. 1991. Long-term changes in the macrofauna of the Puck lagoon. *Acta Ichthyol Piscator* (Suppl.) **21**, 259–64.

Oceanography and Marine Biology: an Annual Review 1998, **36**, 127–78

UCL Press

THE IMPACT OF DREDGING WORKS IN COASTAL WATERS: A REVIEW OF THE SENSITIVITY TO DISTURBANCE AND SUBSEQUENT RECOVERY OF BIOLOGICAL RESOURCES ON THE SEA BED

R. C. NEWELL,[1] L. J. SEIDERER[1] & D. R. HITCHCOCK[2]
[1] *Marine Ecological Surveys Limited, West Country Office, Trewood Cottage, Steeple Lane, St Ives, Cornwall TR26 2PF, UK*
[2] *Coastline Surveys Limited, Bridgend Farmhouse, Stonehouse, Gloucestershire GL10 2AX, UK*

Abstract The present review provides a framework within which the impact of dredging on biological resources that live on the sea bed ("Benthic" communities) can be understood, and places in perspective some of the recent studies that have been carried out in relation to aggregates dredging in European coastal waters. The impact of dredging works on fisheries and fish themselves, and on their spawning grounds is outside the scope of this review. We have, however, shown that empirical models for shelf waters such as the North Sea indicate that as much as 30% of total fisheries yield to man is derived from benthic resources, and that these become an increasingly important component of the food web in near-shore waters where primary production by seaweeds (macrophytes) and seagrasses living on the sea bed largely replaces that by the phytoplankton in the water column. Because dredging works are mainly carried out in near-shore coastal deposits, and these are the ones where benthic production processes are of importance in supporting demersal fish production, our review concentrates on the nature of benthic communities, their sensitivity to disturbance by dredging and land reclamation works, and on the recovery times that are likely to be required for the re-establishment of community structure following cessation of dredging or spoils disposal.

Essentially, the impact of dredging activities mainly relates to the physical removal of substratum and associated organisms from the sea bed along the path of the dredge head, and partly on the impact of subsequent deposition of material rejected by screening and overspill from the hopper. Because sediment disturbance by wave action is limited to depths of less than 30 m, it follows that pits and furrows from dredging activities are likely to be persistent features of the sea bed except in shallow waters where sands are mobile. Recent studies using Acoustic Doppler Current Profiling (ADCP) techniques suggest that the initial sedimentation of material discharged during outwash from dredgers does not, as had been widely assumed, disperse according to the Gaussian diffusion principles used in most simulation models, but behaves more like a density current where particles are held together during the initial phase of the sedimentation process. As a result, the principal area likely to be affected by sediment deposition is mainly confined to a zone of a few hundred metres from the discharge chute.

Our review suggests that marine communities conform with well established principles of ecological succession, and that these allow some realistic predictions on the likely recovery of benthic communities following cessation of dredging. In general, communities living in fine mobile deposits, such as occur in estuaries, are characterized by large populations of a restricted variety of species that are well adapted to rapid recolonization of deposits that are subject to frequent disturbance. Recolonization of dredged deposits is initially by these "opportunistic" species and the community is subsequently supplemented by an increased species variety of long-lived and slow-growing "equilibrium" species that characterize stable undisturbed deposits such as coarse gravels and reefs.

Rates of recovery reported in the literature suggest that a recovery time of 6–8 months is characteristic of many estuarine muds where frequent disturbance of the deposits precludes the establishment of long-lived components. In contrast, the community of sands and gravels may take 2–3 yr to establish, depending on the proportion of sand and level of environmental disturbance by waves and currents, and may take even longer where rare slow-growing components were present in the community prior to dredging. As the deposits get coarser along a gradient of environmental stability, estimates of 5–10 yr are probably realistic for development of the complex biological associations between the slow-growing components of equilibrium communities characteristic of reef structures.

Most recent studies show, however, that biological community composition is not controlled by any one, or a combination of, simple granulometric properties of the sediments such as particle size distribution. It is considered more likely that biological community composition is controlled by an array of environmental variables, many of them reflecting an interaction between particle mobility at the sediment–water interface and complex associations of chemical and biological factors operating over long time periods. Such interactions are not easily measured or analyzed, but the results suggest that the time course of recovery of an equilibrium community characteristic of undisturbed deposits is controlled partly by the process of compaction and stabilisation that occurs following deposition.

Biological community composition thus reflects changes in sediment composition, but is also in equilibrium with seabed disturbance from tidal currents and wave action, both of which show spatial variations and interactions with water depth. The processes associated with compaction and stability of seabed deposits may, therefore, largely control the establishment of long-lived components of equilibrium communities and account for the dominance of opportunistic species in the initial stages of colonization in unconsolidated deposits of recently sedimented material after the cessation of dredging.

Introduction

The importance of benthic communities in marine food webs leading to commercially exploitable yields of fish has been widely recognized. Early models for the North Sea (see Steele 1974) suggested that of net primary production by the phytoplankton, approximately 80% was consumed by pelagic herbivores such as copepods and euphausiids, and 20% fell to the sea bed as a detrital input to the benthic community. At each step of the food web, relatively large amounts (80–90%) of the material entering the consumers is remineralized and returned to the water column to support further primary production by the phytoplankton, leaving a small proportion incorporated into the biomass of the consumer.

Because of the complexity of marine food webs, and the major dissipation of energy at each step of the food chain, the empirical model proposed by Steele (1965, 1974) for the North Sea and shown in Fig. 1 indicates that out of $100\,g\,C\,m^{-2}\,yr^{-1}$ produced at the sea surface as net primary production by the phytoplankton, only $0.3\,g\,C\,m^{-2}\,yr^{-1}$ appears as yield to man through the pelagic food web, and approximately $0.13\,g\,C\,m^{-2}\,yr^{-1}$ from demersal fish. Despite the huge dissipation of materials that occurs at each step in the food web, however, sufficient carbon evidently flows through the detrital food web, even in plankton-based ecosystems such as the North Sea, for as much as 30% of total fish production to be dependent on conversion through the community which lives on the sea bed.

More recent analyses of the trophic structure and fluxes of carbon in shelf waters of the North Sea by Joiris et al. (1982) suggest that as much as 50% of the annual phytoplankton production sinks to the sea bed as detritus and is supplemented by faecal

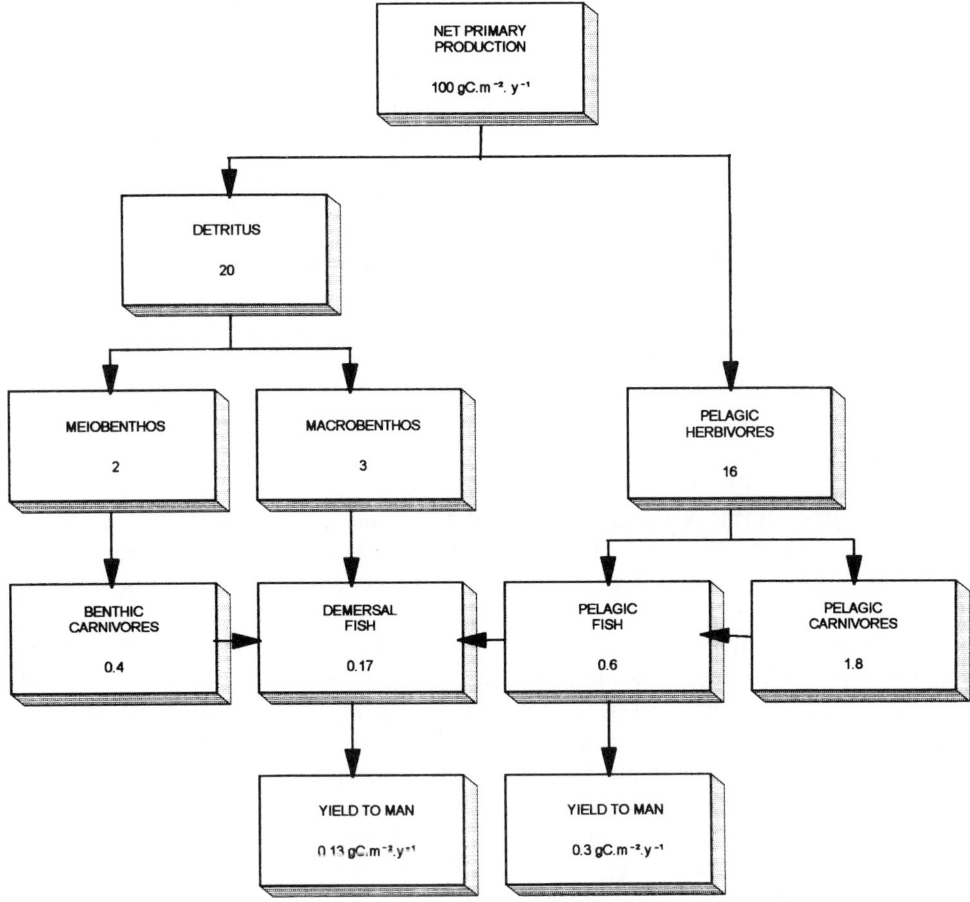

Figure 1 Simplified empirical carbon flow diagram for the phytoplankton-based ecosystem of the shelf waters of the North Sea. Note that, of the $100\,g\,C\,m^{-2}\,yr^{-1}$ of sea surface produced by the phytoplankton, the yield to man through pelagic food webs and pelagic fish is estimated to be $0.3\,g\,C\,m^{-2}\,yr^{-1}$ whereas that through benthic food webs is $0.13\,g\,C\,m^{-2}\,yr^{-1}$, or about 30% of the total exploitable fish yield to man. (Based on Steele 1965).

pellets of the zooplankton (see also Smetacek, 1984). The benthos is thus heavily implicated in carbon flow in coastal systems, and becomes of increasing importance in shallow waters where production by benthic algae (macrophytes) and seagrasses largely replaces that derived from the phytoplankton (see also Taylor & Saloman 1968, Thayer et al. 1975, Mann 1982, Moloney et al. 1986, Newell et al. 1988).

Benthic communities thus play a central role in the transfer of materials from primary production by the phytoplankton, benthic macrophytes and coastal wetlands through the detrital pool into higher levels in the food web, including commercially exploitable fish. Most estimates suggest that even in phytoplankton-based systems such as the North Sea, the yield to man through the benthos to demersal fish stocks is likely to approach 30–40% of that derived through the pelagic system. Partly for this reason, the populations of benthic communities which live on, and in, the deposits on the sea bed have been widely studied in investigations of the integrated effects of disturbance from a variety of natural and other sources.

Early studies include extensive physiological-toxicological work on the potential impact of suspended sediments on commercially significant target organisms (Loosanoff 1962, Sherk 1971, Sherk et al. 1972, 1974, Bright & Ellis 1989, Jokiel 1989, for review, see Moore 1977). Such studies have been extended to include the potential impact of dredging works on the ecology of biological communities in coastal embayments and estuarine ecosystems (Jones & Ellis 1976, Morton 1977, Conner & Simon 1979, Johnston 1981, Ellis & Heim 1985, Ellis & Taylor 1988, Ellis & Hoover 1990, Giesen et al. 1990, Onuf 1994).

Comprehensive studies of the impact of dredging for marine aggregates and sand on marine communities in European waters have been carried out by Millner et al. (1977), Pagliai et al. (1985), Sips & Waardenburg (1989), van Moorsel & Waardenburg (1990, 1991), and Kenny & Rees (1994, 1996). Reviews of the impact of sand and gravel extraction include those of the International Council for the Exploration of the Sea (ICES, 1975, 1977, 1992a,b, 1993), Gayman (1978), de Groot (1986), Nunny & Chillingworth (1986), Hurme & Pullen (1988), Lart (1991) and Charlier & Charlier (1992). A recent review for the Minerals Management Service, US Department of the Interior containing a number of specific case histories on the impact of marine mining has been given in a C-CORE publication (1996; see also Ellis 1987).

Despite the work that has been carried out over the past 30 years, the non-biologist could be forgiven for being bewildered by the diversity of the results and the difficulties of making more than the most general predictions on the effects of dredging activities including the extraction of marine aggregrates on biological resources. Essentially, most studies show that dredging itself is usually accompanied by a significant fall in species numbers, population density and biomass of benthic organisms. The rate of recovery is, however, highly variable depending (among other factors) on the type of community that inhabits the deposits in the dredged area and surrounding deposits, the latitude and the extent to which the community is naturally adapted to high levels of sediment disturbance and suspended particulate load.

In general, rapid rates of initial recolonization have been reported for some coastal deposits where the organisms are mainly mobile "opportunistic" species that have a rapid rate of reproduction and growth. Such organisms may also be able to recolonize the deposits by migration of the adults (see McCall 1976, Conner & Simon 1979, Saloman et al. 1982, Guillou & Hily 1983, Pagliai et al. 1985, van der Veer et al. 1985, Clarke & Miller-Way 1992, Rees & Dare 1993, van Moorsel 1994). In contrast, long-lived and slow-growing species, especially those in high latitudes may take several years before larval recruitment and subsequent growth of the juveniles allows restoration of the original community composition and biomass.

The process of "recovery" following environmental disturbance is generally defined as the establishment of a successional community of species which progresses towards a community that is similar in species composition, population density and biomass to that previously present, or at non-impacted reference sites (C-CORE 1996; see also Ellis & Hoover 1990). Typically, values range from up to one year in fine-grained deposits such as muds and clays (Ellis et al. 1995), although even in the fine deposits that characterize coastal ecosystems such as the Dutch Waddensea, van der Veer et al. (1985) report that recolonization takes 1–3 yr in areas of strong currents but up to 5–10 yr in areas of low current velocity. Longer recovery times are reported for sands and gravels where an initial recovery phase in the first 12 months is followed by a period of several years before pre-extraction population structure is attained (van Moorsel 1994, Kenny & Rees 1996).

Even longer times may be required for biologically-controlled communities that characterize coarse deposits (see Garnett & Ellis 1995), although the evidence is conflicting for coral reef communities. Some studies report long-term damage to coral resources from sedimentation associated with dredging (Dodge & Vaisnys 1977, Bak 1978, Dodge & Brass 1984, Madany et al. 1987, Hodgson 1994; for review see Maragos 1991). Other studies suggest that corals themselves may be tolerant of short-term increases in siltation associated with dredging (Marszalek 1981, Brown et al. 1990) but that modification of community structure of other components of reef communities such as fish species are detectable after multivariate analysis of species composition (Dawson-Shepherd et al. 1992).

Recovery times following disturbance from a variety of sources, including dredging work, may be extended in colder waters at high latitudes where communities typically comprise large slow-growing species that may take many years for recolonization and growth. In a Swedish fjord system, for example, a recovery that was indistinguishable from natural variations was established only after 8 yr following closure of a pulp mill (Rosenberg 1976), whereas de Groot (1979; see also Wright 1977, Aschan 1988) reports that recovery of communities within the Arctic Circle may take more than 12 yr compared with estimates of approximately 3 yr for deposits off the coast of the Netherlands. Similar extended timescales for recolonization by the benthic community have been reported for Antarctic waters by Oliver & Slatterly (1981).

The concept of "recovery" of biological resources is itself not an easy one to define for complex communities whose composition can vary over time, even in areas that remain undisturbed. Whether a community is identical in species composition and population structure following cessation of dredging thus to some extent begs the question of whether the biodiversity would have remained stable over that period in the absence of disturbance by dredging. Probably a more practical approach to the question of "recovery" will be the recognition of the establishment of a community that is capable of maintaining itself and in which at least 80% of the species diversity and biomass has been restored.

This implies a substantial restoration of the carrying capacity of the benthic food webs leading to fish, even though the precise species composition may not be identical to that recorded in the pre-dredged system. This issue of whether biological resources have been restored, and how this should be assessed, is of considerable importance in areas such as Canadian coastal waters where recovery of seabed resources forms part of a Statutory obligation following cessation of mining (D. V. Ellis pers. comm.).

Despite the complexity of the results for specific dredged areas, some firm general principles governing community structure following environmental disturbance have emerged in recent years and these appear to be generally applicable to a wide variety of communities both on the land and on the sea bed. The application of such concepts to coastal communities allows some credible predictions on the scale of impact of environmental disturbance such as that imposed by dredging and dredged spoils disposal and, more important, gives some insight into how long it might take for recovery in dredged areas and the surrounding deposits once dredging has ceased.

The object of the present review is to provide a framework within which the biological impact and subsequent recovery of benthic resources can be understood, with examples drawn mainly from the impact of dredging works in near-shore waters and estuarine systems.

General features of community structure

Most general models of community structure are based on the concept that biological communities do not form a series of distinct groups or assemblages along an environmental gradient, but show a corresponding gradient in community composition. Species that colonize habitats with unpredictable short-term variations in environmental conditions at one end of an environmental gradient of stability are subject to frequent catastrophic mortality. Such conditions occur in many shallow-water, intertidal and estuarine habitats and are characterized by populations which tend to have a high genetic variability that allows at least some components of the population to survive environmental extremes (see Grassle & Grassle 1974, Guillou & Hily 1983). Such organisms are thus selected for maximum rate of population increase, with high fecundity, dense settlement, rapid growth and rather a short life cycle. They are well suited to rapid invasion and colonization of environments where space has been left by a previous catastrophic mortality, whether this has been induced by natural factors or disturbance by man.

Such components have been designated "*r*-strategists" in a pioneer work by MacArthur & Wilson (1967; see also Pianka 1970), although we prefer to use the term "opportunists" for all such early colonizing species. Opportunists rely on a large investment in reproductive effort, rather than on mobility, for success in colonizing habitats made available by the catastrophic destruction of the previous community (see Gadgil & Solbrig 1972, McCall 1976).

Many communities living in unstable environments may comprise small, highly mobile species that are able to take advantage of recently created empty habitats quickly and to colonize them with large populations. These mobile colonizers are often associated with frequently-disturbed habitats (see Osman 1977). We distinguish these as "mobile opportunists" (see also MacArthur 1960, Grassle & Grassle 1974). All such mobile opportunists are *r*-strategists with life-cycle traits of small size, high fecundity, rapid growth and high mortality.

Under the stable conditions that occur at the other end of the environmental continuum, the community is controlled mainly by biological interactions, rather than by extremes of environmental variability. Here the organisms have an "equilibrium strategy" in which they are selected for maximum competitive ability in an environment that is already colonized by many species and in which space for settlement and subsequent growth is limiting. Such organisms are designated "*K*-strategists" or "equilibrium species" and devote a larger proportion of the resources to non-reproductive processes such as growth, predator avoidance and investment in larger adults (MacArthur & Wilson 1967, Gadgil & Bossert 1970, McCall 1976). Between these two extremes are communities whose species may be intermediate between those that occur at the extremes of the environmental gradient and have different relative

Table 1 Table summarizing the population characteristics of *r*-selected opportunists and *K*-selected equilibrium species (based on Pianka 1970, McCall 1976, Rees & Dare 1993, Holt et al. 1995).

Early colonizing species *r*-selected	Equilibrium species *K*-selected
1. Mainly opportunistic species (a) early reproduction (b) many reproductions per year (c) rapid growth (d) early colonizers (e) often catastrophic mortality	1. Equilibrium species (a) delayed reproduction (b) few reproductions per year (c) slow growth (d) late colonizers (e) low death rate
2. Small body size	2. Large mobile animals
3. Generally surface deposit feeders	3. Deposit and suspension feeders
4. Short life span; generally <1 year	4. Long life span; several to many years
5. Population size variable, usually well below carrying capacity of environment and recolonized frequently	5. Fairly constant in time; saturated community in equilibrium with carrying capacity of environment. No recolonization necessary
6. Brood protection with investment of energy into larval food provision (lecithotrophic)	6. No brood protection: larvae widely distributed in the plankton.
Streblospio benedicti *Capitella capitata* *Owenia fusiformis* *Ampelisca abdita* *Scolelepis fuliginosus* *Chaetozone setosa* *Jassa marmorata*	*Nephtys incisa* *Ensis directus* *Sabellaria spinulosa* *Arctica (Cyprina) islandica* *Echinocardium cordatum* *Nephrops norvegicus* *Melinna cristata* *Nucula sp.* *Amphiura filiformis* *Terebellides sp.* *Virgularia mirabilis* *Gari fervensis* *Tellina crassa* *Venerupis rhomboides* *Dosinia exoleta* *Scoloplos armiger* *Abra alba*

proportions of opportunistic *r*-strategists and equilibrium *K*-strategists. The characteristics of *r*-selected and *K*-selected equilibrium species are summarized in Table 1 (based on McCall 1976, Rees & Dare 1993), although it should be emphasized that the distinction is to some extent an arbitrary one, and is blurred in habitats that are subject to only mild environmental disturbance.

Changes in the structure and physical size of the infauna along a gradient of environmental conditions have been described in relation to organic pollution by Pearson & Rosenberg (1978) and in relation to physical disturbance by Rhoads et al. (1978), Oliver et al. (1980) and by Gray & Pearson (1982). These are illustrated in a schematic diagram in Fig. 2. Essentially, such studies show that community composition of benthic infauna (those that live within the deposits) along an environmental gradient is the result of a complex interaction between physico-chemical factors that

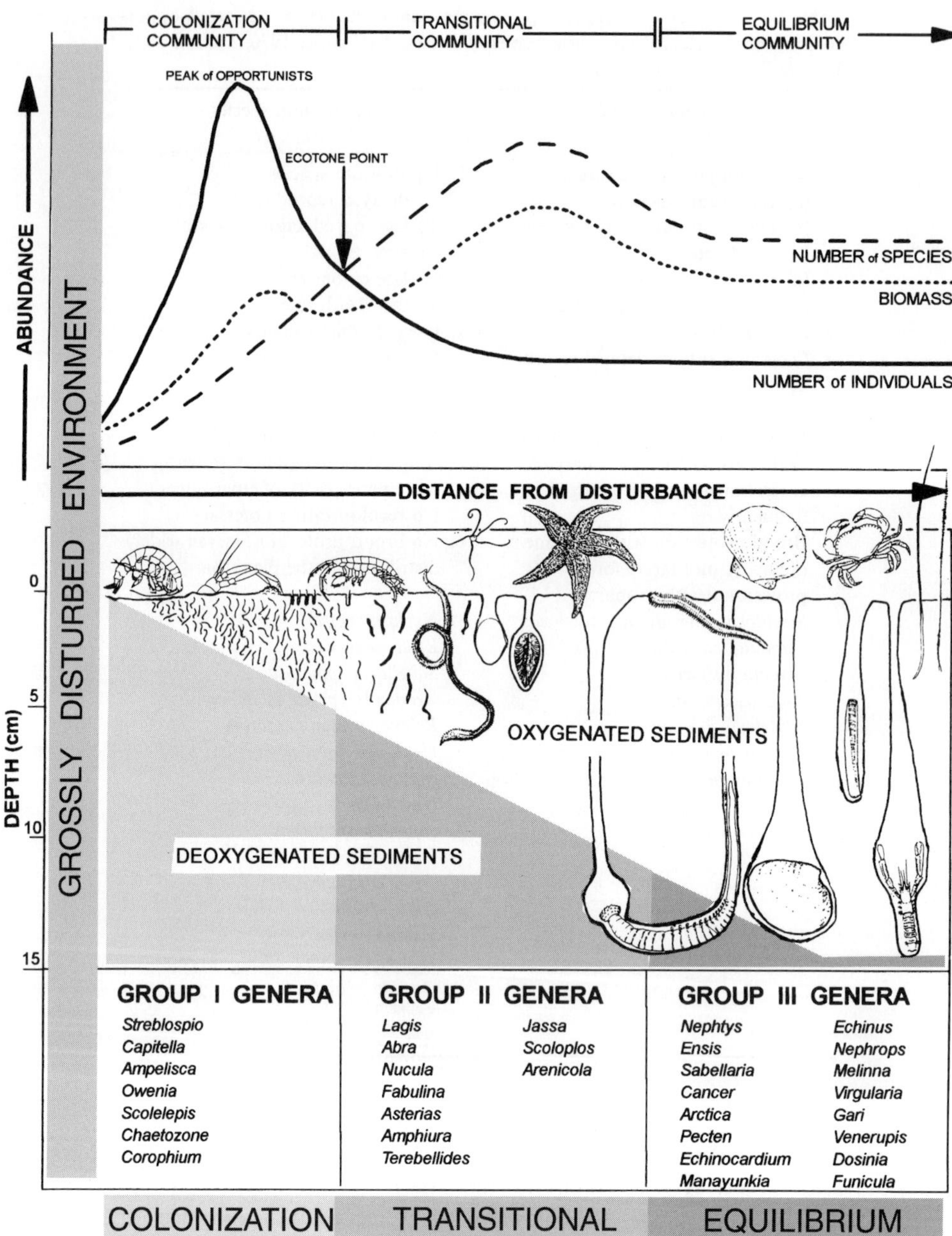

Figure 2 Pictorial diagram showing the ecological succession that characterizes benthic communities through a gradient of environmental disturbance. Note that in highly disrupted environments (on the left side of the diagram) few organisms may be capable of survival. In polluted or semi-liquid muds the sediments are colonized by few (resistant) species but which can attain very high population densities. As the stability of the environment increases, these opportunistic *r*-selected species are replaced by increased species variety, including slower-growing *K*-selected species. Finally in environments of high stability the community is dominated by equilibrium species with complex biological interactions between members of the community. (Based on Pearson & Rosenberg 1978, Rhoads et al. 1978).

operate at one end of the gradient and biologically-controlled interactions under the more uniform environmental conditions that occur in deeper waters (see Sanders 1969, Boesch & Rosenberg 1981).

The large species that comprise the burrowing infauna of stable habitats and those with low organic content maintain oxygen levels in the deposits down to considerable depths (see Flint & Kalke 1986) and often have complex interactions with neighbouring species, including smaller species whose survival depends on their association with large burrowing components (Fig. 2). The importance of bioturbation in both enhancing species diversity and in exclusion of potentially competitive species has been widely documented (Gray 1974, Rhoads 1974, Lee & Swartz 1980, Carney 1981, Rhoads & Boyer 1982, Thayer 1983). Comprehensive reviews by Pearson & Rosenberg (1978) and Hall (1994) summarize the impact of disturbance by a wide variety of factors including storms, dredging, fishing and biological activities on benthic community structure.

Biological interactions may also control community composition on the surface of the deposits. The presence of surface-dwelling bivalves, for example, may allow colonization by barnacles, ascidians and other epifaunal species that would not otherwise occur in the surface of the sediments. In other stable habitats, the activities of suspension-feeding mussels produce consolidated silt deposits that then allow deposit feeders such as the polychaete *Amphitrite*, the burrows of which in turn provide specialized shelter for the commensal scale worm, *Gattyana* (Newell, 1979).

Several studies have shown that the activities of the infauna may also inhibit, rather than facilitate, the occurrence of potential competitors for space. In an important study by Rhoads & Young (1970), it was shown that the benthic environment may be significantly modified by the burrowing and feeding activities of deposit-feeding organisms. This bioturbation results in the production of an uncompacted surface layer of faecal material that may result in the transfer of fine material to the sediment–water interface by turbulent mixing (Wildish & Kristmanson 1979, Snelgrove & Butman 1994) and may lead to the exclusion of potential competitors by deposit feeders (Woodin 1991, Woodin & Marinelli 1991). This inhibition of one type of population by the activities of another has been termed "amensalism" by Odum & Odum (1959) and has since been described in many habitats (Aller & Dodge 1974, Nichols 1974, Driscoll 1975, Eagle 1975, Johnson 1977, Myers 1977a,b, Brenchley 1981, de Witt & Levinton 1985, Brey 1991, Flach 1992).

Loss of these "key species" in *K*-dominated equilibrium communities following disturbance by dredging or other activities can lead to a collapse of the entire biologically-accommodated community even though individual species may be apparently tolerant of environmental disturbance. The colonial polychaete *Sabellaria spinulosa*, for example, provides a complex habitat that is associated with a wide variety of species which would not otherwise occur (see Holt et al. 1995). This polychaete undergoes a natural cycle of accretion and decay along with the associated community with a periodicity of from 5–10 yr (Wilson 1971, Gruet 1986). Disturbance of communities that are dominated by *K*-strategists may therefore take many years for recovery of their full community composition even though recolonization by individual components may occur comparatively rapidly.

As the amount of organic matter in the sediments increases along a gradient towards the fine silts and muds that characterize estuarine habitats, the larger species and deep-burrowing forms are replaced by large numbers of relatively inactive small suspension-feeding and surface deposit feeders including polychaete worms, bivalves and holothurians. This reduction in the species diversity and extent of sediment bioturbation results in an increased sediment stability and a restriction of the oxygenated layer to the surface of the sediments. Species in the intermediate parts of the environmental gradient shown in Figure 2 are thus relatively smaller than their counterparts in deeper waters and comprise a "transitional community" that is confined to a restricted habitat in the surface oxygenated layer of sediment and includes components that have many intermediate characteristics between typical *r*-selected opportunists and *K*-selected equilibrium species.

Because the *K*-selected components in the community live for longer, the individuals must be able to tolerate short-term changes in environmental conditions including siltation. They therefore have generally wider limits of physiological tolerance than *r*-selected opportunistic species that respond to environmental change by selection of genetically adapted components of the population during each of the many reproductive cycles per year.

The transitional community comprises more species than the equilibrium community shown in Figure 2 because of invasion by opportunistic species, but the species variety and mean size rapidly decline as the organisms are increasingly crowded into the upper oxygenated layer at the sediment – water interface. The region between this transitional community and those dominated by large populations of a restricted variety of small opportunists has been referred to as the "ecotone point" by Pearson & Rosenberg (1978) and is shown in Figure 2.

Finally at the extreme end of the physical gradient shown in Figure 2, there is a further restriction of habitat space to the upper oxygenated layer of sediment. This results in a progressive elimination of species and to communities dominated by opportunistic *r*-strategists that are selected for small size, high fecundity and an ability to recolonize rapidly following catastrophic mortality (see Pearson & Rosenberg 1978, Gray & Pearson 1982). Very high population densities of these *r*-selected opportunists can occur (the "Peak of Opportunists" in Figure 2) before these decline as organic pollution or high environmental disturbance eliminates even these rapid colonizers.

A useful tool for determining the extent of impact of environmental impact from a variety of sources is a plot of the proportional contribution of each species in the community to the overall population density of the assemblage as a whole. These curves have been designated "*K*-dominance curves" by Lambshead et al. (1983) and have been widely used in environmental impact studies in recent years (Warwick 1986, Clarke & Warwick 1994). Obviously the equilibrium communities characteristic of undisturbed (or unpolluted) environments have a high species diversity and each component species makes a relatively small contribution to the overall population density. Conversely, as a point source of disturbance is approached the (sensitive) species are replaced by large numbers of those (resistant) members of the community that are capable of survival. This can lead to as much as 80–90% of the population being dominated by only one or two opportunists or *r*-selected species at the Peak of Opportunists shown in Figure 2.

A typical set of results taken from one of our surveys of coastal communities in the eastern English Channel is shown in Fig. 3 (Newell & Seiderer, 1997c). From this it can be seen that as much as 78% of the community in unstable, unconsolidated, mobile deposits at Site 1 was represented by just one species, the opportunist amphipod crustacean *Ampelisca brevicornis*, and that additional species each made only a relatively small contribution to the population. Further along the gradient of sediment stability

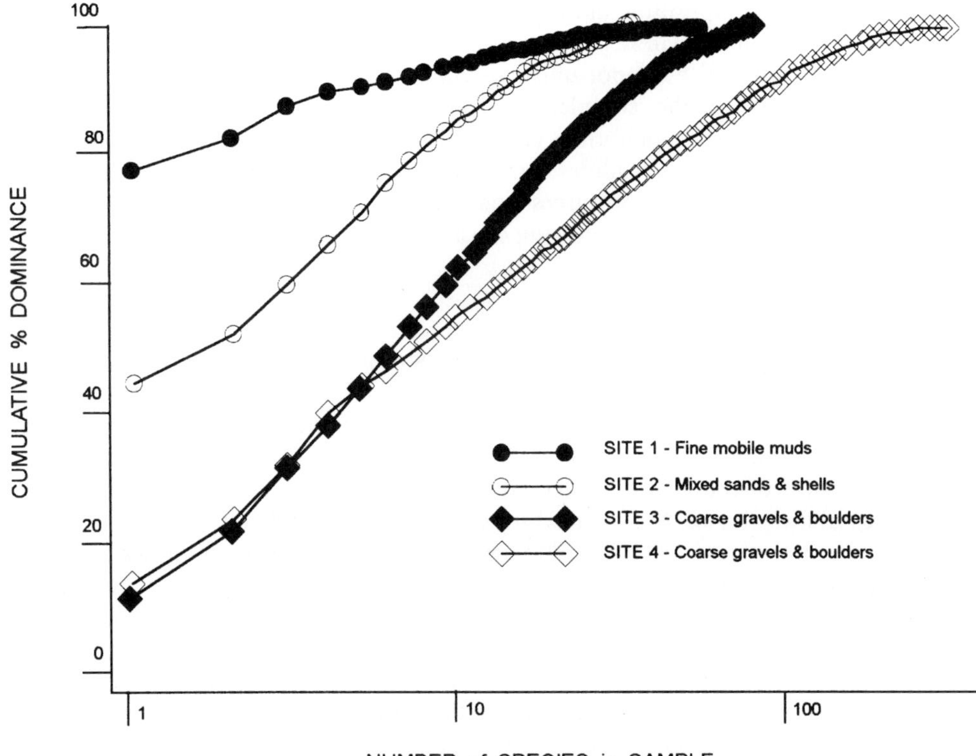

Figure 3 A set of typical *K*-dominance curves showing the proportional contribution of individual species to the overall community in fine mobile muds, in mixed sands and shells and in stable habitats of coarse gravels and boulders off the Kent coast at West Varne in 1996. The fine mobile muds are dominated by the opportunist amphipod crustacean, *Ampelisca brevicornis*, whereas the more stable deposits have a higher total species complement each of which makes a relatively small contribution to the overall population density. (Based on Newell & Seiderer 1997c).

in mixed sands and shells at Site 2, the dominance by one species (*Sabellaria* sp.) alone was approximately 45%. Finally, in the stable environmental conditions of coarse gravels and boulders at Sites 3 and 4 there was a very large species variety of over 300 and a relatively uniform species distribution with dominance values of only 12–15%.

Estimation of *K*-dominance curves adjacent to dredging works and other point sources of environmental disturbance is thus potentially useful because it can be used as a relatively simple index to define the area of immediate impact. It can also be used to determine whether this is enlarging or decreasing with time, without the necessity of the complex analysis of community structure that is required for interpretation of the wider impact on community structure in the transition zone.

These distinctions between the lifestyles and adaptive strategies of opportunists and *K*-selected equilibrium species are of fundamental importance because they go some way towards accounting for the differences in the rate of recovery that has been recorded for biological resources following disturbance by episodic events such as dredging.

Clearly, the species composition and rate of recovery of biological communities following cessation of dredging will depend to a large extent on whether the original communities were dominated by opportunists or equilibrium species, and on the time that is required to develop the complex associations which characterise interactions between the *K*-dominated equilibrium community. Knowledge of the key faunal components and their lifestyle thus allows some predictions on the impact of dredging and spoils disposal on biological resources and on the subsequent rate of recovery of marine community composition following cessation of dredging.

Ecological succession and the recolonization process

These general features of the structure of benthic communities apply not only to successional stages along a gradient of environmental variability, but also to the successive sequence of populations that recolonize deposits after the cessation of environmental disturbance. McCall (1976) and Rees & Dare (1993) have recognized the occurrence of three main types of benthic components of marine communities based on the distinction between *r*-selected opportunistic species and *K*-selected equilibrium species. Group I species comprise those that colonize first after a community has been removed by disturbance. They comprise large populations of small sedentary tube-dwelling deposit feeders that have rapid development, many generations per year, high settlement and death rates. Examples include the polychaete worms *Streblospio*, *Capitella capitata*, and *Owenia fusiformis* as well as the amphipod *Ampelisca*. That is, the Group I community comprises mainly *r*-strategist opportunistic species.

Group II species comprise mainly bivalve molluscs such as *Tellina*, *Nucula* and *Abra*, the tube worm *Lagis* (= *Pectinaria*) and the common starfish (*Asterias rubens*). There is no absolute distinction between this community and the primary colonizers, but the components attain a lower peak abundance than the smaller opportunistic species and have slower recruitment and growth rates. Finally Group III species comprise larger slow-growing *K*-strategist equilibrium species such as the polychaete *Nephtys*, the reef-forming Ross worm (*Sabellaria*), razor shell (*Ensis*), sea urchins such as *Echinocardium* and *Echinus*, scallops (*Pecten*), the ocean quahog (*Arctica islandica*), the edible crab (*Cancer pagurus*) and larger burrowing crustaceans such as *Nephrops* and *Callianassa*.

The changes in species variety, abundance of individuals and biomass during the recolonization process are shown in Fig. 4. Inspection of this figure shows that initially the sediments are almost devoid of benthic macrofauna.

The initial colonizing species are few, but the number of individuals (population density) increase rapidly with time to a peak of (Group I) opportunistic species. As time passes, the short-lived opportunistic species (*r*-strategists) decrease in numbers and biomass as more species invade the area. This transition point where the community is poor in population density and biomass is the same ecotone point shown on the spatial gradient in Figure 2.

Prior to this, the community is characterized by large populations of a few small opportunistic species; after this time the species variety increases, as does the biomass, but the population density declines. This Group II community is a transitional one where the maximum number of species has invaded the newly-available space, and is

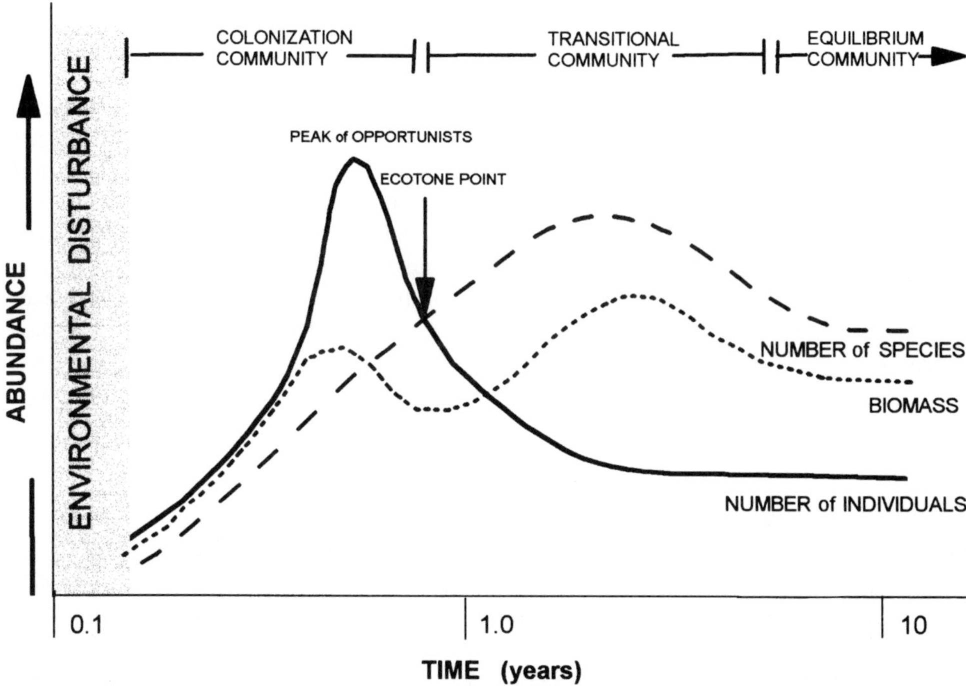

Figure 4 Schematic diagram showing a colonization succession in a marine sediment following cessation of environmental disturbance. Initial colonization is by opportunistic species which reach a peak population density generally within 6 months of a new habitat becoming available for colonization after the catastrophic mortality of the previous community. As the deposits are invaded by additional (larger) species, the population density of initial colonizeis decliners. This ecotone point marks the beginning of a transitional community with high species diversity of a wide range of mixed *r*-selected opportunistic and *K*-selected equilibrium species. This period may last for 1–5 yr depending on a number of environmental factors, including latitude. Provided environmental conditions remain stable, some members of this transition community are eliminated by competition and the community as a whole then forms final equilibrium community comprising larger, long-lived and slow growing species with complex biological interactions with one another. (Based on Pearson & Rosenberg 1978).

followed by a phase where some species are eliminated by competition and the community returns to the (somewhat lower) species composition and biomass characteristic of the undisturbed Group III community.

The sequence shown in Figure 4 indicates that colonization is likely to follow a definite time course of progressive invasion by large numbers of opportunistic species in the first instance, followed by a wider species diversity during the transitional phase and finally by a consolidation phase when competition between the *K*-strategist equilibrium species for the limited space available results in the elimination of some of the transitional colonizers (see also Warwick et al. 1987). The biological diversity in any particular community will then reflect the frequency of disturbance and represent a balance between invasion and subsequent growth of colonizers, and losses by extinction and displacement (see Huston 1994). In areas where environmental disturbance is unevenly distributed, this may lead to a mosaic of communities, each at different stages of the successional sequence shown in Figure 4 (see Johnson 1970, Grassle & Sanders 1973, Whittaker & Levin 1977, Connell 1978), and may partly account for the patchiness of marine communities in dredged areas.

The time taken for recovery of the full species composition and for subsequent exclusion of some of the transition community following the growth of larger *K*-strategist equilibrium species in a particular area will depend largely on the components that occur under natural conditions. In shallow water and estuarine conditions, where the community is in any case dominated by opportunistic species, recovery to the original species composition may be very rapid and coincide with the Peak of Opportunists point in Figure 4. In the stable environmental conditions of deeper waters, the replacement of the initial colonizers in the transitional community following complex biological interactions between the *K*-selected equilibrium species may take several years.

The physical impact of dredging

Impact within the dredged area

The increased exploitation of marine deposits and the physical impact of dredging works has been widely reviewed (see Dickson & Lee 1972, Shelton & Rolfe 1972, Cruikshank & Hess 1975, Eden 1975, Millner et al. 1977, de Groot 1979, van der Veer et al. 1985, Glasby 1986, Lart 1991, Gajewski & Uscinowicz 1993, ICES 1993, Land et al. 1994, Whiteside et al. 1995, Hitchcock & Drucker 1996). Most of the sea-going aggregate dredgers are self-contained and use a centrifugal suction pump to lift the aggregates from the sea bed into a hopper where the material is screened before being transferred to the hold. The *in situ* reserves for economic exploitation normally range from 15–55% gravel. Unless the material is otherwise suitable for direct use as a beach feed or landfill (see Hess 1971), the sand : gravel ratio in the final cargo is adjusted to between 50 : 50 and 65 : 35 depending on customer requirements, local geology and ship performance (A. Hermiston pers.comm).

A proportion of the dredged deposits may therefore be returned to the sea bed through reject chutes when there is a larger proportion of fine material than is required for a commercially viable cargo. In most aggregate deposits the fines comprise only 1–2% of the total and are dominated by the silt fraction although significant quantities of sand may also be discharged in the immediate vicinity of the dredger to increase the gravel component of the cargo. Overspilling of water via spillways from the hopper will also contain some fine sands that are maintained in suspension by the turbulence within the hopper.

Essentially the physical impact of dredging works is dependent partly on the method of dredging, and partly on the amount and grade of deposits rejected by screening (if used) and overspill from the hopper. Two main methods of dredging are used for gravels extraction in European coastal waters. These are anchor hopper dredging and trailer suction hopper dredging and are illustrated in Fig. 5. In anchor dredging the vessel is stationary and dredges the deposits from a sequence of specific points on the sea floor and can therefore leave pits or depressions on the sea bed that may reach as much as 20 m in depth and 75 m in diameter (Cruikshank & Hess 1975, Dickson & Lee 1972).

Because the deposits required for marine aggregates are coarse, and sediment disturbance by wave action is in any case limited mainly to depths of less than 30 m even

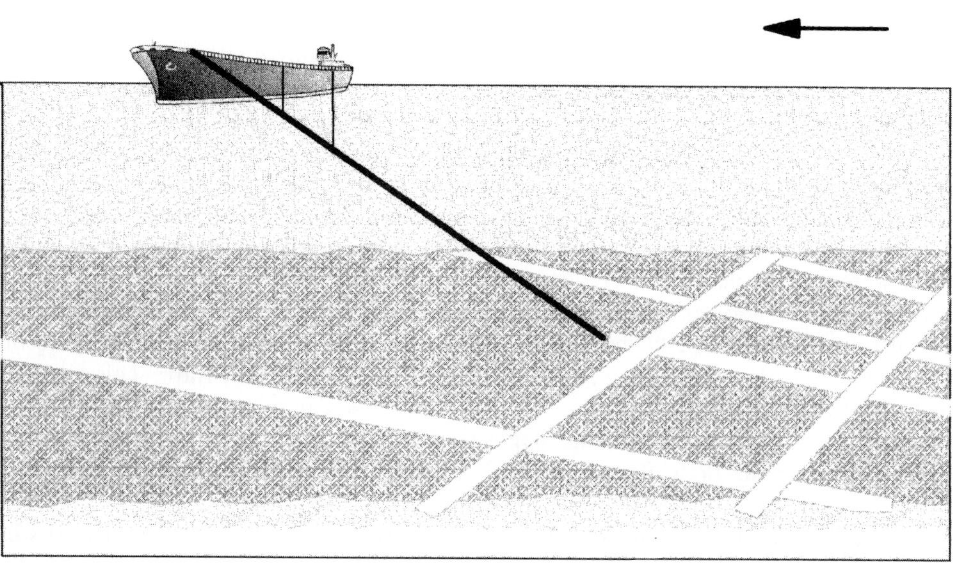

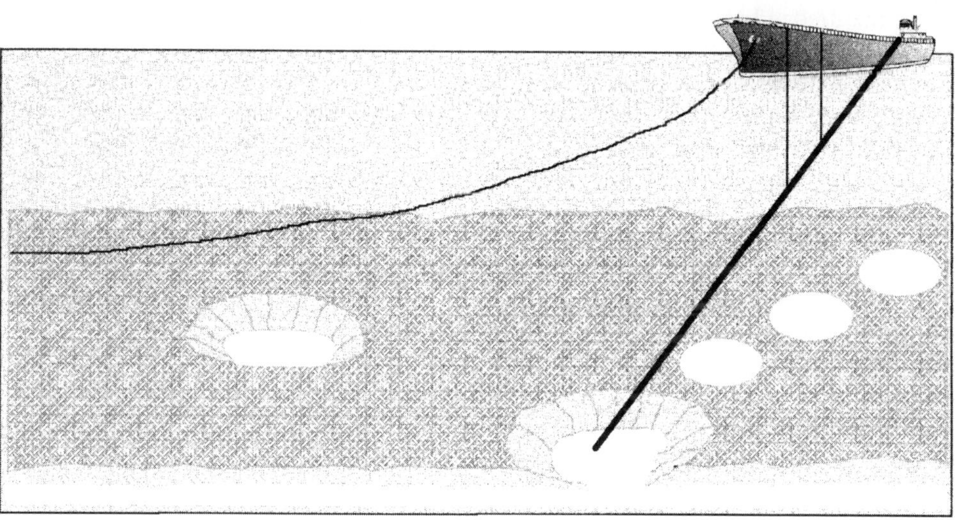

Figure 5 Diagram showing the two principal methods of dredging for marine aggregates in European coastal waters. Upper diagram shows the furrows left on the sea bed by trailer suction dredging while the vessel is under way. In this case the sea bed is crossed by a series of tracks which are 2–3 m wide and up to 50 cm deep. Lower diagram shows the pits left on the sea bed by anchor dredging. In this case the vessel is anchored and the dredged pits may reach as much as 75 m in diameter and 20 m in depth. (Based on Dickson & Lee 1972).

during storm conditions, it follows that not only is the fauna likely to be removed in patches from the dredged areas, but such pits are likely to be persistent features of the sea bed topography for several years except in areas where the sands are mobile (Eden 1975). Dickson & Lee (1972) studied the recovery of test pits dug by anchor dredge in gravel deposits in the Hastings Shingle Bank off the south east coast of England. They found that the pits were very slow to fill and were still visible after two years. In another study, van der Veer et al. (1985) described the recovery of pits in sandy substrata in the Dutch Wadden Sea. They showed that in this instance pits in channels with a high current velocity filled within one year, but those in the lower current velocities which occur in tidal watersheds took 5–10 yr to fill whereas those in tidal flat areas were still visible after 15 yr.

Such sediment movement as does occur is mainly through slumping of the sides of the furrows and subsequent infilling by fine particulates transported by tidal currents into the furrows that reduce current velocity and act as sediment traps. This can lead to heavily anoxic sediments and to colonization by a community which differs considerably from that in the original deposits (Dickson & Lee 1972, Shelton & Rolfe 1972, Kaplan et al. 1975, Bonsdorff 1983, Hily 1983, van der Veer et al. 1985, Hall 1994).

A second method of marine aggregate dredging is for the ship to extract deposits by suction through one or two pipes deployed while the vessel is slowly under way (Figure 5). In this case, side-scan sonar records show that the sea bed within the boundaries of licenced extraction areas in the southern North Sea is crossed by a series of dredge tracks that are 2–3 m wide and up to 50 cm deep (van Moorsel & Waardenberg 1990, Kenny & Rees 1994) although deeper troughs of up to 2 m were recorded from areas where the dredge head had crossed the area several times. Davies & Hitchcock (1992) reported dredge cuts of between 20–55 cm depth and 3–3.8 m width in commercially exploited deposits of the Bristol Channel. Somewhat deeper troughs of up to 70 cm were reported for the Baltic (Gajewski & Uscinowicz 1993). In this case removal of the surface 0.5 m of deposit would be sufficient to eliminate the benthos from the deposits in strips, the total removal depending on the intensity of dredging at a particular worked site.

Despite the shallower depth of removal, the evidence suggests that infilling of the troughs from trailer suction dredging takes at least 12 months in the Baltic and is achieved partly by slumping from the sides and partly by transport of fine material by bottom currents into the sediment traps formed by the dredged furrows (Kaplan et al. 1975, Hily 1983, van der Veer et al. 1985, Gajewski & Uscinowicz 1993). Progressive removal of the original sandy gravel and its replacement by fine sand has also been reported for the sediments off Dieppe by Desprez (1992). In the case of experimental furrows dredged by trailer suction in gravel deposits of the southern North Sea off the Suffolk coast of England, even shallow depressions of only 20–30 cm depth were still visible on side-scan sonar records made up to 4 yr later (Millner et al. 1977). In contrast, dredge furrows in the Bristol Channel have been reported to disappear within 2 to 3 tides because of high sediment mobility (pers. obs.).

Rather unexpectedly, Kenny & Rees (1994, 1996) found an increase in the particle size of deposits in the dredged areas, possibly reflecting the exposure of coarse deposits at depth below the surface gravel layers. In this study, which was carried out in the southern North Sea, the dredged furrows were visible with side-scan sonar even after 2 yr. Similar results have been reported for dredging tracks off the French coast at

Dieppe (Desprez 1992), although winter storms obliterated tracks within a few months on the Klaver Bank in the Dutch sector of the North Sea (Sips & Waardenberg 1989, van Moorsel & Waardenberg 1990, 1991). In general, dredge tracks will persist for varying times depending on the rate of local sediment fluxes, recent measurements suggesting 1–4 days only for the Norfolk Banks, but periods as long as 1–4 yr for more stable deposits off the Owers to the east of the Isle of Wight (A. Hermiston pers.comm.).

Thus both anchor dredging and trailer suction dredging have an important potential impact on the biology of the dredged areas, since no benthos is likely to occur below the dredged depth. This can be expected to lead to a patchy distribution of organisms, reflecting the differences between the dredged furrows and the intervening undredged surfaces. Such recolonization as occurs within the dredged areas is likely to be by migration of adults through transport on tidal currents (Rees et al. 1977, Hall 1994); by transport in sediments slumping from the sides of the pits and furrows (McCall 1976, Guillou & Hily 1983); by the return of some undamaged components through outwash from the chutes and spillways (see Lees et al. 1992, Ministry of Agriculture, Fisheries and Food 1993); and by colonization and subsequent growth of larvae from neighbouring populations. In this case, a clear succession of colonizing species is to be anticipated, leading to the establishment of definite clusters or patches in benthic community composition, depending on the type of deposits that have infilled the dredged areas and the time since the recolonization sequence started.

Impact adjacent to the dredged area

Although a good deal of concern has been expressed about the possible impact of marine aggregate extraction on coastal resources (see ICES 1992a,b), the possible scale of impact outside the immediate dredged area from the settlement on the sea bed of fine material temporarily suspended by marine aggregates dredging is poorly understood.

Hitchcock & Drucker (1996) have summarized values for material lost through the hopper overflow spillways and from the reject chutes during the screening process on a typical modern trailer suction dredger of 4500 t hopper capacity operating in UK waters off the coast of East Anglia. Table 2 shows the size distribution for the material lost through the reject chute and spillway, while the total screened load quantities are

Table 2 Size distribution of overspill and reject material from a typical modern trailer suction dredger of 4500 t hopper capacity (based on Hitchcock & Drucker 1996).

	Proportion in discharge (%)	
Particle size (mm)	Spillway	Reject chute
<0.063	38.0	1.0
0.063–0.125	14.0	0.9
0.125–0.250	5.7	8.9
0.250–0.500	12.9	31.4
0.500–1.000	9.2	27.3
1.000–2.000	3.3	12.0
>2.000	16.9	18.5

summarized in Table 3. These show that during a recorded average loading time of 290 min, 12 158 t of dry solids and 33 356 t of water were pumped by the dredge pump. The data show that 4185 t of dry solids are retained as cargo, while 7223 t of dry solids are returned overboard because of rejection by screening, and a further 750 t from overspill.

It is also clear that some 1338 t of material >2.0 mm (representing 18.5% of the 7235 t in the reject chute – see Table 2) is lost overboard through the reject chute and a further 126 t (representing 16.9% of the 750 t in the spillway – see Table 2) from the spillway. This equates to a loss rate of 76.9 $kg\,s^{-1}$ of particles >2.0 mm from the screening reject chute and 7.2 $kg\,s^{-1}$ from overspill. Assuming that the dredger moves at an average speed of 1 knots, the flux of material >2.0 mm entering the water column is 39.6 $kg\,s^{-1}\,m^{-1}$ from the screening reject chute and 3.7 $kg\,s^{-1}\,m^{-1}$ from the spillways. Much of this material is in the size range 2.0–10.0 mm and falls rapidly to the sea bed with little horizontal displacement during screening. Video recordings during normal loading operations show that such material deposits on the sea bed directly under the dredge vessel (Davies & Hitchcock, 1992).

Finer sand and silt fractions discharged during dredging and screening amount to 5824 t from the reject screening chutes and 338 t from overspill. This is equivalent to a deposition rate of 334 $kg\,s^{-1}$ and 19.4 $kg\,s^{-1}$, respectively. In addition to the sand fraction, up to 213 t (12.2 $kg\,s^{-1}$) of muddy sediment (<0.063 mm diameter) may be lost through the rejection process and 285 t (16.4 $kg\,s^{-1}$) from overspill. The material may be expected to settle more slowly than the sand-sized fraction and has a typical settling velocity of 0.1–1.0 $mm\,s^{-1}$.

In its simplest form, the settlement velocity and residence time of such particles in the water column can be estimated from Stoke's Law. If the residence time of particles in the water column is known, the duration and speed of currents will then determine the excursion pattern before settlement. Estimates of dispersion of fine material based on these Gaussian diffusion principles suggest that very fine sand particles may travel up to 11 km from the dredge site, fine sand up to 5 km, medium sand up to 1 km and coarse sand less than 50 m (H. R. Wallingford 1994, cited in Hitchcock & Drucker 1996).

Similar estimates based on the settlement velocity of fine silt-sized particles (<0.063 mm diameter) suggest that this material could remain in suspension for up to 4–5 tidal cycles and be carried for as much as 20 km on each side of a point source of discharge. Most recent studies made on the dispersion of sediment plumes generated from dredging operations suggest, however, that the area of impact of outwash from dredging activities is smaller than estimates based on Gaussian diffusion models, especially where the proportion of silt and clay in the deposits is low. This appears to be due to complex cohesion properties of the discharged sediment particles that settle to

Table 3 Typical screened load quantities for a suction dredger of 4500 t hopper capacity (based on Hitchcock & Drucker 1996).

	Dry solids (tonnes)	Water (tonnes ±5%)
Quantity pumped	12 158	33 356
Quantity retained	4185	874
Quantity reject through screening	7223	13 499
Quantity lost through overspill	750	21 387
Total losses	7973	34 886

the sea bed as a density current and do not conform to settlement rates based on the specific gravity and size of the component particles themselves.

A detailed study by Gajewski & Uscinowicz (1993) in relation to trailer dredging in the Baltic showed that the width of the plume, as determined by the light extinction in the water, did not exceed 50 m. Settlement of the suspended matter onto the sea floor was measured by a series of sediment traps deployed approximately 1 m above the sea bed. When the dredger was discharging mainly fine sand (0.25–0.125 mm) at a concentration of approximately 11 000 mg l^{-1}, it was found that deposition of 7500–15 000 g m^{-2} were recorded in the troughs during dredging. The settlement in traps deployed 50 m away from the dredger's route was, however, less than 1220 g m^{-2}. At distances greater than 50 m the amount of material settling on the sea floor decreased rapidly (Fig. 6).

More recently, Acoustic Doppler Current Profiling (ADCP) techniques have been used to determine plume dispersion in relation to spoils dispersal from both commercial aggregate dredgers (Hitchcock & Dearnaley 1995, Hitchcock & Drucker 1996)

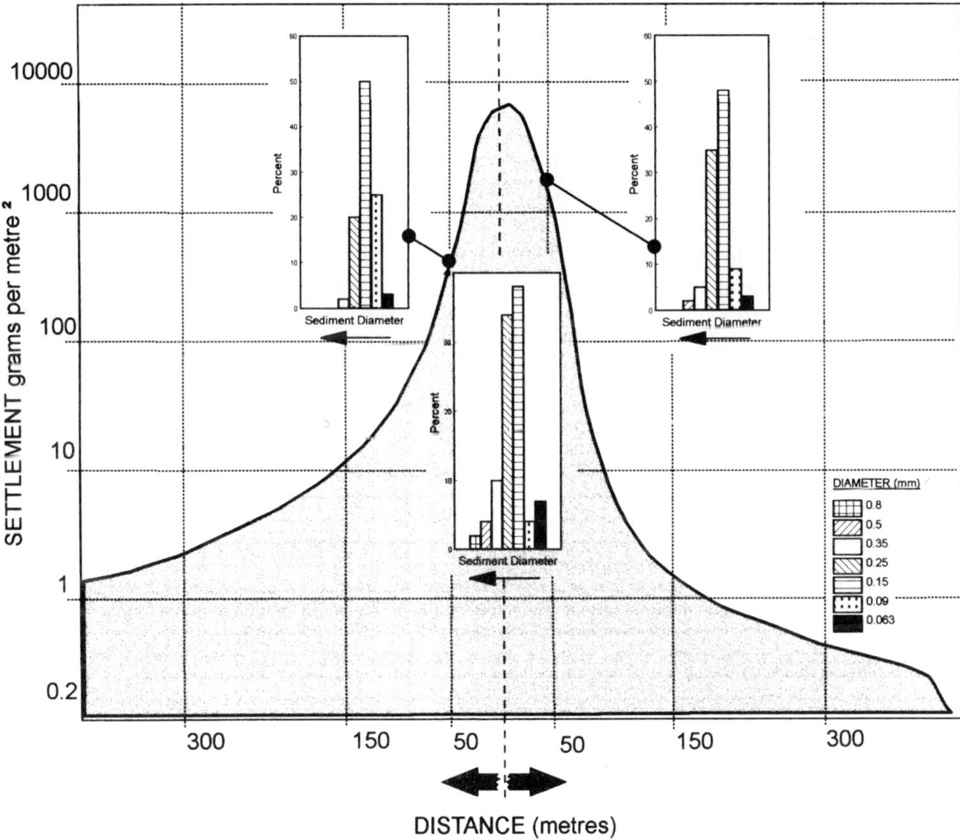

Figure 6 Diagram showing the settlement of sand during dredging operations from trailer dredging in the Baltic. Particle size profiles for the sediments deposited in the track of the dredger and 50 m on each side of the dredger are also shown. Note that the main deposition of sand from flow-off from dredging operations was confined to distances within 150 m on each side of the track of the dredger. (After Gajewski & Uscinowicz 1993).

and in relation to capital dredging works and sand mining (Land et al. 1994, Whiteside et al. 1995). Remote airborne and satellite imagery has also proved to be a useful tool in defining the contours of sediment dispersal (Whiteside et al. 1995).

These studies confirm that the initial sedimentation of material discharged during outwash from dredgers does not, as had been widely assumed, disperse according to the Gaussian diffusion principles used in most simulation models, but behaves more like a density current where particles are held together by cohesion during the initial phase of the sedimentation process. As a result, the principal area likely to be affected by sediment deposition is much less than the "worst case" scenarios predicted from conventional Gaussian diffusion simulation models, and is mainly confined to a zone of a few hundred metres from the discharge chutes.

A recent study by Whiteside et al. (1995; see also Johanson & Boehmer 1975, Gayman 1978) has shown that the behaviour of plumes discharged during sand dredging can best be regarded as comprising an initial "dynamic phase" during which the sediment–water mixture descends rapidly to the sea bed as a density current jet at a rate that depends on the overflow density, the diameter of the discharge pipe, the water depth, the velocity of discharge and the speed of the dredger. During its passage through the water column and following impact with the sea bed the sediment is dispersed into the water and forms a well defined plume astern of the dredger. This second longer phase has been referred to as the "passive phase" of dispersion by Whiteside et al. (1995) and starts approximately 10 min after outflow. During this phase the material behaves in a relatively simple settling mode according to Stokes' Law, the plume then decaying to background levels after a period of 2–3 h.

Their study showed that approximately 100 m (corresponding to approximately 3 min from the overflow) astern of a dredger working in Hong Kong waters the plume surface sediment concentrations were from 75–150 $mg\,l^{-1}$. Levels were halved in 10 min and reduced to 20–30 $mg\,l^{-1}$ after 30 min. This approached the recorded background suspended solids concentration of 10–15 $mg\,l^{-1}$ and indicated that only a relatively small proportion of the fines category ($<63\,\mu m$) remained in the water column at the start of the passive phase of dispersion 10 min after discharge. Even then, their data suggest that the settlement rate of the plume continued to be more rapid than simple particle settlement would suggest.

A plume dispersion model developed by Whiteside et al. (1995) for the surface layer (the upper 8 m of the water column) for up to 40 min after discharge is shown in Fig. 7 and compares well with summed plume decay measurements in the vicinity of the dredger. The contours for sediment deposition evidently remain as a narrow band extending for approximately 100 m on each side of the track of the dredging vessel, much as recorded by Gajewski & Uscinowicz (1993) for Baltic waters.

Very similar rapid rates of deposition and decay of sediment outwash plumes have been recorded by Hitchcock & Drucker (1996), who studied plume generation and decay from four dredge vessels ranging in capacity from 2000–5000 t during normal loading operations off the coast of East Anglia, UK. During the plume tracking exercise peak current velocities reached 0.6 $m\,s^{-1}$ and the water depth was approximately 22 m. The concentration of total suspended sediment discharged was approximately 2500 $mg\,l^{-1}$ comprising mainly sand-sized material and with $<30\,mg\,l^{-1}$ mud (<0.063 mm diameter).

The total concentration of suspended solids in the water column at different depths and distances from the dredger measured by water sampling and optical transmis-

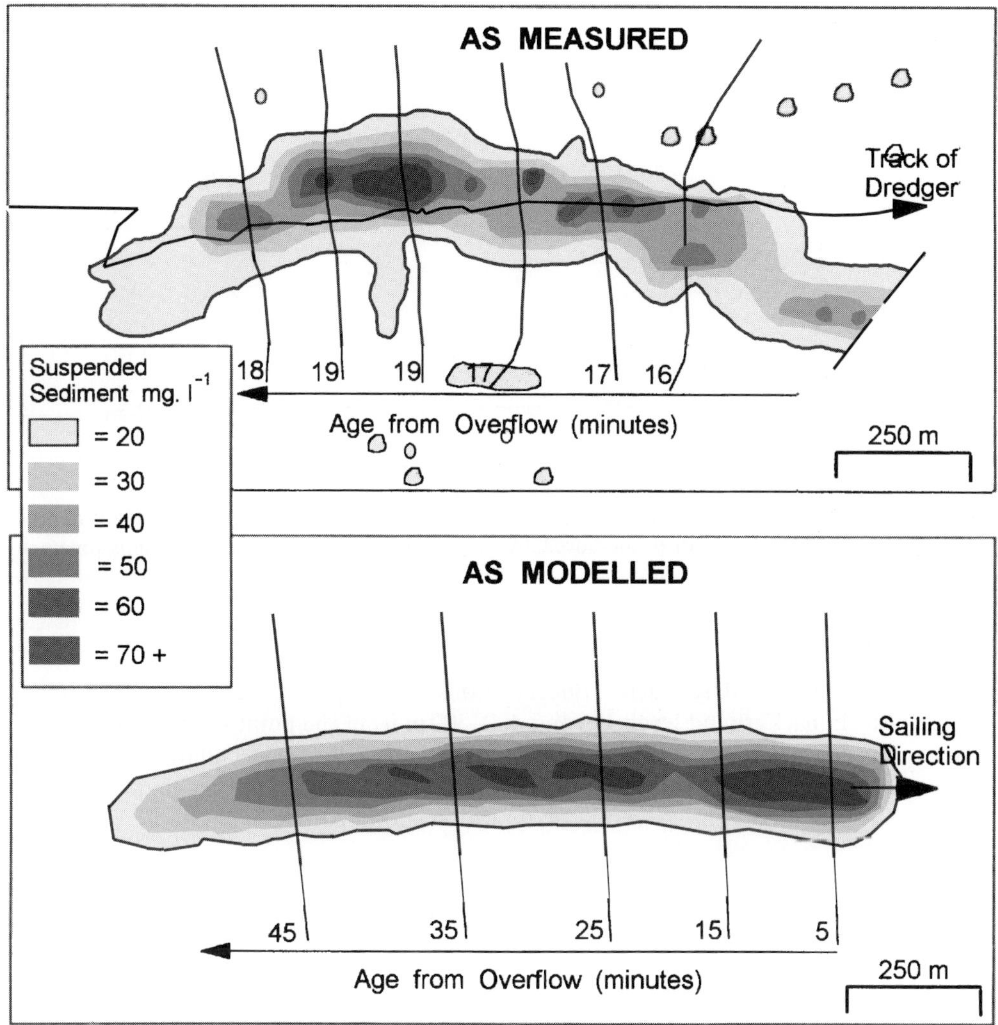

Figure 7 Diagram showing contours for suspended sediment concentrations astern of a trailer dredger operating in Hong Kong waters. Upper diagram shows the contours as measured in the upper 8 m of water across the plume at various arbitrary time intervals during field studies of a sediment plume from 16–19 min after discharge. Lower diagram shows the output of a simulation model developed for sediment dispersion based on rapid initial sedimentation during a dynamic phase and a second longer passive phase which starts approximately 10 min after outflow. (After Whiteside et al. 1995).

someters is summarized in Fig. 8. The corresponding values for silt-sized material (<0.063 mm) are shown in Fig. 9. These data show that concentrations of sand-sized material are reduced to background levels over a distance of only 200–500 m from the point of release into the water column and that the concentration of even silt-sized particles is reduced to background values of 2–5 mg l^{-1} over a similar distance. This very rapid reduction in suspended sediment concentrations is similar to that reported by Whiteside et al. (1995).

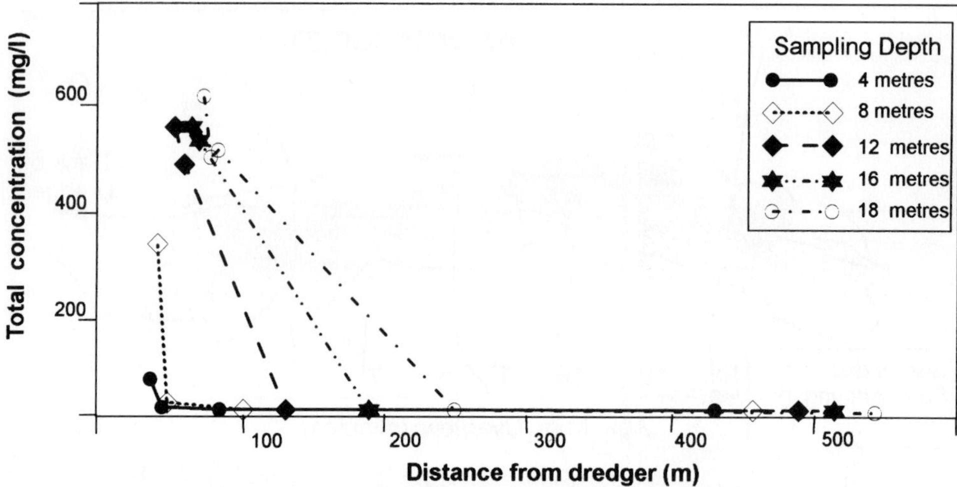

Figure 8 The total concentration of suspended solids in the water column at different depths and distances from the dredger measured by water sampling and optical transmissometers. (After Hitchcock & Drucker 1996).

Although suspended sediment concentrations in the plume were not significantly different from background levels beyond 400–500 m from the point source of discharge using conventional water sampling and optical transmissometer techniques, it is interesting to note that it is possible to track the plume using ADCP techniques over a distance of up to 3.5 km. A series of typical sections across a discharge plume at

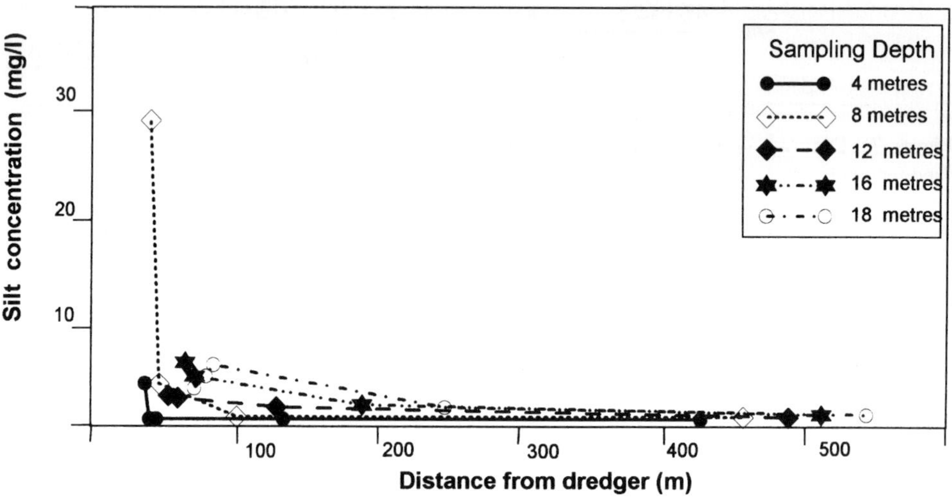

Figure 9 The concentration of silt-sized material (<0.063 mm) in the water column at different depths and distances from the dredger measured by water sampling and optical transmissometers. (After Hitchcock & Drucker 1996).

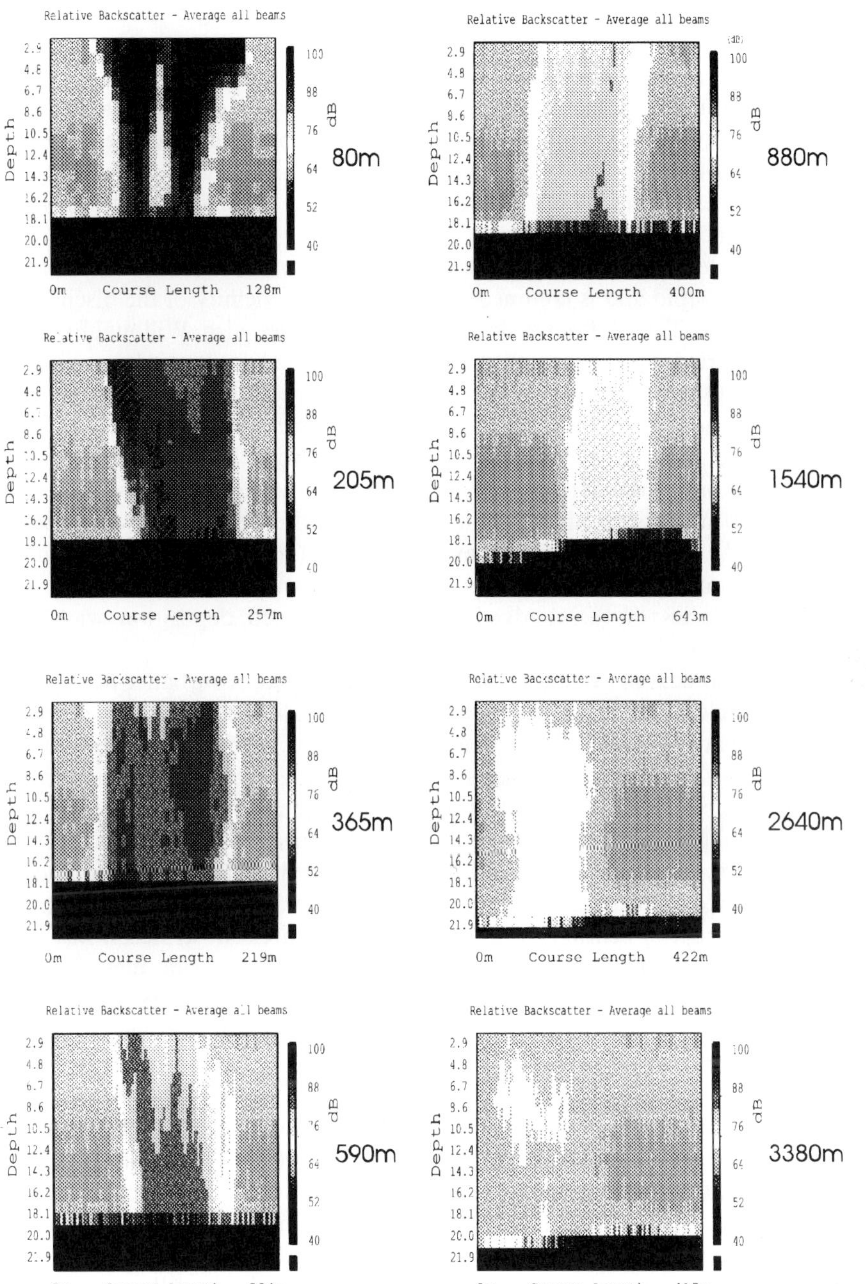

Figure 10 Acoustic Doppler Current Profiler (ADCP) acoustic backscatter across the plume of a dredger at varying distances downstream (m). Water speed $100\,\text{cm}\,\text{s}^{-1}$. Note that although sedimentation is achieved within 880 m, the relative backscatter suggests an additional "plume", perhaps representing entrained air bubbles, biochemical precipitates or organic matter released from the sediment, which extends up to 3380 m astern of the dredger. (After Hitchcock & Drucker 1996).

varying distances from 80 m up to 3335 m away from a dredger is shown in Fig. 10. This shows clearly the decay of the plume to background levels at approximately 400 m, but that a residual impact is detectable up to 3.5 km by ADCP methods. Whether this residual impact is caused by air bubbles or organic matter becoming entrained into the water column during the dredging operation is at present unknown. However, it is noteworthy that there is now a good deal of evidence that suggests that disturbance of marine sediments may release sufficient organic matter into the water column to enhance benthic production.

These results for the dispersion of sediment in the water column thus suggest that sedimentation is rapid and is confined to the immediate vicinity of the discharge. They confirm earlier studies of Poiner & Kennedy (1984; see also Willoughby & Foster 1983) who reported that sediment deposition generated from dredging activities in Moreton Bay, Queensland was confined to the immediate vicinity of the dredging works. Concentrations of suspended sand-sized material were reported to decay to background levels over a distance of only 200–500 m from the point of release into the water column from a commercial aggregate dredger. They estimated that the sediment deposition 500 m outside the boundary of the dredged area was 29.6 kg m^{-2} (23 mm m^{-2}). At 1 km deposition was 21.2 k m^{-2} (16 mm m^{-2}), at 1.5 km it was 15 kg m^{-2} (12 mm m^{-2}), at 2 km it was 10.7 kg m^{-2} (8 mm m^{-2}) and finally at 2.5 km from the boundaries of the dredged area the estimated deposition was less than 7.6 kg m^{-2} (6 mm m^{-2}).

There is a good deal of evidence from other surveys that disturbance of sediments by dredging may release sufficient organic materials to enhance the species diversity and population density of organisms outside the immediate zone of deposition of particulate matter. Disturbance of the sediments may thus enhance benthic production outside the immediate zone of deposition provided that contaminants from polluted sediments are not associated with the disposal of spoils. Stephenson et al. (1978) and Jones & Candy (1981) both document the enhanced diversity and abundance of benthic faunas near to dredged channels. Poiner & Kennedy (1984) showed that there was an enhancement of benthic biota close to dredged areas at Moreton Bay, Queensland and that the level of enhancement decreased with increasing distance from the dredged area up to a distance of approximately 2 km. They ascribe this to the release of organic nutrients from the sediment plume, a process which is well known from other studies (Ingle 1952, Biggs 1968, Sherk 1972, Oviatt et al. 1982; Walker & O'Donnell 1981).

The results reviewed above thus suggest that the impact of dredging activities mainly relates to the physical removal of substratum and associated organisms from the sea bed along the path of the dredge head and to the impact of subsequent deposition of sediment from outwash during the dredging process. The evidence from direct studies on the sedimentation of particulate matter suggests that the impact of sedimentation on biological resources on the sea bed is likely to be confined to distances within a few hundred metres of the dredger where the deposits are sands and gravels. It should be remembered, however, that discharge of dredge spoils from maintenance and capital dredging works in estuaries may result in much larger dispersion plumes that reflect the dominantly fine particles and strong current flows that occur in estuaries, and that the same processes, which result in the release of dissolved organic matter, can also result in the release of bound surface contaminants from the sediments into the water column.

The impact of dredging on biological resources

Sensitivity to Disturbance

The impact of disturbance by the dredge head during marine aggregate dredging has been reviewed on pp. 140–43. The effects of sediment deposition and spoils disposal outside the immediate boundaries of dredged areas in coastal waters has also been widely studied and includes extensive physiological–ecological work on a wide variety of animals including plankton, benthic invertebrates and fish species (for reviews, see Sherk 1971, Moore 1977). Early studies by Loosanoff (1962) showed that different species of commercially significant filter-feeding molluscs were differently affected by suspended sediment. Subsequent studies by Sherk (1971) and Sherk et al. (1974) included both plankton and fish species. They showed that, as in the case of bivalves, fish species have varying tolerances of suspended solids, filter-feeding species being more sensitive than deposit feeders and larval forms being more sensitive than adults (see also Matsumoto 1984).

Many of the macrofauna that live in areas of sediment disturbance are well adapted to burrow back to the surface following burial (see Schafer 1972). Studies by Maurer et al. (1979) showed that some benthic animals could migrate vertically through more than 30 cm of deposited sediment, and this ability may be widespread even in relatively deep waters. Kukert (1991) showed, for example, that approximately 50% of the macrofauna on the bathyal sea floor of the Santa Catalina Basin were able to burrow back to the surface through 4–10 cm of rapidly deposited sediment. A good deal of the apparent recolonization of deposits following dredging or spoils disposal may therefore reside in the capacity of adults to migrate up through relatively thin layers of deposited sediments (see also Ellis & Heim 1985), or to migrate in during periods of storm-induced disturbance (see Hall 1994).

There is good evidence that the activities of filter-feeding bivalves, in particular, can play an important part in controlling the natural phytoplankton and seston loads in the water column (Cloern 1982), to an extent that food may become a limiting resource in the benthic boundary layer at the sediment–water interface (Wildish & Kristmanson 1984, Fréchette et al. 1989, 1993; see also Dame 1993, Snelgrove & Butman 1994) as well as on coral reef flats and in cryptic reef habitats (Glynn 1973, Buss & Jackson 1981). Because the suspension-feeding component is evidently highly effective in removing particulate matter from sea water, the release of large quantities of suspended matter can lead to a loss of suspension-feeding components through clogging of the gills. This has led to a corresponding increase in the community of deposit feeders in some areas such as St Austell Bay off the southwest coast of England (Howell & Shelton 1970).

In general, however, most recent studies of filter feeders that live in coastal waters show that bivalves, in particular, are highly adaptable in their response to increased turbidity such as can be induced by periodic storms, dredging or spoils disposal and can maintain their feeding activity over a wide range of phytoplankton concentrations and inorganic particulate loads (Shumway et al. 1985, 1990, Newell et al. 1989, Newell & Shumway 1993 Iglesias et al. 1996, Navarro et al. 1996, Urrutia et al. 1996).

Although these studies on the physiology of individual species can give some insight into the differing susceptibilities of the macrofauna to increased turbidity, or to burial from dredger outwash, in general it is difficult to make predictions of the impact of dredging on whole communities from the results of studies on individual species. Partly for this reason, and because the interactions between the components of natural populations are complicated in space and time, most recent studies on the impact of spoils disposal and dredging works have been carried out on whole communities, rather than individual species. Such studies have concentrated on three main features of benthic communities, namely the number of individuals (population density), number of species (the diversity) and the biomass (to give an index of the growth following recolonization).

Sampling is conventionally carried out by means of a grab that allows collection of a sediment sample from a known area of seabed deposits, which are then eluted through a 1-mm mesh sieve to extract the macrofauna. Sediment samples from fine deposits such as occur in coastal embayments, lagoons and estuaries are relatively easy to obtain by means of equipment such as the van Veen and Smith-McIntyre grabs, the Ponar grab (Ellis & Jones 1980), or the Day grab whose jaws are held closed by the tension of the wire from which the grab is suspended rather than by a spring-loaded mechanism (see Holme & McIntyre 1984). Sampling of coarser gravel deposits is complicated, however, by the fact that the larger stones become trapped between the jaws of conventional grabs, leading to extensive losses through "washout" from the grab. Partly because of this problem, most work on coarser deposits has been carried out with semi-quantitative dredges such as the Anchor dredge (Forster 1953, Holme 1966, Kenny et al. 1991) or the Ralier du Baty dredge used by Davoult et al. (1988).

More recently, however, Sips & Waardenburg (1989) and Kenny & Rees (1994) have used a Hamon grab for quantitative studies on the fauna of gravels and sands. This grab takes a scoop out of the seabed deposits, rather than relying on the closure of opposing jaws (see Holme & McIntyre 1984). This greatly reduces the problem of fauna losses through "washout" during the sampling process and the Hamon grab is now widely used in the quantitative evaluation of the benthos in coarse sands and gravels.

Such studies emphasize that the macrofauna may vary considerably even over relatively short distances, and that a proper understanding of the distribution of benthic communities is necessary if damage to potentially important communities is to be avoided during dredging operations. Figure 11 shows, for example, the distribution of two important members of the benthic community in mixed gravel, sand and muddy deposits off the coast of East Anglia in August 1996 (Newell & Seiderer 1997b). Inspection of this figure shows that the main population of the reef-building tubeworm, *Sabellaria spinulosa* (called Ross by the fishermen) occurs in the northwestern part of the survey area, and corresponds with a localized patch of coarse stones and cobbles that give sufficient stability to support a rich reef community. This species may be predated upon by the pink shrimp (*Pandalus*) and is potentially important as a feeding ground for a variety of demersal fish species (see Warren, 1973).

In contrast, the populations of the comb worm (*Lagis* = *Pectinaria koreni*) occur in mobile muddy sands in the southwest of the survey area. This species is an important prey item for sole (*Solea solea*), dab (*Limanda limanda*) and plaice (*Pleuronectes platessa*) (see Lockwood 1980, Basimi & Grove 1985, Carter et al. 1991, Horwood 1993, also Peer 1970) and therefore represents a food resource within the survey area that requires conservation.

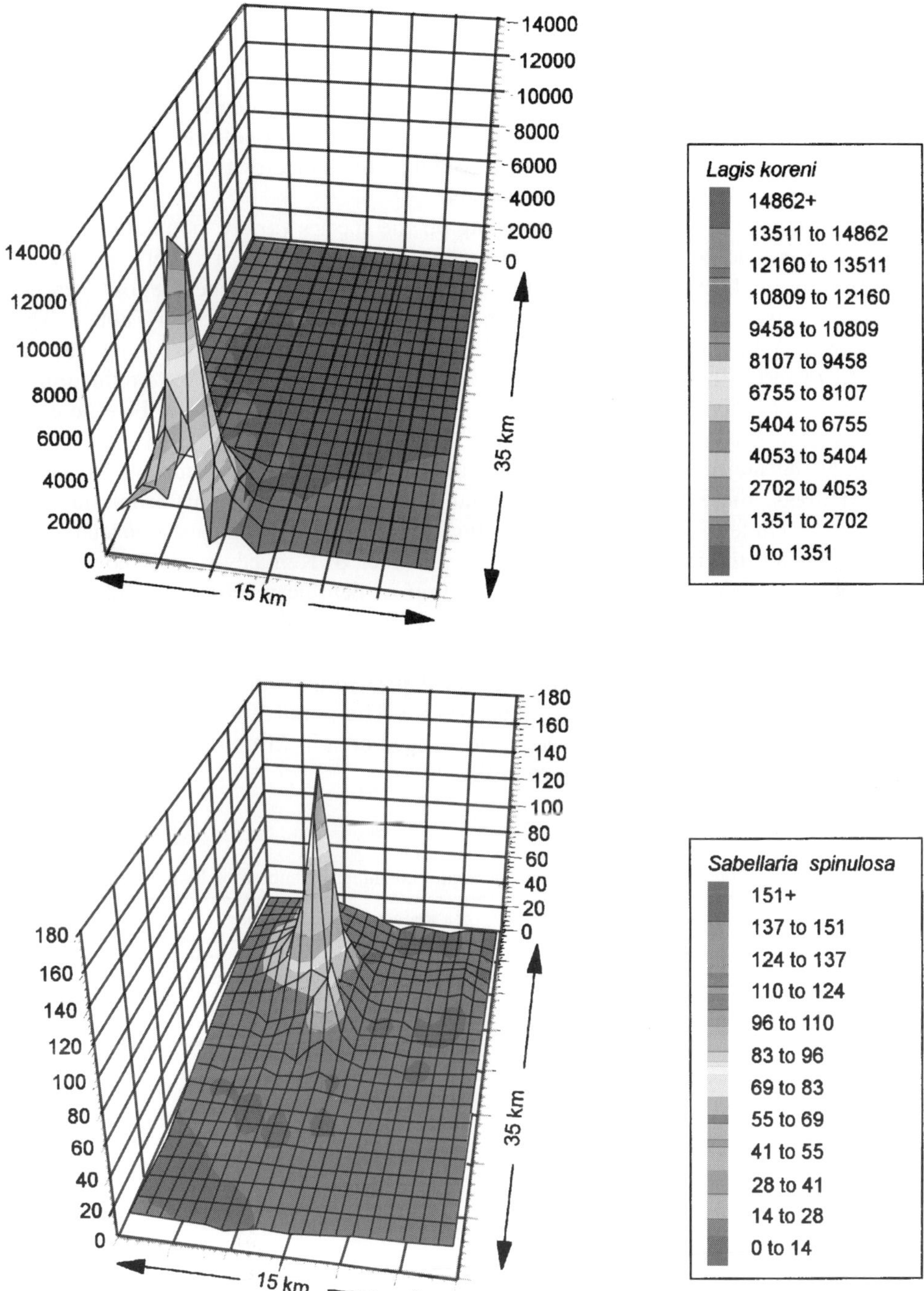

Figure 11 Schematic diagram of a survey area in the southern North Sea off Suffolk showing the distribution of the comb worm (*Lagis koreni*) in fine deposits of the survey area, and that of the colonial "Ross" worm (*Sabellaria spinulosa*) in areas where coarse boulders provide a stable environment for development of reef-forming species. Population density in numbers of individuals per 0.25 m^2 Hamon grab sample. (Newell & Seiderer 1997b).

Impact of dredging on diversity and abundance

The impact of dredging on benthic communities varies widely, depending, among other factors, on the intensity of dredging in a particular area, the degree of sediment disturbance and recolonization by passive transport of adult organisms (see Hall 1994) and the intrinsic rate of reproduction, recolonization and growth of the community that normally inhabits the particular deposits.

Some examples of the impact of dredging on the species variety, population density (number of individuals) and biomass of benthic organisms from a variety of habitats ranging from muds in coastal embayments and lagoons, to oyster shell deposits, and to sands and gravel deposits in the southern North Sea are summarized in Table 4. This shows that both maintenance dredging and marine aggregates dredging can be expected to result in a 30–70% reduction of species diversity, a 40–95% reduction in the number of individuals, and in a similar reduction in the biomass of benthic communities in the dredged area.

Despite the major impact of dredging on benthic community composition within dredged areas, there is little evidence that deposition of sediments from outwash through the chuteways during the dredging process has a significant impact on the benthos outside the immediate dredged area. Poiner & Kennedy (1984) showed that the population density and species composition of benthic invertebrates adjacent to dredging works on sandbanks in Moreton Bay, Queensland, Australia increased rapidly outside the boundaries of the dredged area, as might be anticipated from the relatively small amounts of sediment that are deposited beyond a few hundred metres of the dredger trail (see Figs 6 and 7 and p. 150). The population density and species diversity recorded from a transect across a dredged area in Moreton Bay in July 1982 by Poiner & Kennedy (1984) is shown in Fig. 12.

Table 4 Table showing the impact of dredging on benthic community composition from various habitats.

		% Reduction after dredging			
Locality	Habitat type	Species	Individuals	Biomass	Source
Chesapeake Bay	Coastal embayment muds-sands	70	71	65	Pfitzenmeyer 1970
Goose Creek, Long Island, NY	Shallow lagoon mud	26	79	63–79	Kaplan et al. 1975
Tampa Bay, Florida	Oyster shell	40	65	90	Conner & Simon 1979
Moreton Bay, Queensland, Australia	Sand	51	46	–	Poiner & Kennedy 1984
Dieppe, France	Sands-gravels	50–70	70–80	80–90	Desprez 1992
Klaver Bank, Dutch Sector, North Sea	Sands-gravels	30	72	80	van Moorsel 1994
Lowestoft, Norfolk, UK	Gravels	62	94	90	Kenny & Rees 1994
Hong Kong	Sands	60	60	–	Morton 1996
Lowestoft, Norfolk, UK	Sands-gravels	34	77	92	Newell & Seiderer 1997a

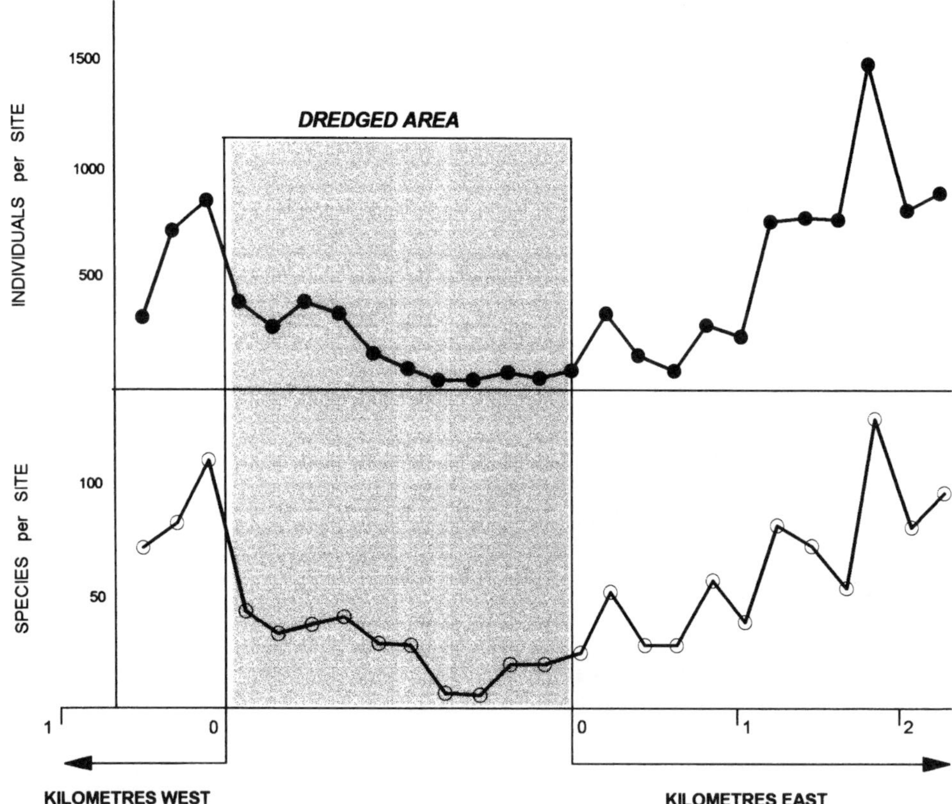

Figure 12 Diagram showing the number of individuals and species of benthos recorded in July 1982 on a transect crossing a dredged area on a sublittoral sandbank in Moreton Bay, Queensland, Australia. Note that species variety and population density increased rapidly outside the immediate boundaries of the dredged area. (Based on Poiner & Kennedy 1982).

Other than this study, there is surprisingly little detailed information on the precise boundaries of biological impact or "footprint" surrounding areas that have been dredged for sands and gravels. The circumstantial evidence from the boundaries of sediment deposition suggest, however, that biological impact is likely to be confined to the immediate vicinity of the dredged area.

One of the problems with assessing the impact of dredging works and the recovery of benthic communities over time is that biological communities are often subject to major changes in population density and community composition, even in areas that are apparently unaffected by dredging. Variations in the population density and species composition of the large bivalve population recorded between 1988 and 1991 in the sand and gravel deposits of the Klaver Bank in the Dutch sector of the North Sea by van Moorsel (1994) are shown in Fig. 13. This shows the major change in population of the bivalve, *Dosinia exoleta* between the summer of 1988 and the spring of 1989 and the loss of the large bivalve, *Arctica islandica* from the deposits even before aggregate extraction had taken place.

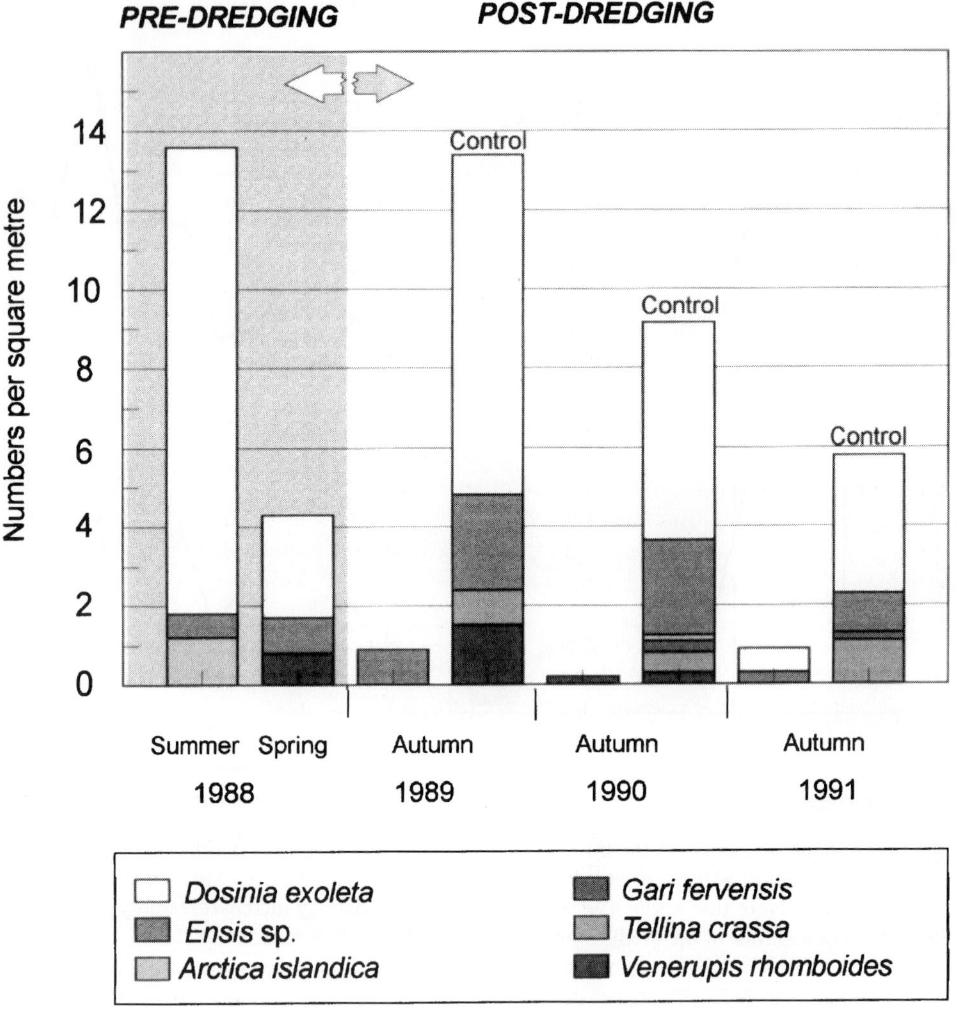

Figure 13 Diagram summarizing the changes in population density and species composition of large bivalves on the Klaver Bank in the southern North Sea between 1988 and 1991. Pre-dredging values in 1988 show major seasonal changes in density and species composition. After dredging in the summer of 1989, large differences in population density and species composition emerged between dredged and control areas, and these differences persisted for at least two years. (After van Moorsel 1994).

A short period of aggregate extraction was carried out in the study area on the Klaver Bank in the summer of 1989. Thereafter, clear differences emerged between the large populations of bivalves in control areas outside the dredged zone and those within the dredged area, despite the natural variations in species composition and population density that evidently occurred in the deposits of the survey area. These differences persisted until the end of the survey period in autumn of 1991, suggesting that this slow-growing component of the benthos remains impacted for at least 2 yr after cessation of dredging.

The process of recolonization and recovery

These complex changes in community structure following dredging, and which occur during the recovery process, are difficult to assess by mere inspection of the data for species composition, population density and biomass. Most recent studies on community structure in relation to environmental gradients, therefore, whether these are natural or induced by man, use relatively sophisticated analytical techniques that incorporate the type of species as well as their individual population densities and biomass to assess changes in community structure. The use of these techniques is beyond the scope of this review, but useful accounts for the biologist are given in Kruskal (1977), Hill (1979), Field et al. (1982), Heip et al. (1988), Magurran (1991), Warwick & Clarke (1991), Clarke & Ainsworth (1993), Clarke & Warwick (1994, and references cited therein).

Probably the most widely-used method is detrended correspondence analysis (DECORANA), an ordination technique that arranges stations along axes according to their similarity in species composition (Hill 1979). This is often used in association with two-way indicator species analysis (TWINSPAN) to identify species that characterize particular parts of an environmental gradient such as might be imposed, for example, by dredging or spoils overspill, or communities in relation to wider spatial gradients (see Eleftheriou & Basford 1989).

A second approach is the use of non-parametric multivariate analyses of community structure as outlined by Field & McFarlane (1968), Field et al. (1982) and Clarke & Warwick (1994). This procedure has recently become available in a convenient software package PRIMER (Plymouth Routines in Multivariate Ecological Research) and is now widely used in the analysis of benthic community structure in European coastal waters.

Despite problems in the interpretation of long-term studies on the abundance and composition of marine communities, studies that are carried out over even relatively short time periods can give important information on the recovery process following cessation of dredging. The most comprehensive analysis of the impact of dredging on community composition and on the process of recolonization and recovery in mixed gravel deposits is that of Kenny & Rees (1994, 1996). They carried out an intensive dredging programme by suction trailer dredger in an experimental area off Lowestoft, Norfolk in the southern North Sea and subsequently monitored the recovery over a period of 8 months in the first instance, although this was increased to 2 yr in an extended study of the recolonization process (Kenny & Rees 1996). Dredging occurred in April 1992, during which the suction dredger SAND HARRIER removed a total of 52 000 t of mixed aggregates from an area measuring 500 m by 270 m, an estimated 70% of the surface deposits down to an average depth of 0.3 m having been removed from the experimental area. The species variety, population density and biomass in the experimentally-dredged site was then compared with that in a reference site nearby over the 8-month period between March and December 1992.

The results from their study are summarized in Fig. 14. This shows that the number of species in the dredged site declined from 38 to only 13 species following dredging, whereas the number of species remained at about 35 during the 8-month period at the reference site. The number of species in the dredged area subsequently increased somewhat in the following 7 months, suggesting that some recolonization occurs even over this relatively short time.

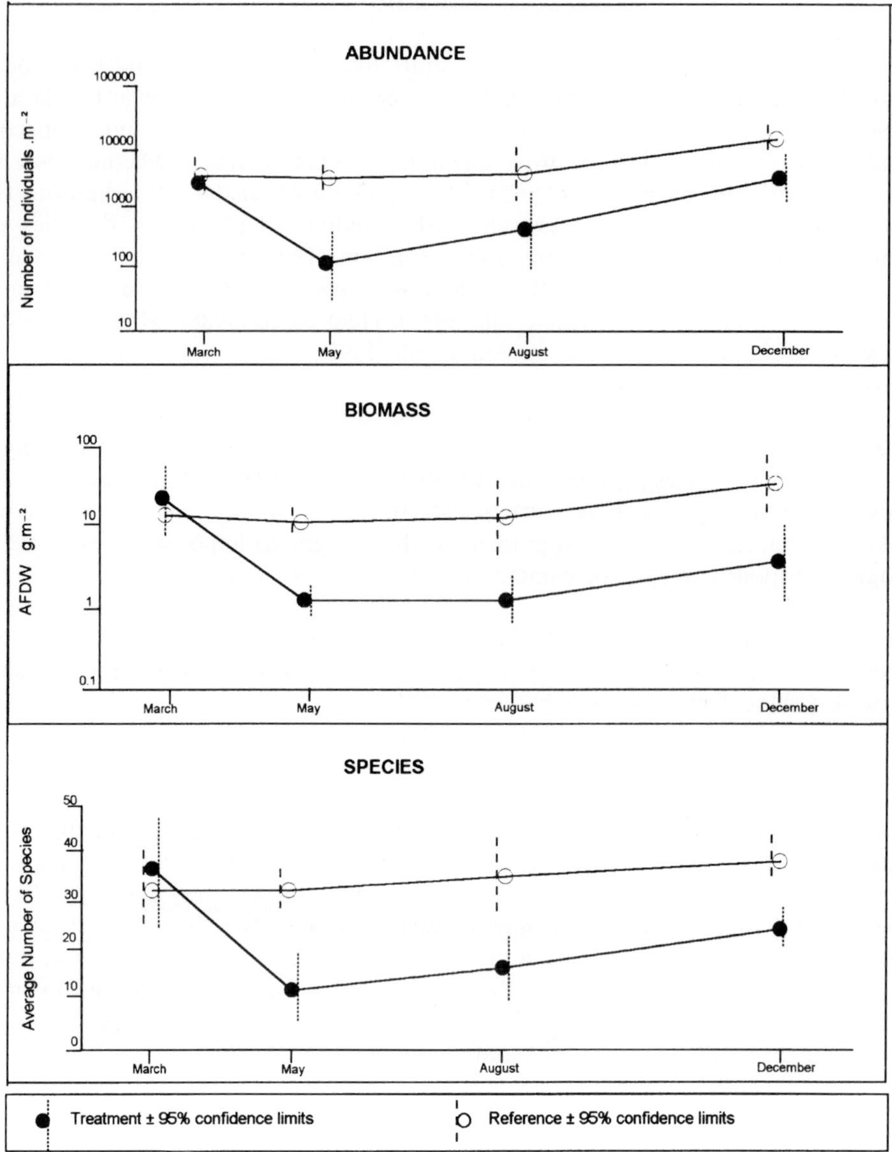

Figure 14 Graphs showing the mean values for the abundance of individuals (No. per m^2) from five Hamon grab samples each of 0.25 m^2 taken in a dredged site and at a reference site. Dredging occurred in April, and samples were taken in the pre-dredged deposits in March 1992 and through to December 1992. Values for the biomass are expressed as g AFDW m^{-2} from five Hamon grab samples. The average number of species in each of the five Hamon grab samples is also shown. The 95% confidence limits are indicated as bars. Note that there was a significant increase in species variety and abundance during the 7-month post-dredging period, but that the biomass increased only slowly. This indicates that recruitment was mainly of small individuals by larval settlement. Despite this recolonization, it is clear that population density, biomass and species variety had not recovered at the end of the 7-month post-dredging period. (After Kenny & Rees 1994).

The average population density for all taxa of 2769 individuals recorded by Kenny & Rees (1994) prior to dredging was reduced after dredging to only 129 ind. m^{-2}, compared with a relatively uniform invertebrate population density of 3300 ind. m^{-2} in the reference site. Again, the population density showed a significant increase in the 7 months after dredging had ceased.

Inspection of Figure 14 shows that the high biomass of 23 g AFDW m^{-2} was reduced to only 1 g AFDW m^{-2} after dredging. This reflects the removal of relatively large macrofaunal species, such as the mussel *Modiolus modiolus*, from the dredged sediments and was followed by a slower rate of increase in the post-dredging period than that recorded for population density. This implies that recolonization was initially by small individuals that then grew relatively slowly during the 7 months after dredging had ceased.

Figure 15 shows the output of a non-metric multidimensional scaling (MDS) ordination (see Kruskal 1977, Kruskal & Wish 1978, Field et al. 1982) of the data for the macrofauna sampled in gravel deposits before dredging of the experimental site off

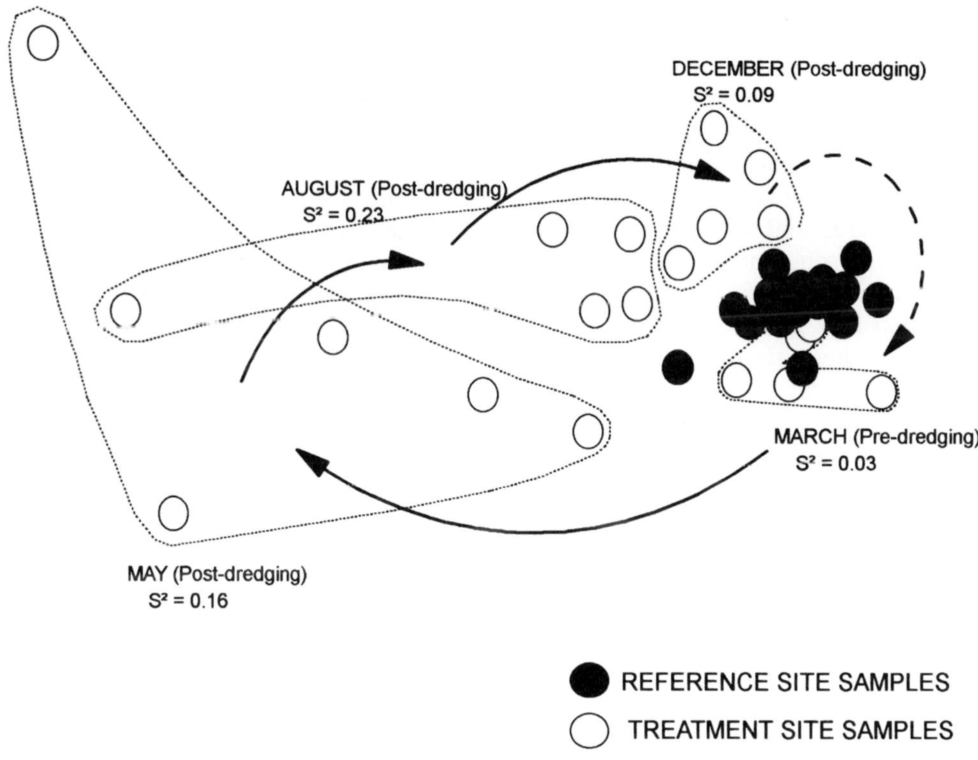

Figure 15 Two-dimensional multidimensional scaling (MDS) ordination for the benthic communities in a survey area off Norfolk in March 1992 prior to dredging, and in May, August and December 1992. Note that dredging of the experimental area resulted in an initial impact on community structure which differed from that in control areas and to that in the deposits prior to dredging. In the following months community structure became more similar to that in the undredged deposits, but was still distinct at the end of the 7-month post-dredging period. (After Kenny & Rees 1994).

Norfolk, and in the 7 months after dredging (after Kenny & Rees 1994). Their multivariate analysis of community structure prior to dredging and in the months following dredging shows a number of important features of the recolonization process that highlight the general principles of succession outlined in Fig. 2 (p. 134).

The first point that is clear from their results is that the community within the dredged site prior to dredging in March 1992 formed a small "cluster" on the MDS ordination. This indicates that the communities sampled within the experimental site were similar to one another, and were also evidently very similar to those in the reference site since they are close together on the MDS ordination.

The experimental area was again sampled in May 1992, 1 month after completion of dredging. At this stage it can be seen from Fig. 15 that dredging had resulted in two important changes in community structure. First, the communities in all the samples from the dredged site were well separated in the MDS plot from those in March and from those in the reference site. This implies a major change in community composition following dredging. Secondly, the communities at each of the sampling sites within the dredged area were different from one another. This is indicated by the fact that they have an increased derived variance (S^2) and no longer form a tight "cluster" on the MDS ordination shown in Fig. 15 (see also Warwick & Clarke 1993). This increased variance would be expected when some samples were taken from the dredged furrows themselves whereas others were from areas between furrows.

One of the interesting features of this study is that it shows that much of the initial process of colonization of the gravel deposits off the Norfolk coast was accomplished within the following 7-month period. Inspection of Fig. 15 shows that the community in the dredged area became more similar to those in the surrounding deposits of the reference area and to those in the pre-dredged site, and also had a closer internal similarity to one another (S^2 reduced to 0.09) in the months following cessation of dredging. This shows that many of the commoner species present in the dredged area in March 1992 prior to dredging had recolonized by December 1992. The clear difference from both the reference site and the community prior to dredging suggests, however, that many of the rarer components of the community had not yet colonized the dredged area in the following 7 months.

The study was then extended to include data for a 2-yr period following dredging. These results are reported by Kenny & Rees (1996). They showed that although recruitment of new species, especially *r*-selected species such as the barnacle *Balanus crenatus* and the ascidian, *Dendrodoa grossularia* had occurred by December 1992, even at the end of a 2-yr period both the average species abundance and biomass for the dredged area were lower than those in the reference site.

It is also clear from their work that the community composition in the dredged area was not restored even 2 yr after dredging. Inspection of Figure 16 shows the tightly clustered samples from the reference site and from the pre-dredged experimental site in March 1992. The marked shift in community composition and the increased variation between samples taken in May 1992 shortly after dredging is shown, as well as data collected in May 1993, 1 yr post-dredging and in May 1994, 2 yr after dredging. It is apparent from Figure 16 that despite the significant recolonization that had evidently occurred within 7 months of dredging, the community in the dredged area remained distinct from that in the reference area and from that in the deposits prior to dredging, even after 2 yr. Whether this reflects residual differences in the nature of the deposits following dredging, or the long time period required for establishment of the rarer

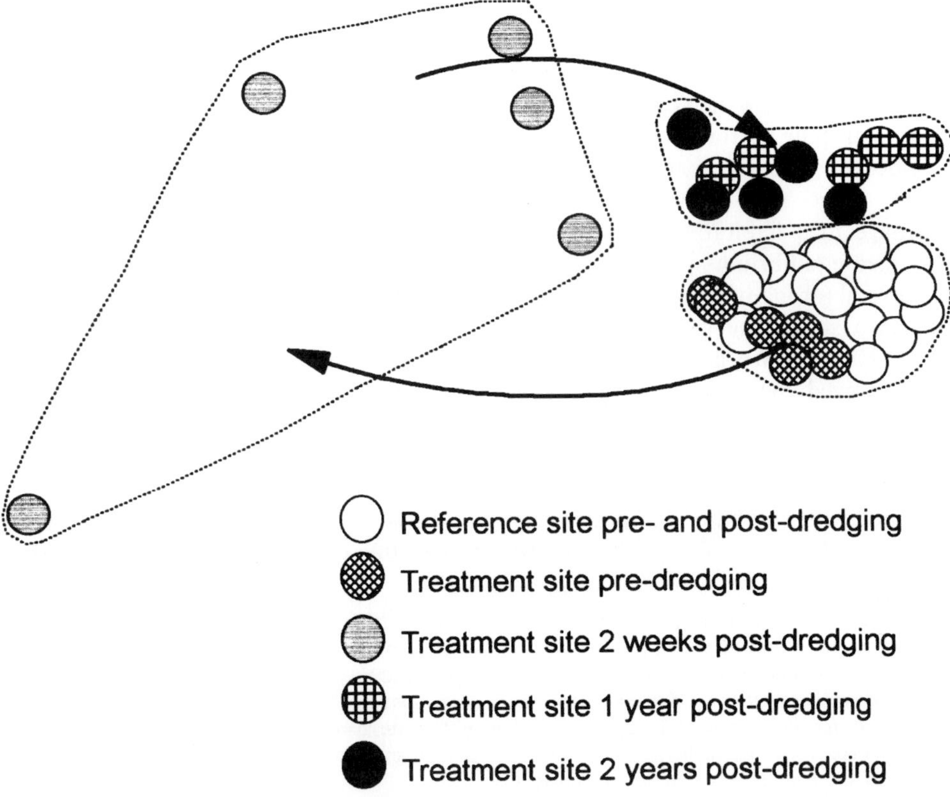

Figure 16 Two-dimensional multidimensional scaling (MDS) ordination for the benthic communities in a survey area off Norfolk in March 1992, and for following 2 yr post-dredging. Note that despite the increasing similarity of the community in the dredged area to those in the surrounding sediments over the 2 yr post-dredging period, recovery had not been fully accomplished even after 2 yr. (After Kenny & Rees 1996).

components of the original community is not yet known.

The results that have been reviewed above thus show that the process of recolonization involves two distinct phases; first, recolonization of species composition and population density by settlement of small individuals as larvae and juveniles; secondly, a period of growth during which the biomass approaches that in the undisturbed deposits. Inspection of Figure 16 shows, however, that in the gravel deposits of the southern North Sea this process had only entered its initial phase of partial restoration of community structure in the 7-month period that followed cessation of dredging, and that full recovery may take several years, much as would be anticipated for typical equilibrium communities on the sea bed (see Figure 4, p. 139).

The rate of recovery of biological resources

The rates of recovery of biological resources following capital and maintenance dredging, disposal of dredged spoils and marine aggregate dredging have been widely studied in other habitats and conform with the general principles of ecological succession shown in Figures 2 and 3. That is, communities that inhabit fine semi-liquid and disturbed muds comprise opportunistic *r*-selected species that have a high rate of recolonization and which can reach high population densities within weeks or months of a catastrophic mortality. Conversely, communities that inhabit less disturbed deposits of deeper waters or coarse substrata have complex associations and are characterized by large slow-growing species that are selected for maximum competitive advantage in a habitat where space is already crowded. These large, slow-growing *K*-selected equilibrium species recolonize only slowly following disturbance and may take several (or many) years for recovery of full species composition and biomass.

Table 5 shows the rates of recovery of the benthic fauna following dredging in various habitats. We have included semi-liquid muds from freshwater tidal areas and have arranged the data along a gradient of increasing environmental stability and predictability through estuarine and coastal muds to sands and gravels and coral reef assemblages. Inspection of the data summarized in Table 5 shows that recovery of the benthic fauna in highly disturbed semi-liquid muds can occur within weeks. This is associated with an ability for species such as *Limnodrilus* spp., *Ilyodrilus*, *Coelotanypus* sp. and *Procladius* to migrate through the surrounding deposits and to recolonize dis-

Table 5 Table showing the rates of recovery of the benthic fauna following dredging in various habitats. Note that highly disturbed sediments in tidal fresh waters and estuaries that are dominated mainly by opportunistic (*r*-strategist) species have a rapid rate of recovery. Recovery times increase in stable habitats of gravels and coral reefs that are dominated by long-lived components with complex biological interactions controlling community structure. Longevity and slow growth are also associated with slow recolonization rates in sub-arctic seas. Examples have been arranged along a gradient from disturbed muds of freshwater-tidal estuarine conditions to stable reef assemblages.

Locality	Habitat type	Recovery time	Source
James River, Virginia	Freshwater semi-liquid muds	±3 wk	Diaz 1994
Coos Bay, Oregon	Disturbed muds	4 wk	McCauley et al. 1977
Gulf of Cagliari, Sardinia	Channel muds	6 months	Pagliai et al. 1985
Mobile Bay, Alabama	Channel muds	6 months	Clarke et al. 1990
Chesapeake Bay	Muds-sands	18 months	Pfitzenmeyer, 1970
Goose Creek, Long Island, NY	Lagoon muds	>11 months	Kaplan et al. 1975
Klaver Bank, Dutch Sector, North Sea	Sands-gravels	1–2 yr (ex-bivalves)	van Moorsel 1994
Dieppe, France	Sands-gravels	>2 yr	Desprez 1992
Lowestoft, Norfolk, UK	Gravels	>2 yr	Kenny & Rees 1994, 1996
Dutch Coastal Waters	Sands	3 yr	de Groot 1979, 1986
Tampa Bay, Florida	Oyster shell (complete defaunation)	>4 yr	US Army Corps of Engineers 1974
Tampa Bay, Florida	Oyster shell (incomplete defaunation)	6–12 months	Conner & Simon 1979
Boca Ciega Bay, Florida	Shells-sands	10 yr	Taylor & Saloman 1968
Beaufort Sea	Sands-gravels	12 yr	Wright 1977
Florida	Coral reefs	>7 yr	Courtenay et al. 1972
Hawaii	Coral reefs	>5 yr	Maragos 1979

turbed muds as adults. A similar recolonization of disturbed deposits in dredged channels may also account for the relatively fast recolonization of some muds and sands in near-shore waters, especially those where tidal currents may transport juveniles into the dredged area (see Hall 1994).

Inspection of the recolonization rates reported in the literature and summarized in Table 5 suggest that a period of 2–4 yr is a realistic estimate of the time required for recovery in gravels and sands, but that this time may be increased to more than 5 yr in coarser deposits, including coral reef areas. Interestingly, the data for areas in Tampa Bay, Florida that had been dredged for oyster shell, suggest that a period of as much as 10 yr may be required for recovery following complete defaunation whereas a recovery time of only 6–12 months is required for recovery following partial dredging and incomplete defaunation (see Benefield 1976, Conner & Simon 1979). This suggests that areas of undisturbed deposits between dredged furrows may provide an important source of colonizing species that enable a faster recovery than might occur solely by larval settlement and growth (see also van Moorsel 1993, 1994).

Other more complex environmental factors also evidently affect the rate of recovery of dredged areas. Studies in the Dutch Wadden Sea by van der Veer et al. (1985) show that the recovery of species composition and biomass of benthic organisms was related to the speed of infilling of dredged pits. These data are summarized in Table 6, which shows that even 16 yr after cessation of dredging no recovery of the benthos had occurred on a tidal flat at Terschelling Sand. On a tidal watershed at Oosterbierum a partial recovery of 85% of the species and 39% of the biomass had occurred after 4 years. This is typical of recolonization by small individuals that were in the process of growth towards the original biomass levels of the undisturbed deposits, a process which would clearly take several more years.

In the tidal channels, both the rate of infill and recolonization were related to the speed of currents. A partial recovery of 57% of the species and 67% of the biomass was recorded after 3 yr in a tidal channel at Paesensrede (see Table 6), with greater recovery and shorter time periods being recorded in areas of faster current. Even then, it will be noted that the species composition had not recovered and that the biomass evidently became dominated by fewer species of relatively large size compared with those in the surrounding deposits.

The likely recolonization rates for the benthic community of estuarine muds, sands, gravels and reef areas have been superimposed onto a generalized colonization succession in Figure 17, which allows some predictions to be made on the rates of recovery of deposits following dredging. The fine muds that characterize coastal

Table 6 Table showing the percentage recovery recorded in a variety of habitats in the estuarine Dutch Wadden Sea following dredging up to 15 yr previously. Based on van der Veer et al. 1985.

Area	Habitat	Time interval since dredging	% Recovery Species	% Recovery Biomass
Terschelling sand	Tidal flat	16 yr	0	0
Oosterbierum	Tidal watershed	4 yr	85	39
Paesensrede	Tidal channel	3 yr	57	67
Holwerderbalg	Tidal channel	2 yr	64	100
Kikkertgat	Tidal channel	1 yr	88	116

embayments, estuaries and lagoons are likely to be colonized by large populations of a relatively restricted variety of opportunistic *r*-selected species, which are capable of rapid colonization within months of space being made available for colonization and growth. Because such deposits are subject to regular disturbance under natural conditions prior to dredging, the ecological succession recovers to the colonization phase shown in Figure 17, but does not proceed to the development of *K*-selected slow-growing equilibrium species within the community. Recovery of the "normal" community in disturbed deposits such as muds, therefore, can be achieved within months of cessation of dredging, or disposal of spoils.

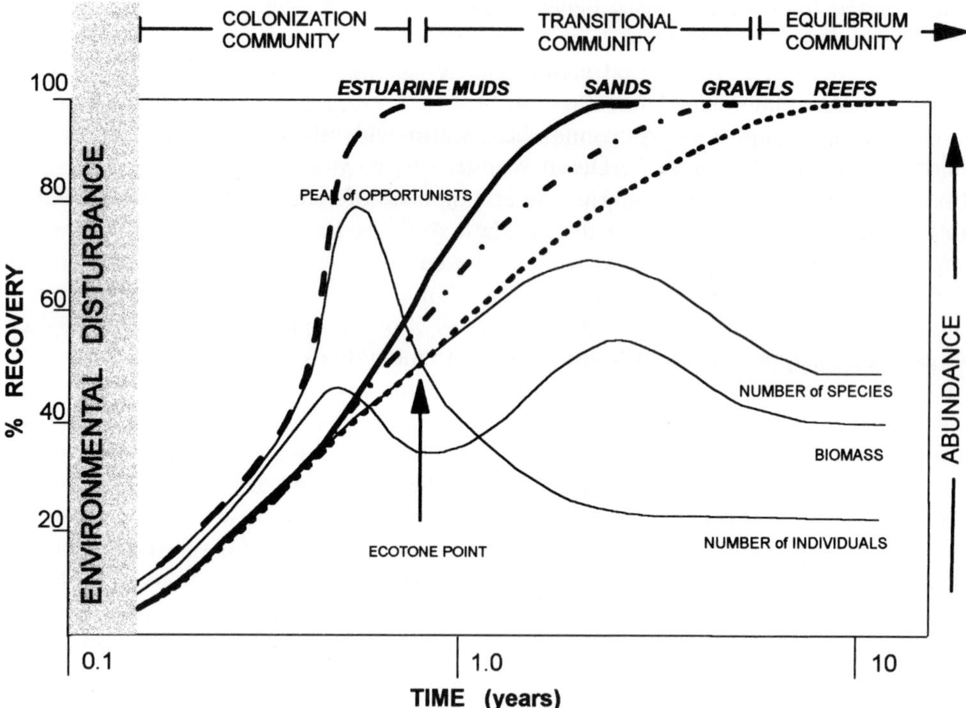

Figure 17 Schematic diagram showing the likely recolonization rates for the benthic community of estuarine muds, sands and reef areas. The curves for recovery have been superimposed onto a generalized colonization succession and allows some predictions to be made on the rates of recovery of deposits following dredging. Note that the fine muds that characterize coastal embayments, estuaries and lagoons are likely to be recolonized by a relatively restricted variety of opportunistic *r*-selected species within months of space being made available for recolonization and growth. Because such deposits are subject to regular disturbance, the succession recovers to the colonization phase, but does not proceed to the development of long-lived slow-growing *K*-selected species. The natural communities of gravels and sands, however, contain varying proportions of slow-growing *K*-selected equilibrium species depending on the degree of disturbance by waves and currents. These communities are held in a transitional state by natural environmental disturbance and are likely to recover within a period of 2–3 yr after cessation of dredging. Finally, the recovery curve for reef communities indicates that a period of 8–10 yr may be required for the long process of establishment and growth of the long-lived and slow-growing *K*-selected species characteristic of equilibrium communities.

The natural communities of gravel and sand deposits, however, contain varying proportions of slow-growing, *K*-selected equilibrium species, depending, among other factors, on the degree of disturbance by waves and the speed of tidal currents. In this case, the tail of the sigmoid recovery curve becomes more pronounced because the rarer components of the equilibrium community may take several years to recolonize the deposits, even after the main components of the community have become established. Where the deposits are sandy, periodic mortality of the long-lived components may result in major seasonal changes in community composition such as occurs in the North Sea on the Klaver Bank (van Moorsel 1994), and as has been reported for the sediments of Liverpool Bay by Eagle (1975). Under these conditions, the community will be held in a transitional state by natural environmental disturbance, and is likely to recover within 2–3 yr after cessation of dredging.

There is good evidence that disturbance of the deposits by man may result in a shift from the equilibrium community characteristic of undisturbed deposits towards the transitional community, which characterizes deposits in areas of natural environmental disturbance. Studies by de Groot (1984) suggest, for example, that the increasingly heavy bottom gear used by trawlers has been associated with a shift in community composition of the benthos of the North Sea, and this also applies to the benthos of the Wadden Sea.

As might be anticipated from the successional sequence shown in Figure 17, long-lived components such as molluscs and larger crustaceans in near-shore waters, such as the Wadden Sea, have decreased in numbers and diversity over the years and have been replaced by larger populations of rapidly growing polychaete species (Reise 1982, Riesen & Reise 1982, Reise & Schubert 1987).

Finally, the community recovery curve for reef communities indicates that a period of 8–10 yr may be required for the process of establishment and growth of the long-lived and slow-growing *K*-selected equilibrium species and for the development of the biological interactions that are familiar to those who have observed the immense diversity and complexity of life on undisturbed reef structures. This long process of establishment of an equilibrium community reflects partly the time required for colonization by rarer components of the community, but is also influenced by the nature and stability of the substratum following cessation of dredging, and the time required for complex stabilization processes involving both physical compaction and biological interactions. The relationship between biological community structure, sediment composition and seabed stability is considered in more detail below.

Community composition and seabed stability

The influence of sediment composition in controlling the nature of communities of animals that live on the sea bed has been widely recognized since the pioneer studies of Petersen (1913), Thorson (1957) and Sanders (1958). Most recent evidence suggests, however, that the precise relationship between biological community composition and specific properties of the sediments is poorly understood. In some estuaries and shallow water coastal embayments, fine grained and silty deposits clearly support an entirely distinct community compared with those from mobile sands or on stable substrata such as rocks and boulders.

On the other hand it is a matter of common observation that although very fine mobile muds may be dominated by opportunistic species such as the amphipod *Ampelisca brevicornis* or the polychaete *Lagis koreni*, the same silts can become consolidated into clays and then support long-lived and sedentary equilibrium species such as the boring piddock bivalves *Pholas dactylus* and *Barnea parva* as well as an epifauna of hydroids, ascidians and other species more characteristic of reefs. Clearly, the stability of the sediment, rather than particle size itself, is of importance in controlling community structure. In other instances it is clear that the deposits on the sea-bed undergo a complex process of consolidation or "armouring" that allows the establishment of communities that are more typical of rocks and reefs reflecting the complex relationships between the physical deposits and biological activities of the animals themselves.

The relationship between community composition and sediment type in deeper waters of the continental shelf is less well documented than that for estuaries and lagoons. Some early studies suggest that macrobenthic communities can be distinguished on a basis of sediment granulometry (Glémarec 1973, Buchanan et al. 1978, Flint 1981) but other studies have shown little correlation (Buchanan 1963, Day et al. 1971). Efforts to identify what physical properties are of greatest importance in controlling the structure of marine communities are often frustrated by the fact that most of the sediment variables obtained from conventional sorting methods are interdependent since they are expressed as a percentage of the total sample (see Weston 1988). A high percentage of silt, for example, is inversely related to the percentage of the other sediment components. Again, many of the physical properties of sediments are linked with other features such as depth of disturbance by wave action, strength and duration of currents, and may themselves be linked with complex biological interactions including the surface area available for microbial food components, and the presence of species that can exclude potential competitors.

Partly for this reason, most recent studies have concluded that the complexity of soft-bottom communities defies any simple paradigm relating to a single factor, and that there should be a shift towards understanding relationships between the distribution of organisms in terms of a dynamic relationship between the sediments and their hydrodynamic environment. According to this view, complex shear forces at the sediment–water interface are considered to play a dominant role in controlling food availability, settlement of larvae, microbial food availability, pore water flow and other environmental features that affect the benthic organisms that inhabit marine deposits. It is therefore considered unlikely that any one factor alone, or even a combination of single granulometric properties, can account for the distribution of animals in most sedimentary habitats (for review, see Snelgrove & Butman 1994).

Despite this emerging view that sediment granulometry itself is unlikely to control the composition and distribution of biological communities on the sea bed, concern has been expressed that dredging for marine aggregates can result in significant changes in sediment composition. Studies off Dieppe, France have shown, for example, a large increase in the proportion of fine sand in deposits that have been intensively worked for marine aggregates (see Desprez 1992, ICES 1992, 1993). Again, the infill of pits and grooves from dredging for marine aggregates is commonly dominated by the fine deposits which are capable of mobilization by shear stress induced by waves and tidal currents (Dickson & Lee 1972, Shelton & Rolfe 1972, Millner et al. 1977).

If sediment composition were of importance in controlling biological community composition, such changes following dredging could potentially prevent subsequent recolonization by communities that were similar to those that occurred in the deposits prior to dredging (see Windom 1976) and could by implication affect the nature and abundance of food organisms for commercial fish stocks.

We have analyzed the relationship between biological community composition and the sediment granulometry in undredged coastal deposits in the English Channel and southern North Sea and find that both biological communities and the sediments fall into relatively distinct groups or communities when analyzed by multivariate techniques (Newell & Seiderer 1997d). However, there is little evidence of any correspondence between the distribution of different sediment types and biological communities in the survey areas. Analysis of the Spearman rank correlation between the similarity of biological communities and any one, or a combination of, particle size indices show that granulometric properties of the sediments are likely to account for a maximum of 45% of the variability of the biological component, leaving approximately 55% determined by other environmental factors.

The conclusion to be drawn from these results is that they support recent views that biological community composition is not controlled by any one, or a combination of simple granulometric properties of the sediments such as particle size distribution. It is considered more likely that biological community composition is controlled by an array of environmental variables, many of them reflecting an interaction between particle mobility at the sediment–water interface and complex associations of chemical and biological factors operating over long periods.

Such interactions are not easily measured or analyzed, but the results clearly suggest that restoration of sediment composition after completion of dredging for marine aggregates is not, within broad limits, a prerequisite for the establishment of marine communities that are comparable with those that occurred in the deposits prior to dredging. What is possibly of more importance in controlling the time course of recovery of an equilibrium community characteristic of undisturbed deposits is the process of compaction and stabilization. This will reflect changes in sediment composition, but is also in equilibrium with seabed disturbance from tidal currents and wave action, both of which show spatial variations and interactions with water depth. The processes associated with compaction and stability of seabed deposits may therefore largely control the establishment of long-lived components of equilibrium communities and account for the dominance of opportunistic species in the initial stages of colonization of recently sedimented material in unconsolidated deposits after the cessation of dredging.

Conclusions

At the outset of this review, we assessed the importance of the benthic community to fisheries production and outlined our intention of providing an ecological framework within which the impact of dredging can be understood. We have shown that systems models for shelf waters such as the North Sea suggest that the flow of materials from primary production by the phytoplankton passes partly through planktonic grazers, but that 20–50% sinks to the sea-bed either from dead and decaying phytoplankton cells, or as faecal material derived from the feeding activities of the grazing zooplankton (Steele 1974, Joiris et al. 1982, Newell et al. 1988). Such material then passes into the benthic food web, whose production in turn forms an important food resource for demersal fish.

It has been estimated from empirical models developed for the North Sea that as much as 30% of total fisheries yield to man is derived from benthic resources (see Fig. 1, p. 129). Production by the benthos is therefore important, not only as a resource in itself, but as a key food resource for demersal fish stocks. It becomes an increasingly important component of the marine food web in near-shore waters where primary production by larger macrophytes and seagrasses living on the sea-bed largely replaces that from the phytoplankton in the water column (for review, see Mann 1982).

From this it is clear that reclamation of large areas of coastal wetlands, coastal embayments or estuaries can have a potentially important effect on the supply of materials and energy to marine food webs, and that even in plankton-based deeper water ecosystems such as the North Sea, fish yields based on benthic production are sufficiently large to warrant proper conservation of benthic resources. Our review has concentrated, therefore, on the nature of benthic communities, their susceptibility to disturbance by dredging and land reclamation works, and on the evidence that is available for the recovery times required for the re-establishment of community structure following dredging or spoils disposal.

Our review of the literature shows that the communities of near-shore habitats are characterized by large populations of a relatively restricted variety of species that are well-adapted to exploit space that has become newly available by episodic catastrophic mortality. Such species are generally small, often mobile, and are selected for maximum rate of population increase, with high fecundity, dense settlement, rapid growth and rather a short life cycle. Such species have been designated "*r*-strategists" (see MacArthur & Wilson 1967, Pianka 1970) and have been referred to in our review as opportunists. Their population characteristics allow a rapid recovery of the initial community structure in deposits that are naturally subjected to high levels of environmental disturbance. It is, not surprising, therefore, to find that there are frequent reports in the literature of community recovery times that range from a few weeks to several months for disturbed deposits such as semi-liquid muds in tidal fresh waters, estuaries, lagoons and dredged channels (see Table 5, p. 162 and Fig. 17, p. 164).

In deeper waters, or where the substratum is sufficiently stable to allow the long-term survival of benthic organisms, the habitat tends to be crowded. Under these conditions, organisms have an "equilibrium strategy" and are selected for maximum competitive ability in an environment in which space for colonization and subsequent growth is limiting. Such species have been designated "*K*-strategists" and devote a larger proportion of their resources to non-reproductive processes such as growth, predator avoidance, and investment in larger adults (MacArthur & Wilson 1967, Gadgil & Bossert 1970). Because the *K*-selected equilibrium species live longer, they tend to have wider limits of physiological tolerance, which allows them to survive those variations in environmental conditions that occur in their habitat. Many have active site selection phases that include chemical recognition of the presence of adults of the same species, a strategy which ensures that environmental conditions have been within the limits of tolerance for long enough to allow survival of other members of the same species (for review, see Newell 1979).

Such *K*-selected equilibrium species develop complex biological associations with other long-lived components of the community, and may alter the environment in such a way as to both allow the presence of many other species that would not otherwise occur, and also inhibit other potential competitors for space. Biological interactions between the components of equilibrium communities that are characteristic of stable substrata thus lead to the development of complex communities that may take many years, or decades, to re-establish following destruction. It is therefore not surprising to find that as one moves along a gradient of increasing sediment stability from muds through sands to gravels and reefs, there is a corresponding increase in the times reported for recovery of community structure (Table 5, p. 162).

Knowledge of the components that comprise the benthic community on the sea bed, whether these are *r*-selected opportunistic species or *K*-selected equilibrium species, thus gives important information not only on key resources that may require protection, but on the likely rate of recovery following dredging. Inspection of the schematic colonization succession shown in Figure 17 (p. 164) suggests that a recovery time of 6–8 months is characteristic for many estuarine muds whereas sands and gravels may take from 2–3 yr depending on the proportion of sand and the local disturbance by waves and currents. As the deposits become coarser, estimates of 5–10 yr are probably realistic for the development of the complex biological associations between the slow-growing components of equilibrium communities characteristic of reef structures.

Our review suggests that processes associated with compaction and stabilization of seabed deposits may largely control the time-course of recovery of these long-lived components of equilibrium communities and account for the dominance of opportunistic species in the initial stages of colonization of recently sedimented material in unconsolidated deposits following the cessation of dredging.

Acknowledgements

This work was partly supported through a co-operative research initiative funded by the Minerals Management Service, US Department of the Interior, Washington DC (Contract #14-35-001-30763 to D. R. Hitchcock) and supported by the following organisations: Messrs ARC Marine Ltd, United Marine Dredging Ltd, South Coast Shipping Ltd, Civil & Marine Ltd and H. R. Wallingford Ltd.

We are grateful for the helpful assistance with the literature given by Professor S. E. Shumway of the University of Long Island, NY, Professor R. I. E. Newell of the University of Maryland and Professor Derek V. Ellis of the University of Victoria, Canada. We also acknowledge with thanks the advice and comments of Mr Tom Matthewson of Oakwood Environmental, Wormley, Surrey, Mr A. Hermiston of ARC Marine, Southampton, Dr A. J. Kenny and Dr H. L. Rees of the Centre for Environment, Fisheries & Aquaculture Science (CEFAS), Burnham-on-Crouch, Essex. Messrs South Coast Shipping and Oakwood Environmental kindly gave permission for use of unpublished data obtained in baseline surveys carried out by us on their behalf off the coasts of Kent and East Anglia during 1996.

References

Aller, R. C. & Dodge, R. E. 1974. Animal-sediment relations in a tropical lagoon. *Journal of Marine Research* **32**(2), 209–32.

Aschan, P. 1988. The effect of Iceland scallop (*Chlamys islandica*) dredging at Jan Mayen and in the Spitsbergen area. *ICES CM* 1988, No. 16.

Bak, R. P. M. 1978. Lethal and sublethal effects of dredging on reef corals. *Marine Pollution Bulletion* **9**, 14–16.

Basimi, R. A. & Grove, D. J. 1985. Estimates of daily food intake by an inshore population of *Pleuronectes platessa* L. off eastern Anglesey, North Wales. *Journal of Fish Biology* **27**, 505–20.

Benefield, R. L. 1976. Shell dredging sedimentation in Galveston and San Antonio Bays 1964–69. *Technical Series, Texas Parks and Wildlife Department* **19**, 1–34.

Biggs, R. B. 1968. Environmental effects of overboard spoils disposal. *Journal of Sanitary Engineering Division* **94**, No. SA3, Proc. Paper 5979, June 1968. Contr. No. 328, 477–87.

Boesch, D. F. & Rosenberg, R. 1981. Response to stress in marine benthic communities. In *Stress effects on natural ecosystems*, G. W. Barrett & R. Rosenberg (eds). Chichester: John Wiley & Sons, 179–98.

Bonsdorff, E. 1983. Recovery potential of macrozoobenthos from dredging in shallow brackish waters. *Fluctuation and succession in marine ecosystems. Proceedings of the 17th European Marine Biology Symposium*, Brest: France, *Oceanologica Acta*, 27–32.

Brenchley, G. A. 1981. Disturbance and community structure: an experimental study of bioturbation in marine soft-bottom environments. *Journal of Marine Research* **39**, 767–90.

Brey, T. 1991. The relative significance of biological and physical disturbance: an example from intertidal and subtidal sandy bottom communities. *Estuarine Coastal and Shelf Science* **33**, 339–60.

Bright, D. A. & Ellis, D. V. 1989. Aspects of histology in *Macoma carlottensis* (Bivalvia: Tellinidae) and *in situ* histopathology related to mine-tailings discharge. *Journal of the Marine Biological Association of the United Kingdom* **69**, 447–64.

Brown, B. E., Le Tissier, M. D. A., Scoffin, T. P. & Tudhope, A. W. 1990. Evaluation of the environmental impact of dredging on intertidal coral reefs at Ko Phuket, Thailand, using ecological and physiological parameters. *Marine Ecology Progress Series* **65**, 273–81.

Buchanan, J. B. 1963. The bottom fauna communities and their sediment relationships off the coast of Northumberland. *Oikos* **14**, 154–75.

Buchanan, J. B., Sheader, M. & Kingston, P. F. 1978. Sources of variability in the benthic macrofauna off the south Northumberland coast. 1971–1976. *Journal of Marine Biological Association of the United Kingdom* **58**, 191–209.

Buss, L. B. & Jackson, J. B. C. 1981. Planktonic food availability and suspension–feeder abundance: evidence of *in situ* depletion. *Journal of Experimental Marine Biology and Ecology* **49**, 151–61.

Carney, R. S. 1981. Bioturbation and biodeposition. In *Principles of Benthic Marine Paleoecology*, A. J. Boucot (ed.). New York: Academic Press, 357–99.

Carter, C. G., Grove, D. J. & Carter, D. M. 1991. Trophic resource partitioning between two co-existing flatfish species off the North coast of Anglesey, North Wales. *Netherlands Journal of Sea Research* **27**, 325–35.

C-CORE. 1996. Proposed mining technologies and mitigation techniques: a detailed analysis with respect to the mining of specific offshore mineral commodities. *Contract Report for U.S Department of the Interior, Minerals Management Service.* C-CORE Publication 96-C15.

Charlier, R. H. & Charlier, C. C. 1992. Environmental, economic and social aspects of marine aggregates exploitation. *Environmental Conservation* **19**, 29–37.

Clarke, D. & Miller-Way, T. 1992. *An environmental assessment of the effects of open-water disposal of maintenance dredged material on benthic resources in Mobile Bay, Alabama.* National Technical Information Service, Springfield, Virginia 22161, USA.

Clarke, D. G., Homziak, J., Lazor, R., Palermo, M. R., Banks, G. E., Benson, H. A., Johnson, B. H., Smith-Dozier, T., Revelas, G. & Dardeau, M. R. 1990. *Engineering design and environmental assessment of dredged material overflow from hydraulically filled hopper barges in Mobile Bay, Alabama.* Miscellaneous Paper D-90-4. Vicksburg, MS. US Army Engineer Waterways Experiment Station.

Clarke, K. R. & Ainsworth, M. 1993. A method of linking multivariate community structure to environmental variables. *Marine Ecology Progress Series* **92**, 205–19.

Clarke, K. R. & Warwick, R. M. 1994. Change in Marine Communities: an approach to statistical analysis and interpretation. Natural Environment Research Council.

Cloern, J. E. 1982. Does the benthos control phytoplankton biomass in South San Francisco Bay? *Marine Ecology Progress Series* **9**, 191–202.

Connell, J. H. 1978. Diversity in tropical rain forests and coral reefs. *Science* **199**, 1302–10.

Conner, W. G. & Simon, J. L. 1979. The effects of oyster shell dredging on an estuarine benthic community. *Estuarine and Coastal Marine Science* **9**, 749–58.

Courtenay, W. R., Hartig, B. C. & Loisel, G. R. 1972. Ecological monitoring of two beach nourishment projects in Broward County, Florida. *Shore and Beach* **40**(2), 8–13.

Cruikshank, M. J. & Hess, H. D. 1975. Marine sand and gravel mining. *Oceanus* **19**, 32–44.

Dame, R. F. 1993. The role of bivalve filter feeder material fluxes in estuarine ecosystems. *NATO ASI Series* G 33, *Bivalve Filter Feeders*; In *Estuarine and coastal ecosystem processes*, R. F. Dame (ed.). Heidelberg Springer–Verlag, 246–68.

Davies, C. M. & Hitchcock, D. R. 1992. Improving the exploitation of marine aggregates by the study of the impact of marine mining equipment. Unpublished Report Ref. GR/G 20059. Marine Technology Directorate UK.

Davoult, D., Dewarumez, J. M., Prygiel, J. & Richard, A. 1988. Benthic communities map of the French part of the North Sea. *Station Marin de Wimereux* URA-CNRS. 1363.

Dawson-Shepherd, A., Warwick, R. M., Clarke, K. R. & Brown, B. E. 1992. An analysis of fish community responses to coral mining in the Maldives. *Environmental Biology of Fishes* **33**, 367–80.

Day, J. H., Field, J. G. & Montgomery, M. P. 1971. The use of numerical methods to determine the distribution of the benthic fauna across the continental shelf of North Carolina. *Journal of Animal Ecology* **40**, 93–126.

de Groot, S. J. 1979. An assessment of the potential environmental impact of large-scale sand-dredging for the building of artificial islands in the North Sea. *Ocean Management* **5**, 211–32.

de Groot, S. J. 1984. The impact of bottom trawling on benthic fauna of the North Sea. *Ocean Managment* **9**, 177–90.

de Groot, S. J. 1986. Marine sand and gravel extraction in the North Atlantic and its potential environmental impact, with emphasis on the North Sea. *Ocean Management* **10**, 21–36.

Desprez, M. 1992. Bilan de dix années de suivi de l'impact biosédimentaire de l'extraction de graves marins au large de Dieppe. Comparaison avec d'autres sites. Rapport Groupe d'tude des Milieux Estuariens et Littoraux GEMEL. St Valery/Somme.

(Cited in Report of the working group on the effects of extraction of marinc sediments on fisheries. ICES Report No. CM 1993/E:7 Marine Environmental Quality Committee, 51–67).

de Witt, T. H. & Levinton, J. S. 1985. Disturbance, emigration and refugia: how the mud snail *Ilyanassa obsoleta* (Say) affects the habitat distribution of an epifaunal amphipod, *Microdeutopus gryllotalpa* (Costa). *Journal of Experimental Marine Biology and Ecology* **92**, 97–111.

Diaz, R. J. 1994. Response of tidal freshwater macrobenthos to sediment disturbance. *Hydrobiologia* **278**, 201–12.

Dickson, R. & Lee, A. 1972. Study of the effects of marine gravel extraction on the topography of the sea bed. *ICES, C.M. 1972*/E: **25**.

Dodge, R. E. & Brass, G. W. 1984. Skeletal extension, density and calcification of a reef coral (*Montastrea annularis*): St Croix, US Virgin Islands. *Bulletin of Marine Science* **34**, 288–307.

Dodge, R. E. & Vaisnys, J. R. 1977. Coral populations and growth patterns: responses to sedimentation and turbidity associated with dredging. *Journal of Marine Research* **35**, 715–30.

Driscoll, E. G. 1975. Sediment-animal-water interaction, Buzzards Bay, Massachusetts. *Journal Marine Research* **33**, 275–302.

Eagle, R. A. 1975. Natural fluctuations in a soft bottom community. *Journal of the Marine Biological Association of the United Kingdom* **55**, 865–78.

Eden, R. A. 1975. North Sea environmental geology in relation to pipelines and structures. *Oceanology International* **75**, 302–9.

Eleftheriou, A. & Basford, D. J. 1989. The macrobenthic infauna of the offshore northern North Sea. *Journal of the Marine Biological Association of the United Kingdom* **69**, 123–43.

Ellis, D. V. 1987. A decade of environmental impact assessment at marine and coastal mines. *Marine Mining* **6**, 385–417.

Ellis, D. V. & Heim, C. 1985. Submersible surveys of benthos near a turbidity cloud. *Marine Pollution Bulletin* **16**(5), 197–203.

Ellis, D. V. & Hoover, P. M. 1990. Benthos recolonising mine tailings in British Columbia fiords. *Marine Mining* **9**, 441–57.

Ellis, D. V. & Jones, A. A. 1980. The Ponar grab as a marine pollution sampler. *Canadian Research* June/July 1980, 23–25.

Ellis, D. V., Pedersen, T. F., Poling, G. W., Pelletier, C. & Horne, I. 1995. Review of 23 years of STD: Island Copper Mine, Canada. *Marine Georesources and Technology*, **13**(1–2), 59–99.

Ellis, D. V. & Taylor, L. A. 1988. Biological engineering of marine tailings beds. In *Environmental Management of Solid Wastes*. W. Salomons & V. Forstner (eds). Heidelberg: Springer–Verlag.

Field, J. G., Clarke, K. R. & Warwick, R. M. 1982. A practical strategy for analyzing multispecies distribution patterns. *Marine Ecology Progress Series* **8**, 37–52.

Field, J. G. & MacFarlane, G. 1968. Numerical methods in marine ecology. I. A quantitative similarity analysis of rocky shore samples in False Bay, South Africa. *Zoologica Africana* **3**, 119–38.

Flach, E. C. 1992. Disturbance of benthic infauna by sediment reworking activities of the lugworm, *Arenicola marina*. *Netherlands Journal of Sea Research* **30**, 81–9.

Flint, R. W. 1981. Gulf of Mexico outer continental shelf benthos. macrofaunal- environmental relationships. *Biological Oceanography* **1**, 135–55.

Flint, R. W. & Kalke, R. D. 1986. Biological enhancement of estuarine benthic community structure. *Marine Ecology Progress Series* **31**, 23–33.

Forster, G. R. 1953. A new dredge for collecting burrowing animals. *Journal of the Marine Biological Association of the United Kingdom* **32**, 193–198.

Fréchette, M, Butman, C. A. & Geyer, W. R. 1989. The importance of the boundary layer flows in supplying phytoplankton to the benthic suspension-feeder, *Mytilus edulis* L. *Limnology and Oceanography* **34**, 19–36.

Fréchette, M., Lefaivre, D. & Butman, C. A. 1993. Bivalve feeding and the benthic boundary layer. *NATO ASI SERIES* G 33. *Bivalve Filter Feeders* In *Estuarine and Coastal Ecosystem Processes*, R. F. Dame (ed.). Heidelberg: Springer-Verlag, 325–64.

Gadgil, M & Bossert, W. H. 1970. Life historical consequences of natural selection. *American Naturalist* **104**, 1–24.

Gadgil, M & Solbrig, O. T. 1972. The concept of *r*- and *K*-selection: evidence from wild flowers and some theoretical considerations. *American Naturalist* **106**, 14–31.

Gajewski, L. S. & Uscinowicz, S. 1993. Hydrologic and sedimentologic aspects of mining marine aggregate from the Slupsk Bank (Baltic Sea). *Marine Georesources and Geotechnology* **11**, 229–44.

Garnett, R. H. T & Ellis, D. V. 1995. Tailings disposal at a marine placer mining operation by West Gold, Alaska. *Marine Georesources and Geotechnology*, **13**(1–2), 41–57.

Gayman, W. 1978. Offshore dredging study: environmental ecological report. *Ocean Management* **4**, 51–104.

Giesen, W. B. J. T. van Katwijk, M. M. & den Hartog, C. 1990. Eelgrass condition and turbidity in the Dutch Wadden Sea. *Aquatic Botany* **37**, 71–85.

Glasby, G. P. 1986. Marine minerals in the Pacific. *Oceanography and Marine Biology: an Annual Review* **24**, 11–64.

Glémarec, M. 1973. The benthic communities of the European North Atlantic continental shelf. *Oceanography and Marine Biology: an Annual Review* **11**, 263–89.

Glynn, P. W. 1973. Ecology of a Caribbean coral reef. The *Porites* flat biotope. Part II. Plankton community with evidence for depletion. *Marine Biology* **22**, 1–21.

Grassle, J. F. & Grassle, J. P. 1974. Opportunistic life histories and genetic systems in marine benthic polychaetes. *Journal of Marine Research* **32**, 253–84.

Grassle, J. F. & Sanders, H. L. 1973. Life histories and the role of disturbance. *Deep-Sea Research* **20**, 643–59.

Gray, J. S. 1974. Animal-sediment relationships. *Oceanography and Marine Biology: an Annual Review* **12**, 223–61.

Gray, J. S. & Pearson, T. H. 1982. Objective selection of sensitive species indicative of pollution-induced change in benthic communities. I. Comparative methodology. *Marine Ecology Progress Series* **9**, 111–19.

Gruet, Y. 1986. Spatio-temporal changes of sabellarian reefs built by the sedentary polychaete *Sabellaria alveolata* (Linne). *P.S.Z.N.I. Marine Ecology*, **7**, 303–19.

Guillou, M. & Hily, C. 1983. Dynamics and biological cycle of a *Melinna palmata* (Ampharetidae) population during the recolonisation of a dredged area in the vicinity of the harbour of Brest (France). *Marine Biology* **73**, 43–50.

Hall, S. J. 1994. Physical disturbance and marine communities: life in unconsolidated sediments. *Oceanography and Marine Biology: an Annual Review* **32**, 179–239.

Heip, C., Harman, P. M. J. & Soetaert, K. 1988. Data processing, evaluation and analysis. In *Introduction to the study of meiofauna*, R. P. Higgins & H. Thiel (eds). Washington, DC: Smithsonian Institution, 197–231.

Hess, H. D. 1971. *Marine sand and gravel mining industry in the United Kingdom.* NOAA Technical Report ERL 213-MMTCI. US Department of Commerce, Boulder, Colorado.

Hill, M. O. 1979. DECORANA- A FORTRAN *Program for Detrended Correspondence Analysis and Reciprocal Averaging.* Ithaca, Cornell University.

Hily, C. 1983. Macrozoobenthic recolonisation after dredging in a sandy mud area of the Bay of Brest enriched by organic matter. In *Fluctuation and Succession in Marine Ecosystems.* Proceedings of the 17th European Marine Biology Symposium, Brest. France. *Oceanologica Acta* 113–20.

Hitchcock, D. R. & Dearnaley, M. P. 1995. Investigation of benthic and surface plumes associated with marine aggregate production in the United Kingdom: overview of year one. *Proceedings of the XVth Information Transfer Meeting, Gulf Coast Region.* INTERMAR, New Orleans, USA.

Hitchcock, D. R. & Drucker, B. R. 1996. Investigation of benthic and surface plumes associated with marine aggregates mining in the United Kingdom. In *The global ocean – towards operational oceanography.* Proceedings of Conference on Oceanology International. Spearhead Publications, Surrey *Conference Proceedings* **2**, 221–84.

Hodgson, G. 1994. The environmental impact of marine dredging in Hong Kong. *Coastal Management of Tropical Asia* **2**, 1–8.

Holme, N. A. 1966. The bottom fauna of the English Channel. Part II. *Journal of the Marine Biological Association of the United Kingdom* **46**, 401–93.

Holme, N. A. & McIntyre, A. D. (eds). 1984. *Methods for the Study of Marine Benthos.* International Biological Programme Handbook. 2nd edn Oxford: Blackwell.

Holt, T. J., Jones, D. R., Hawkins, S. J. & Hartnoll, R. G. 1995. The sensitivity of marine communities to man-induced change – a scoping report. Countryside Council for Wales. Contract No. FC 73-01-96. Port Erin Marine Laboratory, Centre for Coastal Studies, University of Liverpool. Port Erin, Isle of Man UK.

Horwood, J. 1993. The Bristol channel sole (*Solea solea* (L): a fisheries case study. *Advances in Marine Biology* **29**, 215–367.

Howell, B. R. & Shelton, R. G. J. 1970. The effect of china clay on the bottom fauna of St Austell and Mevagissey Bays. *Journal of the Marine Biological Association of the United Kingdom* **50**, 593–607.

Hurme, A. K. & Pullen, E. J. 1988. Biological effects of marine sand mining and fill placement for beach replenishment: lessons for other uses. *Marine Mining*, **7**, 123–36.

Huston, M. 1994. *Biological diversity: the coexistence of species on changing landscapes.* Cambridge: Cambridge University Press.

ICES 1975. *Report of the working group on effects on fisheries of marine sand and gravel extraction.* Co-operative Research Report No. 46.

ICES 1977. *Second report of the ICES working group on effects on fisheries of marine sand and gravel extraction.* Co-operative Research Report No. 64.

ICES 1992a. *Report of the working group on the effects of extraction of marine sediments on fisheries.* Marine Environmental Quality Committee Report Ref C.M. 1992/E:7.

ICES 1992b. *Report of the working group on the effects of extraction of marine sediments on fisheries.* Co-operative Research Report No. 182.

ICES 1993. *Report of the working group on the effects of extraction of marine sediments on fisheries.* Marine Environmental Quality Committee Report, Ref. CM 1993/E:7.

Iglesias, J. I. P., Urrutia, M. B., Navarro, E., Alvarez-Jorna, P., Larretxea, X., Bougrier, S. & Heral, M. 1996. Variability of feeding processes in the cockle, *Cerastoderma edule* (L.) in response to changes in seston concentration and composition. *Journal of Experimental Marine Biology and Ecology* **197**, 121–43.

Ingle, R. M. 1952. Studies on the effect of dredging operations on fish and shellfish. *State of Florida Board of Conservation Technical Series* No. 5. 1–25.

Johanson, E. E. & Boehmer, W. R. 1975. Examination of predictive models for dredged material dispersion in open water. In *1st International Symposium on Dredging Technology* Canterbury: University of Kent. BHRA Fluid Engineering, Bedford, Session D3, 33–46.

Johnson, R. G. 1970. Variations in diversity within benthic marine communities. *American Naturalist* **104**, 285–300.

Johnson, R. G. 1977. Vertical variation in particulate matter in the upper twenty centimetres of marine sediments. *Journal of Marine Research* **35**, 273–82.

Johnston, S. A. 1981. Estuarine dredge and fill activities: a review of impacts. *Environmental Management* **5**, 427–40.

Joiris, C., Billen, G., Lancelot, C., Daro, M. H., Mommaerts, J. P., Bertels, A., Bosicart, M., Nijs, J. & Hecq, J. H. 1982. A budget of carbon cycling in the Belgian coastal zone: relative roles of zooplankton, bacterioplankton and benthos in the utilisation of primary production. *Netherlands Journal of Sea Research* **16**, 260–75.

Jokiel, P. L. 1989. Effects of marine mining dredge spoils on eggs and larvae of a commercially important species of fish, the Mahimahi (*Coryphaena hippurus*). *Marine Mining* **8**, 303–15.

Jones, A. A. & Ellis, D. V. 1976. Sub-Obliterative effects of mine-tailing on marine infaunal benthos. II. Effect on the productivity of *Ammotrypane aulogaster*. *Water, Air and Soil Pollution* **5**, 299–307.

Jones, G. & Candy, S. 1981. Effects of dredging on the macrobenthic infauna of Botany Bay. *Australian Journal of Marine and Freshwater Research* **32**, 379–99.

Kaplan, E. H., Welker, J. R., Kraus, M. G. & McCourt, S. 1975. Some factors affecting the colonisation of a dredged channel. *Marine Biology* **32**, 193–204.

Kenny, A. J. & Rees, H. L. 1994. The effects of marine gravel extraction on the macrobenthos: Early post-dredging recolonisation. *Marine Pollution Bulletin* **28**(7), 442–7.

Kenny, A. J. & Rees, H. L. 1996. The effects of marine gravel extraction on the macrobenthos: Results 2 years post-dredging. *Marine Pollution Bulletin* **32**(8/9), 615–22.

Kenny, A. J., Rees, H. L. & Lees, R. G. 1991. *An inter-regional comparison of gravel assemblages off the English east and south-east coasts: preliminary results.* International Council for the Exploration of the Sea (ICES) Report CM 1991/E:27.

Kruskal, J. B. 1977. Multidimensional scaling and other methods of discovering structure. In *Mathematical methods for digital computers. Vol. 3. Statistical methods for digital computers*, K. Enstein et al. (eds). New York: Wiley-Interscience, 296–339.

Kruskal, J. B. & Wish, M. 1978. Multidimensional scaling. Beverley Hills, CA: Sage Publications.

Kukert, H. 1991. *In situ* experiments on the response of deep sea macrofauna to burial disturbance. *Pacific Science* **45**, 95 only.

Lambshead, P. J. D., Platt, H. M. & Shaw, K. M. 1983. The detection of differences among assemblages of marine benthic species based on an assessment of dominance and diversity. *Journal of National History* **17**, 859–74.

Land, J., Kirby, R & Massey, J. B. 1994. Recent innovations in the combined use of acoustic doppler current profilers and profiling silt meters for suspended solids monitoring. In *Proceedings of 4th Nearshore and Estuarine Cohesive Sediment Transport Conference.* Hydraulic Research Wallingford, UK.

Lart, W. J. 1991. *Aggregate dredging; fisheries perspectives.* Sea Fish Industry Authority: Seafish Technology. Seafish Report No. 404.

Lee, G. F. & Swartz, R. C. 1980. Biological processes affecting the distribution of pollutants in marine sediments. Part II. Biodeposition and bioturbation. In *Contaminants and Sediments*, R. A. Baker (ed.). Ann Arbor: Michigan, 555–605.

Lees, R. G., Kenny, A. & Pearson, R. 1992. The condition of benthic fauna in suction trailer dredger outwash: initial findings. pp 22–28 In *Report of the Working Group on the Effects of Extraction of Marine Sediments on Fisheries. International Council for the Exploration of the Sea.* CM 1992/E:7.

Lockwood, S. J. 1980. The daily food intake of 0-group plaice (*Pleuronectes platessa* L.) under natural conditions. *Journal du Conseil, Conseil International pour l'Exploration de la Mer.* **39**, 745–69.

Loosanoff, V. L. 1962. Effects of turbidity on some larval and adult bivalves. *Proceedings of the Gulf and Caribbean Fisheries Institute*, Nov. 1961 80–94. Institute of Marine Science, University of Miami.

MacArthur, R. H. 1960. On the relative abundance of species. *American Naturalist* **94**, 25–36.

MacArthur, R. H. & Wilson, E. O. 1967. *Theory of island biogeography.* Princeton NJ: Princeton University Press.

Madany, I. M., Ali, S. M. & Akhter, M. S. 1987. The impact of dredging and reclamation in Bahrain. *Journal of Shoreline Management* **3**, 255–68.

Magurran, A. E. 1991. *Ecological diversity and its measurement*, London: Chapman & Hall.

Mann, K. H. 1982. *Ecology of coastal waters: a systems approach.* Studies in Ecology Vol. 8. Oxford: Blackwell Scientific Publications.

Maragos, J. E. 1979. *Environmental surveys five years after offshore marine sand mining operations at Keauhou Bay, Hawaii.* International Report of US Army Engineer Division, Pacific Ocean, Fort Shafter, HI.

Maragos, J. E. 1991. Impact of coastal construction on coral reefs in the US-affiliated Pacific islands. *Coastal Management* **21**, 235–69.

Marszalek, D. S. 1981. Impact of dredging on a subtropical reef community, Southeast Florida, USA. *Proceedings of the 4th International Coral Reef Symposium Manila* **1**, 147–53.

Matsumoto, W. M. 1984. *Potential impact of deep seabed mining on the larvae of tunas and billfishes.* US Dept Commerce, NOAA Technical Memorandum NMFS, NOAA-TM-NMFS- SWFC-44.

Maurer, D. L., Leathem, W., Kinner, P. & Tinsman, J. 1979. Seasonal fluctuations in coastal benthic invertebrate assemblages. *Estuarine and Coastal Marine Science* **8**, 181–93.

McCall, P. L. 1976. Community patterns and adaptive strategies of the infaunal benthos of Long Island Sound. *Journal of Marine Research* **35**, 221–66.

McCauley, J. E., Parr, R. A. & Hancock, D. R. 1977. Benthic infauna and maintenance dredging: a case study. *Water Research* **11**, 233–42.

Millner, R. S., Dickson, R. R. & Rolfe, M. S. 1977. *Physical and biological studies of a dredging ground off the east coast of England.* ICES C.M. 1977/E: 48.

Ministry of Agriculture, Fisheries and Food. 1993. Aquatic Monitoring Report No. 36. Lowestoft: Directorate of Fisheries Research.

Moloney, C. L., Bergh, M. O., Field, J. G. & Newell, R. C. 1986. The effect of sedimentation and microbial nitrogen regeneration in a plankton community: a simulation investigation. *Journal of Plankton Research* **8**, 427–45.

Moore, P. G. 1977. Inorganic particulate suspensions in the sea and their effects on marine animals. *Oceanography and Marine Biology: an Annual Review*, **15**, 225–363.

Morton, B. 1996. The subsidiary impacts of dredging (and trawling) on a subtidal benthic molluscan community in the southern waters of Hong Kong. *Marine Pollution Bulletin* **32**(10), 701–10.

Morton, J. W. 1977. Ecological effects of dredging and dredge spill disposal: a literature review. *US Fish and Wildlife Service, Technical Papers*, No. 94, 1–33.

Myers, A. C. 1977a. Sediment processing in a marine subtidal sandy bottom community. I. Physical aspects. *Journal of Marine Research* **35**, 609–32.

Myers, A. C. 1977b. Sediment processing in a marine subtidal sandy bottom community. II. Biological consequences. *Journal of Marine Research* **35**, 633–47.

Navarro, E., Iglesias, J. I. P., Camacho, P. A., & Labarta, U. 1996. The effect of diets of phytoplankton and suspended bottom material on feeding and absorption of raft mussels (*Mytilus galloprovincialis* Lmk). *Journal of Experimental Marine Biology and Ecology* **198**, 175–89.

Newell, C. R. & Shumway, S. E. 1993. Grazing of natural particulates by bivalve molluscs: a spatial and temporal perspective. *NATO ASI Series* G 33, In *Estuarine and Coastal ecosystem processes*, R. F. Dame (ed.). Heidelberg: Springer–Verlag, 85–148.

Newell, C. R., Shumway, S. E., Cucci, T. L. & Selvin, R. 1989. The effects of natural seston particle size and type on feeding rates, feeding selectivity and food resource availability for the mussel, *Mytilus edulis* Linnaeus, 1758 at bottom culture sites in Maine. *Journal of Shellfish Research* **8**, 187–96.

Newell, R. C. 1979. *Biology of intertidal animals.* Faversham, Kent: Marine Ecological Surveys.

Newell, R. C., Moloney, C. L., Field, J. G., Lucas, M. I. & Probyn, T. A. 1988. Nitrogen models at the community level: plant–animal–microbe interactions. In *Nitrogen cycling in coastal marine environments*, T. H. Blackburn & J. Sorensen (eds). Chichester: SCOPE Publications, John Wiley, 379–414.

Newell, R. C. & Seiderer, L. J. 1997a. Benthic ecology off Lowestoft: Dredging Application Area 454. Report prepared for Oakwood Environmental by Marine Ecological Surveys. Ref SCS/454/1.

Newell, R.C . & Seiderer, L. J. 1997b. Benthic ecology off Suffolk (Shipwash Gabbard): Dredging Application Area 452. Report prepared for Oakwood Environmental by Marine Ecological Surveys. Ref SCS/452/1.

Newell, R. C. & Seiderer, L. J. 1997c. Benthic ecology of West Varne Licensed Dredging Areas 432/1 & 432/2. Report prepared for Oakwood Environmental by Marine Ecological Surveys. Report Ref SCS/432/1.

Newell, R. C. & Seiderer, L. J. 1997d. Influence of sediment granulometry on benthic community composition. Report prepared for Oakwood Environmental by Marine Ecological Surveys.

Nichols, F. H. 1974. Sediment turnover by a deposit-feeding polychaete. *Limnology and Oceanography* **19**, 945–50.

Nunny, R. S. & Chillingworth, P. C. H. 1986. *Marine dredging for sand and gravel.* London: HMSO.

Odum, E. P. & Odum, H. T. 1959. *Fundamentals of Ecology*. Philadelphia: W. B. Saunders.

Oliver, J. S. & Slattery, P. N. 1981. Community complexity, succession and resilience in marine soft-bottom environments. In *6th Biennial International Estuarine Research Conference. Estuaries* **4**(3), 259.

Oliver, J. S., Slattery, P. N., Hulberg, L. W. & Nybakken, J. W. 1980. Relationships between wave disturbance and zonation of benthic invertebrate communities along a sub-tidal high energy beach in Monterey Bay, California. *Fisheries Bulletin of the United States* **78**, 437–54.

Onuf, C. P. 1994. Seagrasses, dredging and light in Laguna Madre, Texas, U.S.A. *Estuarine, Coastal and Shelf Science* **39**, 75–91.

Osman, R. W. 1977. The establishment and development of a marine epifaunal community. *Ecological Monographs*. **47**, 37–63.

Oviatt, C. A., Hunt, C. D., Vargo, G. A. & Kopchynski, K. W. 1982. Simulation of a storm event in marine microcosms. *Journal of Marine Research* **39**, 605–18.

Pagliai, A. M. B., Varriale, A. M. C., Crema, R., Galletti, M. C. & Zunarelli, R. V. 1985. Environmental impact of extensive dredging in a coastal marine area. *Marine Pollution Bulletin* **16**(12), 483–8.

Pearson, T. H. & Rosenberg, R. 1978. Macrobenthic succession in relation to organic enrichment and pollution of the environment. *Oceanography and Marine Biology: an Annual Review* **16**, 229–311.

Peer, D. L. 1970. Relation between biomass, productivity and loss to predators in a population of a marine benthic polychaete, *Pectinaria hyperborea*. *Journal of the Fisheries Research Board of Canada* **27**, 243–53.

Petersen, C. G. J. 1913. Valuation of the sea. II. Animal communities of the sea bottom and their importance for marine zoogeography. *Report of the Danish Biology Station to the Board of Agriculture* **21**, 1–44.

Pfitzenmeyer, H. T. 1970. *Gross physical and biological effects of overboard spoil disposal in Upper Chesapeake Bay*. N.R.I Special Report No. 3 26–35. Chesapeake Biological Laboratory, Solomons, Maryland. Contr. No. 397.

Pianka, E. R. 1970. On *r*- and *K*-selection. *American Naturalist* **104**, 592–7.

Poiner, I. R. & Kennedy, R. 1984. Complex patterns of change in the macrobenthos of a large sandbank following dredging. *Marine Biology* **78**, 335–52.

Rees, E. I. S., Nicholaidou, A & Laskaridou, P. 1977. The effects of storms on the dynamics of shallow water benthic associations. In *The Biology of Benthic Organisms*. B. F. Keegan et al (eds). New York: Pergamon, 465–74.

Rees, H. L. & Dare, P. J. 1993. *Sources of mortality and associated life-cycle traits of selected benthic species: a review*. Fisheries Research Data Report, No. 33.

Reise, K. 1982. Long-term changes in the macrobenthic invertebrate fauna of the Wadden Sea: are polychaetes about to take over? *Netherland Journal of Sea Research* **16**, 29–36.

Reise, K. & Schubert, A. 1987. Macrobenthic turnover in the subtidal Wadden Sea: the Norderaue revisited after 60 years. *Helgoländer. Meeresuntersuchungen* **41**, 69–82.

Rhoads, D. C. 1974. Organism-sediment relationships on the muddy sea floor. *Oceanography and Marine Biology: an Annual Review* **12**, 263–300.

Rhoads, D. C. & Boyer, L. F. 1982. The effects of marine benthos on physical properties of sediments: A successional perspective. In *Animal-sediment relations*, P. L. McCall & M. J. S. Tevesz (eds). New York: Plenum Press, 3–52.

Rhoads, D. C., McCall, P. L. & Yingst, J. Y. 1978. Disturbance and production on the estuarine seafloor. *American Scientist* **66**, 577–86.

Rhoads, D. C. & Young, D. K. 1970. The influence of deposit-feeding organisms on sediment stability and community structure. *Journal of Marine Research* **28**, 150–78.

Riesen, W. & Reise, K. 1982. Macrobenthos of the subtidal Wadden Sea: revisited after 55 years. *Helgoländer Meeresuntersuchungen*. **35**, 409–23.

Rosenberg, R. 1976. Benthic faunal dynamics during succession following pollution abatement in a Swedish estuary. *Oikos* **27**, 414–27.

Saloman, C. H., Naughton, S. P. & Taylor, J. L. 1982. Benthic community response to dredging borrow pits, Panama City beach, Florida. M. R. 82-3. US Army Corps of Coastal Engineers Research Center, Fort Belvoir, Va.

Sanders, H. L. 1958. Benthic studies in Buzzards Bay. I. Animal sediment relationships. *Limnology and Oceanography* **3**, 245–58.

Sanders, H. L. 1969. Marine benthic diversity and the stability-time hypothesis. In *Brookhaven Symposia in Biology* (*22*), *Diversity and Stability in Ecological Systems*. Upton, New York: Brookhaven National Lab. 71–81.

Schafer, W. 1972. *Ecology and Palaeoecology of Marine Environments*. Chicago: University of Chicago Press.

Shelton, R. G. & Rolfe, M. S. 1972. The biological implications of aggregate extraction: Recent studies in the English Channel. *ICES C.M. 1972*/E: 26.

Sherk Jr, J. A. 1971. The effects of suspended and deposited sediments on estuarine organisms. *Contrib. No. 443 Chesapeake Biological Laboratory, Solomons, Maryland.*

Sherk Jr, J. A. 1972. Current status of the knowledge of the biological effects of suspended and deposited sediments in Chesapeake Bay. *Chesapeake Science* **13** (Suppl.), 137–44.

Sherk Jr, J. A., O'Connor, J. M. & Neumann, D. A. 1972. Effects of suspended and deposited sediments on estuarine organisms. Phase II. *Ref. No. 72-9E Deptartment of Environmental Research, Chesapeake Biological Laboratory, University of Maryland.*

Sherk Jr, J. A., O'Connor, J. M., Neumann, D. A., Prince, R. D. & Wood, K. V. 1974. Effects of suspended and deposited sediments on estuarine organisms. Final Report. *Ref. No. 74-20 Department of Environmental Research, Chesapeake Biological Laboratory, Prince Frederick, Maryland.*

Shumway, S. E., Cucci, T. L. Newell, R. C. & Yentsch, C. L. 1985. Particle selection, ingestion and absorption in filter-feeding bivalves. *Journal of Experimental Marine Biology and Ecology* **91**, 77–92.

Shumway, S. E., Newell, R. C., Crisp, D. J. & Cucci, T. L. 1990. Particle selection in filter-feeding bivalve molluscs: a new technique on an old theme. In *The Bivalvia- Proceedings of a Memorial Symposium in Honour of Sir Charles Maurice Yonge, Edinburgh. 1986*. B. Morton (ed.). Hong Kong: University Press, 151–65.

Sips, H. J. J. & Waardenburg, H. W. 1989. The macrobenthic community of gravel deposits in the Dutch part of the North Sea (Klaverbank). Ecological impact of gravel extraction. Reports of the Bureau Waadenburg bv, Culemborg, The Netherlands.

Smetacek, V. 1984. The supply of food to the benthos. In: *Flows of energy and materials in marine ecosystems*, M. I. R. Fasham (ed.). 517–48. New York: Plenum Press.

Snelgrove, P. V. R. & Butman, C. A. 1994. Animal-sediment relationships revisited: cause versus effects. *Oceanography and Marine Biology: an Annual Review* **32**, 111–77.

Steele, J. H. 1965. Notes on some theoretical problems in production ecology. *Memoire dell'Istituto Italiano di Idrobiologia Dott Marco de Marchi* **18** (Suppl.), 383–98.Steele, J. H. 1974. *The Structure of Marine Ecosystems.* Cambridge, Mass: Harvard University Press.

Stephenson, W. W. T., Cook, S. D. & Newlands, S. J. 1978. The macrobenthos of the Middle Banks area of Moreton Bay. *Memoirs of the Queensland Museum* **18**, 95–118.

Taylor, P. M. & Saloman, C. H. 1968. Some effects of dredging and coastal development in Boca Ciega Bay, Florida. *United States, Fish and Wildlife Service, Fishery Bulletin* **67**(2), 213–41.

Thayer, C. W. 1983. Sediment-mediated biological disturbance and the evolution of marine benthos. In *Biotic interactions in recent and fossil benthic communities.* M. J. S. Tevesz & P. L. McCall (eds). New York: Plenum Press, 479–625.

Thayer, C. W., Wolfe, D. A. & Williams, R. B. 1975. The impact of man on seagrass systems. *American Scientist* **63**, 288–96.

Thorson, G. 1957. Bottom communities (sublittoral or shallow shelf). *Geological Society of American Memoir* **67**, 461–534.

Urrutia, M. B., Iglesias, J. I. P., Navarro, E. & Prou, J. 1996. Feeding and absorption in *Cerastoderma edule* under environmental conditions in the Bay of Marennes-Oleron (Western France). *Journal of the Marine Biological Association of the United Kingdom* **76**, 431–50.

U.S. Army Corps of Engineers, 1974. *Draft Environmental Statement – Oyster Shell Dredging Tampa & Hillsborough Bays, Florida, USA.* Army Engineer District, Jacksonville, Florida.

van der Veer, H. W., Bergman, M. J. N & Beukema, J. J. 1985. Dredging activities in the Dutch Waddensea: effects on macrobenthic infauna. *Netherlands Journal of Sea Research* **19**, 183–90.

van Moorsel, G. W. N. M. 1993. Long-term recovery of geomorphology and population development of large molluscs after gravel extraction at the Klaverbank (North Sea). *Rapport Bureau Waardenburg bv, Culemborg, The Netherlands.*

van Moorsel, G. W. N. M. 1994. The Klaver Bank (North Sea), geomorphology, macrobenthic ecology and the effect of gravel extraction. *Rapport Bureau Waardenburg and North Sea Directorate (DNZ)*, Ministry of Transport, Public Works & Water Management, The Netherlands.

van Moorsel, G. W. N. M. & Waardenburg, H. W. 1990. Impact of gravel extraction on geomorphology and the macrobenthic community of the Klaverbank (North Sea) in 1989. *Rapport Bureau Waardenburg bv, Culemborg, The Netherlands.*

van Moorsel, G. W. N. M & Waardenburg, H. W. 1991. Short-term recovery of geomorphology and macrobenthos of the Klaverbank (North Sea) after gravel extraction. *Rapport Bureau Waardenburg bv, Culem-*

borg, The Netherlands.

Walker, T. A. & O'Donnell, G. 1981. Observations on nitrate, phosphate and silicate in Cleveland Bay, Northern Queensland. *Australian Journal of Marine and Freshwater Research* **32**, 877–87.

Warren, P. J. 1973. The fishery for the pink shrimp *Pandalus montagui* in the Wash. MAFF, Laboratory Leaflet No. 28.

Warwick, R. M. 1986. A new method for detecting pollution effects on marine macrobenthic communities. *Marine Biology* **92**, 557–62.

Warwick, R. M. & Clarke, K. R. 1991. A comparison of methods for analysing changes in benthic community structure. *Journal of the Marine Biological Association of the United Kingdom.* **71**, 225–44.

Warwick, R. M. & Clarke, K. R. 1993. Increased variabilty as a symptom of stress in marine communities. *Journal of Experimental Marine Biology and Ecology* **172**, 215–26.

Warwick, R. M., Pearson, T. H. & Ruswahyuni 1987. Detection of pollution effects on marine macrobenthos: further evaluation of the species abundance/biomass method. *Marine Biology* **95**, 193–200.

Weston, D. P. 1988. Macrobenthos-sediment relationships on the continental shelf off Cape Hatteras, North Carolina. *Continental Shelf Research* **8**(3), 267–86.

Whiteside, P. G. D, Ooms, K. & Postma, G. M. 1995. Generation and decay of sediment plumes from sand dredging overflow. Proceedings of the 14th World Dredging Congress. Amsterdam, The Netherlands: World Dredging Association (WDA), 877–92.

Whittaker, R. H. & Levin, S. A. 1977. The role of mosaic phenomena in natural communities. *Theoretical Population Biology* **12**, 117–39.

Wildish, D. J. & Kristmanson, D. D. 1979. Tidal energy and sublittoral macrobenthic animals in estuaries. *Journal of the Fisheries Research Board of Canada* **36**, 197–206.

Wildish, D. J. & Kristmanson, D. D. 1984. Importance to mussels of the benthic boundary layer. *Canadian Journal of Fisheries and Aquatic Sciences* **41**, 1618–25.

Willoughby, M. A. & Foster, D. N. 1983. *Middle Banks Humber River plume mapping and hydrographic surveys. 18th June to 3rd July 1982.* Technical Report, Univesity of New South Wales Water Research Laboratory, Australia, No. 83 1–13.

Wilson, D. P. 1971. *Sabellaria* colonies at Duckpool, North Cornwall, 1961–70. *Journal of the Marine Biological Association of the United Kingdom.* **51**, 509–80.

Windom, H. L. 1976. Environmental aspects of dredging in the coastal zone. *Critical Reviews in Environmenal Control* **6,** 91–110.

Woodin, S. A. 1991. Recruitment and infauna: positive or negative cues? *American Zoologist* **31**, 797–807.

Woodin, S. A. & Marinelli, R. 1991. Biogenic habitat modification in marine sediments: the importance of species composition and activity. *Symposia of the Zoological Society of London* **63**, 231–50.

Wright, D. G. (Coordinator) 1977. Artificial islands in the Beaufort Sea. A review of potential impacts. *Department of Fishery and Environment.* Winnipeg, Manitoba. Sept 1977.

Oceanography and Marine Biology: an Annual Review 1998, **36**, 179–215

UCL Press

OBELIA (CNIDARIA, MEDUSOZOA, HYDROZOA): PHENOMENON, ASPECTS OF INVESTIGATIONS, PERSPECTIVES FOR UTILIZATION

S. D. STEPANJANTS
Zoological Institute of the Russian Academy of Sciences
199034 Saint-Petersburg, Russia

Abstract About eight species are joined together into the genus named *Obelia* (Cnidaria, Medusozoa, Hydrozoa). However there are some species (about 10) attributed by some investigators to *Obelia*, and by others to three independent genera: *Gonothyraea*, *Laomedea* and *Hartlaubella*, which are nevertheless supposed to be closely related to each other and to *Obelia*. All the specialists agree that all these genera belong to a subfamily Obeliinae, family Campanulariidae. This review covers the topics: taxonomy, cell composition and cytogenetics, sexual reproduction and early developmental stages, growth and coloniality problems, feeding and digestion, other ecological problems, horizontal and vertical distribution, luminescence. It aims to demonstrate results gained by the specialists dealing with this group and to show the general trend of the continuing investigations. An appendix gives information concerning the Workshop, "*Obelia* as a dominant in epibiotic communities", which took place in September 1996 in St Petersburg.

Introduction

The Cnidaria are divided into four discrete classes; Scyphozoa, Cubozoa, Hydrozoa and Anthozoa. The evolutionary based status of each of them within the cnidarian system is viewed by different taxonomists in different ways (Hadzi 1963, Uchida 1963, Rees 1966, Werner 1973, Petersen 1979, 1990, Stepanjants 1988, Grasshoff 1991a,b, 1997). However, the discussion of this problem is beyond the scope of the present review.

In any case the first three classes could clearly make up a subtype Medusozoa (Petersen 1979) as life histories of the species include two stages: polyps and medusae. Within different groups of Medusozoa these two stages might be expressed differently: both might be well developed; the medusoid stage might be partly or almost completely missed out; or the polypoid stage might be partly or almost completely missed out. However, with the predominance of one of the stages the other one is by no means suppressed.

When the polyp stage predominates medusae exist as attached medusoids at different degrees of reduction or even as gonads (*Hydra*, *Rhysia*), whose development into polyp tissues follows the pattern observed in "medusa nodule" – "entocodon" (Bouillon & Werner 1965, Brien 1965, Brinckmann 1965, Stepanjants 1988). It testifies to the medusoid nature of gonads. In Scyphomedusae, Cubomedusae and Trachymedusae, when the predominance of medusa stage occurs, polyps could be observed through the species' life histories, and although some investigators view them as larvae (Rees 1966, Werner 1973, etc.) they must surely be considered polyps morphologically, even if they are found not on bottom, but within the water column (Stepanjants 1988). The second cnidarian group – Anthozoa – is actually the second subtype (Petersen 1979). It differs because of the absence of a medusa stage in the life history. In addition, both of these groups have some other principle characteristics: Medusozoa are mostly radially symmetrical, while Anthozoa are mostly bilaterally symmetrical; and both groups differ radically in the morphology of nematocysts and in the cnidom composition (composition of nematocysts) (Bozhenova 1988, Bozhenova et al. 1988).

After this short general introduction it is possible to switch over to the subject of our investigation – *Obelia*, which belongs to one of the four listed medusozoan classes, that is, the Hydrozoa.

Taxonomy

About eight species belong without doubt under the generic name *Obelia* Peron et Lesueur 1810 (Cnidaria, Medusozoa, Hydrozoa), but taxonomists disagree concerning the validity of some of these species. For instance Professor P. F. S. Cornelius, a great authority on *Obelia* taxonomy, who has studied the family Campanulariidae, including *Obelia*, for many years, names only four valid *Obelia* species: *O. bidentata*, *O. dichotoma*, *O. geniculata* and *O. longissima* (Cornelius 1975, 1982a, 1990a,b, 1995, in press). Some other investigators agree with this interpretation (Ostman – in most of her conclusions 1982a,b, 1983, 1989, in press, Morri & Boero 1986, Calder 1991). Ostman illustrates morphological differences between the subtypes of nematocysts in these species by means of brilliant electronic photographs.

In different regional investigations, however, more species are listed as valid: *O. plicata* was mentioned as a valid species in the North Atlantic (Vervoort 1985), and *O. chinensis*, *O. oxidentata*, *O. plana* and *O. bicuspidata* (=*O. bidentata*) from the Sea of Japan (Hirohito 1995) also appear to be valid. Kubota (in press) has the same opinion about the listed species except for *O. plana*, which he supposes to be a synonym of *O. longissima*. Other authors compared life histories of *O. longissima* from the White Sea and Barents Sea with that of *O. plana* from the Sea of Japan and concluded that the two are synonymous (Stepanjants et al. 1993), but more recently Stepanjants et al. (in press) returned to the previous view that both *O. longissima* and *O. plana* are valid species.

Not only Japanese investigators supposed *O. oxydentata* to be a valid species. The French investigator Gravier-Bonnet (1972, 1979, in press) studied this species from Madagascar. Although Cornelius (1982a) and Calder (1991) included *O. australis* in the list of synonyms of *O. dichotoma* one should not ignore the opinion of Ralph (1957), who worked all her life on the New Zealand hydroids and supposed *O. australis* to be an independent species inhabiting New Zealand, Australian, and Tasmanian shallow waters. Thus, by now it makes sense to accept eight or nine presumably valid species of *Obelia*: *O. australis*, *O. bidentata*, *O. chinensis*, *O. dichotoma*, *O. geniculata*, *O. longissima*, *O. oxydentata*, *O. plana* (and possibly, *O. plicata*).

There are, however, some species (about 10) that are attributed by some investigators to *Obelia*, and by some others to three independent genera. These genera are *Gonothyraea*, *Laomedea* and *Hartlaubella*, but they are supposed to be related very closely both to each other and to *Obelia*. Major specialists agree that all these genera belong to a subfamily Obeliinae Haeckel 1889, family Campanulariidae (Cornelius 1990a,b, 1995, in press, Calder 1991, etc.). Some also include *Clytia* into Obeliinae (Boero et al. 1996). I do not agree with this opinion because *Clytia* species form an isolated group with distinct characters of either polyps or medusae, which make *Clytia* markedly different from the other representatives of Obeliinae (Cornelius, in press).

Laomedea Lamouroux 1812 All the specialists who accept the necessity of raising the genus and including such species as *Laomedia angulata* Hincks 1868, *L. calceolifera* (Hincks 1871), *L. exigua* M.Sars 1857, *L. flexuosa* Alder 1857, and *L. neglecta* Alder 1856 into it, find practically no substantial differences in this group of species and those included in *Obelia*, except for the free medusa stage occurring in *Obelia* species. Medusoids of *Laomedea* species are reduced to sporosacs (Splettstoesser 1924, Cornelius 1977a,b, 1982a,b,c, 1985, 1995, Antsulevitch 1987b).

Gonothyraea Allman 1864 This genus contains one or three species: *Gonothyraea loveni* (Allman1859), *G. hyalina* Hincks 1866(?) and *G.inornata* Nutting 1901(?). The last species was described from Alaska, but its generic status is uncertain (may be *Laomedea*?). However, this species was found in the Okchotsk Sea, off the Kamchatka shore (Olga Sheiko pers. comm) and it was included in the species list for American shores (Cairns et al. 1991). There are no other differences between the species of this genus and those of *Obelia*, except the absence of the free medusa stage; their only principal difference from *Laomedea* lies in the morphology of the medusoids: in *Gonothyraea* they are fixed medusae without manubriums (meconidiums) (Boero & Bouillon 1987, 1989, Boero & Sara 1987, Boero et al. 1996).

Hartlaubella Poche 1914 This genus is represented by only one species *Hartlaubella gelatinosa* (Pallas 1766). It is also difficult to find principle characters to distinguish this genus from the other three genera except for the free medusa stage being absent. Medusoids are present in the form of sporosacs.

It should be noted that Professor P. F. S. Cornelius, whose recent two-volume monograph (1995) contained the up-to-date taxonomy of Thecaphora Hydroidea, accepted all the four genera of Obeliinae, but did not give an identification key for these genera, while for practically all the other genera and the rest of the families there are keys in his monograph. This fact emphasizes the difficulties of exact differentiation of the generic characters in Obeliinae. Hydranth morphological analysis that Cornelius has used for diagnostic purposes in other subfamilies (hypostom form, number of tentacles, their length, type of nematocyst distribution on tentacles, and others) fails to give convincing generic characters in the Obeliinae (Cornelius 1987). Thus presence or absence of the free medusa stage and the degree of medusoid development appear to be the only characters for defining the Obeliinae genera: that is, *Obelia* – free medusae; *Gonothyraea* – reduced medusae – meconidium, eumedusoid; *Laomedea* and *Hartlaubella* – extreme medusa reduction – sporosac (Kuhn 1913).

Although major taxonomists recognize the type of medusa development to be a generic character (Boero et al. 1996), many examples are known where within the same genus one may find species with different degrees of medusa development *(Bimeria, Hydractinia, Amphysbetia, Sertularella, Aglaophenia,* and others). Obviously, while considering a taxa's natural system and setting off genera it seems to be impossible to restrict ourselves to a sole character.

To illustrate the presented point of view one might try to construct several versions of identification keys for the genera of the Obeliinae by means of previously published diagnoses and characters (Cornelius 1995, Boero et a1. 1996).

In order to make clear the morphological details being used in the keys presented below it would have been desirable to give morphological descriptions of the colonies of polyps and organization of the medusae but this would have considerably extended the present review. However, indications of the characters used in keys (Millard 1975, Cornelius 1982a, 1995, Ostman 1982b, Cornelius & Ostman 1987, Stepanjants et al. 1993) are given in Figures 1 to 4.

Keys published for the first time.

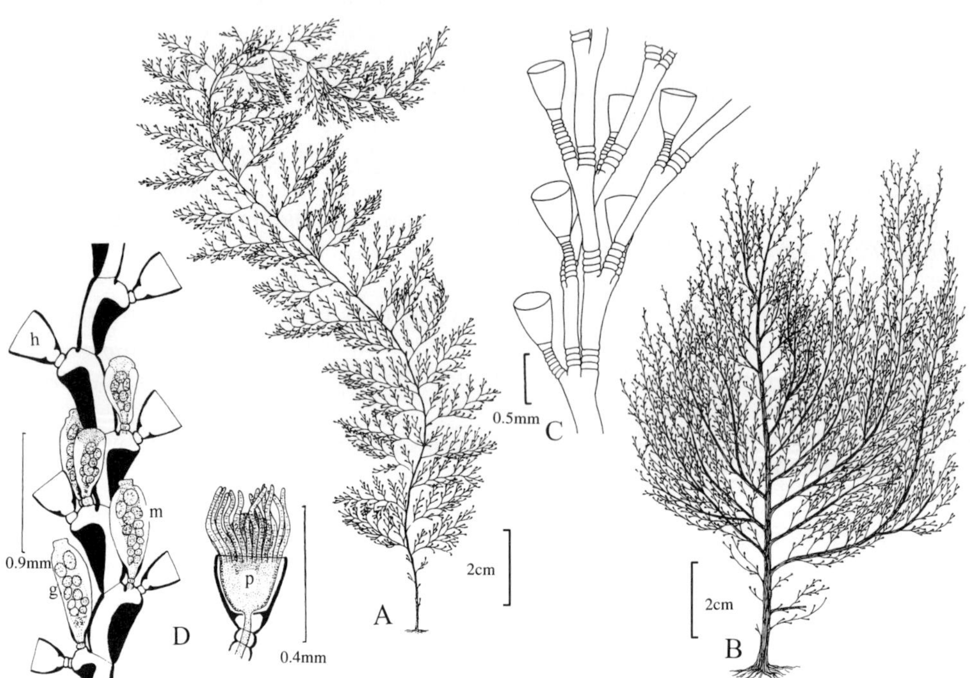

Figure 1 *Obelia* colonies and their fragments. A, *Obelia longissima* (Pallas 1766) branching colony with monosiphonic stem, in natural size (after Cornelius 1995). *B, Obelia dichotoma* (Linnaeus 1758) branching colony with polysiphonic stem, in natural size (after Cornelius 1995). C, Detail of *O. dichotoma* colony: hydrothecae rim is even (after Cornelius 1995). D, *Obelia geniculata* (Linnaeus, 1758) unbranching colony with hydrothecae (h), gonothecae (g), medusae into every gonotheca (m) and polyp into hydrotheca (p) (after Millard 1975). The typical perisarcal thickenings on one side of every internode for this species are shown in black.

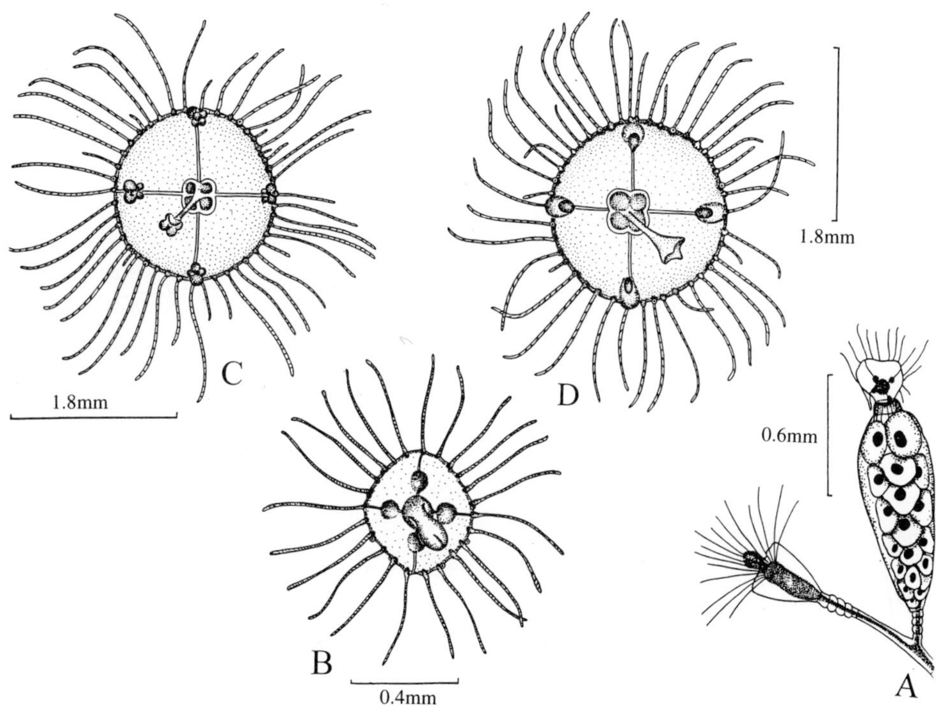

Figure 2 *Obelia longissima* (Pallas, 1766) medusae. A, young medusa at the moment of its leaving of gonotheca (the Barents Sea). B, medusa of one day old (the Barnets Sea). C, D, medusa female (C) and male (D) of 22 days old (the White Sea). (After Stepanjants et al. 1993).

First version:

1 (2). There are free medusae ... genus *Obelia*
2 (1). Free medusae are absent
3 (6). Medusoid is of sporosac type
4 (5). The adult colony stem is polysiphonic Hydrotheca margin is denticulate ... genus *Hartlaubella*
5 (4). The adult colony stem is monosiphonic. Hydrotheca margin is smooth; if stem is bisiphonic, hydrotheca rim is denticulate (*Laomedea neglecta*) ... genus *Laomedea*
6 (3). Medusoid is in the form of a meconidium (eumedusoid) type ... genus *Gonothyraea*

What can be seen from this version of the key?

(1) Identification of sterile colonies is impossible from the key.
(2) It seems rather doubtful that colonies in *Gonothyraea* and *Laomedea* are of absolutely monosiphonic character: according to Cornelius (1995: 284), the *L. neglecta* stem might be bisiphonic.

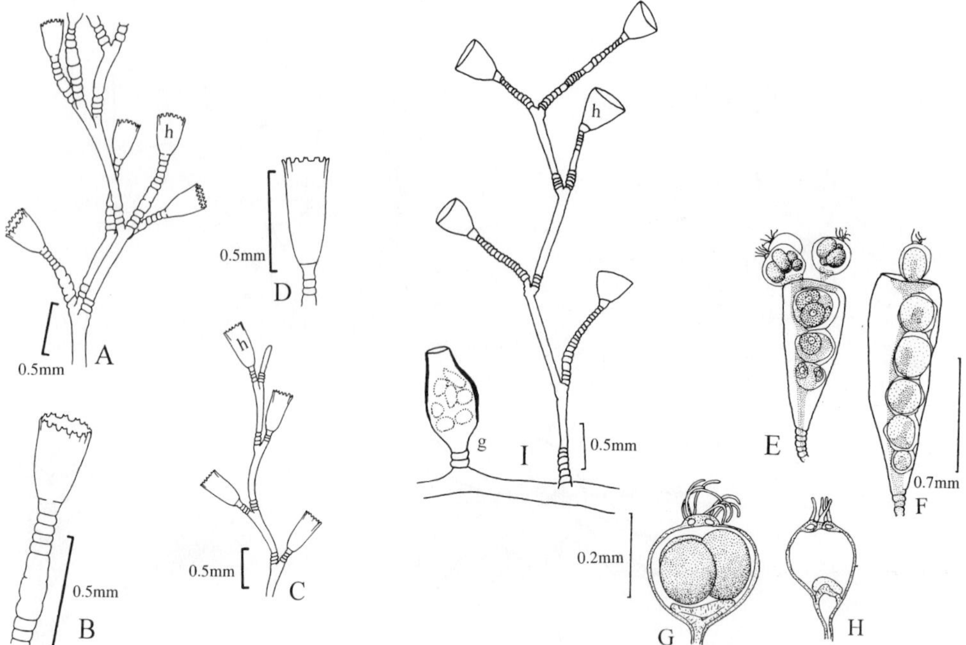

Figure 3 Colonies and its fragments of different Obeliinae. A, B, Fragments of colony of *Hartlaubella gelatinosa* (Pallas 1766) (A) and hydrotheca (B, h) with the denticulate bicuspid rim (after Cornelius 1995). C, D, Fragment of colony of *Gonothyraea loveni* (Allman 1859) (C) and hydrotheca (D, h) with the denticulate bicuspid rim (after Cornelius 1995). E–H, Gonothecae of *Gonothyraea loveni* (Allman 1859) (E, F) with eumedusoids, female (G) and male (H) eumedusoids (after Millard 1975). I, *Laomedea angulata* Hincks 1861 colony with the monosiphonic stem, hydrothecae (h) with the even rim female sporosac (g) (after Cornelius 1982a).

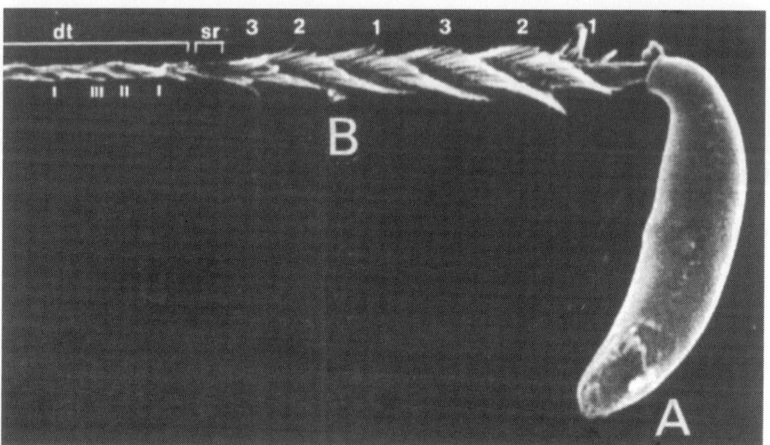

Figure 4 *Laomedea flexuosa* Alder 1857 (after Ostman 1982b). Scanning electron micrograph of the discharged b-rhabdoid nematocyst capsule. A, capsule; B, tube; 1, 2, 3 the long paired spines; sr, the short spineless region; dt, spined distal tube; I, II, III, three rows of minute spines on the distal part of tube.

(3) The denticulate rim of the hydrotheca in *L. neglecta* permits the assignment of this species to *Hartlaubella* rather than to *Laomedea.*

Other versions of the key will bring new difficulties of the same kind to the identification of genera.

Second version:

1	(4). The hydrotheca rim is always denticulate, every denticle is bicuspid. Hydrothecae form varies from cylindrical to conic.	
2	(3). The stem of adult colonies is polysiphonic ...	genus *Hartlaubella*
3	(2). The stem of adult colonies is momosiphonic ...	genus *Gonothyraea*
4	(1). The hydrotheca rim is mostly smooth or slightly crenated. Hydrotheca as a rule is bell-shaped. If there are hydrothecal marginal bicuspidal denticles, hydrothecae as a rule are cylindrical (*Laomedea neglecta; Obelia bidentata*).	
5	(6). There are free medusae ...	genus *Obelia*
6	(5). There are no free medusae. There are medusoids (sporosacs) attached to colony ...	genus *Laomedea*

The second version of the key is based on the marginal hydrothecal dentelation – a character that might seem good and strong. But then it would have been reasonable to place such species as *L. neglecta* and *Obelia bidentata* into *Gonothyraea* or *Hartlaubella*, especially as nobody has seen *O. bidentata** free medusae. In *Laomedia neglecta* and *Obelia bidentata* the form of the hydrothecae might also suggest that this species be assigned to *Hartlaubella* or *Gonothyraea*. In this case, as in the first version, the sterile colonies of *Obelia* and *Gonothyraea* could not be differentiated.

To summarize, the position of those taxonomists integrating *Obelia*, *Laomedea*, and *Gonothyraea* into one genus *Obelia* (Naumov 1960) is clear enough. The same might be said about the necessity to abolish the genus *Hartlaubella* (Boero et al. 1996).

Cell composition and cytogenetics

It is well known that Cnidaria, being lower invertebrates, have the most simple structure whether viewed on cellular, tissue or morphological levels.

Except for some publications dating back to the end of the last century (Fraipont 1880, Huxley & Beer 1923, Beadle & Booth 1938) and several investigations on the cellular composition of other representatives of the Cnidaria, Hydrozoa and Campanulariidae (Aizu 1967, David 1983, Bode et al. 1986, Schmidt & David 1986), in the

* The information concerning adult free medusae being unknown for *O. bidentata* may be found in Cornelius's two-volume monograph (1995, Vol. 2: 292) and in the monograph by Migotto (1995) with reference to Vannucci (1955). In a personal communication, A. Migotto confirmed that he had never found free medusae of this species.

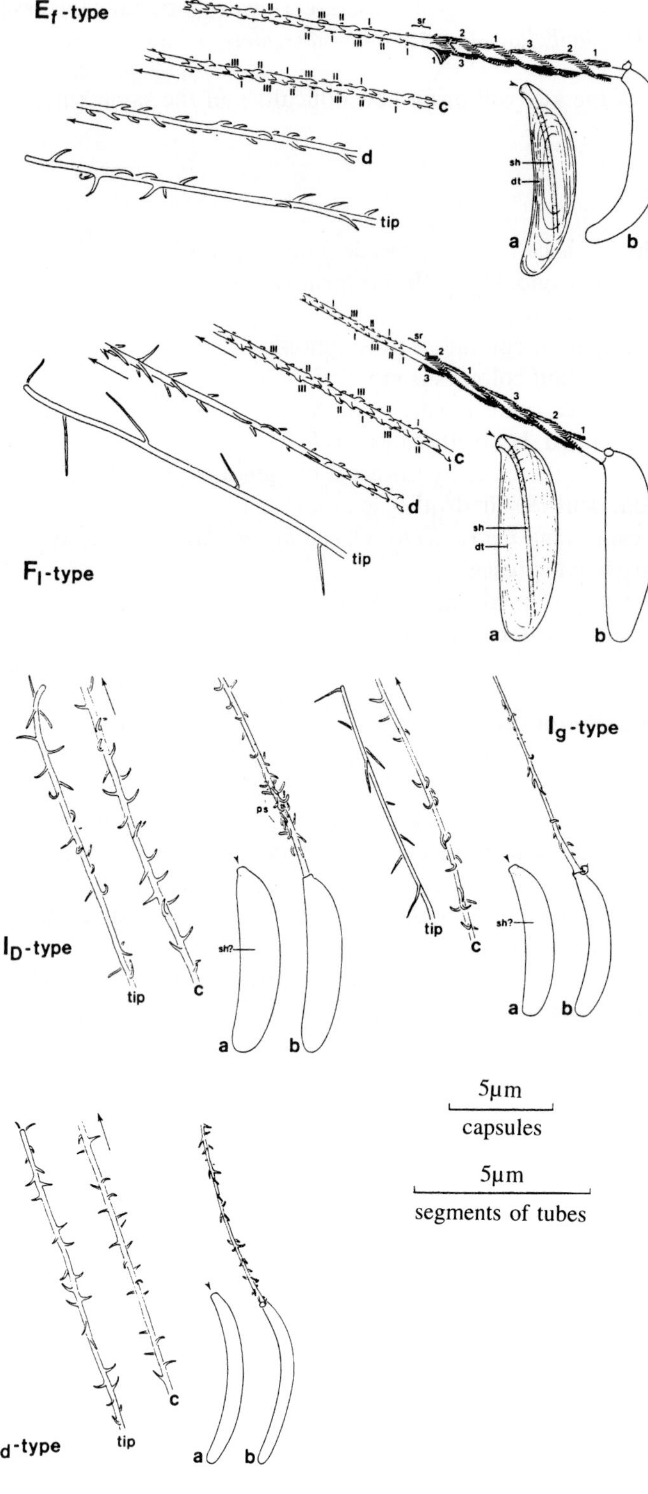

E_f-type
sr
sh
dt
a
b
c
d
tip
F_l-type
I_D-type
I_g-type
ps
sh?
I_d-type
5μm
capsules
5μm
segments of tubes

Obeliinae exploration into the cellular structure of the tissues (i.e. cell differentiation and dedifferentiation, specificity of some clones cells being formed, and specificity of migration of cells from one cell layer into another) began in the 1960s to 1970s (Knight 1965, 1968a,b, 1970a,b, 1971) and especially the 1980s, mainly by Russian scientists (Polteva 1972, 1975, 1981, 1986, Dondua & Dondua 1975, Polteva & Aizenstadt 1980a,b, Aizenstadt & Polteva 1981, 1982, 1986, 1989, Donakov 1985, 1987, 1988, 1989, Aizenshtadt 1986, Makarenkova & Seravin 1986, Polteva et al. 1987, 1988, Donakov & Makarenkova 1989, Donakov et al. in press). The listed publications allow us to understand the principles of tissue formation in evolutionarily more advanced organisms, that is, the Obeliinae might be used as model objects in such fundamental biological studies.

These investigations were conducted mainly on the species that might be named *Obelia loveni* (=*Gonothyraea loveni*) according to Naumov's system (Naumov 1960). Interesting data concerning the cytogenetic potential of cells during the period of sexual maturation and after, up to their winter reduction to the hydrorhyza state, were obtained on White Sea colonies of this species. Similar investigations on *Laomedea* and other close relatives of *Obelia* will allow us to compare the peculiarities of the cytogenetic processes in different Obeliinae.

Information concerning the chromosome composition of Obeliinae species is scarce (Faulkner 1929, Anonymous 1930).

Nematocysts

Like other representatives of the Cnidaria, Obeliinae contain adhesive and penetrative capsules – nematocysts (cnidae) – in their tissues and are characterized by a specific set of nematocysts of different types, named cnidome. Nematocysts of all the Cnidaria, including *Obelia*, have been studied for a long time (Weill 1934a,b, Carlgren 1940, Cutress 1955, Ito & Inoue 1962, Westfall 1966a,b,c, Mariskall 1974, Kubota 1976, Ostman 1979a,b, 1982b, 1983, Bouillon 1985, Watson & Wood 1989, and others). Meanwhile the greatly improved properties of present-day optical equipment allow much improved investigation of the morphology of nematocysts (Hessinger & Lenhoff 1989). Ostman studied the morphological and functional peculiarities of the nematocysts of the Campanulariidae for many years (Ostman 1979a,b, 1982b, 1983, 1987, 1989, in press, Cornelius & Ostman 1987, Ostman et al. 1987). Now, the morphology of uncharged capsulae and the tubes armature of discharged capsulae can be investigated in great detail by means of scanning (SEM) and light (LM) microscopy. It was stated that cnidoms of all the investigated Campanulariidae included capsulae of two types: (a) pseudo-microbasic b-mastigophores, and (b) isorhizous haplonemes (Fig. 5). However, more detailed studies on the morphology of nematocysts allows us to dis-

Figure 5 Nematocysts (b-rhabdoid = microbasic b-mastigophore and atrichous isorhiza types) of *Laomedea flexuosa* (subtype Ef), *Obelia longissima* (subtype Fl), *Obelia dichotoma* (subtypes ID and Id) and *Obelia geniculata* (subtype Ig); a, undischarged capsulae; b, discharged capsulae with the proximal part of tubes; c, d, tip, distal part of tubes; sh, shaft; dt, coils of distal tube; sr, spinless region of tube; ps, proximal spines; 1, 2, 3, separate strands of long paired spines forming the right-handed triple helix in the proximal armature; I, II, III, rows of minute hook-shaped spines (after Ostman 1982b).

tinguish "A", "B" etc. subtypes in each of the listed types. Ostman distinguished six subtypes within the first type and four within the second type. These subtype investigations appear to be rather promising in so far as they allow us to make clear the details of the morphology of the nematocysts and to make use of them for the taxonomic analysis of the group at the species level (Ostman in press, Migotto 1995).

Isoenzymes

Polyacrylamide gel electrophoresis (PAGE) investigations on some *Campanularia, Clytia, Obelia, Laomedea,* and *Gonothyraea* species have shown the possibilities of isoenzymens being used for taxonomic studies of the Obeliinae, mostly at the species level (Ostman 1982a).

Sexual reproduction, early development stages

Publications concerning the sexual reproduction of Obeliinae (Miller 1969, 1970, Lunger, 1971) are not very numerous. Weismann (1883), Goette (1907), and Kuhn (1913) deal with general problems of gametogenesis including the Obeliinae. Some publications contain information concerning gonangium formation, and eggs and sperm ripening in *Laomedea flexuosa* (Hargitt 1913, O'Rand 1972a,b, 1974, O'Rand & Miller 1974) and *L. angulata* (Babic 1912). Information on the correlation of these processes with the cycles of the moon has been given by Elmhirst (1925).

The early embryogenesis of Obeliinae has been little studied (Merezkovsky 1883, Wulpert 1902, Muller-Cale 1913, Teissier 1930, 1932, Kubota 1981, Dementjev & Minichev 1985, Dementjev 1989a,b). During recent years more attention was drawn to these questions (Krauss & Rodimov, in press).

Life history

Representatives of the Medusozoa, including the species of *Obelia* and closely related genera, are characterized by life histories that include a bottom-living colonial vegetative polypoid stage and a sexual medusoid stage, which may be represented by either free-living medusae or medusoids attached to the colonies. The latter cannot release themselves and to live free in the plankton. This peculiarity of the Medusozoa life history and its manifestation in Obeliinae ontogenesis is very important and its investigation might provide interesting potential for evolutionary studies.

At the present time there are several different points of view concerning the evolutionary trends observed in the Campanulariidae. The general issue upon which all the different evolutionary versions agree is that the free medusa of *Obelia* has quite an unusual morphology compared with that of the other hydromedusae (Cornelius 1995, in press, Boero et al. 1996). This is expressed in the following characters of the medusa: (a) the mesogloea on the umbrella is very thin; (b) the marginal velum is nearly absent; (c) the ring canal is nearly absent; (d) there are very elastic solid marginal tentacles, that are not extensile*; (e) the manubrium is fairly short and lacks a peduncle; (f) cirrae are absent; (g) there are closed adradial marginal vesicles with 1–2 concretions; (h) feeding behaviour and movements in water are atypical.

Nonetheless the significance of these unusual morphological and behavioural characteristics of the *Obelia* medusa has been interpreted differently by different specialists. One of the points of view (Cornelius 1990b, 1995, in press) supposes the morphological peculiarities of *Obelia* medusae to be the result of gradual regressive evolution or the secondary development of some characters from earlier degenerated forms (e.g. from *Gonothyraea* eumedusoids). Boero & Sara (1987), following unpublished data of Edwards et al., supposed that *Obelia* free medusae have developed secondarily from reduced ones, and have regained some characters that had disappeared from the phenotype but were coded in the genotype. This phenomenon was called (after McKinney & McNamara 1991) "atavistic evolution" of *Obelia* medusae. According to an alternative theory (Boero & Bouillon 1989, Boero et al. 1996), *Obelia* medusae could be derived by secondary evolution from polyps. Authors following this theory supposed polyps to be larvae and considered this phenomenon to be a demonstration of paedomorphosis. Cornelius (in press) seems to be right when he states that at present the data are not sufficient to consider the questions of evolution of the *Obelia* medusa solved.

When discussing peculiarities of the life history of Obeliinae, one should not forget that the morphology of the free medusa of *Obelia* does not allow the different *Obelia* species to be distinguished (Kramp 1961). This topic has already been discussed repeatedly (Cornelius 1995), but until recently there was no chance to use the characters of the medusa for constructing identification keys. In the early 1980s, however, Kubota (1981) proposed a method for breeding medusae of his *O. plana* in experimental conditions and measuring growth parameters throughout the duration of the medusa stage (about 30 days). Later, this method was used to investigate the life history of *O. longissima* in the White and Barents Seas (Stepanjants & Letunov 1989, Stepanjants et al. 1993), in the Black Sea (Beloussova 1991), and in the South Vancouver Island area (the last case concerns life histories of *O. longissima* and *O. dichotoma* investigated by S. Stepanjants). Measurements were made during 7–30 days of the medusa stage, the following parameters being noted: umbrella diameter, number of marginal tentacles, gonad diameter, distance between the middle of the gonad and the middle of the stomach, measured along the radial canal where the gonads are placed.

Subsequent statistical investigations using the computer data base drawn from these studies demonstrated the possibility of finding species specific characteristics of medusae if the distribution of pairs of characters (such as – umbrella diameter and marginal tentacle number; gonad diameter and distance between the middle of the gonad and the middle of the stomach) rather than individual characters is compared (Stepanjants et al. in press). This method allows not only species differences among *Obelia* medusae to be determined, but also the species status of *Obelia* medusae in different areas of the ocean.

* The statement of these authors concerning the marginal tentacles of the medusa of *Obelia* being incapable of contraction is doubtful. Repeated observations on living *Obelia* medusae feeding in the laboratory indicate the opposite. If a needle with a piece of food is brought to a medusa's tentacle (during the first days of feeding especially), the tentacle shortens strongly – as strongly as under rapid fixation with formalin (Stepanjants et al. 1993).

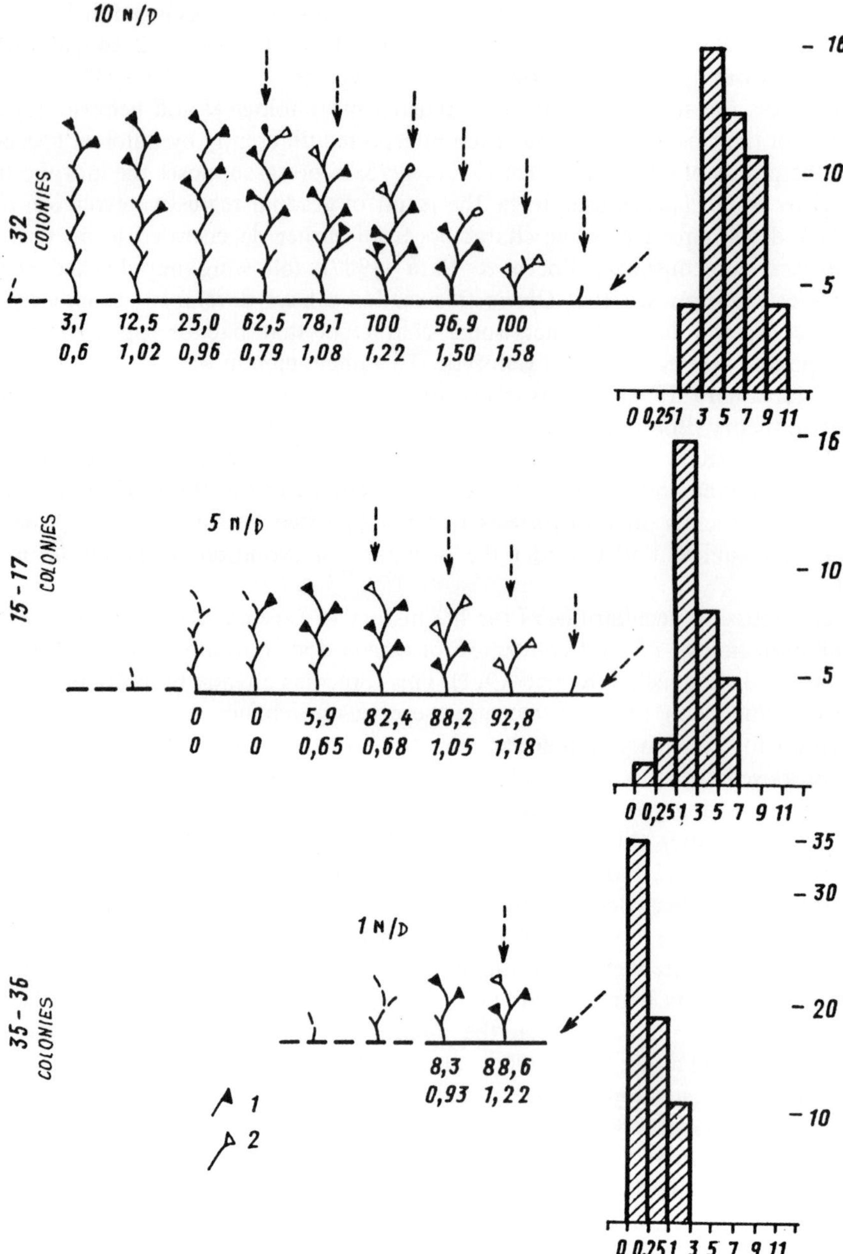

Figure 6 Schematic drawing the colonies *Gonothyraea loveni* under three controlled rations (10, 5 and one nauplii of *Artemia salina* per day) during the stage of "inadequate feeding". Food was been given to proximal (oldest) upright stalks. Left, number of colonies. Right, variation rows of 2 days increase of stolon length under corresponding rations, mm. Centre, drawing of colonies: 1, old hydranths; 2, arising for the last 2 days hydranths; arrows indicate the growing tips. Under each scheme of the colony: top row numbers – % of growing stalks for corresponding position in a colony; lower row – stalks increase in length for 2 days, mm. (After Marfenin 1993a).

Growth and coloniality

Problems of growth in *Obelia* colonies and the effect of amino- and nucleid-acid components were investigated first by Hammett and his co-workers (Hammett & Padis 1935, Hammett & Chapman 1938, Hammett & Barbour 1939, Hammett 1940, 1942, 1943, 1946, 1950, 1951a,b,c,d,e, Hammett & Comee 1940, Hammett & Rivard 1940a,b, Hammett & Hammett 1945). The growth and polymorphic transformations of *Obelia* colonies were investigated by Berrill (1948, 1949, 1950, 1961). Some experimental investigations of variation in colonial "organs" were initiated by Russian investigators (see Korsakova 1950). Some investigations carried out on different hydroids, including the Campanulariidae, dealt with the dynamics of stolon growth and the inter-relationship between these processes and nutrition (Crowell & Rusk 1950, Crowell 1953, 1956, 1957, Crowell et al. 1955, Crowell & Wyttenbach 1957, Crowell & Hartman 1960, Wyttenbach 1968, 1969, Wyttenbach et al. 1973, Suddith 1974). Researches undertaken by Stebbing (1976, 1980, 1981a,b,c, 1982, 1987) addressed growth kinetics and other problems of colonial growth in the Campanulariidae. Ryland & Warner (1986) dealt with the peculiarities of colony growth and building-up of colonies in so-called "modular" organisms, and with the specificity of the polyps' distribution and their size.

Developmental rhythms were investigated first by Brock and his co-workers (Brock et al. 1967, Brock 1970, 1974, 1975a,b,c, 1976, 1979). The morphogenetic line, elaborated by the Moscow Scientific School under Professor L. Beloussov's leadership, might be considered a comparatively new direction for investigations of Obeliinae. Growth pulsations of colonies were originally discovered in White Sea hydroids by Saint-Hilaire (1925, 1930), but first came under intensive consideration in the 1960s (Beloussov 1961, 1968, 1973, 1980, 1991, L. V. Beloussov unpubl., Beloussov et al. 1972, 1984, 1986, Beloussov & Dorfman 1974, Labas et al. 1981, 1989, Badenko et al. 1984, N. I. Kazakova unpubl., and others). Their conclusions concerning the mechanisms and functions of the growth pulsations, with "the co-operative cells agility" being in fact real mechanisms in the formation of *Obelia* colonies, were summarized by Beloussov et al. (1988). Later, five successive levels of the vertical growth of *Obelia* stems were recognized: (a) growth pulsations, (b) annulation, (c) annulation sequences, (d) sequences "annulate + smooth stem sections", (e) stem as a whole with hydranth (L. V. Beloussov unpubl.).

Colonial morphology and interactions between colonial and local processes in colony growth were also investigated by the Moscow School (Kosevich 1983, 1984, 1988, 1989, 1990, 1991, 1992, Marfenin & Kosevich 1984a,b, Kosevich & Marfenin 1986, Kosevich et al. 1990).

Following the investigations of growth processes in *Obelia* colonies, studies on the regulation of the distributive system functioning in colonies were started (Marfenin 1985, 1988, Marfenin & Khomenko 1988, Maljutin & Marfenin 1988, Leontovich & Marfenin 1990). At the same time, the effect of ecological factors on a colony's growth (Burykin 1980, in press) and, somewhat later, the ionic aspects of colonies' morphogenesis came under consideration (Kosevich 1992, I. A. Kosevich et al. unpubl.).

Monographic descriptions of coloniality principles of hydroids, including those of *Obelia*, (Marfenin 1993a,b) might be considered one of the results of the physiologists of the Moscow School. The first of these monographs discusses general principles underlying coloniality as a whole with an emphasis on hydroids. The analysis is provided with some examples including the Obeliinae. Harmony of the proportional

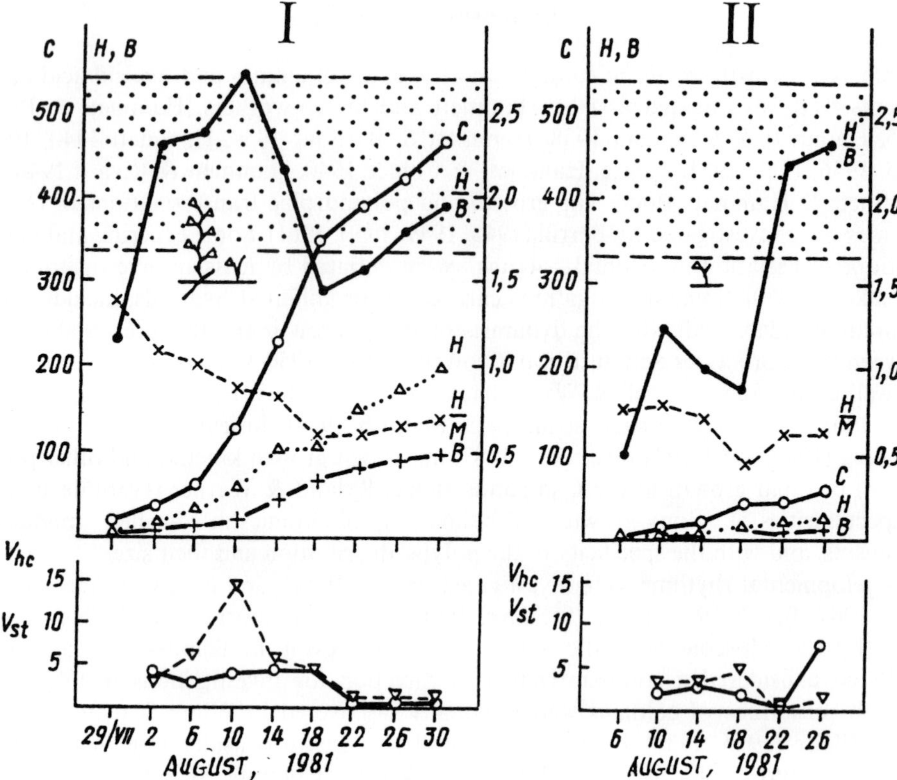

Figure 7 Two variants (I and II) of autoregulation, the proportional relationship between the parts of the colony *Obelia geniculata*. C, the summary hydrophyton length, mm; H, the number of active hydranths; M, the summary length of all upright stalks; Vhc, mean 4 days increase of upright stalks tips, mm; Vst, mean 4 days increase of stolons, mm; dotted area – the normal value of the overcolony structure parameter H/B. Diagrams of the colonies show them at the first day of observation (after Marfenin 1993a).

development of the colonies and some internal factors responsible for the differences in their morphology, or dependence of growth of colonies on food ration and diet, are shown, with *Gonothyraea loveni* as an example (Fig. 6). Self-regulating aspects of the proportional growth of colonies is demonstrated, with *Obelia geniculata* taken as an example (Fig. 7). The second monograph (Marfenin 1993b) concerns only functional morphology of colonies, including the Obeliinae. A chapter devoted to morphology and functions of sympodial colonies is based on Obeliinae material, and another chapter dealing with the relationship between colony growth and hydrodynamics seem to be worth attention. The author's conclusions concern mostly morphological and physiological integration of hydroid colonies. Following his predecessors, Marfenin, with many specific examples, discusses the coloniality phenomenon and thus demonstrates the depth of colonial integration in hydroids. He confirms that a colony is a unitary organism by its origin.

In his monographs Marfenin (1993a,b) postulates two hypotheses:

(1) Phylogenetically it is more likely that coloniality arose independently from vegetative reproduction, as a parallel ecological strategy.

(2) The term "individual" might be applied to a colony's zooids, but from the standpoint of a comparative-anatomical aspect only. A hydroid colony is more likely to be a "multimouth" organism, and the zooids are its organs.

The present writer cannot agree with these hypotheses; one of the points that contradicts them is the observed budding of medusoids on colonies and particularly their capability to release free medusae.

The investigations dealing with the problems of growth, senescence, resorption and restoration of colonies in the Campanulariidae, including the Obeliinae, are of great interest. They were first performed by Huxley & Beer (1923), Crowell (1951, 1953, 1956, 1957, 1960, 1961), and Strehler & Crowell (1961) and later elaborated at a new level by Palinscar (1965a,b, 1969), Toth (1969a,b), and Hughes (1987). The full list of publications on this subject may be found in the review by Gili & Hughes (1995).

Problems of frustulation, and formation of the other dormant and dispersal stages, in some Obeliinae species are discussed in numerous publications (Billard 1904, Reisinger 1937, Berrill 1961, McClary 1961, Cornelius 1982a, 1995, Makarenkova et al. 1985, Makrushin 1986, Chapligina 1992).

Settlement of planulae

Settlement peculiarities in different representatives of the Hydroidea have been described in several publications (Nishihira 1965, Shepherd & Watson 1970, Hughes 1977, Aldrich et al. 1980, Schmidt & Warner 1991, Sommer 1992). More recently, publications have appeared that deal with specificity of settlement of the planulae in different Obeliinae (Chikadze & Railkin 1992, in press, Orlov 1994, 1996, in press, Orlov et al. 1994, Orlov & Marfenin 1995, Dobretzov in press) and with the effect of different factors on this process, the most important being microbial films formed on the natural substrata used by *Obelia* (Railkin 1994, 1995, Chikadze & Railkin in press, Dobretzov in press).

Feeding and digestion

There is little information concerning diets of Obeliinae species (Jennings 1973). Being mostly passive predators (Letunov 1981, Letunov & Stepanjants 1986) *Obelia* species consume fast moving small planktonic organisms. Marfenin (1993b, Marfenin & Khomenko 1989) subdivided hydroids according to their digestion spectrum and placed representatives of the Obeliinae into a group whose food bolus contains tintinnids, rotatorians, harpacticids, cladocerans, ostracods, crustacean nauplii, polychaetaes, molluscan larvae, fish eggs and larvae, and even diatoms. Experimental feeding of some *Obelia* with infusorians appeared to be successful (Marfenin 1993b). More or less similar information on diets of the Obeliinae may be found in the ecological review of Gili & Hughes (1995). Some authors report interesting correlations between size of food and size of nematocysts for some species (Purcell & Mills 1989, Coma 1994).

The digestive processes in Obeliinae colonies are closely related to the specifity of colonial organization and peculiarities of the functioning of the distributive system in these colonies (see above, p. 191). The relationship between structure and feeding behaviour of colonies and separate zooids is discussed in several publications (Warner 1977, Hughes 1980, Ryland, & Warner 1986, Hunter 1989, Gili & Hughes 1995).

The digestive behaviour and the role of factors stimulating digestion processes in *Obelia* colonies were investigated by some Russian specialists (Letunov & Andreev 1980, Letunov 1987, Seravin & Gudkov 1990). Results of studies dealing with intracellular digestion are outlined in publications by Lunger 1963, Jenings 1973, and Lenhoff 1974. For representatives of the Obeliinae this process was investigated by Makarenkova (1988a,b,c, 1989, 1990), who showed the role of digestive cells and mesogloea in digestive and food transport processes.

Other ecological questions

The effect of temperature has been most fully investigated for *O. geniculata*. Ralph (1956) showed that the morphology of the colony and perisarc thickness in *O. geniculata* vary in different parts of its range of distribution and she attributed this variation to differences in water temperature. It is well known that *O. geniculata* does not penetrate into the high Arctic. Its more or less rich populations in the White Sea and Barents Sea might be considered examples of warm-water elements penetrating into Atlantic water (Stepanjants 1989).

Interesting information is known concerning intracellular ice formation in some hydroids and Obeliinae (Ushakov 1925, Jankowsky et al. 1969, Stepanjants 1979). There are data available on the capability of Obeliinae to tolerate reduced salinity (Genzano 1988, Cornelius 1995). A series of publications has demonstrated the effects of heavy metals and other toxic elements on the growth of *Laomedea flexuosa* and *Gonothyraea loveni* colonies (Theede et al. 1979, Stebbing 1980, 1981a,b,c, Stebbing & Santiago-Fandino 1983, Karlsen & Aristarkhov 1986, Karlsen & Marfenin 1987). There is interesting published information concerning the relationship of Obeliinae representatives with other living substrata (Leloup 1931, Zirpolo 1939, 1940).

The important role of different species of the Obeliinae in epibiotic communities is undoubted (Turpaeva 1967, 1969, 1982, Gorin 1975, Zevina et al. 1975, Chapligina 1980, 1993, Oshurkov & Seravin 1983, Oshurkov 1986, 1993, Maslennikov et al.1987, Khomenko 1989, and others). As typical representatives of epibiotic fauna, some of these species may dominate in these communities. Settlements of some species (e.g. *O. longissima*) on piles and on the bottoms of ships are suggested as one of the most probable ways of extending the range of distribution. According to Chapligina (1992), this was a possible means for the initial introduction of *Laomedea flexuosa* and *L. calceolifera* into the Sea of Japan.

As to the biomass of some Obeliinae, in the White Sea *Laminaria saccharina*-biocenosis, for example, *Obelia longissima* has a biomass of about $700\,g\,m^{-2}$ (Letunov & Stepanjants 1986); in the Barents Sea, on navigation buoys, this species reaches a biomass of $40\,g\,m^{-2}$ (Kusnetzova & Zevina 1967, 1978); in the White Sea, on the mussel artificial substratum, the biomass of *O. longissima* colonies is as high as $5\,g\,m^{-2}$ (Kunin in press); and, according to long-term investigations, the biomass of the species arbitrarily identified as *O. longissima* in the near-shore area off Kamchatka (Oshurkov 1993) may reach more than $2\,kg\,m^{-2}$ (Khomenko 1989, Oshurkov 1993).

All this allows us to suppose that colonies of this and other allied species, cultivated on artificial substrata, might be used as a raw material for the creation of some valuable materials: for example, chitin from the colonies' skeleton, prostaglandins, polyunsaturated fatty acids, etc. (Letunov & Stepanjants 1986).

Data show that the peak of liberation of *O. longissima* medusae into the plankton may effect the abundance of White Sea herring, because poison from the nematocysts of the medusae appears to be detrimental to herring larvae. Where artificial breeding grounds for herring are used they should not be sited near collectors for mussel aquaculture. Under White Sea conditions, abundant settlements of *Obelia*, which always accompany artificial mussel settlements, might provoke peaks of abundance of the release of medusae into the plankton, a process which takes place just at the time of hatching of the larvae of White Sea herring. During years when abundant medusae are released they may adversely affect the abundance of the herring population (Ivanchenko 1997, in press).

In Marfenin's monographs, attention is drawn to the role of hydroids including the Obeliinae in ecosystem energy balance. Marfenin refers to some publications (Turpaeva 1967, 1969, 1982, Partaly 1974) and gives a list of animals that consume hydroids, for example, Nudibranchia and Caprellida are permanently present on *O. longissima* colonies in the White Sea. There are some publications that describe the relationships between Nudibranchia and hydroids (Cattaneo-Vietty & Boero 1989), but there is no documented information concerning the role of representatives of the Obeliinae in general energy flow. There is some information on the role of *Obelia* in regulating the production in some species populations (Sullivan & Banzon 1990) and in Marfenin's opinion this role is much more significant than one might suppose.

Horizontal distribution

The problem of the distribution of *Obelia* and related genera in the ocean is of special interest. The genus *Obelia* may be considered cosmopolitan, since species belonging to the genus appear to be abundant from the Arctic to the Antarctic (Calder 1970, Stepanjants 1979, 1980, Cornelius 1982a, in press, Antsulevitch 1987a, Panteleeva in press). But, as was discussed earlier, none of the known individual species of *Obelia*, *Gonothyraea* and *Laomedea* can be considered cosmopolitan; rather they are widespread species but with different ecological natures, which prevents their distribution from being worldwide, as shown below.

O. longissima is a cold water species that is widespread in both northern and southern hemispheres, including the Black Sea (Beloussova 1991, Terentjev in press), but is absent in tropical areas. It may therefore be designated a bipolar species (Stepanjants, 1979, 1980, Stepanjants et al. 1996, 1997) (Fig. 8A).

O. plana, if valid, has a distribution range close to that of *O. longissima*. It was described from the North Atlantic, and has been found in the North and South Atlantic, Mediterranean Sea and the Sea of Japan, but is unknown from the Arctic (Kubota 1981, Morri 1984, Hirohito 1995).

O. dichotoma has a range of distribution similar to that of *O. longissima*, but it is more a warm-water species, being absent in the Arctic, but found in tropical waters (Rao et al. 1972, Stepanjants 1980).

O. geniculata is also a widespread species (Ralph 1957, Blanco & Morris 1977, Cornelius 1995), but it is absent in the major Arctic and far-east seas (Stepanjants 1980).

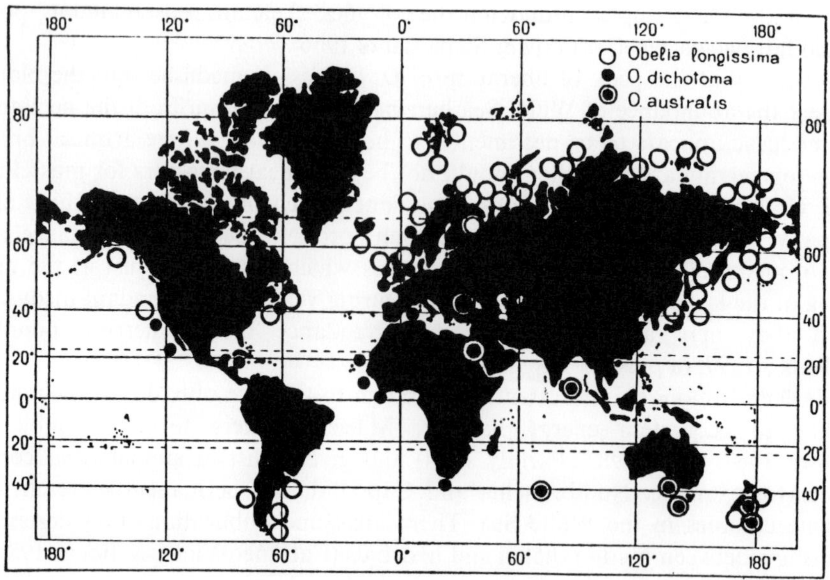

A

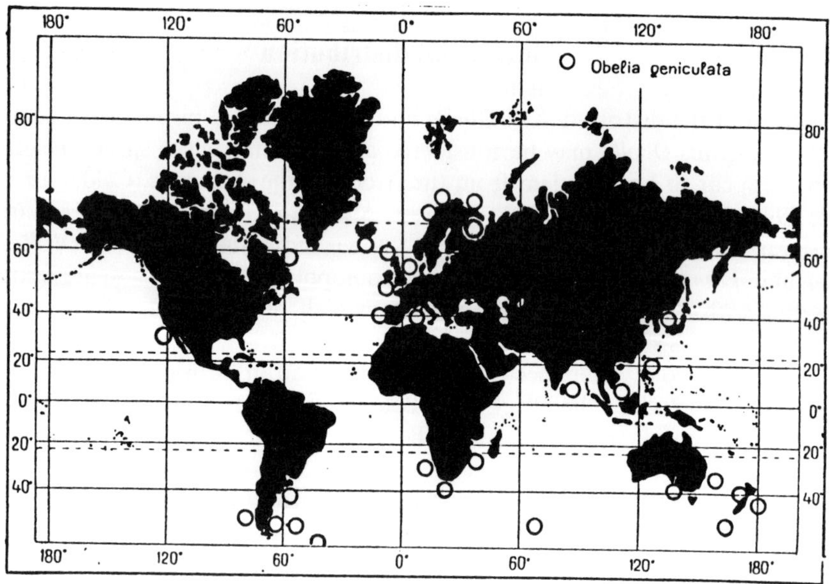

B

Figure 8 The maps of geographical distribution *Obelia longissima, O. dichotoma, O. australis* (A) and *O. geniculata* (B). (After Stepanjants 1980).

The fact that it is widely distributed both in the White and Barents Seas and found in the tidal zone of the Sea of Japan might be considered an example of penetration of warm-water fauna into the Arctic (Stepanjants 1980) (Fig. 8B).

O. bidentata has a distribution range close to that of *O. geniculata*; it also might be classified as a warm-water species (Cornelius 1995).

O. oxydentata, *O. australis*, and *O. chinensis* are more warm-water species than the foregoing: they are found in subtropical and even in tropical waters (Ralph 1957, Hirohito 1995, Gravier-Bonnet in press).

Gonothyraea species are widely distributed; their range includes the Arctic, but they are not known in the tropics, and up to now have not been found in far-east seas, except for *Gonothyraea* (*Laomedea*?) *inornata* (Nutting 1901), described for the Alaska near-shore area and probably living in the Kamchatka near-shore area (see p. 181). Chapligina (1992) was the first to record the spread of *Laomedia flexuosa* and *L. calceolifera* into the Sea of Japan; it is possible to predict the import of further representatives of *Gonothyraea* and *Laomedea* into these waters on shipping.

Laomedea species are widely distributed through the Northern Atlantic, southwards to the Mediterranean and the Azores, and *L. calceolifera* is also known from South African waters (Millard 1975). They penetrate into the White and Barents Seas to the north and, as noted above, have recently been found in the Sea of Japan (Antsulevitch 1987b). None of the listed species could be classified as cosmopolitan as they are absent from the tropics and from the high Arctic. Being warm-water species *Laomedea* representatives may be found in the tidal and subtidal zones of Arctic seas and may penetrate to the North Pacific.

Hartlaubella gelatinosa, judging from the distribution characteristics given by Cornelius (1995), may be considered a bipolar species.

Thus, following the classical concept, there are no individual species among the Obeliinae that might conform to the term "cosmopolitan", but among the genera *Obelia* certainly appears to be cosmopolitan.

The problem first touched upon by Cornelius (1995) and later posed by him more plainly (Cornelius, in press) in the question – "Cosmopolitan morphospecies or allopatric siblings?" – is of great interest, and this approach to species problems and biogeographic specificities at the species and generic levels, with hydroids used as an example, appears to be rather perceptive although very untraditional. In addition to references to this problem presented by Cornelius (Sole-Cava & Thorpe 1989, 1991, Thorpe et al. 1992) it would be useful to cite some other publications (Grebel'nyi 1988, 1989).

Vertical distribution

Analysis of available information indicates that representatives of the Campanulariidae including the Obeliinae are found preferentially in the littoral and sublittoral (and not deeper than 200 m) in all oceans (Mori 1978, Stepanjants 1979, 1989, Boero & Fresi 1986, Zamponi 1987, Zamponi & Genzano 1988, 1990a,b, Piraino & Morri 1990, El Beshbeeshy 1991, Cornelius 1995, Hirohito 1995). However some species have been found in deeper water: the expeditions "BIOGAS" and "BALGIM" in the Bay of Biscay and Gibraltar recorded *Obelia bidentata* at a depth of 681 m; *O. dichotoma* at 150–

521 m (2400 m?); an unidentified *Obelia* species at 521 m; and *O.* cf. *plicata* (at ? m) (Vervoort 1985, Ramil & Vervoort 1992). According to El Beshbeeshy (1991), in Patagonian waters *O. dichotoma* and *O. longissima* were found in the littoral between 300 m and 510 m and *O. geniculata* between 1040–1200 m (although the latter seems doubtful).

Luminescence

The luminescence exhibited by members of the Obeliinae is one of their most interesting features. The process of luminescence in polyps and medusae is still only partly understood (Morin et al. 1968, Morin & Reynolds 1969, 1970, 1974, Morin & Cooke 1971a,b,c, Morin & Hastings 1971a,b, Morin 1974). This character is known only for some *Obelia* species (Campbell 1974a,b, 1985, Letunov & Vysotski 1988a,b, Gitel'son et al. 1990, Bondar et al. 1992, Vysotski in press, E. S. Vysotski et al. unpubl.).

The nature of the luminescence in these organisms is now known; it is provided by a specific Ca^{++} dependent photoprotein, the molecules of which are contained in specific cells – the photocytes (Morin & Reynolds 1974, Freeman & Ridgway 1986). Different *Obelia* species have different zones of photocyte concentration in the tissues of their polyps or medusae (Vishniakov 1989a,b, in press). The luminescence reaction is associated with the interaction of photoprotein and Ca++ ions incorporated in the free state into cells, or on the outside. The photoprotein responds to the lowest Ca++ concentrations by breaking the oxygen bridge in its molecule, thus releasing the free oxygen, which causes the luminescence.

The capability to luminescence, associated with Ca-dependent proteins, is being studied most intensively (E. S. Vysotski et al. unpubl.), because of its potential for use in medicine and pharmacology. A Ca-dependent photoprotein separated from *Obelia* species, and therefore named obelin, appears to be useful, even in trace amounts, for the early diagnosis of malignant diseases; microinjection of obelin into diseased cells of an organism causes them to luminesce. Investigation of the morphological and biochemical aspects of luminescence in *Obelia* species will establish the principles that may allow utilization of these marine invertebrates as a raw material for the creation of the valuable medicine to be developed.

From the foregoing summary, the reasons for the interest taken by biologists in different lines of investigation of this marine invertebrate group become clear. Obeliinae have been used as a model object in different biological laboratories, while university courses in some countries involve studying these animals (Cornelius in press).

Acknowledgements

I am grateful to the editors of *Oceanography and Marine Biology: an Annual Review* who invited me to write this review. I wish to acknowledge my colleagues who kindly commented and corrected this review in spite of other work, especially Paul Cornelius and Carina Ostman. The brilliant bibliography of Leptolida by Wim Vervoort (1995), and the ecological review by J. M. Gili and R. G. Hughes (1995) distinctly lightened my work on this review, and I thank them for it.

References

Aizenshtadt, T. V. 1986. The functional role of hydrorhyza during the sexual reproduction of marine hydroid polyps of the genus *Obelia*. *Ontogenez* **17**(1), 20–6. (In Russian with English summary).

Aizenshtadt, T. V. & Polteva, D. G. 1981. Origin of germ cells and early stages of oogenesis in a marine hydroid polyp *Obelia*. *Ontogenez* **12**(3), 243–50. (In Russian with English summary).

Aizenshtadt, T. V. & Polteva, D. G. 1982. Vitellogenesis in hydroid polyps *Obelia flexuosa* and *Obelia loveni*. *Ontogenez* **13**(1), 28–33. (In Russian with English summary).

Aizenshtadt, T. V. & Polteva, D. G. 1986. The functional role of hydrorhyza during the sexual reproduction of marine hydroid polyps of the genus *Obelia*. An ultrastructural study. *Ontogenez* **17**(1), 14–21. (In Russian with English summary).

Aizenshtadt, T. V. & Polteva, D. G. 1989. The sea hydropolyp hydrorhyza, genus *Obelia*. In *The fundamental investigations of the recent Porifera and Cnidaria*, V. M. Koltun et al. (eds). USSR: Academy of Science, 24–5. (In Russian with English summary).

Aizu, S. 1967. Transformation of ectoderm to entoderm in *Obelia*. *Science Reports of the Tohoku University* **4**, 33(3–4), 375–82.

Aldrich, I. C., Crowe, W., Fitzgerald, M., Murphy, M., McManus, C., Magennis, B. & Murphy, D. 1980. Analysis of environmental gradients and patchiness in the distribution of the epiphytic marine hydroid *Clava squamata*. *Marine Ecology Progress Series* **2**(4), 293–301.

Anonymous, 1930. Observation on living chromosomes in *Obelia*. *Nature* **125**, 67 only.

Antsulevitch, A. E. 1987a. Hydroids from the shelf waters of Kurile Islands. *Zoological Institute, Academy of Science USSR (Leningrad)*. (In Russian with English summary).

Antsulevitch, A. E. 1987b. Hydroids of the genus *Laomedea* Lamouroux, 1812 in the fauna of the USSR. *Vestnik Leningradskoga Universiteta, Biologiya* **4**(24), 11–18. (In Russian).

Babic, K. 1912. Dimorphismus der gonangien bei *Laomedea angulata* Hincks. *Zoologischer Anzeiger* **39**(13–14), 457–60.

Badenko, L. A., Beloussov, L. V., Zelinchenok, R. I., Labas, Y. A. & Letunov, V. N. 1984. Analysis of hydroid *Obelia longissima* growth pulse. *Biofizika* **29**(6), 1014–17. (In Russian with English summary).

Beadle, L. C. & Booth, F. A. 1938. The reorganization of tissue masses of *Cordylophora lacustris* and the effect of oral cone grafts, with supplementary observations on *Obelia gelatinosa*. *Journal of Experimental Biology* **15**(3), 303–26.

Beloussov, L. V. 1961. Vital observations on cell displacement in the hydroid polyp *Obelia flexuosa*. *Doklady Akademii Nauk SSSR* **133**(6), 1490–93. (In Russian).

Beloussov, L. V. 1968. Interpretation of real order of morphogenesis in hydroid polyp *Obelia loveni*. *Biologicheskie Nauki (Moscow)* **7**(55), 21–7. (In Russian).

Beloussov, L. V. 1973. Growth and morphogenesis of some marine Hydrozoa according to histological data and time-lapse studies. In *Recent Trends in Research in Coelenterate Biology*. T. Tokioka & S. Nishimura (eds). *Publications of Seto Marine Biological Laboratory* **20**, 315–66.

Beloussov, L. V. 1980. Morphogenesis and species form of hydroid polyps. In *The theoretical and practical importance of the coelenterates*, D. V. Naumov & S. D. Stepanjants (eds). Leningrad: Zoological Institute of Academy of Science. 10–15 (In Russian).

Beloussov, L. V. 1991. Basic morphogenetic processes in Hydrozoa and their evolutionary implications: an exercise in rational taxonomy. In *Coelenterate Biology: recent research on Cnidaria and Ctenophora*, R. B. Williams et al. (eds). Dordrecht: Kluwer Academic, 61–8.

Beloussov, L. V., Badenko, L. A., Kachurin, A. L. & Kurilo, F. 1972. Cell movements in morphogenesis of hydroid polyps. *Journal of Embryology and Experimental Morphology* **27**(2), 317–33.

Beloussov, L. V. & Dorfman, Y. G. 1974. Mechanisms of growth and morphogenesis in hydroid polyps. *Ontogenez* **5**(5), 437–45. (In Russian).

Beloussov, L. V., Labas, Y. A. & Badenko, L. A. 1984. Growth pulsation and rudiments shape in hydroid polyps. *Zhurnal Obshchei Biologii* **45**(6), 796–805. (In Russian with English summary).

Beloussov, L. V., Labas, Y. A., Badenko, L. A., Kazakova, N. I. & Zinchenko, V. P. 1986. Cytophysiology of the pulsatorial morphogenesis in hydroid polyps. *Abstracts of the 10th Soviet Embryologists Conference*, 140 only. (In Russian).

Beloussov, L. V., Labas, Y. A., Kazakova, N. I. & Badenko, L. A. 1988. Growth pulsations in hydroid polyps. In *Porifera and Cnidaria. Modern and perspective investigations.*

V. M. Koltun & S. D. Stepanjants (eds). Leningrad: Zoological Institute of Academy of Science, 47–57. (In Russian with English summary).

Beloussova, N. P. 1991. The first finding of *Obelia* medusae (Hydrozoa, Campanulariidae) in the Black Sea. *Zoolicheskii Zhurnal* **70**(3), 133–4. (In Russian with English summary).

Berrill, N. J. 1948. A new method of reproduction in *Obelia. Biological Bulletin* **95**(1), 94–9.

Berrill, N. J. 1949. The polymorphic transformation of *Obelia. Quarterly Journal of Microscopical Science* **90**(3), 235–64.

Berrill, N. J. 1950. Growth and form in calyptoblastic hydroids. II. Polymorphism within the Campanulariidae. *Journal of Morphology* **87**, 1–26.

Berrill, N. J. 1961. *Growth, development and pattern.* San Francisco: Freeman.

Billard, A. 1904. Contribution a l'étude des Hydroides (multiplication, regeneration, greffes, variations). *Annales des Sciences Naturelles, Zoologie et Biologie Animale* **20**, 1–251.

Blanco, O. M. & Morris, M. R. 1977. Nueva cita para *Obelia geniculata* (L.) (Hydroida – Campanulariidae). *Neotropica* **23**(69), 91–3.

Bode, H. R., Dunne, J., Heimfeld, S., Huang, L., Javois, L., Koizumi, O., Westerfield, J. & Yaross, M. 1986. Transdifferentiation occurs continuously in adult *Hydra*. In *Commitment and instability in cell differentiation. Current Topics in Developmental Biology* **20**, 257–80.

Boero, F. & Bouillon, J. 1987. Inconsistent evolution and paedomorphosis among the hydroids and medusae of Athecatae/Anthomedusae and the Thacatae/Leptomedusae (Cnidaria, Hydrozoa). In *Modern trends in the systematics, ecology and evolution of Hydroids and Hydromedusae*, J. Bouillon et al. (eds). London: Clarendon Press, 229–50.

Boero, F. & Bouillon, J. 1989. An evolutionary interpretation of anomalous medusoid stages in the life cycles of some Leptomedusae (Cnidaria). In *Reproduction, genetics and distributions of marine organisms. Proceedings of the 23rd European Marine Biology Symposium*, J. S. Ryland & P. A. Tyler (eds). Fredensborg, Denmark: Olsen & Olsen, 37–41.

Boero, F. & Fresi, E. 1986. Zonation and evolution of a rocky bottom hydroid community. *P. S. Z. N. I.: Marine Ecology* **7**(2), 123–50.

Boero, F. & Sara, M. 1987. Motile sexual stages and evolution of Leptomedusae (Cnidaria). *Bollettino di Zoologia* **54**(2), 131–9.

Boero, F., Bouillon, J. & Piraino, S. 1996. Classification and phylogeny in the Hydroidomedusae (Hydrozoa, Cnidaria). In *Advances in hydrozoan biology*. S. Piraino et al. (eds). *Scientia Marina* **60**(1), 17–33.

Bondar, V. S., Trofimov, K. P. & Vysotski, E. S. 1992. Physico-chemical properties of the hydroid polyp *Obelia longissima* photoprotein. *Biokhimiya* **57**(10), 1481–90. (In Russian with English summary).

Bouillon, J. 1985. Essai de classification des Hydropolypes – Hydromeduses (Hydrozoa-Cnidaria). *Indo-Malayan Zoology* **2**(1), 29–243.

Bouillon, J. & Werner, B. 1965. Production of medusae buds by the polyps of *Rathkea octopunctata* (M. Sars) (Hydroida Athecata). *Helgoländer Wissenschafliche Meeresuntersuchungen* **12**(1–2), 137–48.

Bozhenova, O. V. 1988. Present views on the classification of the nematocysts of Cnidaria. In *Porifera and Cnidaria. Modern and perspective investigations*, V. M. Koltun & S. D. Stepanjants (eds). Leningrad: Zoological Institute of Academy of Science, 57–71. (In Russian with English summary).

Bozhenova, O. V., Grebel'nyi, S. D. & Stepanjants, S. D. 1988. The possible ways of the Cnidaria nematocysts evolution. In *Porifera and Cnidaria. Modern and perspective investigations*, V. M. Koltun & S. D. Stepanjants (eds). Leningrad: Zoological Institute of Academy of Science, 71–4. (In Russian with English summary).

Brien, P. 1965. L'embryogenese et la senescence de l'hydre d'eau douce. *Memoires de l'Academie Royale de Belgique, Science* **36**(1), 1–113.

Brinckmann, A. 1965. The biology and development of *Rhysia automnalis* gen. n. sp.n. (Anthomedusae/Athecatae, Rhysiidae fam.n.). *Canadian Journal of Zoology* **43**, 941–52.

Brock, M. A. 1970. Ultrastructural studies on the life cycle of a short-lived metazoan, *Campanularia flexuosa.* II. Structure of the old adult. *Journal of Ultrastructure Research* **32**(1–2), 118–41.

Brock, M. A. 1974. Growth, developmental and longevity rhythms in *Campanularia flexuosa*. In *The developmental biology of the Cnidaria*. R. L. Miller & C. R. Wyttenbach (eds). *American Zoologist* **14**(2), 757–71.

Brock, M. A. 1975a. Circannual rhythms. 1. Free-running rhythms in growth and development of the marine cnidarian *Campanularia flexuosa*. *Comparative Biochemistry and Physiology* (A) **51**(2), 377–83.

Brock, M. A. 1975b. Circannual rhythms. 2. Temperature-compensated free-running rhythms in growth and development of marine cnidarian *Campanularia flexuosa*. *Comparative Biochemistry and Physiology* (A) **51**(2), 385–90.

Brock, M. A. 1975c. Circannual rhythms. 3. Rhythmicity in the longevity of hydranths of the marine cnidarian, *Campanularia flexuosa*. *Comparative Biochemistry and Physiology* (A) **51**(2), 391–8.

Brock, M. A. 1976. Free running rhythmicity in the life spans of hydranths of the marine cnidarian, *Campanularia flexuosa*. In *VIth International Interdisciplinary Cycle Research Symposium*. *Journal of Interdisciplinary Cycle Research* **7**(4), 269–78.

Brock, M. A. 1979. Differential sensitivity to temperature steps in the circannual rhythm of hydranth longevity in the marine cnidarian, *Campanularia flexuosa*. *Comparative Biochemistry and Physiology* (A) **64**(3), 381–90.

Brock, M. A., Strehler, B. I. & Brandes, D. 1967. Ultrastructural studies on the life of a short-lived metazoan, *Campanularia flexuosa*. I. Structure of the young adult. *Journal of Ultrastructure Research* **21**(3–4), 281–312.

Burykin, Y. B. 1980. Regulatory role of some ecological factors in the processes of growth and integration of colonial hydroids. In *The theoretical and practical importance of the coelenterates*, D. V. Naumov & S. D. Stepanjants (eds). Leningrad: Zoological Institute of Academy of Science, 16–19. (In Russian).

Burykin, Y. B. in press. Role in pulsations in metamorphosis of *Gonothyraea loveni* (Allm., 1859) planula. In Obelia (*Cnidaria, Medusozoa, Hydrozoa*). *Phenomenon. Aspects of investigations. Perspectives of employment. Zoosystematica Rossica.* (Suppl.). S. D. Stepanjants (ed.).

Cairns, S. D., Calder, D. R., Brinckmann-Voss, A., Castro, C. B., Pugh, P. R., Cutress, C. E., Laap, W. C., Fautin, D. G., Larson, R. J., Harbison, G. R., Arai M. N. & Opresko, D. M. 1991. Common and scientific names of aquatic invertebrates from the United States and Canada: Cnidaria and Ctenophora. *American Fisheries Society, Special Publication* No. 22, 1–75.

Calder, D. R. 1970. Thecate hydroids from the shelf waters of northern Canada. *Journal of the Fisheries Research Board of Canada* **27**(9), 1501–47.

Calder, D. R. 1991. Shallow-water hydroids of Bermuda. The Thecate, exclusive of Plumularioidea. *Life Science Contributions Royal Ontario Museum* **154**, 1–140.

Campbell, A. K. 1974a. Studies on Obelin Ca2 + -activated bioluminescent protein from hydroid *Obelia geniculata*. *Biochemical Society Transactions* **2**(5), 995–6.

Campbell, A. K. 1974b. Extraction, partial purification and properties of Obelin, calcium-activated luminescent protein from the hydroid *Obelia geniculata*. *Biochemical Journal* **143**(2), 411–18.

Campbell, A. K. 1985. *Intracellular calcium*. Chichester: Wiley.

Carlgren, O. 1940. A contribution to the knowledge of the structure and distribution of the cnidae in the Anthozoa. *Lund Universitets Årsskrift Avdelningen 2, N.F. 36*, 1–62.

Cattaneo-Vietty, R. & Boero, F. 1989. Relationships between eolid (Mollusca, Nudibranchia) radular morphology and their cnidarian prey. *Bollettino Malacologico* **24**, 215–22.

Chapligina, S. F. 1980. Hydroidea in aggregations of the north-western part of the Sea of Japan. In *Ecology of fouling in the north-western part of the Pacific Ocean*, V. Kudryashov (ed.). *Sbornik Rabot Instituta Biologiya Morya* **20**, 56–71. (In Russian).

Chapligina, S. F. 1992. Introduction of two hydroid species *Laomedea flexuosa* and *L. calceolifera* (Cnidaria, Hydroidea, Campanulariidae) into the Sea of Japan. *Zoologicheskii Zhurnal* **71**(9), 5–10. (In Russian with English summary).

Chapligina, S. F. 1993. Hydroids in the fouling of mariculture installations in Peter the Great Bay of the Sea of Japan. *Biologiya Morya* **2**, 29–36. (In Russian with English summary).

Chikadze, S. Z. & Railkin, A. I. 1992. Glucose inhibits settlement, attachment and metamorphosis of Hydrozoa (*Obelia loveni*). *Vestnik Sankt Petersburgskogo Universiteta Seriya Biologiya* **3**(17), 29–32. (In Russian with English summary).

Chikadze, S. Z. & Railkin, A. I. in press. Interactions between larvae *Obelia loveni* and macroalgae are mediated by bacterial-algal films. In Obelia (*Cnidaria, Medusozoa, Hydrozoa*). *Phenomenon. Aspects of investigations. Perspectives of employment. Zoosystematica Rossica.* (*Suppl.*). S. D. Stephanjans (ed.).

Coma, R. 1994. *Evaluacion del balance energetico de dos especies de cnidarios bentonicos marinos*. PhD thesis, University of Barcelona.

Cornelius, P. F. S. 1975. The hydroid species of *Obelia* (Coelenterata, Hydrozoa: Campanulariidae) with notes on the medusa stage. *Bulletin of the British Museum (Natural History), Zoology* **28**(6), 249–53.

Cornelius, P. F. S. 1977a. The linking of polyp and medusa stages in *Obelia* and other coelenterates. *Biological Journal of the Linnean Society* **9**, 45–57.

Cornelius, P. F. S. 1977b. The gradual discovery of the relation between medusa and polyp in *Obelia* and other coelenterates. *Journal of the Quekett Microscopical Club* **33**(3), 172–4.

Cornelius, P. F. S. 1982a. Hydroids and medusae of the family Campanulariidae recorded from the eastern North Atlantic, with a world synopsis of genera. *Bulletin of the British Museum (Natural History), Zoology* **42**(2), 3–148.

Cornelius, P. F. S. 1982b. Rediscovery in Britain of the hydroid *Laomedea angulata*. A request for records. *Porcupine Newsletter* **2**(2), 113–17.

Cornelius, P. F. S. 1982c. Further notes on genus-group names in the hydroid family Campanulariidae Johnston, 1836. Z.N. (S) 2326. *Bulletin of Zoological Nomenclature* **39**, 222–5.

Cornelius, P. F. S. 1985. Results of the Porcupine Newsletter *Laomedea angulata* enquiry. *Porcupine Newsletter* **3**(3), 88 only.

Cornelius, P. F. S. 1987. Taxonomic characters from the hydranths of thecate hydroids. In *Modern trends in the systematics, ecology and evolution of hydroids and hydromedusae*. J. Bouillon et al. (eds). Oxford: Clarendon Press, 29–42.

Cornelius, P. F. S. 1990a. European *Obelia* (Cnidaria, Hydroida). *Journal of Natural History* **24**(3), 535–78.

Cornelius, P. F. S. 1990b. Evolution in leptolid life-cycles. (Cnidaria: Hydroida). *Journal of Natural History* **24**(3), 579–94.

Cornelius, P. F. S. 1995. North-West European thecate hydroids and their medusae. Parts 1, 2. Synopses of the British Fauna (New Series). R. S. K. Barnes & J. H. Crothers (eds) No. 50, Part 1 1–346, Part 2 1–389.

Cornelius, P. F. S. in press. A changing taxonomic paradigm: studies on *Obelia* and some other Campanulariidae (Cnidaria: Hydrozoa). In Obelia *(Cnidaria, Medusozoa, Hydrozoa). Phenomenon. Aspects of investigations. Perspectives of employment. Zoosystematica Rossica.* (Suppl.) S. D. Stepanjants (ed.).

Cornelius, P. F. S. & Ostman, C. 1987. Redescription of *Laomedea exigua* M. Sars, a hydroid new to Scandinavia, with comments on its nematocysts, life cycle and feeding movements. *Zoologica Scripta* **16**(1), 1–8.

Crowell, S. 1951. Additions to our knowledge of *Obelia* and *Campanularia*. *Proceedings of the Indiana Academy of Science* **60**, 309 only.

Crowell, S. 1953. The regression-replacement cycle of hydranths of *Obelia* and Campanulariidae. *Physiological Zoology* **26**, 319–27.

Crowell, S. 1956. Responses of apical growth zones of hydroids to differences in age, temperature, and nutrition. *Anatomical Record* **125**, 620–21.

Crowell, S. 1957. Differential responses of growth zones to nutritive level, age and temperature in the colonial hydroid *Campanularia*. *Journal of Experimental Zoology* **134**(1), 63–90.

Crowell, S. 1960. Non-regulative differentiation in the thecate hydroid *Campanularia*. *Anatomical Record* **138**, 341–2.

Crowell, S. 1961. Developmental problems in *Campanularia*. In *The biology of* Hydra *and of some other coelenterates*, H. M. Lenhoff & W. F. Loomis (eds). Coral Cables, Florida: University of Miami Press, 297–316.

Crowell, S. & Hartman, W. 1960. Reorganization capacities of dissociated tissues of *Campanularia flexuosa*. *Anatomical Record* **138**(3), 342 only.

Crowell, S. & Rusk, M. 1950. Growth of *Campanularia* colonies. *Biological Bulletin* **99**(2), 357 only.

Crowell, S. & Wyttenbach, C. 1957. Factors affecting terminal growth in the hydroid *Campanularia*. *Biological Bulletin* **113**(2), 233–44.

Crowell, S., Wyttenbach, C. & Wilt, F. 1955. Effects of the level of nutrition on the growth pattern in *Campanularia* colonies. *Biological Bulletin* **109**(3), 345 only.

Cutress, C. E. 1955. An interpretation of the structure and distribution of cnidae of Anthozoa. *Systematic Zoology* **4**, 120–37.

David, C. N. 1983. Stem cell proliferation and differentiation in *Hydra*. In *Stemcells. Their identification and characterization*, C. S. Potten (ed.). Edinburgh: Churchill Livingstone, 12–27.

Dementjev, A. E. 1989a. The embryogeny variability of *Obelia* and its connection with the colonies inhabiting. In *The fundamental investigations of the recent Porifera and Coelenterata*, V. M. Koltun et al. (eds). Leningrad: Zoological Institute of Russian Academy of Sciences, 39–41. (In Russian).

Dementjev, A. E. 1989b. Embryology of *Obelia loveni* (Peculiarities of the Black Sea cleavage). *Vestnik Leningradskogo Universiteta Biologiya* **3**(4), 90–2. (In Russian).

Dementjev, A. E. & Minichev, Y. S. 1985. Peculiarities of cleavage in a hydroid polyp *Obelia loveni* (Hydrozoa, Leptolida). *Zoologicheskii Zhurnal* **64**(8), 1133–9. (In Russian with English summary).

Dobretzov, S. V, in press. Macroalgae and microbial film determine substrate selection by the *Gonothyraea loveni* planulae. In Obelia (*Cnidaria, Medusozoa, Hydrozoa*). *Phenomenon. Aspects of investigations. Perspectives of employment. Zoosystematica Rossica.* (Suppl.). S. D. Stepanjants (ed.).

Donakov, V. V. 1985. Sistema stvolovykh interstitialnykh kletok i gametogenesu gidroidnogo polipa *Obelia loveni.* In *Organizm v onto- i filogeneze.* L., 2–7. Ms deposited in Institute of Scientific-Technical Information, No. 463–85 Dep.

Donakov, V. V. 1987. Osennie perestroiki v sisteme stvolovykh interstitialnykh kletok gidroidnogo polipa *Obelia loveni*, Hydrozoa, Leptolida, Thecaphora. In *Problemi reproduktsii, differentsirovki i funktsionirovaniya kletok.* L., 12–20. Ms deposited in Institute of Scientific-Technical Information, No. 3107–V87).

Donakov, V. V. 1988. Spermatogenesis in a colonial hypogenetic hydroid polyp *Obelia loveni* (Hydrozoa, Leptolida). *Ontogenez.* **19**(5), 513–22. (In Russian with English summary).

Donakov, V. V. 1989. Ultrastructural aspects of spermatogenesis in hydroid *Obelia loveni* (Hydrozoa, Thecaphora). In *The fundamental investigations of the recent Porifera and Coelenterata.* V. M. Koltun et al. (eds). Leningrad: Zoological Institute, 41–3. (In Russian).

Donakov, V. V. & Makarenkova, E. P. 1989. The cells reproduction and differentiation in hydropolyp *Obelia loveni* (Hydrozoa, Campanulariidae). In *The fundamental investigations of the recent Porifera and Coelenterata*, V. M. Koltun et al. (eds). Leningrad: Zoological Institute, 43–6. (In Russian).

Donakov, V. V., Polteva D. G. & Genikovich, G. in press. Hydranth development in Campanulariidae hydroids: origin, resorption, regeneration. In Obelia (*Cnidaria, Medusozoa, Hydrozoa*). *Phenomenon. Aspects of investigations. Perspectives of employment. Zoosystematica Rossica.* (Suppl.). S. D. Stepanjants (ed.).

Dondua, A. K. & Dondua, O. A. 1975. Effect of actinomycin D on transformation of planulae *Obelia loveni* (Allman). *Ontogenez* **6**(6), 627–30. (In Russian).

El Beshbeeshy, M. 1991. *Systematische, morphologische und zoogeographysche untersuchungen an den Thecaten Hydroiden des Patagonischen Schelfs.* Dissertation, Universität Hamburg.

Elmhirst, R. 1925. Lunar periodicity in *Obelia. Nature* **116**, 358–9.

Faulkner, G. H. 1929. The early prophases of the first oocyte division as seen in life, in *Obelia geniculata. Quarterly Journal of Microscopical Science* **73**(2), 225–42.

Fraipont, P. J. 1880. Recherches sur l'organisation histologique et le development de la *Campanularia angulata. Archives de Zoologie Expérimentale et Générale* **8**, 433–66.

Freeman, G. & Ridgway, F. B. 1986. Endogenous photoproteins, calcium channels and calcium transients during in hydrozoans metamorphosis. *Wilhelm Roux's Archives of Developmental Biology* **196**(1), 30–50.

Genzano, G. N. 1988. Hallazgo de *Gonothyraea loveni* (Allman, 1859) (Hydrozoa: Campanulariidae) en el estuario del Rio de la Plata. *Spheniscus* **7**, 11–12.

Gili, J.-M. & Hughes, R. G. 1995. The ecology of marine benthic hydroids. *Oceanography and Marine Biology: an Annual Review* **33**, 351–426.

Gitel'son, G. I., Tugai, V. A. & Zakharchenko, A. N. 1990. Production of Obelin, a calcium activated photoprotein from *Obelia longissima* and its application for registration of the calcium outflux from the fragmented sarcoplasmic reticulum of skeletal muscles. *Ukrainskil Biokhimicheskil Zhurnal* **62**(2), 69–76.

Goette, A. 1907. Vergleichende Entwicklungsgeschichte der Geschlechsindividuen der Hydropolypen. *Zeitschrift für Wissenschaftliche Zoologie* **87**, 1–335.

Gorin, A. N. 1975. Relationship between the distribution of the main epibiotic organisms of the Sea of Japan and some environmental factors. In *Epibioses of the Japan and Okhotsk Seas.* Vladivistok: Far East Scientific Centre, 21–44. (In Russian).

Grasshoff, M. 1991a. Die evolution der Cnidaria. I. Die Entwicklung zur Anthozoen-Konstruction. *Natur und Museum* **121**(8), 225–36.

Grasshoff, M. 1991b. Die Evolution der Cnidaria. II. Solitare und coloniale Anthozoen. *Natur und Museum* **121**(9), 269–82.

Grasshoff, M. 1997. Outlines of Coelenterate evolution based on principles and constructional morphology. *Proceedings of the 6th International Conference of Coelenterate Biology*, J. C. den Hartog (ed.). Leiden: National Natuurlijk Museum, 195–208.

Gravier-Bonnet, N. 1972. Hydroides epiphytes de trois phanerogames marines en provenance de Nossi-Be (NW de Madagascar). *Tèthys* (Suppl.) **3**, 3–10.

Gravier-Bonnet, N. 1979. Hydraires semi-profonds de Madagascar, (Coelenterata Hydrozoa), étude systematique et ecologique. *Zoologische Verhandelingen* (*Leiden*) **169**, 3–76.

Gravier-Bonnet, N. in press. *Obelia* (Cnidaria Hydrozoa) in sea-grass beds of Madagascar (Indian Ocean). In Obelia (*Cnidaria, Medusozoa, Hydrozoa*). *Phenomenon. Aspects of investigations. Perspectives of employment. Zoosystematica Rossica.* (Suppl.). S. D. Stepanjants (ed.).

Grebel'nyi, S. D. 1988. The indirect evidences of the existence of the asexual sporiferous generation in the life cycle of some low invertebrates. In *Porifera and Cnidaria. Modern and perspective investigations.* V. M. Koltun & S. D. Stepanjants (eds). Leningrad: Zoological Institute of Academy of Science, 75–80. (In Russian with English summary).

Grebel'nyi, S. D. 1989. On the life cycle of the low invertebrates. In *The fundamental investigations of the recent Porifera and Coelenterata,* V. M. Koltun et al. (eds). Leningrad: Zoological Institute of Russian Academy of Science 38–9.

Hadzi, J. 1963. The evolution of the Metazoa. *International Series of Monographs on Pure and Applied Biology* **16**, 1–499.

Hammett, F. S. 1940. Synthetic l-leucine in developmental growth of *Obelia geniculata. Growth* **4**, 81–7.

Hammett, F. S. 1942. The role of amino acids and nucleic acid components in developmental growth. I. The growth of an *Obelia* hydranth. *Growth* **6**, 59–123.

Hammett, F. S. 1943. The role of amino acids and nucleic acid components in developmental growth. Part One. The growth of an *Obelia* hydranth. Chapter One. Description of *Obelia* and its growth. *Growth* **7**, 331–9.

Hammett, F. S. 1946. Correlations between the hydranth components of *Obelia* colony populations. *Growth* **10**, 89–172.

Hammett, F. S. 1950. Quantitative growth in *Obelia* colonies in culture. I. Rates of basic activities. *Growth* **14**, 263–4.

Hammett, F. S. 1951a. Quantitative growth in *Obelia* colonies in culture. II. Seasonal changes in rates of basic activities. A. Initiation. *Growth* **15**, 23–37.

Hammett, F. S. 1951b. Quantitative growth in *Obelia* colonies in culture. II. Seasonal changes in rates of basic activities. B. Proliferation. *Growth* **15**, 125–40.

Hammett, F. S. 1951c. Quantitative growth in *Obelia* colonies in culture. II. Seasonal changes in rates of basic activities. C. Differentiation. *Growth* **15**, 189–200.

Hammett, F. S. 1951d. Quantitative growth in *Obelia* colonies in culture. II. Seasonal changes in rates of basic activities. D. Organization. *Growth* **15**, 201–10.

Hammett, F. S. 1951e. Quantitative growth in *Obelia* colonies in culture. II. Seasonal changes in rates of basic activities. E. Maintenance, regression, catabolism, injury. *Growth* **15**, 225–39.

Hammett, F. S. & Barbour, H. G. 1939. The retardation of growth by deuterium oxide (*Obelia geniculata*). *Growth* **3**, 403–8.

Hammett, F. S. & Chapman, S. S. 1938. Free amino acid localization in *Obelia geniculata. Growth* **2**, 223 only.

Hammett, F. S. & Comee, R. 1940. The growth reaction of *Obelia geniculata* to D(-)-threonine. *Growth* **4**, 139–43.

Hammett, F. S. & Hammett, D. W. 1945. Seasonal changes in *Obelia* colony composition. *Growth* **9**, 54–144.

Hammett, F. S. & Padis, N. 1935. Correlation between developmental status of *Obelia* hydranths and direction of hydroplasmic streaming within their pedicels. *Protoplasma* **23**(1), 81–5.

Hammett, F. S. & Rivard, D. 1940a. The influence of extra-alkaline sea water on the developmental growth of *Obelia geniculata. Growth* **4**, 101–5.

Hammett, F. S. & Rivard, D. 1940b. The effect of centrifugation on hydranth and stolon regeneration and growth in *Obelia geniculata. Growth* **4**, 111–22.

Hargitt, G. T. 1913. Germ cells of coelenterates. I. *Campanularia flexuosa. Journal of Morphology* **24**(3), 1–21.

Hessinger, D. A. & H. M. Lenhoff (eds), 1989. *The biology of nematocysts.* San Diego: Academic Press.

Hirohito, Emperor Showa, annotated by M. Yamada, 1995. The hydroids of Sagami Bay. II. Thecata, Biology Laboratory of the Imperial Household, Tokyo. I–V, 1–355; 1–244. (In Japanese).

Hughes, R. G. 1977. Aspects of the biology and life history of *Nemertesia antennina* (L.). (Hydrozoa: Plumulariidae). *Journal of the Marine Biological Association of the United Kingdom* **57**, 641–57.

Hughes, R. G. 1980. Current induced variations in the growth and morphology of hydroids. In *Developmental and cellular biology of coelenterates,* P. Tardent & R. Tardent (eds). *Proceedings of the IVth International Coelenterate Conference,* 179–84.

Hughes, R. G. 1987. The loss of hydranths of *Laomedea flexuosa* Alder and other hydroids, with reference to hydroid senescence. In *Modern trends in the systematics, ecology and evolution of hydroids and Hydromedusae,* J. Bouillon et al. (eds). Oxford: Clarendon Press, 171–84.

Hunter, T. 1989. Suspension feeding in oscillating flow: the effect of colony morphology and flow regime on plankton capture by the hydroid *Obelia longissima*. *Biological Bulletin* **176**(1), 41–9.

Huxley, J. S. & de Beer, G. R. 1923. Studies in dedifferentiation. IV. Resorption and differential inhibition in *Obelia* and *Campanularia*. *Quarterly Journal of Microscopical Science, New Series* **67**, 473–95.

Ito, T. & Inoue, K. 1962. Systematic studies on the nematocysts of Cnidaria. I. Nematocysts of Gymnoblastea and Calyptoblastea. *Memoires of the Ehime University IIB* **4**, 445–60.

Ivanchenko, O. F. 1997. *Obelia* as density regulator for early White Sea herring larvae. In *Ecological investigations of the White Sea organisms*, V. Y. Berger & A. D. Naumov (eds). Materials of International Conference, 16–19 July, 1997, St Petersburg, 30–2.

Ivanchenko, O. F. in press. *Obelia* as density regulator for early White Sea organisms. In Obelia (*Cnidaria, Medusozoa, Hydrozoa*). *Phenomenon. Aspects of investigations. Perspectives of employment. Zoosystematica Rossica.* (Suppl.). S. D. Stepanjants (ed.).

Jankowsky, H. D., Laudien, H. & Precht, H. 1969. Ist eine intrazellulare Eisbildung by Tieren todlich? Versuche mit Polypen der Gattung *Laomedea*. *Marine Biology* **3**(1), 73–7.

Jennings, J. B. 1973. *Feeding, digestion and assimilation in animals.* London: Macmillan Press.

Karlsen, A. G. & Aristarkhov, V. M. 1986. Change in toxic action of zinc chloride solutions on colonial hydroids of *Obelia loveni* and *Obelia flexuosa*, caused by constant magnetic field. *Izvestiya Akademii Nauk SSSR, Seriya Biologicheskaya* **2**, 230–37. (In Russian with English summary).

Karlsen, A. G. & Marfenin, N. N. 1987. The copper and zinc ions actions on some hydroids in laboratory. *Biologicheskii Nauki* **1**, 71–3.

Khomenko, A. V. 1989. The morphology, ecology and peculiarities of the life cycle of the hydroid polyp *Obelia longissima* (Pallas, 1766) (Hydrozoa, Thecaphora). In *Hydrobiological Research in the Avachinsk Bay*, O. G. Kussakin (ed.). Vladivostok: Biologiya Morya Institut 59–68.

Knight, D. P. 1965. Behavioural aspects of emergence in the hydranth of *Campanularia flexuosa* (Hincks). *Nature* **206**, 1170–71.

Knight, D. P. 1968a. *The cellular basis for quinone tanning in the hydroid* Campanularia flexuosa. Doctoral thesis, University of Cambridge.

Knight, D. P. 1968b. Cellular basis for quinone tanning of the perisarc in the thecate hydroid *Campanularia* (= *Obelia*) *flexuosa*. *Nature* **218**, 584–6.

Knight, D. P. 1970a. Tanning cell in a thecate hydroid *Campanularia flexuosa*. *Proceedings of the Challenger Society* **4**, 60–1.

Knight, D. P. 1970b. Sclerotization of the perisarc of the calyptoblastic hydroid, *Laomedea flexuosa*. 1. Identifications and localization of dopamine in the hydroid. *Tissue and Cell* **2**, 467–77.

Knight, D. P. 1971. Sclerotization of the perisarc of the calyptoblastic hydroid, *Laomedea flexuosa*. 2. Histochemical demonstration of phenol oxidase and attempted demonstration of peroxidase. *Tissue and Cell* **3**(1), 57–64.

Korsakova, G. F. 1950. The problem of the variation of organs in the integrity of colonial hydroids (experiments with *Obelia longissima* and *Laomedea flexuosa*). *Zoologicheskii Zhurnal* **29**(4), 307–22. (In Russian).

Kosevich, I. A. 1983. Dimorfizm pobegov v kolonii gidroida *Obelia longissima* (Pallas, 1766) (Hydrozoa, Thecaphora, Campanulariidae). In Molecular biology and general directions of modern biology development; 14th Conference of young scientists of the Biological Faculty of Moscow University. Ms deposited in Institute of Scientific-Technical Information, No. 1507-84. (In Russian).

Kosevich, I. A. 1984. Movements of colonies of *Obelia loveni* (Allman) (Hydrozoa, Thecaphora, Campanulariidae) along a substrate. In *Modern problems of oceanology*. D. G. Sedov & S. S. Ivanov (eds). Moscow: Academy of Science, USSR, Shirshov Institute of Oceanology, 92–3. (In Russian).

Kosevich, I. A. 1988. Interaction of local and colonial processes in colony growth of *Obelia loveni* (Allm.). In *Porifera and Cnidaria. Modern and perspective investigations.* V. M. Koltun & S. D. Stepanjants (eds). Leningrad: Zoological Institute of Russian Academy of Science, 85–90. (In Russian with English summary).

Kosevich, I. A. 1989. Interaction of local and colonial processes in colony growth of *Obelia loveni* (Allm.). In *The fundamental investigations of the recent Porifera and Coelenterata.* V. M. Koltun et al. (eds). Leningrad: Zoological Institute of Russian Academy of Science, 51–2. (In Russian).

Kosevich, I. A. 1990. Developments of stolon's and stem's internodes in hydroid genus *Obelia* (Campanulariidae). *Vestnik Moskovskogo Universiteta Seriya Biologiya* 1990 (3), 26–32. (In Russian).

Kosevich, I. A. 1991. Control of the structure of giant shoots in the colonial hydroid *Obelia longissima* Pallas, 1766 (Campanulariidae). *Ontogenez* 22(3), 308–15. (In Russian with English summary).

Kosevich, I. A. 1992. Colonial hydroids sensibility to the increased concentration of copper and zinc ions. *Ecologiya, Sverdlovsk* 1992 (1), 89–91. (In Russian).

Kosevich, I. A. & Marfenin, N. N. 1986. Colonial morphology of the hydroid *Obelia longissima* (Pallas, 1766) (Campanulariidae). *Vestnik Moskovskogo Universiteta Seriya Biologiya* 1986 (3), 44–52. (In Russian).

Kosevich, I. A., Marfenin, N. N. & Chudakov, L. I. 1990. The automatic registration of the rhythmic pulsations of hydroids. *Biologicheskie Nauki* 1990 (4), 132–9. (In Russian).

Kramp, P. L. 1961. Synopsis of the medusae of the world. *Journal of the Marine Biological Association of the United Kingdom* **40**, 1–469.

Krauss, Y. A. & Rodimov, A. A. (in press). The embryonic development of the *Obelia* species with sessile. In Obelia *(Cnidaria, Medusozoa, Hydrozoa). Phenomenon. Aspects of investigations. Perspectives of employment. Zoosystematica Rossica.* (Suppl.). S. D. Stepanjants (ed.).

Kubota, S. 1976. Notes on the nematocysts of Japanese hydroids, I. *Journal of the Faculty of Science Hokkaido University*, Series VI, Zoology **20**(2), 230–43.

Kubota, S. 1981. Life-history and taxonomy of an *Obelia* species (Hydrozoa: Campanulariidae) in Hokkaido, Japan. *Journal of the Faculty of Science Hokkaido University*, Series VI, Zoology **22**(4), 379–99.

Kubota, S. in press. Fauna of *Obelia* (Cnidaria, Hydrozoa) in Japanese waters, with special reference to life cycle of *Obelia dichotoma.* In Obelia *(Cnidaria, Medusozoa, Hydrozoa). Phenomenon. Aspects of investigations. Perspectives of employment. Zoosystematica Rossica.* (Suppl.). S. D. Stepanjants (ed.).

Kuhn, A. 1913. Entwicklungsgeschichte und Verwandtschaftsbeziehungen der Hydrozoen. I Teil: Die Hydroiden. Ergebenisse und Fortschrifte der Zoologie **4**(1–2), 1–284.

Kunin, B. l. in press. Long-term variation in hydroid mass of *Obelia longissima* (Pallas) (Hydrozoa, Thecaphora: Campanulariidae) on early stages of development of artificial community in mussel *Mytilus edulis* L. mariculture at the White Sea. In Obelia *(Cnidaria, Medusozoa, Hydrozoa). Phenomenon. Aspects of investigations. Perspectives of employment. Zoosystematica Rossica.* (Suppl.). S. D. Stepanjants (ed.).

Kuznetsova, I. A. & Zevina, G. B. 1967. Obrastaniys v rayonach stroitel'stva prilivnich electrostanziy na Barenzevom i Belom morjach. *Trudy Instituta Okeanologii Akademii Nauk SSSR* **85**, 18–28.

Kusnetzova, I. A. & Zevina, G. B. 1978. Sostav obrastaniya severnoy chasti Kol'skogo poluostrova. In *Biopovredgdeniya materialov i zaschita ot nich.* Moscow Academy of Science, USSR, Shirshov Institute of Technology, 47–53.

Labas, Y. A., Beloussov, L. V., Badenko, L. A. & Letunov, V. N. 1981. Growth pulsations of Metazoa. *Doklady Akademii Nauk SSSR* **25**(2), 1247–50.

Labas, Y. A., Zinchenko, V. P., Beloussov, L. V. & Kazakova, N. I. 1989. The dynamics and the role of free cytoplasmic calcium in the autorhythmical growth pulsations in developing rudiments of the colonial hydroids. *Zhurnal Obshchei Biologii* **50**(4), 504–15. (In Russian with English summary).

Leloup, E. 1931. Un cas d'epibiose de l'hydropolype, *Laomedea geniculata* (Linne). *Bulletin du Musée Royal d'Histoire Naturelle de Belgique* **7**(24), 1–3.

Lenhoff, H. M. 1974. On the mechanism of action and evolution of receptors associated with feeding and digestion. In *Reviews and new perspectives. Coelenterate biology.* L. Muscatine & H. N. Lenhoff (eds). London: Academic Press, 211–43.

Leontovich, A. A. & Marfenin, N. N. 1990. Connection of major intercolonial processes at branching in colonial hydroids. *Zhurnal Obshchei Biologii* **51**(3), 353–62. (In Russian with English summary).

Letunov, V. N. 1981. Pishevoye povedeniye, pishchevaritel'no- raspredelitel'niy apparat i morphogenez u kolonialnikh gidroidnikh polipov otr. Leptolida. Abstract of dissertation, 1–16. (In Russian).

Letunov, V. N. 1987. Growth and morphogenesis in colonial hydroids – mass macro-foulers. In *The study of the process of marine bio-fouling and the elaboration of methods to counteract it*, O. A. Scarlato (ed.). Leningrad: Zoology Institute, 44–7. (In Russian).

Letunov, V. N. & Andreev, A. P. 1980. The role of physical and chemical stimulation in induction of food and defence behaviour reactions on hydranths of *Obelia loveni.* In *The theoretical and practical importance of the Coelenterata.* A. V. Naumov & S. D. Stepanjants (eds). Leningrad: Zoological Institute of Academy of Science, 50–55.

Letunov, V. N. & Stepanjants, S. D. 1986. A study of *Obelia longissima* (Pallas, 1766) (Hydrozoa, Thecaphora, Campanulariidae) and characteristics of autoecology of this species in the White Sea. In *Ecological investigations of the benthic organisms of the White Sea.* V. V. Fedjakov & V. V. Lukanin (eds). Leningrad, Zoological Institute, 17–29. (In Russian).

Letunov, V. N. & Vysotsky, E. S. 1988a. Bioluminescence activity and Ca^{++} dependent photoprotein content of colonial hydroids *Obelia longissima* (Pallas, 1766) (Hydrozoa, Thecaphora, Campanulariidae). *Zhurnal Obshchei Biologii* **49**, 381–7. (In Russian).

Letunov, V. N. & Vysotsky, E. S. 1988b. The bioluminescence of Cnidaria and Ctenophora. In *Porifera and Cnidaria. Modern and perspective investigations*, V. M. Koltun & S. D. Stepanjants (eds). Leningrad: Zoological Institute of Academy of Science, 90–93. (In Russian with English summary).

Lunger, P. D. 1963. Fine-structural aspects of digestion in a colonial hydroid. *Journal of Ultrastructure Research* **9**(3,4), 362–80.

Lunger, P. D. 1971. Early stages of spermatozoan development in the colonial hydroid *Campanularia flexuosa. Zeitschrift für Zellforschung und Mikroskopische Anatomie* **116**, 37–51.

Makarenkova, E. P. 1988a. Digestion of colonial hydroids: horseradish peroxidase absorption in the gastrodermal cells of *Obelia loveni* (Campanulariidae, Hydrozoa). In *Porifera and Cnidaria. Modern and perspective investigations.* V. M. Koltun & S. D. Stepanjants (eds). Leningrad: Zoological Institute of Academy of Science, 94–8.

Makarenkova, E. P. 1988b. A study of the gastrodermal epithelium of hydroid polyp *Obelia loveni* and *O. longissima* using light and electron microscopy. *Tsitologiya* **30**(5), 539–47. (In Russian with English summary).

Makarenkova, E. P. 1988c. Determination of hydroplasma motion velocity in the colonial hydroids of the genus *Obelia* using fluorescent dyes. *Doklady Akademii Nauk SSSR* **302**(5), 1275–7. (In Russian).

Makarenkiva, E. P. 1989. Digestion of colonial hydroids: horseradish peroxydase absorption in the gastrodermal cells of *Obelia loveni.* In *The fundamental investigations of the recent Porifera and Coelenterata.* V. M. Koltun et al. (eds). Leningrad: Zoological Institute of Russian Academy of Science, 65–6. (In Russian).

Makarenkova, E. P. 1990. Horseradish peroxidase absorption in the gastrodermal epithelial cells of the colonial hydroid *Obelia loveni. Tsitologiya* **32**(1), 21–27. (In Russian with English summary).

Makarenkova, E. P., Makarenkov, S. N. & Letunov, V. N. 1985. Frustulation in hydroids of the genus *Obelia* (Leptolida, Campanulariidae) as a modified form of stoloniferous growth. *Zoologicheskii Zhurnal* **64**(11), 1614–20. (In Russian with English summary).

Makarenkova, E. P. & Seravin, L. N. 1986. Cell contacts and their position in the hydranth bud of *Obelia loveni* (Thecaphora). *Tsitologiya* **28**(7), 677–82. (In Russian with English summary).

Makrushin, A. V. 1986. Resting stages of *Clava multicornis, Obelia loveni* and *O. flexuosa* (Hydroidea, Leptolida). *Zoologicheskii Zhurnal* **65**(8), 1254–8. (In Russian with English summary).

Maljutin, O. I. & Marfenin, N. N. 1988. In search of indices for the water stream's action on hydroids research. In *Porifera and Cnidaria. Modern and perspective investigations.* V. M. Koltun & S. D. Stepanjants (eds). Leningrad: Zoological Institute of Academy of Science, 98–103. (In Russian with English summary).

Marfenin, N. N. 1985. The functioning of the pulsatory-peristaltic type transport system in colonial hydroids. *Zhurnal Obshchei Biologii* **46**(2), 153–64. (In Russian with English summary).

Marfenin, N. N. 1988. A transport system activity in the hydroid colony: a new method and events. In *Porifera and Cnidaria. Modern and perspective investigations.* V. V. Koltun & S. D. Stepanjants (eds). Leningrad: Zoological Institute of Academy of Science, 111–16. (In Russian with English summary).

Marfenin, N. N. 1993a. *The phenomenon of coloniality.* Moscow: Moscow State University. (In Russian with English summary).

Marfenin, N. N. 1993b. *Functional morphology of the colonial hydroids.* Zoological Institute of Russian Academy of Sciences, St Petersburg. (In Russian with English summary).

Marfenin, N. N. in press. Functional morphology differences between the hydroids: *Obelia longissima, O. geniculata, Gonothyraea loveni,* and *Laomedea flexuosa,* occurred at the common biotops in the White Sea. In Obelia (*Cnidaria, Medusozoa, Hydrozoa*). *Phenomenon. Aspects of investigations. Perspectives of employment. Zoosystematica Rossica.* (Suppl.). S. D. Stepanjants (ed.).

Marfenin, N. N. & Khomenko, A. V. 1988. The colony morphology and signs of the morphofunctional adaptation of the hydroid *Obelia geniculata* (L.) to the life in a strong stream. *Nauchnye Doklady Vysshei Shkoly Biolicheskie Nauki* 1988 (5), 35–44. (In Russian with English summary).

Marfenin, N. N. & Khomenko, T. N. 1989. The feeding spectrum of the White Sea hydroids dominant species. In *The fundamental investigations of the recent Porifera and Coelenterata*, V. M. Koltun et al. (eds). Leningrad: Zoological Institute of Russian Academy of Sciences, 80–3. (In Russian).

Marfenin, N. N. & Kosevich, I. A. 1984a. Colonial morphology of the hydroid *Obelia loveni* (Allm.) (Campanulariidae). *Vestnik Moscovskogo Universiteta Seriya Biologiya* 1984 (2), 37–46. (In Russian with English summary).

Marfenin, N. N. & Kosevich, I. A. 1984b. Colony formation, behaviour and life cycle of hydranths, reproduction. *Vestnik Moscovskogo Universiteta Seriya Biologiya* 1984 (3), 16–24. (In Russian with English summary).

Mariskall, R. N. 1974. Nematocysts. In *Review and new perspectives. Coelenterate Biology*. L. Muscatine and H. M. Lenhoff (eds). New York: Academic Press, 1–17.

Maslennikov, S. I., Kashin, I. A. & Fadeev, V. I. 1987. Formirovaniye obrastaniya gidroidnogo tipa na sadkach promishlennich ustanovok dlya kultivirovaniya grebeshka. 3 Vsesoyusn. *Konfer. Biopovredgdeniy. Moscow, Nauka*, 262–3. (In Russian).

McClary, A. 1961. Experimental studies on bud development in *Craspedacusta sowerbii. Transactions of the American Microscopical Society* **80**(3), 343–53.

McKinney, M. L. & McNamara, K. 1991. *Heterochrony: the evolution of ontogeny*. New York: Plenum Press.

Merezkovsky, C. 1883. Histoire du developpement de la meduse *Obelia. Bulletin de la Société Zoologique France* **8**, 18–49.

Migotto, A. E. 1995. Benthic shallow-water hydroids (Cnidaria, Hydrozoa) of the coast of Sao Sebastiano, Brazil, including a checklist of Brazilian hydroids. *Zoologische Verhandelingen Rijksmuseum van Natuurelijke Historie Leiden* **306**, 1–128.

Millard, N. A. H. 1975. Monograph on the Hydroida of southern Africa. *Annals of the South African Museum* **68**, 1–513.

Miller, R. L. 1969. Preliminary observations on maturation of the female gonangium of *Campanularia flexuosa. Biological Bulletin* **137**, 409–10.

Miller, R. L. 1970. Sperm migration prior to fertilization in the hydroid *Gonothyraea loveni. Journal of Experimental Zoology* **175**(4), 493–504.

Morin, J. G. 1974. Coelenterate bioluminescence. In *Reviews and new perspectives. Coelenterate Biology*, L. Muscatine & H. M. Lenhoff (eds). New York: Academic Press, 397–438.

Morin, J. G. & Cooke, I. M. 1971a. Behavioural physiology of the colonial hydroid *Obelia*. I. Spontaneous movements and correlated electrical activity. *Journal of Experimental Biology* **54**(3), 689–706.

Morin, J. G. & Cooke, I. M. 1971b. Behavioural physiology of the colonial hydroid *Obelia*. II. Stimulus-initiated electrical activity and bioluminescence system. *Journal of Experimental Biology* **54**(3), 707–21.

Morin, J. G. & Cooke, I. M. 1971c. Behavioural physiology of the colonial hydroid *Obelia*. III. Characteristics of the bioluminescent system. *Journal of Experimental Biology* **54**(3), 723–35.

Morin, J. G. & Hastings, J. W. 1971a. Biochemistry of the bioluminescence of colonial hydroids and other coelenterates. *Journal of Cell Physiology* **77**(3), 305–12.

Morin, J. G. & Hastings, J. W. 1971b. Energy transfer in a bioluminescent system. *Journal of Cell Physiology* **77**(3), 313–18.

Morin, J. G. & Reynolds, G. T. 1969. Fluorescence and time distribution of photon emission of bioluminescent photocytes in *Obelia geniculata. Biological Bulletin* **137**, 410 only.

Morin, J. G. & Reynolds, G. T. 1970. Repetitive flashing of luminescence in hydroid. *American Zoologist* **10**, 505–6.

Morin, J. G. & Reynolds, G. T. 1974. The cellular origin of bioluminescence in the colonial hydroid *Obelia. Biological Bulletin* **147**(2), 397–410.

Morin, J. G., Reynolds, G. T. & Hastings, J. W. 1968. Excitatory physiology and localization of bioluminescence in *Obelia. Biological Bulletin* **135**, 429–30.

Morri, C. 1978. Contributo all conoscenza degli idrozoi lagunari italiani: idropolipi di alcuni laghi costieri mediotirrenici. *Annali del Museo Civico di Storia Naturele di Genova* **82**, 163–71.

Morri, C. 1984. Contributo alla conoscenza degli Idrozoi lagunari italiani: Idropolipi di alcune lagune costiere sarde et note sulla distribuzione del genere *Obelia* nelle lagune italiane. *Rendiconti Seminario Facolta Scienze, Universita Cagliari* **54**, (Suppl.), 41–8.

Morri, C. & Boero, F. 1986. Marine fouling hydroids. In *Catalogue of the main marine fouling organisms*, **7**. *Hydroids*, 1–91.

Muller-Cale, K. 1913. Zur Entwicklungsgeschichte einiger Thecaphoren. *Zoologische Jahrbücker, Abteilung für Anatomie und Ontogenie der Tiere* **37**, 83–112.

Naumov, D. V. 1960. Gidroidi i gydromedusy morskikh, solonovatovodnykh i presnovodnykh basseinov SSSR. Opredelitely po faune SSSR, izdav. Zoological Institute of Russian Academy of Science, Nauk SSSR 70, 1–627. English transl. by Israel Progr., 1969: Hydroids and Hydromedusae of the USSR, No. 5108, 1–631.

Nishihira, M. 1965. The association between Hydrozoa and their attachment substrata with special reference to algal substrata. *Bulletin of the Marine Biological Station of Asamushi*, **12**(2,3), 75–92.

O'Rand, M. G. 1972a. A soluble cell surface material required for spermatozoan-epithelial cell interaction during fertilization in *Campanularia flexuosa. Biological Bulletin* **143**, 472 only.

O'Rand, M. G. 1972b. In vitro fertilization capacitation-like interaction in the hydroid *Campanularia flexuosa*. *Journal of Experimental Zoology* **182**(3), 299–305.

O'Rand, M. G. 1974. Gamete interaction during fertilization in *Campanularia* – The female epithelial cell surface. In *The developmental biology of the Cnidaria*, R. L. Miller & C. R. Wyttenbach (eds). *American Zoologist* **14**(2), 487–93.

O'Rand, M. G. & Miller, R. L. 1974. Spermatozoan vesicle loss during penetration of the female gonangium in the hydroid, *Campanularia flexuosa*. *Journal of Experimental Zoology* **188**(2), 179–93.

Orlov, D. V. 1994. Larval settlement in colonial hydroids. PhD dissertation, Moscow State University. (In Russian).

Orlov, D. V. 1996. The role of biological interactions in settlement of hydrozoan planulae. *Zoologicheskii Zhurnal* **75**(6), 8–19. (In Russian with English summary).

Orlov, D. V. in press. Differential larval settlement in two related colonial hydroids. In Obelia (*Cnidaria, Medusozoa, Hydrozoa*). *Phenomenon. Aspects of investigations. Perspectives of employment. Zoosystematica Rossica*. (Suppl.). S. D. Stepanjants (ed.).

Orlov, D. V. & Marfenin, N. N. 1995. Behaviour and settlement of the planula larvae of *Gonothyraea loveni* (Allman) (Hydroidea, Thecaphora) at the White Sea intertidal zone. In *Cnidaria. Modern and perspective investigations. II*. S. D. Stepanjants (ed.). *Trudy Zoologicheskogo Institute Akademii Nauk* **261**, 103–20. (In Russian with English summary).

Orlov, D. V., Marfenin, N. N. & Railkin, A. I. 1994. The settlement of the planulae of the White Sea hydroids *Gonothyrea loveni* (Allman) and *Laomedia flexuosa* (Hincks) on the slim bacterial film from the surface of intertidal algae. *Vestnik Moskovskogo Universiteta Seriya Biologiya* **3**(16), 47–51. (In Russian).

Oshurkov, V. V. 1986. Development and structure of some fouling communities of Avachinskiy Bay. *Biologiya Morya* **5**, 20–7. (In Russian).

Oshurkov, V. V. 1993. Succession and dynamic of fouling communities of the upper sublittoral. Dissertation Abstracts, 1–44. (In Russian).

Oshurkov, V. V. & Seravin, L. N. 1983. Fouling biocenosis formation of Chupa Bay (the White Sea). *Vestnik Leningradovskogo Universiteta Seriya Biologiya* **3**, 37–46. (In Russian).

Ostman, C. 1979a. Two types of nematocysts in Campanulariidae (Cnidaria, Hydrozoa) studied by light and scanning electron microscopy. *Zoologica Scripta* **8**(1), 5–12.

Ostman, C. 1979b. Nematocysts in the Phialidium medusae of *Clytia hemisphaerica* (Hydrozoa, Campanulariidae) studies by light and scanning electron microscopy. *Zoon Uppsala* **7**, 125–42.

Ostman, C. 1982a. Isoenzymes and taxonomy in Scandinavian hydroids (Cnidaria, Campanulariidae). *Zoologica Scripta* **11**(3), 155–63.

Ostman, C. 1982b. Nematocysts and taxonomy in *Laomedea*, *Gonothyraea* and *Obelia* (Hydrozoa, Campanulariidae). *Zoologica Scripta* **11**(4), 227–41.

Ostman, C. 1983. Taxonomy of Scandinavian hydroids (Cnidaria, Campanulariidae) a study based on nematocysts morphology and isoenzymes. *Acta Universitatus Upsaliensis* (Abstracts of Dissertation from the Faculty of Science) No. 672, 1–22.

Ostman, C. 1987. New techniques and old problems in hydrozoan systematics. In *Modern trends in the systematics, ecology and evolution of hydroids and hydromedusae*, J. Bouillon et al. (eds). Oxford: Clarendon Press, 67–82.

Ostman, C. 1989. Nematocysts as taxonomic criteria within the family Campanulariidae, Hydrozoa. In *The biology of nematocysts*. D. A. Hessinger & H. M. Lenhoff (eds). New York: Academic Press, 501–17.

Ostman, C. in press. Nematocysts and their values as taxonomic parameters within the Campanulariidae (Hydrozoa). A review based on light and scanning electron microscopy. In Obelia (*Cnidaria, Medusozoa, Hydrozoa*). *Phenomenon. Aspects of investigations. Perspectives of employment. Zoosystematica Rossica*. (Suppl.). S. D. Stepanjants (ed.).

Ostman, C., Piraino, S. & Roca, I. 1987. Nematocyst comparison between some Mediterranean and Scandinavian campanulariids (Cnidaria, Hydrozoa). In *Modern trends in the systematics, ecology and evolution of hydroids and hydromedusae*, J. Bouillon et al. (eds). Oxford: Clarendon Press, 607–13.

Palincsar, E. E. 1965a. The effect of fluoride ion on the longevity of *Campanularia flexuosa* hydranth. *Bulletin of the Mount Desert Island Biological Laboratory* **5**(1), 60–1.

Palincsar, E. E. 1965b. Respiratory rate and senescence in *Campanularia flexuosa*. *Bulletin of the Mount Desert Island Biological Laboratory* **5**(1), 61–2.

Palinscar, E. E. 1969. The fine structure of senescence in the regression-replacement cycle of *Campanularia flexuosa*. *Bulletin of the Mount Desert Island Biological Laboratory* **8**, 48 only.

Panteleeva, N. N. in press. *Obelia longissima* (Pallas,1766) and *Obelia geniculata* (L., 1758) (Hydrozoa, Thecaphora, Campanulariidae) in the Barents Sea. Morphology, distribution, ecology and special features of the life history. In Obelia (*Cnidaria, Medusozoa, Hydrozoa*). *Phenomenon. Aspects of investigations. Perspectives of employment. Zoosystematica Rossica.* (Suppl.). S. D. Stepanjants (ed.).

Partaly, E. M. 1974. Effects of planktonic crustaceans on a population of the hydroid *Perigonimus megas* in a fouling community. *Oceanologiya* (Oceanology) 14(4), 719–23. (In Russian).

Petersen, K. W. 1979. Development of coloniality in Hydrozoa. In *Biology and systematics of colonial organisms*, C. Larwood & B. Rosen (eds). *Symposia of the Systematics Association* **11**, 105–39.

Petersen, K. W. 1990. Evolution and taxonomy in capitate hydroids and medusae. *Zoological Journal of the Linnean Society* **100**, 101–231.

Piraino, S. & Morri, C. 1990. Zonation and ecology of epiphytic hydroids in a Mediterranean coastal lagoon the "Stagnone" of Marsala (north-west Sicily). *P.S.Z.N.I.: Marine Ecology* **11**(1), 43–60.

Polteva, D. G. 1972. The morphogenetic role of the gastrodermis in somatic embryogenesis in the hydroid polyp *Obelia* geniculata. *Vestnik Leningradskogo Universiteta Seriya Biologiya* 1972 (1), 27–36. (In Russian with English summary).

Polteva, D. G. 1975. Cells transformation in morphogenetic processes in hydroids of genus *Obelia*. Voprosi Stranitel'noy i eksperimental'noy morfobgii morskih organizmov Apatity, 61–74. (In Russian).

Polteva, D. G. 1981. Cells and reparative morphogenesis in Coelenterata. *Arkhiv Anatomii Gistologii Embriologii* **80**(2), 5–17. (In Russian with English summary).

Polteva, D. G. 1986. Morphogenesis and changes in hydroid cell populations. *Ontogenez* **17**(4), 434–5. (In Russian).

Polteva, D. G. & Aizenshtadt, T. V. 1980a. Cell differentiation in the embryogenesis of *Obelia*. Origination of interstitial cells in early larval development of *Obelia flexuosa* and *O.loveni*. In *The theoretical and practical importance of the coelenterates.* D. V. Naumov & S. D. Stepanjants (eds). Leningrad: Zoological Institute of Academy of Science, 86–91. (In Russian).

Polteva, D. G. & Aizenshtadt, T. V. 1980b. The fine structure of marine hydroid *Obelia* larvae (Hydrozoa) interstitial cells differentiation during early larval development. *Tsitologiya* **22**(3), 271–6. (In Russian with English summary).

Polteva, D. G., Donakov, V. V. & Aizenshtadt, T. V. 1987. Seasonal changes in cell populations of hydroid *Obelia* loveni in hypogenetic colony. *Zoologicheskii Zhurnal* **66**(8), 1135–47. (In Russian with English summary).

Polteva, D. G., Donakov, V. V. & Markova, L. G. 1988. Morphogenesis and cell processes in hydroid polyps *Obelia loveni* and *Clava multicornis*. In *Porifera and Cnidaria. Modern and perspective investigations*, V. M. Koltun & S. D. Stepanjants (eds). Leningrad: Zoological Institute of Academy of Science, 111–16. (In Russian with English summary).

Purcell, J. E. & Mills, C. E. 1989. The correlation between nematocyst types and diets in pelagic Hydrozoa. In *The biology of nematocysts*, D. A. Hessinger & H. Lenhoff (eds). New York: Academic Press, 462–85.

Railkin, A. I. 1994. Behavioural and physiological reactions of hydroids and bivalves to some antifouling substances. *Zoologicheskii Zhurnal* **73**(7,8), 22–30. (In Russian with English summary).

Railkin, A. I. 1995. The negative chemotaxis and the suppression of the settlement of hydroid planulae by the effect of bacterial repellent. In *Cnidaria. Modern and perspective investigations. II.* S. D. Stepanjants (ed.). Trudy Zoologicheskogo Instituta Academii Nauk, **261**, 121–9. (In Russian with English summary).

Ralph, P. M. 1956. Variation in *Obelia geniculata* (Linnaeus, 1758) and *Silicularia bilabiata* (Coughtrey, 1875) (Hydroida, f. Campanulariidae). *Transactions of the Royal Society of New Zealand* **84**(2), 279–96.

Ralph, P. M. 1957. New Zealand thecate hydroids. I. Campanulariidae and Campanulinidae. *Transactions of the Royal Society of New Zealand* **84**(4), 811–54.

Ramil, F. & Vervoort, W. 1992. Report on the Hydroida collected by the "BALGIM" expedition in and around the Strait of Gibraltar. *Zoologische Verhandelingen Rijksmuseum van Natuurlijke Historie Leiden* **277**, 1–262.

Rao, K. V. Rama & Rurya Rao, K. V. 1972. On the occurence of the hydroid *Obelia commissuralis* McCrady along the west coast of India. *Journal of the Marine Biological Association of India* **13**(1,2), 144–5.

Rees, W. J. 1966. The evolution of the Hydrozoa. In *Cnidaria and their evolution.* W. J. Rees (ed.). *Symposia of the Zoological Society of London* **16**, 199–222.

Reisinger, E. 1937. Der Entlandungsvorgang der Nesselkapsel. *Verhandlungen der Deutschen Zoologischen Gesellschaft* **39**, 311–15.

Ryland, J. S. & Warner, G. F. 1986. Growth and form in modular animals: ideas on the size and arrangement of zooids. *Philosophical Transactions of the Royal Society of London* Series B, **313**, 53–76.

Saint-Hilaire, C. 1925. Morphogenese der periderms by den Hydroiden (*Gonothyraea loveni* Allman). Bulleten Nauchnogo Obschestra pri Voronjezskom Universitets **1**, 20 only.

Saint-Hilaire, C. 1930. Morphogenetische Untersuchungen der, nichtzellularen Gebilde bei Tieren. Periderm der Hydroiden. *Zoologische Jahbücher, Abteiling für Allegmeine Zoologie und Physiologie der Tiere* **47**(4), 511–622.

Schmidt, G. H. & Warner, G. F. 1991. The settlement and growth of *Sertularia cupresina* (Hydrozoa, Sertulariidae) in Langstone Harbour, Hampshire, UK. *Hydrobiologia* **216/217**, 215–19.

Schmidt, T. & David, C. N. 1986. Gland cells in *Hydra*: cell cycle kinetics and development. *Journal of Cell Science* **85**, 197–215.

Seravin, L. N. & Gudkov, A. V. 1990. Significance of chemical and physical irritans in feeding behaviour of zooids of the hydroid polyp *Obelia loveni*. *Biologiya Morya* **16**(4), 207–11.

Shepherd, S. A. & Watson, J. E. 1970. The sublittoral ecology of West Island, South Australia. 2. The association between hydroids and algal substrate. *Transactions of the Royal Society of South Australia* **94**, 139–46.

Sole-Cava, A. M. & Thorpe, J. P. 1989. Biochemical correlates of genetic variation in marine lower invertebrates. *Biochemical Genetics* **27**(5,6), 303–12.

Sole-Cava, A. M. & Thorpe, J. P. 1991. High levels of genetic variation in natural populations of marine lower invertebrates. *Biological Journal of the Linnean Society* **44**(1), 65–80.

Sommer, C. 1992. Larval biology and dispersal of *Eudedrium racemosum* (Hydrozoa, Eudendriidae). In *Aspects of hydrozoan biology*, J. Bouillon et al. (eds). *Scientia Marina*, **56**(2–3), 205–11.

Splettstoesser, W. 1924. Beitrage zur Kenntniss der Gattung *Laomedea* (sensu Broch). *Zoologische Jahbücher Systematik* **48**(5,6), 367–433.

Stebbing, A. R. D. 1976. The effects of low metal levels on a clonal hydroid. *Journal of the Marine Biological Association of the United Kingdom* **56**, 977–94.

Stebbing, A. R. D. 1980. Increase in gonozooid frequency as an adaptive response to stress in *Campanularia flexuosa*. In *Developmental and cellular biology of coelenterates. Proceedings of the IVth International Coelenterate Conference* P. Tardent & R. Tardent (eds). Amsterdam: Elsevier/North-Holland Biomedical Press, 27–32.

Stebbing, A. R. D. 1981a. The kinetics of growth control in a colonial hydroid. *Journal of the Marine Biological Association of the United Kingdom* **61**, 35–63.

Stebbing, A. R. D. 1981b. Effects of reduced salinity on colonial growth and membership in a hydroid. *Journal of Experimental Marine Biology and Ecology* **55**, 233–41.

Stebbing, A. R. D. 1981c. Hormesis stimulation of colony growth in *Campanularia flexuosa* by copper, cadmium andother toxicants. *Aquatic Toxicology* **1**(3–4), 227–38.

Stebbing, A. R. D. 1982. Hormesis. The stimulation of growth by low levels of inhibitors. *Science of Total Environment* **22**, 213–34.

Stebbing, A. R. D. 1987. Growth hormesis, a by-product of control. *Health Physics* **52**(5), 543–7.

Stebbing, A. R. D. & Santiago-Fandino, V. J. R. 1983. The combined and separate effects of copper and cadmium on the growth of *Campanularia flexuosa* (Hydrozoa) colonies. *Aquatic Toxicology* **3**(3), 183–93.

Stepanjants, S. D. 1979. Hydroids of the Antarctic and Subantarctic waters. Biological Results Soviet Antarctic Expedition 6. *Issledovaniya Fauny Morei* XXII (XXX), 1–199. (In Russian).

Stepanjants, S. D. 1980. On the cosmopolitism in hydroids. In *The theoretical and practical importance of the coelenterates*, D. V. Naumov & S. D. Stepanjants (eds). Leningrad: Zooligical Institute of Academy of Science, 114–22. (In Russian).

Stepanjants, S. D. 1988. Cnidaria origin and the possible way of the Hydrozoa evolution. In *Porifera and Cnidaria. Modern and perspective investigations.* V. M. Koltun & S. D. Stepanjants (eds). Leningrad: Zoological Institute of Academy of Science, 130–44. (In Russian with English summary).

Stepanjants, S. D. 1989. Hydrozoa of the Eurasian arctic seas. In *The Arctic Seas. Climatology, oceanography, geology, and biology.* Y. Herman (ed.). New York: Van Ostrand Reinhold, 397–430.

Stepanjants, S. D. & Letunov, V. N. 1989. Life-history of the White Sea population of *Obelia longissima* (medusae development). In *The fundamental investigations of the recent Porifera and Coelenterata.* V. M. Koltun et al. (eds). Leningrad: Zoological Institute of Russian Academy of Science, 115–17. (In Russian).

Stepanjants, S. D., Lobanov, A.L. & Dianov, M.B. in press. New approaches to search of the interspecies differences in genus *Obelia* frames. I. Analysis of the morphological characters of Obelian polyps colonies with use of computer packages BIKEY. II. Analysis of the growth and development parameters of the different species *Obelia* free medusae from the different areas of the ocean (experimental conditions). In Obelia (*Cnidaria, Medusozoa, Hydrozoa*). *Phenomenon. Aspects of investigations. Perspectives of employment. Zoosystematica Rossica.* (Suppl.). S. D. Stepanjants (ed.).

Stepanjants, S. D., Panteleeva, N. N. & Beloussova, N. P. 1993. The Barents Sea medusae development in the laboratory condition. In *Marine plankton. Taxonomy, ecology, distribution. II*. S. D. Stepanjants (ed.). *Issledovanija Fauni Morey* **45**(53), 106–30.

Stepanjants, S. D., Svoboda, A. & Vervoort, W. 1996. The problem of bipolarity, with emphasis on the Hydroidea (Cnidaria Hydrozoa). *Russkiy Gydrobiol Ogicheskig Zhurmal*, Special Issue, 5–34. (In Russian with English summary).

Stepanjants, S. D., Svoboda, A & Vervoort, W. 1997. The problem of bipolarity, with emphasis on the Medusozoa (Cnidaria: Anthozoa excepted). In *Proceedings of the 6th International Conference on Coelenterate Biology*, J. C. den Hartog (ed.). Leiden: National Natuurhistorisch Museum, 455–64.

Strehler, B. L. & Crowell, S. 1961. Studies on the comparative physiology of aging. I. Function vs. age of *Campanularia flexuosa. Gerantologia* **5**(1), 1–8.

Suddith, R. L. 1974. Cell proliferation in the terminal regions on the internode and stolons of the colonial hydroid *Campanularia flexuosa*. In *The developmental biology of the Cnidaria*, R. L. Miller & C. R. Wytten (eds). *American Zoologist* **14**(2), 745–55.

Sullivan, B. K. & Banzon, P. V. 1990. Food limitation and benthic regulations of population of the copepod *Acartia hudsonica* Pinhey in nutrient-limited systems. *Limnology and Oceanography* **35**, 1618–31.

Teissier, G. 1930. Polyembrionnie accidentele de *Gonothyraea loveni* Allm. *Bulletin de la Société Zoologique France* **55**, 61–82.

Teissier, G. 1932. Developpement embryonaire d'*Obelaria gelatinosa* (Pallas). *Bulletin de la Société Zoologique France* **57**, 228–33.

Terentjev, A. S. in press. Distribution of *Obelia longissima* (Pallas, 1766) in the open area of the north-western Black Sea. In Obelia (*Cnidaria, Medusozoa, Hydrozoa*). *Phenomenon. Aspects of investigations. Perspectives of employment. Zoosystematica Rossica.* (Suppl.). S. D. Stepanjants (ed.).

Theede, H., Scholtz, N. & Fisher, H. 1979. Temperature and salinity effects on the acute toxicity of cadmium to *Laomedea loveni* (Hydrozoa). *Marine Ecology Progress Series* **1**(11), 13–19.

Thorpe, J. P., Ryland, J. S., Cornelius, P. F. S. & Beardmore, J. A. 1992. Genetic divergence between branched and unbranched forms of the thecate hydroid *Aglaophenia pluma. Journal of the Marine Biological Association of the United Kingdom* **72**, 887–94.

Toth, S. E. 1969a. Inhibition of hydroid aging in *Campanularia flexuosa. Science* **166**, 619–20.

Toth, S. E. 1969b. Aging and regression in the colonial marine hydroid *Campanularia flexuosa* with special reference to senescence in hydroids. *International Review of General and Experimental Zoology* **4**, 49–79.

Turpaeva, E. P. 1967. To the question of the species relations into fouling biocenosis. *Trudy Instituta Okeanologii Akademii Nauk SSSR* **85**, 43–7. (In Russian).

Turpaeva, E. P. 1969. Symphysiological connections into fouling oligomixing biocenosis. *Doklady Akademii Nauk SSSR* **189**(2), 415–18. (In Russian).

Turpaeva, E. P. 1982. Experience of the system investigations of marine animals associations. In *System Investigations. Methodological problems.* Annual issue from Institute of Oceanology Moscow, Nauka, 360–82. (In Russian).

Uchida, T. 1963. The systematic position of the Hydrozoa. *Japanese Journal of Zoology* **14**(1), 1–14.

Ushakov, P. V. 1925. Seasonal changes on the littoral of the Kola Bay. *Trudy Leningradskogo Obschestva Estestvoispytatelii* **65**(1), 4–71. (In Russian).

Vannucci, M. 1955. On the newly liberated medusa of *Obelia hyalina* Clarke, 1879. *Dusenia* **6**(1,2), 55–60.

Vervoort, W. 1985. Deep-water hydroids. In *Peuplements profonds du golfe de Gascogne*. L. Laubier & C. Monniot (eds). IFREMER, Service Documentations Publications, Brest France, 267–97.

Vishniakov, A. E. 1989a. The photocytes ultrastructure of hydroid polyp *Obelia longissima* (Pallas) (Hydroidea, Thecaphora). In *The fundamental investigations of the recent Porifera and Coelenterata*, V. M. Koltun et al. (eds). Leningrad: Zoological Institute of Russian Academy of Sciences, 35–6. (In Russian).

Vishniakov, A. E. 1989b. The photocytes distribution in colonies of *Obelia geniculata* (L.) (Hydroidea, Thecaphora). In *The fundamental investigations of the recent Porifera and Coelenterata*, V. M. Koltun et al. (eds). Leningrad: Zoological Institute of Russian Academy of Sciences, 37–8. (In Russian).

Vishniakov, A. E. in press. Morphofunctional peculiarities of photocytes of *Obelia longissima* hydropolyp. In Obelia (*Cnidaria, Medusozoa, Hydrozoa*). *Phenomenon. Aspects of investigations. Perspectives of employment. Zoosystematica Rossica.* (Suppl.). S. D. Stepanjants (ed.).

Vysotski, E. S. in press. Calcium-regulated photoprotein of hydroid polyp *Obelia* longissima. In Obelia (*Cnidaria, Medusozoa, Hydrozoa*). *Phenomenon. Aspects of investigations. Perspectives of employment. Zoosystematica Rossica.* (Suppl.). S. D. Stepanjants (ed.).

Warner, G. F. 1977. The shapes of passive suspension feeders. In *The biology of benthic organisms*. B. P.

Keegan et al. (eds). Oxford: Pergamon Press, 567–76.

Watson, G. M. & Wood, R. L. 1989. Colloquium on terminology. In *The biology of nematocysts*, D. A. Hessinger & H. M. Lenhoff (eds). New York: Academic Press, 21–3.

Weill, R. 1934a. Contribution a l'étude des Cnidaires et de leurs nematocystes. I. Recherches sur les nematocystes. Morphologie – Physiologie – Developement. *Travaux de la Station Zoologique de Wimereux* **10**, 1–347.

Weill, R. 1934b. Contribution a l'étude des Cnidaires et de leurs nematocystes. II. Valeur taxonomique du cnidome. *Travaux de la Station Zoologique de Wimereux* **11**, 349–701.

Weismann, A. 1883. *Die Entstehung der sexualzellen bei den Hydromedusen. Zugleich als Beitrag zur Kenntnis des Baues und der Lebenserscheinungen dieser Gruppe.* Jena: Gustav Fischer Verlag.

Werner, B. 1973. New investigations on systematics and evolution of the class Scyphozoa and the phylum Cnidaria. In *Recent trends in research in coelenterate biology*, T. Tokioka & S. Nishimura (eds). *Publications of the Seto Marine Biological Laboratory* **20**, 35–61.

Westfall, J. A. 1966a. Morphology and development of nematocysts and associated structures in the Cnidaria. Dissertation Abstracts International (B) 27(3), 1002.

Westfall, J. A. 1966b. The differentiation of nematocysts and associated structures in the Cnidaria. *Zeitschrift für Zellforschung und Mikroskopische Anatomie* **75**(2), 381–406.

Westfall, J. A. 1966c. Fine structure and evolution of nematocysts. *Proceedings of the 6th International Congress of Electron Microscopy Kyoto*, 235–6.

Wulpert, J. 1902. Embrionalentwicklung von *Gonothyraea loveni. Allgemeine Zeitschrift für Wissenschaftliche Zoologie* **71**, 296–326.

Wyttenbach, C. R. 1968. The dynamics of stolon elongation in the hydroid, *Campanularia flexuosa. Journal of Experimental Zoology* **167**(3), 333–51.

Wyttenbach, C. R. 1969. Genetic variation in the mode of stolon growth in the hydroid *Campanularia flexuosa. Biological Bulletin* **137**, 547–56.

Wyttenbach, C. R., Crowell, S. & Suddith, R. L. 1973. Variations in the mode of stolon growth among different genera of colonial hydroids and their evolutionary implications. *Journal of Morphology* **139**(3), 363–75.

Zamponi, M. O. 1987. Ciclos biologicos de celenterados litorales. II. Las formas polipo y medusa de *Obelia dichotoma* Hincks, 1868 (Leptomedusae, Campanulariidae). *Neotropica* **33**(89), 3–9.

Zamponi, M. O. & Genzano, G. N. 1988. The validity of *Obelia longissima* (Pallas, 1766) (Leptomedusae, Campanulariidae). *American Zoologist* **28**(4), 10A.

Zamponi, M. O. & Genzano, G. N. 1990a. Ciclos biologicos de celenterados litorales. IV. La validez de *Obelia longissima* (Pallas,1766) (Leptomedusae,Campanulariidae). *Spheniscus* **8**, 1–7.

Zamponi, M. O. & Genzano, G. N. 1990b. The use of nematocysts for identification of the common medusa-stage of the genus *Obelia* Peron et Lesueur, 1810 (Leptomedusae, Campanulariidae) from the subantarctic region. *Plankton Newsletter* **13**, 21–3.

Zevina, G. B., Kamenskaya, O. E. & Kubanin, A. A.1975. Colonizers of the Sea of Japan foulings. The complex investigations of the Ocean nature. Moscow, Moscow State University **5**, 240–49. (In Russian)

Zirpolo, G. 1939. Caso di epibiosi di *Obelia geniculata* su *Hippocampus guttatus. Annuarr del Museo di Zoologie del Universtadt, Napoli, N.S.* **7**(8), 1–8.

Zirpolo, G. 1940. Nuovo caso di associazione di idroidi e pesci con revisione critica dei casi *Bollettino della Societa Naturalisti di Napoli* **50**, 127–39.

Appendix

The Workshop "*Obelia* as a dominant in epibiotic communities" took place within the framework of the 31st European Marine Biology Symposium (31st EMBS, St Petersburg, 9–13 September 1996) and had the following goals.

1. To bring together all the information on the Obeliinae.
2. To elaborate an unified programme for their investigation.
3. To prepare species collections of *Obelia* and closely related genera which are to be accumulated in some leading Museums of the world (i.e. British Museum, and the Zoological Museums in St Petersburg and some other cities).

Twenty-three presentations were prepared for this Workshop. However, unfortunately not all colleagues were able to participate in the Conference.

The Organizing Committee decided not to include *Obelia* presentations in the Proceedings of the 31st EMBS, but to publish a separate volume "*Obelia* (Cnidaria, Medusozoa, Hydrozoa). Phenomenon. Aspects of investigation. Perspectives of employment". This volume will include papers not only from the participants in the Workshop, but also from colleagues from different countries who could not participate in the Conference but had been working with *Obelia* for many years and were kind enough to agree to send contributions for the volume.

The 19 papers included may be grouped into topics according to their different fields of investigation.

1. Literary–historical section

"A changing taxonomic paradigm: studies on *Obelia* and some other Campanulariidae (Cnidaria: Hydrozoa)", by P. Cornelius (British Museum Natural History, London, UK).

Cornelius has been investigating *Obelia* for more than 20 years. Now he is one of the most authoritative Hydrozoa taxonomists, specifically of thecate hydroids. This article of his is not just a review of taxonomic and evolutionary concepts on Campanulariidae. Some controversial problems on *Obelia* and related genera taxonomy are discussed. Moreover, the problems of evolutionary aspects, the interrelationships of genera, the origin of the free medusa stage in *Obelia*, the relationships among different Campanulariidae subfamilies, perspectives of investigation of representatives of the Campanulariidae, and other questions are treated in this article.

2. Morphological–taxonomic section, regarding the different approaches to investigation of *Obelia* taxonomy

"Nematocyst comparisons as taxonomic tools within Campanulariidae (Hydrozoa). A review study based on light and scanning electron microscopy", by C. Ostman (Uppsala, Sweden).

"Fauna of *Obelia* (Cnidaria: Hydrozoa) in Japanese waters, with special reference to life cycle of *Obelia dichotoma*", by S. Kubota (Seto Marine Biological Laboratory, Kyoto University, Japan).

"New approaches to search of the interspecies differences in Genus *Obelia* frames.

1. Analysis of the morphological characters of *Obelia* polyps colonies with use of computer packages BIKEY.
2. Analysis of the growth and development parameters of the different species *Obelia* free medusae from the different areas of the ocean (experimental conditions)", by S. Stepanjants, A. Lobanov & M. Dianov (St. Petersburg, Russia).

3. Faunistical–ecological section

"*Obelia longissima* (Pallas 1766) and *Obelia geniculata* (L. 1758) (Hydrozoa, Thecaphora, Campanulariidae) in the Barents Sea. Morphology, distribution, ecology and special features of the life history", by N. Panteleeva (Dalniye Zelentsy, Russia).

"*Obelia* (Cnidaria, Hydrozoa) in sea-grass beds of Madagascar (Indian Ocean)", by N.Gravier-Bonnet (Réunion, France).

"*Obelia longissima* (Pallas 1766). Distribution in the open part of the north-western Black Sea", by A. Terentjev (The Crimea, Ukraine).

"Long-term variation in hydroid mass of *Obelia longissima* (Pallas) (Hydrozoa, Thecaphora, Campanulariidae) on early stages of development of artificial community in mussel *Mytilus edulis* l. Mariculture at the White Sea", by B. Kunin (St Petersburg, Russia).

"*Obelia* as a regulator of the quantity of the early larvae of the White Sea herring", by O. Ivanchenko (St Petersburg, Russia).

4. Ethological–ecological section

The works of the Moscow scientific school.

"Functional morphology differences between the hydroids: *Obelia longissima, O.geniculata, Gonothyraea loveni* and *Laomedea flexuosa*, occurred at the common biotopes in the White Sea", by N. Marfenin (Moscow, Russia).

"Differential larval settlement in two related colonial hydroids", by D. Orlov (Moscow, Russia).

"Role of pulsations in metamorphosis of *Gonothyraea loveni* (Allman 1859) planula", by Y. Burykin (Moscow, Russia).

The works of the St Petersburg scientific school:

"Interactions between larvae *Obelia loveni* and macroalgae are mediated by bacterial-algal films", by S. Chikadze & A. Railkin (St Petersburg, Russia)

"Macroalgae and microbial film determine substrate selection by the *Gonothyraea loveni* planulae", by S. Dobretzov (St Petersburg, Russia)

5. Cyto-embryological section

These investigations were performed by the colleagues of St Petersburg and Moscow Universities.

"Hydranth development in campanulariid hydroids: origin, resorption, regeneration", by V. Donakov, D. Polteva & G. Genikhovich (St Petersburg, Russia)

"The embryonic development of the *Obelia* species with sessile gonophores", by Y. Krauss (Moscow) & A.Rodimov (St Petersburg, Russia)

6. Cyto–biochemical section

"Calcium-regulated photoprotein of hydroid polyp *Obelia longissima*", by E. Vysotski (Krasnoyarsk, Russia).

"Photogenic cells of hydropolyps *Obelia longissima* and *Obelia geniculata*", by A. Vishniakov (St Petersburg, Russia).

Oceanography and Marine Biology: an Annual Review 1998, **36**, 217–340

UCL Press

SYMBIOTIC POLYCHAETES: REVIEW OF KNOWN SPECIES

D. MARTIN[1] & T. A. BRITAYEV[2]

[1] *Centre d'Estudis Avançats de Blanes (CSIC), Camí de Santa Bàrbara s/n, 17300-Blanes (Girona), Spain*

[2] *A. N. Severtzov Institute of Ecology and Evolution (RAS), Laboratory of Marine Invertebrates Ecology and Morphology, Leninsky Pr. 33, 129071 Moscow, Russia*

Abstract Although there have been numerous isolated studies and reports of symbiotic relationships of polychaetes and other marine animals, the only previous attempt to provide an overview of these phenomena among the polychaetes comes from the 1950s, with no more than 70 species of symbionts being very briefly treated. Based on the available literature and on our own field observations, we compiled a list of the mentions of symbiotic polychaetes known to date. Thus, the present review includes 292 species of commensal polychaetes from 28 families involved in 713 relationships and 81 species of parasitic polychaetes from 13 families involved in 253 relationships. When possible, the main characteristic features of symbiotic polychaetes and their relationships are discussed. Among them, we include systematic accounts, distribution within host groups, host specificity, intra-host distribution, location on the host, infestation prevalence and intensity, and morphological, behavioural and/or physiological and reproductive adaptations. When appropriate, the possible directions for further research are also indicated.

Introduction

Living organisms are always connected in some way to other individuals of the same species and to individuals of other species living in the same area. Many of these established interactions are known as direct relationships (i.e. sex recognition, competition, predation). In these cases, the involved organisms are known as free-living. However, some species have become closely associated, often to mutual benefit (frequently referred to as "symbiosis"). These relationships usually convey a high degree of specificity, to the extent that at least one of the involved partners can no longer be considered a free-living organism.

Among the polychaetous annelids, which for the most part are free-living, crawling, burrowing and tube-dwelling, the setting-up of close associations with other marine invertebrates is a rather common phenomenon. This is not surprising because the polychaetes are probably the most frequent and abundant marine metazoans in benthic environments. They may be numerically less important on hard substrata, and some molluscs and crustaceans may co-dominate on soft bottoms, but of all metazoans only the nematodes are more ubiquitous (Fauchald & Jumars 1979). However, despite the obvious importance and increasing knowledge on the ecological role of polychaetes, the literature on symbiotic polychaetes remains largely anecdotal, with the likely exception of some widely known examples of commensalism practised by some species of the family Polynoidae. Moreover, the information is often scattered among studies dealing with many different subjects (e.g. systematics, descriptive ecology, biology).

The aim of the review is, therefore, to attract the attention of scientists to – and to encourage further studies on – the ecology of this particular and diverse group of symbionts. We make an attempt not only to summarize the available data on symbiotic polychaetes, but also to outline the lacunae in the knowledge of their ecology. However, an exception to these dispositions are the Myzostomida, a highly specialized Order of polychaetes that includes more than 130 species (Pettibone 1982) which live in association with echinoderms (Barel & Kramers 1977, Jangoux 1987). The aberrant characteristics of these worms and their special and intimate associations with echinoderms merit a specific approach that falls outside the scope of the present paper. Thus, the whole Order has been omitted, for inclusion in a further review. On the other hand, some associations previously considered as symbiotic, such as those between the orbinids *Orbinia latreilli* and *Phylo foetida* and the holothuroid *Leptosinapta gallieni* (Barel & Kramers 1977), appear to be fortuitous ecological associations (Beauchamp & Zachs 1913). As far as possible, these fortuitous associations have been excluded from the present review. Similarly, the presence of epibionts (e.g. on the carapace of crabs, see Okuda 1934, Abelló et al. 1990), endobionts (e.g. inside the aquiferous canals of sponges, see Pearse 1932, Bacescu 1971, Frith 1976, Pansini & Daglio 1980–81, Westinga & Hoetjes 1981, Alós et al. 1982, Voultsiadou-Koukoura et al. 1987, Koukouras et al. 1992, 1996, Pascual et al. 1996) or simple associates to other organisms (viz. bryozoans in Morgado & Amaral 1981a,b,c, 1984, 1985, vermetids in Ben Eliahu 1975a,b, 1976, corals in Arvanitidis & Koukouras 1994, wood-boring molluscs in Kirkegaard & Santhakumaran 1967) is not regarded as symbiosis without any demonstration of specificity, because most of the involved organisms will inhabit a wide variety of substrata and the presence or absence of the so-called "host" is irrelevant.

The present review provides a list of known symbiotic polychaetes arranged into several huge tables, including the symbiotic species from each polychaete family, the host group and species name, the type of association and the source of information. There are three main tables: non-boring commensals, non-boring parasites, and symbiotic borers. The main body of the review is divided into three main sections according to these three tables. Each section is subdivided into several parts, starting with the analyses of the respective list of relationships. Then follow the information on morphological, biological, ecological and behavioural trends characterizing the symbiotic polychaetes.

Background: previous reviews

Although the oldest known review of commensal organisms is probably that by Beneden (1869), the first – and certainly the only known – reviews entirely dedicated to symbiotic polychaetes are those by Paris (1955), Clark (1956) and Britayev (1981). The first is a collection of some of the typical commensal and parasitic relationships involving polychaetes known to date. The second includes a very full list of symbiotic polychaetes and a description of one new association. The last is the first modern attempt to summarize the main trends of polychaete as symbionts. There have also been a number of reviews in the past that have addressed particular aspects of symbiotic polychaetes. Dales (1957) reviewed the commensal interactions in marine organisms and mentioned several associations involving polychaetes. The extensive review of symbionts of Echinodermata from the NE Atlantic by Barel & Kramers (1977) included an annotated check list of symbiotic polychaetes, whereas Alvà & Jangoux (1989) updated the known relationships between polychaetes and ophiuroids. Overestreet (1983), in his review on the metazoan symbionts of crustaceans, briefly mentioned some polychaetes. Britayev (1981) described two new polychaete species commensal with coelenterates, and reviewed the bibliography on their associated polychaetes. The symbiotic invertebrates associated with the sea urchin *Echinus sculentus* L. and with the polychaete *Chaetopterus appendiculatus* (Grube) were reviewed by Comely & Ansell (1988) and Petersen & Britayev (1997), respectively. The symbiotic relationships within benthic organisms from South California, polychaetes included, were reviewed by Fox & Ruppert (1985). The symbiotic polychaetes from more or less restricted geographical areas have more often been reviewed within the framework of systematic papers (e.g. Gibbs 1969 for the Solomon Islands, Okuda 1936a, 1950, and Uchida 1975 for the Japanese waters), whereas some systematic reviews referred to groups whose members included well-known symbiotic species, such as the families Caobangidae (Jones 1974), Oenonidae (Pettibone 1957), Histriobdellidae (Moyano et al. 1993), Iphitimidae (Fage & Legendre 1933, Gaston & Benner 1981) and Polynoidae (e.g. Hanley 1989, Pettibone 1993, 1996). The shell-boring spionid species of *Polydora* and related genera were reviewed by Blake (1969, 1971, 1996). Finally, the physiology of commensal polychaetes was discussed by Davenport (1966a,b).

All these review articles, though comprehensive, address individually only particular aspects of the complex symbiotic interactions involving polychaetes. There has been a more or less steady dribble of information on symbiotic polychaetes over the years since Paris (1955) and Clark (1956). Therefore, it is now becoming possible to begin a synthesis of the available information. The present review will attempt to summarize current knowledge on symbiotic polychaetes as well as discuss the main systematic, biological, ecological and behavioural trends of these associations. When appropriate, possible directions for further research will also be indicated.

Before considering the inter-relationships between polychaetes and their associated hosts, it is necessary to define the different concepts that are most often used when describing symbiotic associations. However, it would not be appropriate here to discuss extensively the different conceptual approaches because this aspect has been largely discussed (e.g. Dales 1957, Rhode 1981, Boucher et al. 1982, Margolis et al. 1982, Addicott 1984). The particular terms adopted for the present review and the reasons for this adoption will, however, be considered in more detail.

Definition of terms

Although many attempts to classify the inter-relations between organisms have been proposed, distinctions between different types of associations are not always clear. There are many factors that further define the nature of the interactions (e.g. the degree of association among the species, the degree to which the association is necessary for survival, the kinds of benefits, the extent of reciprocal specialization by one species in response to the other, the temporal pattern, the stage of the life cycle at which the interaction occurs, the location of the guest on or inside the host). Moreover, there are almost as many definitions as there are authors writing about close associations. In practice, it is difficult to obtain unequivocal demonstrations of all factors affecting a given interaction: the terms should preferably be used as stepping stones in helping to understand the real relationships in any particular association.

A partnership in which the advantage is wholly on one side, usually at the expense of the other partner, will be referred to as "parasitism" (Dales 1957). Where an association is clearly to the advantage (not necessarily from a trophic point of view) of one of the members without seriously inconveniencing or harming the other, the relationship will be referred to as "commensalism" (Dales 1957). Any interaction in which two (or more) species reciprocally benefit from the presence of the other species is referred to as "mutualism" (Addicott 1984). There is also a typical system of symbols to describe these three types of relationships: "+", "−" and "0" indicate positive, negative and neutral effects for a given associate, respectively. Therefore, the associations can be represented as "++" (mutualism), "+0" (commensalism), and "+−" (parasitism).

The term "symbiosis" has been considered as synonymous with mutualism (e.g. Rhode 1981). However, it may also refer to the extent to which two species live in close association. Thus, in a wider sense, symbiosis is more often used to describe all kinds of organism associations (Henry 1966). In this review, we preferred to retain the generic sense of symbiosis, which will thus include parasitism, commensalism and mutualism. In this case, the partners involved in "symbiotic" associations will be referred to as "the symbiont" (or "the guest") and "the host". The host harbours the symbiont, which is usually smaller than the host and always derives benefits from the association. In addition, the symbionts tend to have a more active role in the formation of associations than the hosts. In the case of commensalisms and parasitisms, however, the symbiont may be more specifically referred to as "the commensal" and "the parasite", respectively.

Because of the difficulties of applying the experimental method to such a complex network of interactions, we often know little about their nature. Strict borderlines can not be easily drawn between the various types of associations. For instance, a parasite may become a predator by killing the host, or it may become beneficial to the host species, that is, the relationship may become mutualistic. A commensal may affect the host and sometimes damage it, and thus become a parasite. In practice, it is difficult to infer unequivocal demonstrations of benefit (be it reciprocal or not) from the analysis of an association or from the existing literature. Moreover, the interpretations of the relationships become peculiarly susceptible to anthropomorphic bias: the parasites simply seem "bad", the mutualists and commensals seem "good". However, many associations that were initially considered as parasitisms have been, with increasing knowledge, redefined as commensalistic or even mutualistic. Some examples will be mentioned in coming chapters. For the purposes of this review, we will thus consider two broad groups of associations, parasitisms and commensalisms. The cases of mutualism involving polychaetes are very few, so we have included them within the commensalisms, with indications of the bi-directional trends if known. Some species will be labelled as parasites or commensals according to the original criteria of their descriptors, without it being possible to give much more information so that the character of the interactions can be determined. However, when possible, special care will be taken to pinpoint the known relevant trends of the associations, thus allowing the readers to judge. Moreover, a special chapter will be dedicated to the polychaetes that bore into living organisms (i.e. those that produce burrows mainly, but not exclusively, in organisms with calcareous structures), which are often difficult to catalogue into the two types of associations.

Other labels that will be frequently used to define the nature of an association are: "ecto-" and "endo-" (for guests living outside or inside the host, respectively), "obligatory" and "facultative" (for guests that can not survive without the host or that can also be free living, respectively), "permanent" and "temporary" (for guests living as symbionts during their whole life or only during one phase of their life cycle, respectively), and "monoxenous" and "polyxenous" (for symbionts inhabiting one or a few hosts and those inhabiting many different hosts, respectively). Moreover, by considering the close associations as infestations, some specific terminology applied (Margolis et al. 1982). The relationship between the number of infested hosts and the total number of hosts will be referred to as "prevalence", whereas "intensity" will be the number of symbionts present in each infested host, "mean intensity" will be the mean number of individuals of a particular symbiotic species per infested host in a sample, and "abundance" or "relative density" will be the mean number of symbiont individuals per host examined (i.e. equal to prevalence multiplied by mean intensity).

Commensal polychaetes

Commensalisms are the most abundant relationships among symbiotic polychaetes. This statement may reflect a real dominance of "+0" associations. However, the low level of knowledge and scattered available information on the biology of the species involved may artificially exaggerate their relevance. Most known symbiotic polychaetes have been considered as "commensals" as their associations with other organisms lack clear "parasitic" features. However, their status may be further modified with the appraisal of new information on host-symbiont relationships. For example, the relationships involving the scaleworms *Arctonoe vittata* (with its gastropod host *Diodora aspera*) and *Gastrolepidia clavigera* (with its holothuroid hosts), rather than commensalisms, are closer to mutualism and parasitism, respectively (Dimock & Dimock 1969, T. A. Britayev pers. obs.). However, we decided to include within the commensals those species often considered as parasites but lacking specific behaviour descriptions. For example, the highly modified species of the family Spintheridae are often described as feeding on their host sponges (Pettibone 1963, 1982). However, whether this feeding activity damages the host or benefits it by cleaning its surface has not yet been demonstrated. A high morphological specialization is not in itself a demonstration of parasitism, as indicated by the re-evaluation of the histriobdellids as commensals of their host crustaceans (Jennings & Gelder 1976, Cannon & Jennings 1987).

Taxonomical distribution of commensal polychaetes

In the present review, 292 species of commensal polychaetes belonging to 28 families are reported to be involved in 713 different commensal relationships (Tables 1 and 8, p. 306; Fig. 1). The total number of polychaete families may range between 84 (Fauchald

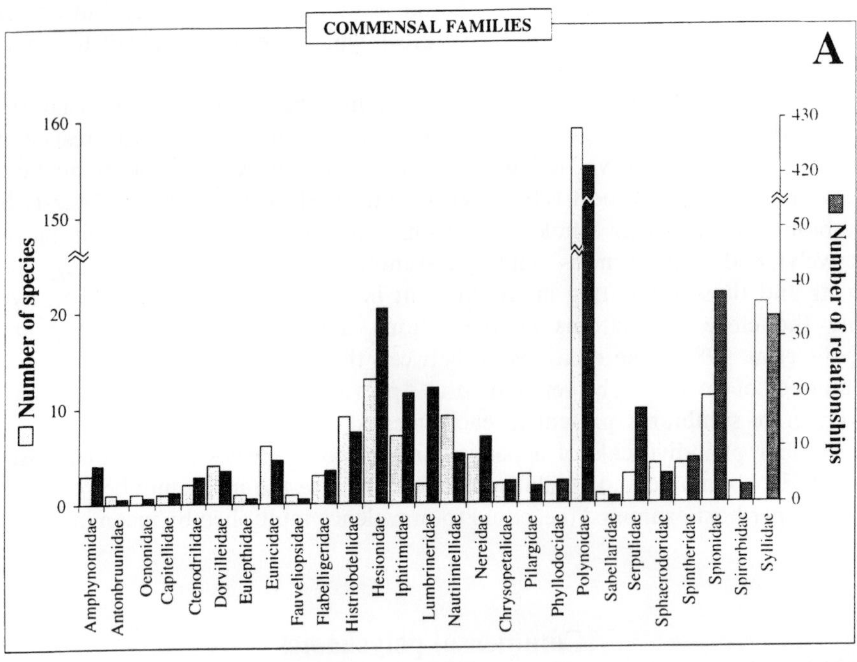

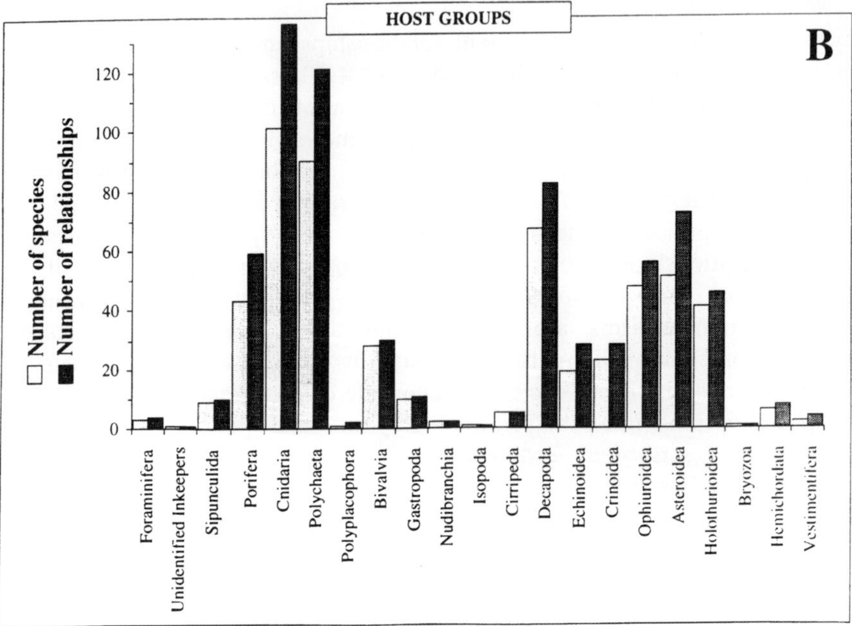

Figure 1 A, Number of commensal relationships and number of commensal species from each polychaete family including commensal forms. B, Number of species from each taxonomic group of host harbouring commensal polychaetes and number of relationships on which they are involved.

Table 1 List of commensal polychaete species (boring species excluded). Type: Type of relationship (O, obligatory; F, facultative).

Commensal species	Authority	Host group	Species	Type	References
Amphynomidae					
Hipponoe gaudichaudi	Aldouin & M. Edwards	Cirriped	*Lepas anatifera*	O	Paris 1955
Hipponoe gaudichaudi		Cirriped	*Lepas anserifera*	O	Britayev & Memmy 1989
Hipponoe gaudichaudi		Cirriped	*Lepas pectinata*	O	Pettibone 1963
Hipponoe gaudichaudi		Cirriped	*Lepas* sp.	O	Okuda 1950, Day 1967, Nuñez et al. 1991
Hipponoe gaudichaudi		Cirriped	*Dosima fascicularis*	O	Britayev & Memmy 1989
Benthoscolex cubanus	Hartman	Echinoid	*Archeopneustes hystrix*	O	Hartman 1942, Emson et al. 1993
Paramphinome pulchella	Sars	Sponge	Unidentified	F	Pettibone 1963
Antonbruunidae					
Antonbruunia viridis	Hartman & Boss	Bivalve	*Lucinia fosteri*	O	Hartman & Boss 1965
Oenonidae					
Oenone diphyllidis		Sponge	*Speciospongia vespera*	O	Pearse 1932
Capitellidae					
Capitella capitata	Fabricius	Decapod	*Pinnixia fava*	F	Clark 1956
Capitella capitata		Bivalve	*Schizothaeus capax*	F	Clark 1956
Ctenodrilidae					
Ctenodrilus serratus	Schmidt	Echinoid	*Centrostephanus coronatus*	F	Fauchald & Jumars 1979
Ctenodrilus serratus		Echinoid	*Strongylocentrotus franciscanus*	F	Fauchald & Jumars 1979
Ctenodrilus sp.		Holothuroid	*Leptosinapta inhaerens*	F	Barel & Kramers 1977
Ctenodrilus sp.		Hologhuroid	*Labidoplax digitata*	F	Barel & Kramers 1977
Ctenodrilus sp.		Hologhuroid	*Holothuria tubulosa*	F	Barel & Kramers 1977
Dorvilleidae					
Dorvillea pseudorubrovittata	Berkeley	Polychaete	*Phyllochaetopterus prolifica*	?	Petersen & Britayev 1997
Ophryotrocha geryonicola	(Esmark)	Decapod	*Cancer borealis*	O	Wesenber-Lund 1938, Gaston & Benner 1981, Pfannestiel et al. 1982
Ophryotrocha geryonicola		Decapod	*Geryon quinquedens*	O	Gaston & Benner 1981
Ophryotrocha geryonicola		Decapod	*Geryon tridens*	O	Gaston & Benner 1981
Ophryotrocha mediterranea	Martin, Abelló & Cartes	Decapod	*Geryon longipes*	O	Desportes et al. 1977, Mori & Belloni 1985, Martin et al. 1991
Ophryotrocha puerilis	Claparède & Metschnikov	Holothuroid	*Cucumaria planci*	F	Monticelli 1892, Barel & Kramers 1977, Jangoux 1987
Eulepthidae					
Grubeulepis geayi	(Fauvel)	Polychaete	*Polyodontes melanotus*	F	Day 1962, Pettibone 1969e

Table 1 *Continued*

Commensal species	Authority	Host group	Species	Type	References
Eunicidae					
Eunice antennata ?	(Savigny)	Cnidarian	Unidentified coral	F	Gardiner 1976
Eunice filamentosa	Grube	Sponge	Unidentified (small)	F?	Treadwell 1921
Eunice floridana	Pourtales	Cnidarian	*Lophohelia prolifera*	F	Britayev 1981
Eunice floridana		Cnidarian	*Amphihelia oculata*	F	Britayev 1981
Eunice harassii	Aldouin & M. Edwards	Bivalve	*Ostrea edulis*	F	Gravier 1900
Eunice marovoi	Gibbs	Gastropod	*Cerithium vertagus*	F	Gibbs 1969, 1971
Eunice pennata	(O. F. Muller)	Cnidarian	*Madrepora pertusa*	F	Fauchald 1992
Eunice pennata		Cnidarian	Madreporid coral	F	Fauvel 1923
Fauveliopsidae					
Fauveliopsis olgae	Hartmann-Schröder	Foraminifer	*Bathisiphon rusticum*	O	Hartmann-Schröder 1983
Flabelligeridae					
Flabelligera affinis	Sars	Echinoid	*Echinus acutus*	F	Barel & Kramers 1977
Flabelligera affinis		Echinoid	*Echinus esculentus*	F	Fauvel 1923, Clark 1956, Barel & Kramers 1977
Flabelligera affinis		Echinoid	*Psammechinus miliaris*	F	Barel & Kramers 1977
Flabelliderma commensalis	(Moore)	Echinoid	*Centrostephanus coronatus*	O	Spies 1977, Fauchald & Jumars 1979
Flabelliderma commensalis		Echinoid	*Strongylocentrotus purpuratus*	O	Clark 1956
Phalacrostemma cidariophilum	Marenzeller	Echinoid	*Cidaris cidaris*	O	Marenzeller 1895, Bellan 1964
Histriobdellidae					
Histriobdella homari	Van Beneden	Decapod	*Homarus americanus*	O	Uzmann 1967, Simon 1967, 1968, Jennings & Gelder 1976, Brettey & Campbell 1985
Histriobdella homari		Decapod	*Homarus vulgaris*	O	Moyano et al. 1993
Stratiodrilus aeglaphilus	Vila & Bahamonde	Decapod	*Aegla laevis talcahuano*	O	Vila & Bahamonde 1985, Moyano et al. 1993
Stratiodrilus haswelli	Harrison	Decapod	*Astacoides madagascariensis*	O	Harrison 1928
Stratiodrilus novaehollandeae	Haswell	Decapod	*Cherax dispar*	O	Moyano et al. 1993
Stratiodrilus novaehollandeae		Decapod	*Cherax punctatus*	O	Moyano et al. 1993
Stratiodrilus platensis	Cordero	Decapod	*Aegla abtao*	O	Gacitua 1992 in Moyano et al. 1993
Stratiodrilus platensis		Decapod	*Aegla bahamondei*	O	Gacitua 1992 in Moyano et al. 1993
Stratiodrilus platensis		Decapod	*Aegla laevis laevis*	O	Cordero 1927
Stratiodrilus pungnaxi	Vila & Bahamonde	Decapod	*Parastacus pugnax*	O	Vila & Bahamonde 1985, Moyano et al. 1993
Stratiodrilus tasmanicus	Haswell	Decapod	*Astacopsis franklinii*	O	Haswell 1900, Harrison 1928
Stratiodrilus cirolanae	Führ	Isopod	*Cirolana venusticauda* var. *simplex*	O	Führ 1971
Stratiodrilus sp.		Decapod	Unknown	O	Moyanao et al. 1993

Table 1 *Continued*

Commensal species	Authority	Host group	Species	Type	References
Hesionidae					
Gyptis brevipalpa	Hartmann-Schröder	Polychaete	*Glycera robusta*	F	Gardiner 1976
Gyptis brevipalpa		Holothuroid	*Leptosynapta tenuis*	F	Gardiner 1976
Gyptis ophiocomae	Storch & Niggemann	Ophiuroid	*Ophiocoma scolopendrina*	O	Storch & Niggemann 1967
Gyptis vittata	Webster & Benedict	Polychaete	*Notomastus lobatus*	?	Pettibone 1963, Gardiner 1976
Hesionella mccullochae	Hartman	Polychaete	*Lumbrineris brevicirra*	F	Hartman 1941
Microphthalmus aberrans	Webster & Benedict	Polychaete	*Lysilla alba*	F	Pettibone 1963
Microphthalmus aberrans		Polychaete	*Enoplobranchus sanguineus*	F	Pettibone 1963
Microphthalmus hamosus	Westheide	Sipunculid	*Sipunculus nudus*	F	Westheide 1982
Microphthalmus sczelkowii	Meckznicov	Polychaete	*Arenicola marina*	F	Reise & Ax 1979
Ophiodromus angustifrons	(Grube)	Asteroid	*Pentaceros hedemanni*	O	Jones 1964
Ophiodromus flexuosus	(Delle Chiaje)	Asteroid	*Astropecten aranciacus*	F	Barel & Kramers 1977
Ophiodromus flexuosus		Asteroid	*Astropecten bispinosus*	F	Barel & Kramers 1977
Ophiodromus flexuosus		Asteroid	*Astropecten platyacanthus*	F	Barel & Kramers 1977
Ophiodromus flexuosus		Asteroid	*Astropecten irregularis*	F	Barel & Kramers 1977
Ophiodromus flexuosus		Asteroid	*Louidia ciliaris*	F	Barel & Kramers 1977
Ophiodromus flexuosus		Balanoglossid	*Balanoglossus* sp.	F	Clark 1956
Ophiodromus flexuosus		Holothuroid	*Leptosynapta* sp.	F	Barel & Kramers 1977
Ophiodromus flexuosus		Polychaete	*Amphitrite edwardsi*	F	Barel & Kramers 1977
Ophiodromus flexuosus		Polychaete	*Euclymene lumbricoides*	F	Barel & Kramers 1977
Ophiodromus obscurus	Verrill	Polychaete	*Lysilla alba*	F	Pettibone 1963
Ophiodromus obscurus		Echinoid	*Lytechinus* sp.	F	Hartman 1951
Ophiodromus obscurus		Holothuroid	*Thyone* sp.	F	Pettibone 1963
Ophiodromus pugettensis	(Johnson)	Asteroid	*Patiria miniata*	F	Bartel & Davenport 1956, Davenport et al. 1960, Lande & Reish 1968, Ricketts et al. 1985
Ophiodromus pugettensis		Asteroid	*Louidia foliolata*	F	Davenport et al. 1960
Ophiodromus pugettensis		Asteroid	*Louidia magnifica*	F	Storch & Rosito 1981
Ophiodromus pugettensis		Asteroid	*Pteraster tesselatus*	F	Storch & Niggemann 1967
Ophiodromus pugettensis		Asteroid	*Orcaster occidentalis*	F	Steinbeck & Ricketts 1941
Ophiodromus pugettensis		Asteroid	*Pisaster ochraceus*	F	Davenport & Hickok 1957, Hickok & Davenport 1957
Ophiodromus pugettensis		Asteroid	*Petalaster foliolata*	F	Stewart 1970, Hilbig 1994
Ophiodromus pugettensis		Holothuroid	*Protankyra bidentata*	F	Okuda 1936a
Ophiodromus pugettensis		Decapod	*Eupagurus* sp.	F	Hickok & Davenport 1957
Ophiodromus pugettensis		Gastropod	*Aletes* sp.	F	Storch & Niggemann 1967

Table 1 *Continued*

Commensal species	Authority	Host group	Species	Type	References
Ophiodromus pugettensis		Bivalve	*Chama* sp.	F	Storch & Niggermann 1967
Ophiodromus pugettensis		Echinoid	*Clypeaster humilis*	F	Storch & Niggermann 1967
Parahesione luteola	(Webster)	Decapod	*Upogebia affinis*	F	Pettibone 1956
Parasyllidea humesi	Pettibone	Bivalve	*Tellina nimphalis*	O	Pettibone 1961
Iphitimidae					
Iphitime cuenoti	Fauvel	Decapod	*Liocarcinus depurator*	O	Fage & Legendre 1925, 1933, Abelló 1985, Belloni & Mori 1985, Abelló et al. 1988, Comely & Ansell 1889
Iphitime cuenoti		Decapod	*Liocarcinus corrugatus*	O	Comely & Ansell 1989
Iphitime cuenoti		Decapod	*Liocarcinus puber*	O	Comely & Ansell 1989
Iphitime cuenoti		Decapod	*Carcinus maenas*	O	Comely & Ansell 1989
Iphitime cuenoti		Decapod	*Cancer pagurus*	O	Comely & Ansell 1989
Iphitime cuenoti		Decapod	*Hyas araneus*	O	Comely & Ansell 1989
Iphitime cuenoti		Decapod	*Inachus dorsettensis*	O	Comely & Ansell 1989
Iphitime cuenoti		Decapod	*Macropipus tuberculatus*	O	Abelló 1985, Mori & Belloni 1985, Abelló et al. 1988
Iphitime cuenoti		Decapod	*Goneplax rhomboides*	O	Fage & Legendre 1925, 1933, Abelló et al. 1988
Iphitime cuenoti		Decapod	*Macropodia tenuirostris*	O	Fage & Legendre 1925, 1933, Hartnoll 1962
Iphitime cuenoti		Decapod	*Maia squinado*	O	Fauvel 1914
Iphitime doderleinii	Marenzeller	Decapod	*Macrocheria kaempferi*	O	Marenzeller 1902
Iphitime hartmanae	Kirkegaard	Decapod	*Hyas araneus*	O	Kirkegaard 1977
Iphitime hartmanae		Decapod	*Hyas coarctatus*	O	Kirkegaard 1977
Iphitime holobranchiata	Pilger	Decapod	*Cancer antennarius*	O	Pilger 1971
Iphitime loxorhinchii	Hertman	Decapod	*Loxorhinchus grandis*	O	Hartman 1952
Iphitime paguri	Fage & Legendre	Decapod	*Eupagurus bernhardus*	O	Fage & Legendre 1933, Moore & Gorzula 1973, Comely & Ansell 1989
Iphitime paguri		Decapod	*Eupagurus prideauxi*	O	Spooner et al. 1957, Comely & Ansell 1989
Iphitime sartoral	Paiva & Nonato	Decapod	*Libinia spinosa*	O	Paiva & Nonato 1991
Iphitime sartoral		Decapod	*Portunus spinicarpus*	O	Paiva & Nonato 1991

Table 1 *Continued*

Commensal species	Authority	Host group	Species	Type	References
Lumbrineridae					
Lumbrineris flabellicola	(Fage)	Cnidarian	*Acanthocyathus spiniger*	O	Miura & Shirayama 1992
Lumbrineris flabellicola		Cnidarian	*Balanophyllia cellulosa*	O	Zibrowius et al. 1975
Lumbrineris flabellicola		Cnidarian	*Balanophyllia* sp.	O	Zibrowius et al. 1975
Lumbrineris flabellicola		Cnidarian	*Caryophyllia sarsiae*	O	Zibrowius et al. 1975
Lumbrineris flabellicola		Cnidarian	*Caryophyllia smithii*	O	Zibrowius et al. 1975
Lumbrineris flabellicola		Cnidarian	*Caryophyllia* sp.	O	Zibrowius et al. 1975
Lumbrineris flabellicola		Cnidarian	*Ceratotrochus duodecimcostatus*		Zibrowius et al. 1975
Lumbrineris flabellicola		Cnidarian	*Dendrophyllia cornigera*	O	Zibrowius et al. 1975
Lumbrineris flabellicola		Cnidarian	*Dendrophyllia cornucopia*	O	Zibrowius et al. 1975
Lumbrineris flabellicola		Cnidarian	*Desmophyllum cristagalli*	O	Zibrowius et al. 1975
Lumbrineris flabellicola		Cnidarian	*Flabellum magnificum*	O	Zibrowius et al. 1975
Lumbrineris flabellicola		Cnidarian	*Flabellum avicola*	O	Zibrowius et al. 1975
Lumbrineris flabellicola		Cnidarian	*Flabellum chunii*	O	Marenzeller 1904, Fage 1936, Zibrowius et al. 1975
Lumbrineris flabellicola		Cnidarian	*Flabellium fornasinii*	O	Zibrowius et al. 1975
Lumbrineris flabellicola		Cnidarian	*Flabellium inconstans*	O	Zibrowius et al. 1975
Lumbrineris flabellicola		Cnidarian	*Flabellium pavoninum*	O	Zibrowius et al. 1975
Lumbrineris flabellicola		Cnidarian	Unidentified hydroids	O	Zibrowius et al. 1975
Lumbrineris flabellicola		Cnidarian	Scleractinian coral		Miura & Shirayama 1992
Lumbrineris flabellicola		Cnidarian	*Stephanocyathus moseleyanus*	O	Zibrowius et al. 1975
Lumbrineris flabellicola		Cnidarian	Unidentified zoantharian	O	Zibrowius et al. 1975
Lumbrineris coccinea	(Renier)	Cnidarian	Unidentified coral	F	Gardiner 1976
Nautiliniellidae					
Flascarpia alvinae	Blake	Bivalve	Unidentified	?	Blake 1993
Laubierus mucronatus	Blake	Bivalve	Unidentified mytilid	O	Blake 1993
Mytilidiphila enseiensis	Miura & Hasimoto	Bivalve	Unidentified mytilid	O	Miura & Hashimoto 1993
Mytilidiphila okinavaensis	Miura & Hasimoto	Bivalve	Unidentified mytilid	O	Miura & Hashimoto 1993
Natsushima bifurcata	Miura & Laubier	Bivalve	*Solemya* sp.	O	Miura & Laubier 1990
Nautiliniella calyptogenicola	Miura & Laubier	Bivalve	*Calyptogena phaseoliformis*	O	Miura & Laubier 1989, 1990
Petrecca thyasira	Blake	Bivalve	*Thyasira insignis*	O	Blake 1990
Shinkai longipedata	Miura & Ohta	Bivalve	*Calyptogena* sp.	O	Miura & Ohta 1991
Shinkai sagamiensis	Miura & Laubier	Bivalve	*Calyptogena soyoae*	O	Miura & Laubier 1990

Table 1 *Continued*

Commensal species	Authority	Host group	Species	Type	References
Nereidae					
Cheilonereis cyclurus	(Harrington)	Decapod	*Pagurus armatus*	O	Harrington 1897, Clark 1956, Hickok & Davenport 1957
Cheilonereis cyclurus		Decapod	*Pagurus branchiomastus*	O	Buzhinskaja 1967
Cheilonereis cyclurus		Decapod	*Pagurus ochotoensis*	O	Buzhinskaja 1967, Wu et al. 1985
Cheilonereis cyclurus		Decapod	*Pagurus* sp.	O	Okuda 1950
Nereis acuminata	Ehlers	Decapod	*Clibanarius vittatus*	F	Gardiner 1976
Nereis fucata	Savigny	Decapod	*Anapagurus laevis*	O	Spooner et al. 1957
Nereis fucata		Decapod	*Eupagurus bernhardus*	O	Spooner et al. 1957, Clark 1956, Gilpin-Brown 1969, Goerke 1971, Jensen & Bender 1973
Nereis fucata		Decapod	*Pagurus cuanensis*	O	Spooner et al. 1957
Nereis fucata		Decapod	*Pagurus prideauxi*	O	Clark 1956, Spooner et al. 1957
Nereis fucata		Decapod	*Pagurus variabilis*	O	Gilpin-Brown 1969
Nereis grayi	Pettibone	Polychaete	*Asychis elongatus*	O	Pettibone 1982
Perinereis cultrifera	(Grube)	Bivalve	*Pteria martensi*	F	Okuda 1950
Chrysopetalidae					
Bhawania goodei	Webster	Sipunculid	*Aspidosiphon elegans*	F	Gibbs 1969
Bhawania goodei		Sipunculid	*Cleosiphon aspergillum*	F	Gibbs 1969
Bhawania goodei		Sipunculid	*Phascolosoma albolineatum*	F	Gibbs 1969
Bhawania potsiana	(Horst)	Polychaete	*Eurithoe complanata*	F	Gibbs 1969, 1971
Pilargidae					
Ancistrosyllis commensalis	Gardiner	Polychaete	*Notomastus lobatus*	O	Gardiner 1976
Ancistrosyllis groenlandica	McIntosh	Holothuroid	*Molpadia* sp.	F	Ganapati & Radhakrishna 1962
Pilargis berkeleyae	Monro	Polychaete	*Chaetopterus cautus*	F	Buzinskaja et al. 1980, Britaeyev 1993, Petersen & Britaeyev 1997
Phyllodocidae					
Mystides bathysiphonicola	Hartmann-Schröder	Foraminifer	*Bathysiphon rusticum*	O	Hartmann-Schröder 1983
Mystides bathysiphonicola		Foraminifer	*Bathysiphon* sp.	O	Hartmann-Schröder 1983
Mystides bathysiphonicola		Foraminifer	*Hormosina normani*	O	Hartmann-Schröder 1983
Protomystides hatsushimensis	Miura	Vetimentiferan	Unidentified	O	Miura 1988
Polynoidae					
Acholoe astericola	(Delle Chiaje)	Asteroid	*Astropecten aranciacus*	O	Barel & Kramers 1977, D. Martin pers. obs.
Acholoe astericola		Asteroid	*Astropecten bispinosus*	O	Marenzeller 1874
Acholoe astericola		Asteroid	*Astropecten hupferi*	O	Pettibone 1996
Acholoe astericola		Asteroid	*Astropecten irregularis*	O	McIntosh 1900, Cuenot 1912, Davenport 1953a, Barel & Kramers 1977

Table 1 *Continued*

Commensal species	Authority	Host group	Species	Type	References
Acholoe astericola		Asteroid	*Astropecten pentacanthus*	O	Pettibone 1996
Acholoe astericola		Asteroid	*Astropecten platyacanthus*	O	Barel & Kramers 1977
Acholoe astericola		Asteroid	*Astropecten* sp.	O	Barel & Kramers 1977
Acholoe astericola		Asteroid	*Louidia ciliaris*	O	Davenport 1953a, 1954, Barel & Kramers 1977
Adyte assimilis	(McIntosh)	Echinoid	*Echinus esculentus*	F	McIntosh 1900, Pettibone 1969a, Barel & Kramers 1977, Comely & Ansell 1988
Adyte assimilis		Echinoid	*Echinus acutus*	F	Barel & Kramers 1977
Alentiana aurantiaca	(Verril)	Cnidarian	*Bolocera tuediae*	F	Pettibone 1963
Antinoella sarsi	(Malmgren)	Echinoid	*Spatangus purpureus*	F	Barel & Kramers 1977
Antipathipolyeunoa nuttingi	Pettibone	Cnidarian	*Antipathes tanacetum*	O	Pettibone 1991b
Arcteobia anticostiensis	(McIntosh)	Polychaete	*Pista flexuosa*	F	Pettibone 1954, 1963
Arcteobia anticostiensis		Polychaete	Unidentified maldanid	F	Pettibone 1954, 1963
Arctonoe fragilis	(Baird)	Asteroid	*Evasterias troschelii*	O	Davenport & Hickok 1951, Britayev 1991, Ruff 1995
Arctonoe fragilis		Asteroid	*Leptasterias aequalis*	O	Britayev 1991
Arctonoe fragilis		Asteroid	*Leptasterias hexactis*	O	Britayev 1991
Arctonoe fragilis		Asteroid	*Louidia foliolata*	O	Ruff 1995
Arctonoe fragilis		Asteroid	*Orthasterias koehleri*	O	Britayev 1991
Arctonoe fragilis		Asteroid	*Pisaster giganteus*	O	Hanley 1989
Arctonoe fragilis		Asteroid	*Pisaster ochraceus*	O	Britayev 1991
Arctonoe fragilis		Asteroid	*Solaster dawsoni*	O	Britayev 1991
Arctonoe fragilis		Asteroid	*Stylasterias forreri*	O	Britayev 1991
Arctonoe pulchra	(Johnson)	Asteroid	*Dermasterias imbricata*	O	Britayev 1991
Arctonoe pulchra		Asteroid	*Louidia foliolata*	O	Britayev 1991
Arctonoe pulchra		Asteroid	*Pteraster tesselatus*	O	Britayev 1991
Arctonoe pulchra		Asteroid	*Solaster stimpsoni*	O	Britayev 1991
Arctonoe pulchra		Gastropod	*Megathura crenulata*	O	Britayev 1991
Arctonoe pulchra		Holothuroid	*Stichopus californicus*	O	Davenport & Hickok 1951, Britayev 1991
Arctonoe pulchra		Holothuroid	*Stichopus parvimensis*	O	Davenport 1950, Britayev 1991
Arctonoe pulchra		Polychaete	*Loimia grubei*	O	Britayev 1991, Ruff 1995
Arctonoe pulchra		Polyplacophor	*Cryptochiton stelleri*	O	Britayev 1991

Table 1 *Continued*

Commensal species	Authority	Host group	Species	Type	References
Arctonoe vittata	(Grube)	Asteroid	*Aphelasterias japonica*	O	Britayev 1991
Arctonoe vittata		Asteroid	*Asterias amurensis*	O	Okuda 1936b, Britayev 1991
Arctonoe vittata		Asteroid	*Asterias rathbunae*	O	Britayev et al. 1989, Britayev 1991
Arctonoe vittata		Asteroid	*Crossaster papposus*	O	Britayev 1991
Arctonoe vittata		Asteroid	*Dermasterias imbricata*	O	Pettibone 1953, Wagner et al. 1979, Britayev 1991, Ruff 1995
Arctonoe vittata		Asteroid	*Evasterias echinosoma*	O	Britayev 1991
Arctonoe vittata		Asteroid	*Henricia leviuscula*	O	Britayev 1991
Arctonoe vittata		Asteroid	*Leptasterias kamtchatica*	O	Britayev 1991
Arctonoe vittata		Asteroid	*Louidia foliolata*	O	Britayev 1991
Arctonoe vittata		Asteroid	*Pteraster tesselatus*	O	Britayev 1991
Arctonoe vittata		Asteroid	*Solaster dawsoni*	O	Britayev 1991
Arctonoe vittata		Asteroid	*Solaster endeca*	O	Britayev 1991
Arctonoe vittata		Asteroid	*Solaster papposus*	O	Barel & Kramers 1977
Arctonoe vittata		Asteroid	*Solaster stimpsoni*	O	Britayev 1991
Arctonoe vittata		Cnidarian	*Metridium senile*	O	Britayev 1991
Arctonoe vittata		Echinoid	*Cidarina cidaris*	O	Britayev 1991
Arctonoe vittata		Gastropod	*Acmaea mitra*	O	Britayev 1991
Arctonoe vittata		Gastropod	*Acmaea pallida*	O	Britayev 1991
Arctonoe vittata		Gastropod	*Diodora aspera*	O	Pettibone 1953, Gerber & Stout 1968, Dimock & Dimock 1969
Arctonoe vittata		Gastropod	*Diodora aspera*	O	Hanley 1989, Britayev 1991, Ruff 1995
Arctonoe vittata		Gastropod	*Haliotis kamchatkana*	O	Okuda 1936b, Britayev 1991
Arctonoe vittata		Gastropod	*Patelloida* sp.	O	Okuda 1936b
Arctonoe vittata		Gastropod	*Puncturella cucullata*	O	Britayev 1991
Arctonoe vittata		Gastropod	*Puncturella multistriata*	O	Britayev 1991
Arctonoe vittata		Holothuroid	*Parastichopus californicus*	O	Britayev 1991
Arctonoe vittata		Holothuroid	*Stichopus japonicus*	O	Britayev 1991
Arctonoe vittata		Polychaete	*Neoamphitrite robusta*	O	Pettibone 1953, Britayev 1991
Arctonoe vittata		Polychaete	*Neoamphitrite* sp.	O	Britayev 1991
Arctonoe vittata		Polychaete	*Thelepus crispus*	O	Pettibone 1953, Britayev 1991
Arctonoe vittata		Polyplacophor	*Cryptochiton stelleri*	O	Webster 1968, Britayev 1991, Ruff 1995
Asterophilia carlae	Hanley	Asteroid	*Culcita novaeguineae*	O	T. A. Britayev pers. obs.
Asterophilia carlae		Asteroid	*Linckia laevigater*	O	Hanley 1989
Asterophilia carlae		Asteroid	*Protoreaster nodosus*	O	T. A. Britayev pers. obs.
Asterophilia carlae		Asteroid	Unidentified	O	Hanley 1989
Asterophilia carlae		Crinoid	Unidentified	O	T. A. Britayev pers. obs.

Table 1 *Continued*

Commensal species	Authority	Host group	Species	Type	References
Australaugeneria michelseni	Pettibone	Cnidarian	Unidentified alcyonarian	F	Pettibone 1969a
Australaugeneria michelseni		Cnidarian	Unidentified gorgonian	F	Pettibone 1969a
Australaugeneria pottsi	(Potts)	Cnidarian	Unidentified gorgonian	O	Pettibone 1969a, 1969d
Australaugeneria rutilans	(Grube)	Cnidarian	*Xenia* sp.	F	Okuda 1950, Pettibone 1969a
Bathynoe cascadiensis	Ruff	Asteroid	*Astrocles actinodetus*	O	Ruff 1991
Bathynoe cascadiensis		Asteroid	*Astrolirus panamensis*	O	Ruff 1991
Bathynoe tuberculata	(Treadwell)	Asteroid	*Brisinga alberti*	O	Hanley 1989, Ruff 1991
Bayerpolynoe floridensis	Pettibone	Cnidarian	*Antipathes columnaris*	O	Pettibone 1991b
Benhamipolynoe anthipathicola	(Benham)	Cnidarian	*Parantipathes tenuispina*	O	Pettibone 1970, 1989a
Benhamipolynoe anthipathicola		Cnidarian	*Antipathes columnaris*	O	Pettibone 1970, 1989a
Benhamipolynoe anthipathicola		Polychaete	Unidentified eunicid	O	Pettibone 1970
Benhamipolynoe cairnsi	Pettibone	Cnidarian	*Conopora adeta*	O	Pettibone 1989a, Hanley & Burke 1991b
Branchipolynoe seepensis	Pettibone	Bivalve	Unidentified Mytilid	O	Pettibone 1986a
Branchipolynoe symmitylida	Pettibone	Bivalve	*Bathymodiolus thermophilus*	O	Pettibone 1984a
Branchipolynoe pettiboneae	Miura & Hashimoto	Bivalve	*Adula* sp.	O	Miura & Hashimoto 1991
Branchipolynoe pettiboneae		Bivalve	*Bathymodiolus* sp.	O	Miura & Hashimoto 1991
Brychionoe karenae	Hanley & Burke	Cnidarian	*Leiopathes* sp.	O	Hanley & Burke 1991a
Capitulatinoe cupisetis	Hanley & Burke	Asteroid	*Astropecten granulatus*	O	Hanley & Burke 1989
Disconatis accolus	(Estcourt)	Polychaete	*Abarenicola affinis*	O	Hanley & Burke 1988
Disconatis contubernalis	Hanley & Burke	Polychaete	Unidentified maldanid	O	Hanley & Burke 1988
Enipo gracilis	Verrill	Polychaete	*Nicomache lumbricalis*	F	Berkeley & Berkeley 1942, Pettibone 1953, 1954
Eunoe bathydomus	Ditlevsen	Holothuroid	*Bathyplotes natans*	O	Wesenberg-Lund 1941
Eunoe depressa	Moore	Decapod	Unidentified pagurid	F	Moore 1905, 1908
Eunoe depressa		Decapod	*Paralithodes kamtchatica*	F	Hartman 1948
Eunoe laetmogonensis	Kirkegaard & Billet	Holothuroid	*Laetmogone violacea*	O	Kirkegaard & Billett 1980
Gastrolepidia clavigera	Schmarda	Holothuroid	*Actinopyga echinites*	O	T. A. Britayev pers. obs.
Gastrolepidia clavigera		Holothuroid	*Actinopyga mauritana*	O	J. R. Hanley pers. comm.
Gastrolepidia clavigera		Holothuroid	*Actinopyga* sp.	O	Potts 1910
Gastrolepidia clavigera		Holothuroid	*Bohadschia argus*	O	Gibbs 1969, Hanley 1989
Gastrolepidia clavigera		Holothuroid	*Bohadschia graffei*	O	Gibbs 1969, T. A. Britayev pers. obs.
Gastrolepidia clavigera		Holothuroid	*Holothuria atra*	O	Day 1967, Gibbs 1969, Hanley 1989
Gastrolepidia clavigera		Holothuroid	*Holothuria edulis*	O	J. R. Hanley pers. comm.
Gastrolepidia clavigera		Holothuroid	*Holothuria gyrifer*	O	Hartman 1954
Gastrolepidia clavigera		Holothuroid	*Holothuria leucospilota*	O	T. A. Britayev pers. obs.

Table 1 *Continued*

Commensal species	Authority	Host group	Species	Type	References
Gastrolepidia clavigera		Holothuroid	*Holothuria maculata*	O	Potts 1910
Gastrolepidia clavigera		Holothuroid	*Holothuria scabra*	O	J. R. Hanley pers. comm.
Gastrolepidia clavigera		Holothuroid	*Microthele axiologa*	O	J. R. Hanley pers. comm.
Gastrolepidia clavigera		Holothuroid	*Microthele nobilis*	O	J. R. Hanley pers. comm.
Gastrolepidia clavigera		Holothuroid	*Stichopus chloronotus*	O	Gibbs 1971, T. A. Britayev pers. obs.
Gastrolepidia clavigera		Holothuroid	*Stichopus hermeni*	O	Hanley 1989
Gastrolepidia clavigera		Holothuroid	*Stichopus horrens*	O	Hartman 1954
Gastrolepidia clavigera		Holothuroid	*Stihopus mollis*	O	Hanley 1993
Gastrolepidia clavigera		Holothuroid	*Stichopus variegatus*	O	Hanley 1989
Gastrolepidia clavigera		Holothuroid	*Thelenota ananas*	O	Gibbs 1969
Gastrolepidia clavigera		Holothuroid	*Thelenota anax*	O	J. R. Hanley pers. comm.
Gattyana cirrosa	(Pallas)	Polychaete	*Amphitrite edwardsi*	F	Bellan 1959
Gattyana cirrosa		Polychaete	*Amphitrite johnstoni*	F	McIntosh 1900, Orton & Smith 1935, Pettibone 1955, 1963, Bellan 1959
Gattyana cirrosa		Polychaete	*Arenicola* sp.	F	McIntosh 1900, Newell 1954, Pettibone 1955, 1963
Gattyana cirrosa		Polychaete	*Chaetopterus sarsii*	F	Storm 1881
Gattyana cirrosa		Polychaete	*Chaetopterus pergamentaceus*	F	Euders 1905, Paris 1955
Gattyana cirrosa		Polychaete	*Chaetopterus variopedatus*	F	Southern 1914, Southward 1956, Bellan 1959, Mettam 1980
Gattyana aff. cirrosa		Polychaete	*Chaetopterus* sp. n. Peters.	?	Petersen & Britayev 1997
Gattyana mossambica	Day	Polychaete	*Eunice tubifex*	F	Day 1960
Gaudichaudius cimex	(Quatrefages)	Decapod	*Diogenes custos*	O	Achari 1977, Pettibone 1986b
Gaudichaudius cimex		Decapod	*Diogenes diogenes*	O	Achari 1977, Pettibone 1986b
Gorgoniapolynoe bayeri	Pettibone	Cnidarian	*Narella clavata*	O	Pettibone 1991a
Gorgoniapolynoe caeciliae	(Fauvel)	Cnidarian	*Acanthogorgia aspera*	O	Pettibone 1991a
Gorgoniapolynoe caeciliae		Cnidarian	*Candidella imbricata*	O	Pettibone 1991a
Gorgoniapolynoe caeciliae		Cnidarian	*Corallium johnsoni*	O	Fauvel 1913
Gorgoniapolynoe caeciliae		Cnidarian	*Corallium niobe*	O	Hartmann-Schröder 1985, Pettibone 1991a
Gorgoniapolynoe caeciliae		Cnidarian	*Corallium tricolor*	O	Hartmann-Schröder 1985
Gorgoniapolynoe caeciliae		Cnidarian	*Pleurocorallium johnsoni*	O	Fauvel 1913, Hartmann-Schröder 1985, Pettibone 1991a
Gorgoniapolynoe cairnsi	Pettibone	Cnidarian	*Allopora eguchii*	O	Pettibone 1991a

Table 1 *Continued*

Commensal species	Authority	Host group	Species	Type	References
Gorgoniapolynoe corralophila	(Day)	Cnidarian	*Allopora bithalamus*	O	Day 1967
Gorgoniapolynoe corralophila		Cnidarian	*Stylaster* sp.	O	Stock 1986
Gorgoniapolynoe corralophila		Cnidarian	*Conopora* sp.	O	Stock 1986
Gorgoniapolynoe corralophila		Cnidarian	*Cryptohelia* sp.	O	Stock 1986
Gorgoniapolynoe galapagensis	Pettibone	Cnidarian	*Narella ambigua*	O	Pettibone 1991a
Gorgoniapolynoe guadalupensis	Pettibone	Cnidarian	*Corallium imperiale*	O	Britayev 1981, Pettibone 1991a
Gorgoniapolynoe muzikae	Pettibone	Cnidarian	*Acanthogorgia* sp.	O	Pettibone 1991a
Gorgoniapolynoe muzikae		Cnidarian	*Candidella helminthophora*	O	Pettibone 1991a
Gorgoniapolynoe muzikae		Cnidarian	*Corallium* sp.	O	Pettibone 1991a
Gorgoniapolynoe uschakovi	Britayev	Cnidarian	*Allopora* sp. 1	O	Britayev 1981
Gorgoniapolynoe uschakovi		Cnidarian	*Callogorgia* sp.	O	Britayev 1981, Pettibone 1991a
Grubeopolynoe semenovi	(Annenkova)	Polychaete	Unidentified ampharetid	F	Pettibone 1969c, Uschakov 1982
Grubeopolynoe tuta	(Grube)	Polychaete	*Amphitrite robusta*	F	Pettibone 1969c
Grubeopolynoe tuta		Polychaete	*Thelepus crispus*	F	Gravier 1905a, 1905b, Pettibone 1969c
Halosydna brevisetosa	Kinberg	Decapod	*Paguristes bakeri*	F	Clark 1956
Halosydna brevisetosa		Nudibranch	*Melibe leonina*	F	Ajeska & Nibakkanen 1976
Halosydna brevisetosa		Polychaete	*Amphitrite robusta*	F	Davenport & Hickok 1951
Halosydna brevisetosa		Polychaete	*Pista pacifica*	F	Clark 1956
Halosydna brevisetosa		Polychaete	*Platynereis agassizi*	F	Clark 1956
Halosydna brevisetosa		Polychaete	*Thelepus crispus*	F	Pettibone 1953
Halosydna johnsoni	(Darboux)	Polychaete	*Pista fratrella*	F	McGinitie & McGinitie 1949
Harmothoinae sp.		Polychaete	*Polyodontes lupinus*	O	Fox & Rupert 1985, Pettibone 1989b
Harmothoe acanellae	(Verril)	Cnidarian	*Acanella arbuscula*	F	Verrill 1881, Ditlevsen 1917
Harmothoe acanellae		Cnidarian	*Acanthogorgia armata*	F	Verril 1881
Harmothoe acanellae		Cnidarian	*Anthomastus grandifloris*	F	Ditlevsen 1917
Harmothoe acanellae		Cnidarian	*Pennatula grandis*	F	Pettibone 1963
Harmothoe areolata	Grube	Polychaete	*Eupolymnia nebulosa*	O	McIntosh 1900
Harmothoe areolata		Polychaete	*Chaetopterus* sp.	O	McIntosh 1900
Harmothoe brevipalpa	Bergstrom	Polychaete	*Phyllochaetopterus* sp.	F	Cañete et al. 1993
Harmothoe brevipalpa		Polychaete	*Pectinaria chilensis*	F	Cañete et al. 1993
Harmothoe coeliaca	Saint-Joseph	Decapod	*Eupagurus bernhardus*	O	Dales 1957
Harmothoe commensalis	Rozbaczylo & Cañete	Bivalve	*Gari solida*	O	Rozbaczylo & Cañete 1993
Harmothoe commensalis		Bivalve	*Semele solida*	O	Rozbaczylo & Cañete 1993
Harmothoe hyalonemae	Martín, Rosell & Uriz	Sponge	*Hyalonema thomsoni*	O	Martín et al. 1992
Harmothoe hyalonemae		Sponge	*Hyalonema lusitanicum*	O	Martín et al. 1992
Harmothoe hyalonemae		Sponge	*Hyalonema toxeras*	O	Martín et al. 1992
Harmothoe hyalonemae		Sponge	*Hyalonema infundibulum*	O	Martín et al. 1992

Table 1 *Continuws*

Commensal species	Authority	Host group	Species	Type	References
Harmothoe imbricata	Linnaeus	Asteroid	*Asterias amurensis*	F	Okuda 1936b
Harmothoe imbricata		Decapod	*Pagurus ochotoensis*	F	Pettibone 1963
Harmothoe imbricata		Polychaete	*Amphitrite robusta*	F	Berkeley & Berkeley 1948
Harmothoe imbricata		Polychaete	*Chaetopterus cautus*	F	Britayev 1993
Harmothoe imbricata		Polychaete	*Chaetopterus* sp.	F	McIntosh 1900, Fauvel 1923
Harmothoe imbricata		Polychaete	*Diopatra ornata*	F	Pettibone 1953
Harmothoe imbricata		Polychaete	*Eupolymnia nebulosa*	F	McIntosh 1900
Harmothoe imbricata		Polychaete	*Polycirrus* sp.	F	McIntosh 1900
Harmothoe imbricata		Polychaete	*Thelepus crispus*	F	McIntosh 1900
Harmothoe impar	(Johnson)	Holothuroid	*Cucumaria planci*	F	Fauvel 1923, Barel & Kramers 1977
Harmothoe ljungmani	(Malmgren)	Polychaete	*Lanice conchilega*	F	Fauvel 1923
Harmothoe longisetis	(Grube)	Polychaete	*Amphitrite edwardsi*	F	Fauvel 1923
Harmothoe longisetis		Polychaete	*Arenicola marina*	F	Fauvel 1923
Harmothoe longisetis		Polychaete	*Chaetopterus insignis*	F	Hornell 1891
Harmothoe longisetis		Polychaete	*Chaetopterus pergamentaceus*	F	Euders 1905, Fauvel 1923, Paris 1955
Harmothoe longisetis		Polychaete	*Chaetopterus variopedatus*	F	Fauvel 1923, Paris 1955
Harmothoe melanicornis	Britayev	Cnidarian	*Allopora* sp. 2	F	Britayev 1981
Harmothoe spinifera	(Ehlers)	Polychaete	*Amphitrite gracilis*	F	Beauchamp & Zachs 1913, Davenport 1953b
Harmothoe spinifera		Polychaete	*Polycirrus caliendrum*	F	Davenport 1953b
Harmothoe spongicola	Hanley & Burke	Sponge	*Aulochone clathroclada*	O	Hanley & Burke 1991b
Harmothoe tenebricosa	Moore	Asteroid	*Heterozonias alternatus*	O	Pettibone 1969b
Harmothoe tenebricosa		Asteroid	*Hoppasteria californica*	O	Pettibone 1969b
Harmothoe tenebricosa		Asteroid	*Solaster borealis*	O	Pettibone 1969b
Harmothoe vinogradovae	Averincev	Cnidarian	Unidentified hydrocoral	F	Britayev 1981
Harmothoe sp.		Polychaete	*Amphitrite gracilis*	F	Clark 1956
Harmothoe sp.		Polychaete	*Nereis* spp.	F	Clark 1956
Harmothoe sp.		Polychaete	*Polycirrus aurantiacus*	F	Clark 1956
Harmothoe sp.		Polychaete	*Eupolymnia nebulosa*	F	Clark 1956
Hartmania moorei	Pettibone	Polychaete	*Nereis virens*	F	Pettibone 1955, 1963
Hartmania moorei		Polychaete	*Praxilella gracilis*	F	Pettibone 1963
Hemilepedia verslusii	Horst	Cnidarian	*Thouarella hilgendorfi*	F	Horst 1915, Britayev 1981
Hermadion fauveli	Gravier	Sponge	*Sarcostegia oculata*	F	Gravier 1918, Paris 1955, Clark 1956, Barel & Kramers 1977
Hermadion hyalinus	Sars	Echinoid	*Echinus acutus*	O	Barel & Kramers 1977
Hermenia verruculosa	Grube	Ophiuroid	*Ophiocoma pusilla*	F	Devaney 1974, Pettibone 1975

Table 1 *Continued*

Commensal species	Authority	Host group	Species	Type	References
Hesperonoe adventor	(Skolgsberg)	Echinoid	*Urechis caupo*	F	Dales 1957
Hesperonoe complanata	(Johnson)	Decapod	*Callianassa californiensis*	F	Clark 1956
Hesperonoe complanata		Decapod	*Callianassa gigas*	F	Clark 1956
Hesperonoe hwanghaiensis	Uschacov & Wu	Decapod	*Callianassa* sp.	F	Uschakov & Wu 1959, Uschakov 1982
Hesperonoe hwanghaiensis		Decapod	*Upogebia* sp.	F	Uschakov & Wu 1959, Uschakov 1982
Hesperonoe sp.		Decapod	*Upogebia* sp.	F	Clark 1956
Hololepidella alba	Hartman-Schröder	Asteroid	*Acanthaster planci*	O	Hartmann-Schröder 1981
Hololepidella comatula	Uchida	Crinoid	*Comanthus parvicirrus*	O	Uchida 1975
Hololepidella comatula		Crinoid	*Comanthina schlegeli*	O	Uchida 1975
Hololepidella comatula		Crinoid	*Lamprometra palmata*	O	Uchida 1975
Hololepidella commensalis	Willey	Ophiuroid	*Macrophiotrix koehleri*	O	Willey 1905, Gibbs 1969
Hololepidella commensalis		Ophiuroid	*Ophiarthrum pictum*	O	Gibbs 1969
Hololepidella commensalis		Echinoid	*Clypeaster humilis*	O	Pettibone 1969c
Hololepidella commensalis		Echinoid	*Peronella lesueuri*	O	Augener 1922, Gibbs 1969
Hololepidella lobata	Hartman-Schröder	Asteroid	*Protoreaster* sp.	O	Hartmann-Schröder 1984
Hololepidella nigropunctata	(Horst)	Asteroid	*Acanthaster planci*	O	Devaney 1967, Uchida 1975
Hololepidella nigropunctata		Asteroid	*Linckia multiflora*	O	Pettibone 1993
Hololepidella nigropunctata		Asteroid	*Mithrodia* sp.	O	Pettibone 1963
Hololepidella nigropunctata		Asteroid	*Pentagonaster regulus*	O	Hendler & Meyer 1982
Hololepidella nigropunctata		Cnidarian	*Fungia scutaria*	O	Pettibone 1993
Hololepidella nigropunctata		Crinoid	*Tropiometra afra*	O	Hanley 1992
Hololepidella nigropunctata		Echinoid	*Diadema savigny*	O	Hendler & Meyer 1982
Hololepidella nigropunctata		Nudibranch	*Phyllidia varicosa*	O	Hanley 1992
Hololepidella nigropunctata		Ophiuroid	*Comanthrus annulatus*	O	Gibbs 1969
Hololepidella nigropunctata		Ophiuroid	*Macrophiotrix hisurta*	O	MacNae & Kalk 1962, Hendler & Meyer 1982
Hololepidella nigropunctata		Ophiuroid	*Macrophiotrix belli*	O	Hendler & Meyer 1982
Hololepidella nigropunctata		Echinoid	Unidentified	O	Pettibone 1993
Hololepidella nigropunctata		Ophiuroid	*Ophiarthrum elegans*	O	Gibbs 1969
Hololepidella nigropunctata		Ophiuroid	*Ophiocoma anaglyptica*	O	Devaney 1967
Hololepidella nigropunctata		Ophiuroid	*Ophiocoma brevipes*	O	Devaney 1967, Gibbs 1969
Hololepidella nigropunctata		Ophiuroid	*Ophiocoma dentata*	O	Devaney 1967
Hololepidella nigropunctata		Ophiuroid	*Ophiocoma doederleini*	O	Pettibone 1993
Hololepidella nigropunctata		Ophiuroid	*Ophiocoma erinaceus*	O	Devaney 1967
Hololepidella nigropunctata		Ophiuroid	*Ophiocoma* sp.	O	Hanley & Burke 1991b
Hololepidella nigropunctata		Sponge	*Latrunculia magnifica*	O	Pettibone 1993

Table 1 *Continued*

Commensal species	Authority	Host group	Species	Type	References
Hololepidella obtusa	Hartmann-Schröder	Crinoid	*Echinometra* sp.	O	Hartmann-Schröder 1984
Hololepidella ophiuricola	Gibbs	Ophiuroid	*Macrophiotrix koehleri*	O	Gibbs 1969, 1971, Pettibone 1993
Hololepidella ophiuricola		Ophiuroid	*Ophiarthrum pictum*	O	Gibbs 1969, 1971
Hololepidella sp.	Day	Ophiuroid	*Macrophiotrix hirsuta cheneyi*	O	Pettibone 1993
Hyperhalosydna striata	(Kinberg)	Polychaete	*Eunice* ssp.	F	Hanley & Burke 1991b
Intoshella euplectellae	McIntosh	Sponge	*Euplectella aspergillium*	O	Uschakov 1982, Pettibone 1996
Intoshela holothuricola	(Izuka)	Holothuroid	Unidentified	O	Izuka 1912, Okuda 1936a, Gallardo 1967, Uschakov 1982
Lagisca irritans	Marenzeller	Cnidarian	*Errina macrogasra*	O	Marenzeller 1904, Zibrowius 1981
Lagisca irritans		Cnidarian	*Stenohelia robusta*	O	Marenzeller 1904, Zibrowius 1981
Lagisca zibrowii	Hartmann-Schröder	Cnidarian	*Pseudosolanderia* sp.	O	Hartmann-Schröder 1992
Lepidasthenia alba	Hartman	Polychaete	*Phyllochaetopterus verrilli*	O	Hartman 1966
Lepidasthenia alba		Polychaete	*Phyllochaetopterus major*	O	Rullier 1972
Lepidasthenia accolus	Estcourt	Polychaete	Unidentified arenicolid	O	Estcourt 1967
Lepidasthenia argus	Hodgson	Polychaete	*Amphitrite edwardsi*	F	Clark 1956, Davenport 1953a, b, Paris 1955
Lepidasthenia digueti	Gravier	Balanoglossid	*Balanoglossus gigas*	O	Gravier 1905a
Lepidasthenia berkeleyae	Pettibone	Polychaete	*Praxilella affinis*	F	Clark 1956
Lepidasthenia commensalis	Webster	Polychaete	*Amphitrite ornata*	O	Gravier 1905b, Pettibone 1963, Gardiner 1976
Lepidasthenia commensalis		Polychaete	*Diopatra cuprea*	O	Pettibone 1963
Lepidasthenia commensalis		Polychaete	*Thelepus setosus*	O	Pettibone 1963
Lepidasthenia commensalis		Sponge	*Ircinia campana*	?	Dauer 1974
Lepidasthenia elegans	(Grube)	Polychaete	Unidentified terebellid	O	Day 1967
Lepidasthenia elegans		Polychaete	*Terebellides* sp.	O	Gibbs 1969
Lepidasthenia elegans		Polychaete	*Mesochaetopterus sagittarius*	O	Gibbs 1971
Lepidasthenia elegans		Sponge	Unidentified	O	Rullier 1974
Lepidasthenia gigas	(Johnson)	Polychaete	*Amphitrite* sp.	O	Gravier 1905b, Essenberg 1918
Lepidasthenia guadalcanalis	Gibbs	Balanoglossid	*Balanoglossus carnosus?*	O	Gibbs 1971
Lepidasthenia longissima	(Izuka)	Polychaete	*Thelepus setosus*	F	Okuda 1936b, Clark 1956
Lepidasthenia ocellata	(McIntosh)	Polychaete	*Spiochaetopterus chaellengeri*	F	Okuda 1936b, Clark 1956
Lepidasthenia ohshimai	Okuda	Holothuroid	*Protankyra bidentata*	F	Okuda 1936a, b, Clark 1956
Lepidasthenia maculata		Polychaete	*Chaetopterus* sp.	O	Gibbs 1969
Lepidasthenia maculata		Polychaete	*Phyllochaetopterus herdmani*	O	Gibbs 1971
Lepidasthenia maculata		Polychaete	*Phyllochaetopterus socialis*	F	Fauvel 1913

Table 1 *Continued*

Commensal species	Authority	Host group	Species	Type	References
Lepidasthenia maculata		Polychaete	*Phyllochaetopterus* sp.	O	Gibbs 1969
Lepidasthenia maculata		Polychaete	*Dasybranchus caducus*	O	Gibbs 1969
Lepidasthenia microlepis	Potts	Sipunculid	*Siphonostoma vastus*	O	Gibbs 1969, 1971
Lepidasthenia microlepis		Sipunculid	*Paraspidosiphon cumingi*	O	Gibbs 1969, 1971
Lepidasthenia microlepis		Sipunculid	Unidentified	O	Hanley & Burke 1991b
Lepidasthenia mossambica	Day	Balanoglossid	*Balanoglossus carnosus*	O	Day 1962, Gibbs 1969
Lepidasthenia stylolepis	Willey	Sipunculid	*Siphonostoma vastus*	O	Gibbs 1969, 1971
Lepidasthenia varia	Treadwell	Polychaete	*Notomastus lobatus*	O	Gardiner 1976
Lepidastheniella comma	(Thomson)	Polychaete	Unidentified Terebellid	O	Munro 1924
Lepidonotopodium fimbriatum	Pettibone	Polychaete	*Alvinella pompejana*	F	Pettibone 1984b
Lepidonotopodium fimbriatum		Bivalve	*Calyptogena magnifica*	F	Pettibone 1984b
Lepidonotopodium fimbriatum		Vetimentiferan	*Riftia pachyptila*	F	Pettibone 1984b
Lepidonotopodium riftense	Pettibone	Vetimentiferan	*Riftia pachyptila*	F	Pettibone 1984b
Lepidonotopodium riftense		Polychaete	*Alvinella pompejana*	F	Pettibone 1984b
Lepidonotopodium riftense		Bivalve	*Calyptogena magnifica*	F	Pettibone 1984b
Lepidonotopodium riftense		Bivalve	Unidentified mytilid (deep-sea)	F	Pettibone 1984b
Lepidonotopodium villiamsae	Pettibone	Bivalve	*Calyptogena magnifica*	F	Pettibone 1984b
Lepidonotopodium villiamsae		Bivalve	Unidentified mytilid (deep-sea)	F	Pettibone 1984b
Lepidonotopodium villiamsae		Polychaete	*Alvinella pompejana*	F	Pettibone 1984b
Lepidonotopodium villiamsae		Vetimentiferan	*Riftia pachyptila*	F	Pettibone 1984b
Lepidonopsis humilis	(Augener)	Echinoid	*Mellita quinquiesperforata*	F	Pettibone 1993
Lepidonopsis humilis		Ophiuroid	*Ophiothrix lineata*	F	Pettibone 1993
Lepidonopsis humilis		Ophiuroid	*Micropholis gracillima*	F	Gardiner 1976
Lepidonotus elongatus	Marenzeller	Polychaete	*Thelepus setosus*	F	Okuda 1936b, Clark 1956
Lepidonotus glaucus	(Peters)	Polychaete	*Eunice* sp.1	F	Hanley & Burke 1990
Lepidonotus glaucus		Polychaete	*Eunice* sp.2	F	Hanley & Burke 1990
Lepidonotus glaucus		Polychaete	*Eunice* sp.3	F	Hanley & Burke 1990
Lepidonotus glaucus		Polychaete	*Eunice* sp.4	F	Hanley & Burke 1990
Lepidonopsis humilis	(Augener)	Echinoid	*Mellita quinquiesperforata*	F	Pettibone 1993
Lepidonopsis humilis		Ophiuroid	*Ophiothrix lineata*	F	Pettibone 1993
Lepidonopsis humilis		Ophiuroid	*Micropholis gracillima*	F	Pettibone 1993
Lepidonotus melanogrammus	Haswell	Ophiuroid	Unidentified	F	Hanley & Burke 1990
Lepidonotus sublevis	Verrill	Decapod	*Pagurus pollicaris*	F	Pettibone 1963
Lepidonotus sublevis		Decapod	*Pagurus pollicaris*	F	Gardiner 1976
Lepidonotus sublevis		Decapod	*Pagurus impressus*	F	Gardiner 1976
Lepidonotus sublevis		Decapod	*Clibanarius vittatus*	F	Gardiner 1976

Table 1 *Continued*

Commensal species	Authority	Host group	Species	Type	References
Lepidonotus sp.		Holothuroid	*Labidoplax digitata*	F	Barel & Kramers 1977
Lepidonotus variabilis	Webster	Sponge	*Ircinia campana*	F	Dauer 1974
Lepidonotus variabilis		Sponge	Unidentified	F	Rullier 1974
Malmgreniella andreapolis	(McIntosh)	Holothuroid	*Leptosynapta bergenensis*	F	Pettibone 1993
Malmgreniella andreapolis		Holothuroid	*Labidoplax digitata*	F	Pettibone 1993
Malmgreniella andreapolis		Holothuroid	*Leptosynapta gallienni*	F	Orton 1923, Spooner et al. 1957
Malmgreniella andreapolis		Holothuroid	Unidentified synaptid	F	Cazaux 1968
Malmgreniella andreapolis		Ophiuroid	*Acrocnida brachiata*	F	Orton 1923, Spooner et al. 1957, Pettibone 1953
Malmgreniella arenicola	(Saint-Joseph)	Polychaete	*Arenicola marina*	F	Saint-Joseph 1888, Spooner et al. 1957, Pettibone 1993
Malmgreniella arenicola		Polychaete	*Arenicola* sp.	F	Spooner et al. 1957, Pettibone 1993
Malmgreniella arenicola		Polychaete	*Notomastus latericeus*	F	Spooner et al. 1957, Pettibone 1993
Malmgreniella arenicola		Polychaete	*Amphitrite johnstoni*	F	Spooner et al. 1957, Pettibone 1993
Malmgreniella arenicola		Polychaete	*Loimia medusa*	F	Spooner et al. 1957, Pettibone 1993
Malmgreniella arenicola		Polychaete	*Lanice conchilega*	F	Hamond 1966, Pettibone 1993
Malmgreniella arenicola		Sipunculid	*Golfingia elongata*	F	Spooner et al. 1957, Pettibone 1993
Malmgreniella arenicola		Sipunculid	*Golfingia vulgaris*	F	Spooner et al. 1957, Pettibone 1993
Malmgreniella baschi	Pettibone	Ophiuroid	*Ophiopsila califonica*	F	Pettibone 1953
Malmgreniella baschi		Polychaete	*Chaetopterus variopedatus*	F	Ruff 1995
Malmgreniella castanea	(McIntosh)	Echinoid	*Spatangus purpureus*	F	Clark 1956, Barel & Kramers 1977, Pettibone 1993
Malmgreniella castanea		Echinoid	*Spatangus raschi*	F	Barel & Kramers 1977
Malmgreniella castanea		Asteroid	*Astropecten irregularis*	F	Barel & Kramers 1977
Malmgreniella capensis	(McIntosh)	Echinoid	*Spatangus capensis*	F	Pettibone 1993
Malmgreniella dicirra	Hartman	Cnidarian	*Allopora eguchii*	O	Pettibone 1993
Malmgreniella dicirra		Cnidarian	*Calyptopora reticulata*	O	Pettibone 1993
Malmgreniella dicirra		Cnidarian	*Conopora pauciseptata*	O	Pettibone 1993
Malmgreniella dicirra		Cnidarian	*Lepidopora* sp.	O	Pettibone 1993
Malmgreniella dicirra		Cnidarian	*Stylaster densicaulis*	O	Pettibone 1993
Malmgreniella galataensis	Pettibone	Ophiuroid	*Amphiura* cf. *fibulata*	O	Pettibone 1993

Table 1 *Continued*

Commensal species	Authority	Host group	Species	Type	References
Malmgreniella galataensis		Ophiuroid	*Ophiophragmus septus*	O	Pettibone 1993
Malmgreniella hendleri	Pettibone	Ophiuroid	*Amphiura* cf. *fibulata*	O	Pettibone 1993
Malmgreniella hendleri		Ophiuroid	*Ophionereis limicola*	O	Pettiobne 1993
Malmgreniella inhacaensis	Pettibone	Balanoglossid	*Balanoglossus* sp.	O	Day 1962, Pettibone 1993
Malmgreniella inhacaensis		Balanoglossid	*Ptychodera* sp.	O	Pettibone 1993
Malmgreniella lunulata	(Delle Chiaje)	Balanoglossid	Unidentified balanoglossids	O	Pettibone 1993
Malmgreniella cf. "*lunulata*"		Polychaete	*Mesochaetopterus rickettsi*	?	Petersen & Britayev 1997
Malmgreniella maccraryae	Pettibone	Ophiuroid	Unidentified	F	Pettibone 1993
Malmgreniella maccraryae		Ophiuroid	*Amphiopholis gracillima*	F	Pettibone 1993
Malmgreniella maccraryae		Ophiuroid	*Amphioplus sepultus*	F	Pettibone 1993
Malmgreniella macginitiei	Pettibone	Decapod	*Callianassa californiensis*	O	Berkeley & Berkeley 1941, Pettibone 1993, Ruff 1995
Malmgreniella macginitiei		Polychaete	*Axiotella rubrocincta*	O	Kudenov 1975, Pettibone 1993, Ruff 1995
Malmgreniella macginitiei		Ophiuroid	*Amphiodia urtica*	O	Pettibone 1993, Ruff 1995
Malmgreniella marphysae	(McIntosh)	Polychaete	*Marphysa sanguinea*	O	Pettibone 1993
Malmgreniella marphysae		Polychaete	*Lagis koreni*	O	Pettibone 1993
Malmgreniella murrayensis	Pettibone	Sponge	Unidentified	O	Pettibone 1993
Malmgreniella nigralba	(Berkeley)	Holothuroid	*Leptosynapta clarki*	O	Berkeley 1924, Pettibone 1953, 1993, Ruff 1995
Malmgreniella panamensis	Pettibone	Ophiuroid	*Ophiopsila* cf. *polysticta*	O	Pettibone 1993
Malmgreniella pettiti	Pettibone	Cnidarian	Unidentified alcyonarian	O	Pettibone 1993
Malmgreniella pierceae	Pettibone	Ophiuroid	*Amphiura* cf. *fibulata*	F	Pettibone 1993
Malmgreniella puntotorensis	Pettibone	Ophiuroid	*Amphiodia trychna*	O	Pettibone 1993
Malmgreniella puntotorensis		Ophiuroid	*Ophionephthys limicola*	O	Pettibone 1993
Malmgreniella puntotorensis		Ophiuroid	*Ophiophragmus cubanus*	O	Pettibone 1993
Malmgreniella puntotorensis		Ophiuroid	*Ophiophragmus pulcher*	O	Pettibone 1993
Malmgreniella puntotorensis		Ophiuroid	*Ophiophragmus septus*	O	Pettibone 1993
Malmgreniella puntotorensis		Ophiuroid	Unidentified	O	Pettibone 1993
Malmgreniella scriptoria	(Moore)	Echinoid	*Briaster latifrons*	F	Moore 1910, Pettibone 1953, 1993, Ruff 1995
Malmgreniella taylori	Pettibone	Ophiuroid	*Amphiodia atra*	F	Pettibone 1993
Malmgreniella taylori		Ophiuroid	*Amphiopholis gracillima*	F	Pettibone 1993
Malmgreniella variegata	(Treadwell)	Ophiuroid	*Ophionereis reticulata*	F	Millot 1953, Pettibone 1993
Malmgreniella variegata		Ophiuroid	*Ophionereis annulata*	F	Pettibone 1993
Malmgreniella sp.		Holothuroid	Unidentified synaptid	O	D. Martin pers. obs.
Minusculisquama hughesi	Pettibone	Polychaete	Unidentified maldanid	F	Pettibone 1983

Table 1 *Continued*

Commensal species	Authority	Host group	Species	Type	References
Neohololepidella murrayi	Pettibone	Sponge	calcareous sponge	O	Pettibone 1969c
Ophthalmonoe pettiboneae	Petersen & Britayev	Polychaete	*Chaetopterus appendiculatus*	O	Petersen & Britayev 1997
Parabathynoe brisinga	Pettibone	Asteroid	*Brisinga* sp.	O	Pettibone 1990
Paractonoella aphthalma	Gallardo	Ophiuroid	Unidentified	F?	Gallardo 1967, Uschacov 1982, Pettibone 1996
Paradyte crinoidicola	(Potts)	Asteroid	*Astropecten* sp.	O	Hanley 1984, Hanley & Burke 1991b
Paradyte crinoidicola		Crinoid	*Capillaster multiradiatus*	O	Thesunov et al. 1989, Hanley 1984, Hanley & Burke 1991b
Paradyte crinoidicola		Crinoid	*Colobometra persispinosa*	O	Thesunov et al. 1989
Paradyte crinoidicola		Crinoid	*Comantheria rotula*	O	Hanley 1984, Hanley & Burke 1991b
Paradyte crinoidicola		Crinoid	*Comanthina belli*	O	Hanley 1984, Hanley & Burke 1991b
Paradyte crinoidicola		Crinoid	*Comanthina schelegi*	O	Hanley 1984, Hanley & Burke 1991b
Paradyte crinoidicola		Crinoid	*Comanthus annulatus*	O	Thesunov et al. 1989
Paradyte crinoidicola		Crinoid	*Comanthus bennetti*	O	Hanley 1984, Hanley & Burke 1991b
Paradyte crinoidicola		Crinoid	*Comanthus parvicirrus*	O	Hanley 1984, Hanley & Burke 1991b
Paradyte crinoidicola		Crinoid	*Comantula purpurea*	O	Hanley 1984, Hanley & Burke 1991b
Paradyte crinoidicola		Crinoid	*Comatella stelligera*	O	Hanley 1984, Hanley & Burke 1991b
Paradyte crinoidicola		Crinoid	*Himenometra robustispina*	O	Hanley 1984, Hanley & Burke 1991b
Paradyte crinoidicola		Crinoid	*Lamprometra klunzingeri*	O	Hanley 1984, Hanley & Burke 1991b
Paradyte crinoidicola		Crinoid	*Lamprometra palmata*	O	Hanley 1984, Hanley & Burke 1991b
Paradyte crinoidicola		Crinoid	*Oligometra carpenteri*	O	Hanley 1984, Hanley & Burke 1991b
Paradyte crinoidicola		Crinoid	*Oligometra serripinna*	O	Hanley 1984, 1992, Hanley & Burke 1991b
Paradyte crinoidicola		Crinoid	*Petasometra helianthoides*	O	Hanley 1984, Hanley & Burke 1991b
Paradyte crinoidicola		Crinoid	*Stephanometra indica*	O	Hanley 1984, Hanley & Burke 1991b
Paradyte crinoidicola		Crinoid	*Tropiometra carinata*	O	Day 1962, Pettibone 1969a, Hanley 1984, Hanley & Burke 1991b
Paradyte crinoidicola		Crinoid	*Zygometra comata*	O	Hanley 1984, Hanley & Burke 1991b
Paradyte crinoidicola		Crinoid	*Zygometra elegans*	O	T. A. Britayev pers. obs.
Paradyte crinoidicola		Crinoid	Unidentified	O	Hanley 1992
Paradyte laevis	(Day)	Cnidarian	*Dendronephthya* sp.	F	Okuda 1950

Table 1 *Continued*

Commensal species	Authority	Host group	Species	Type	References
Paradyte tentaculata	(Horst)	Cnidarian	*Dendronephthya* sp.	O	Hanley 1992
Paradyte tentaculata		Cnidarian	*Nephthea* sp.	O	Hanley 1992
Parahalosidnopsis tubicola	(Day)	Polychaete	*Loimia medusa*	O	Day 1973, Pettibone 1977, Gaikwad 1993
Phyllosheila wigleyi	McIntosh	Asteroid	*Brisinga* sp.	O	Pettibone 1961
Polyeunoa laevis	McIntosh	Cnidarian	*Thouarella variabilis*	F	Pettibone 1969c
Polyeunoa maculata	(Day)	Polychaete	*Mesochaetopterus japonicus*	O	Day 1973, Hartmann-Schröder 1981, Gaikwad 1993
Polynoe antarctica	Kinberg	Polychaete	Unidentified terebellid	F	Uschakov 1962, Hartmann-Schröder 1989
Polynoe antarctica		Cnidarian	*Thuiria* sp.	F	Hartmann-Schröder 1989
Polynoe scolopendrina	Savigny	Polychaete	*Eupolymnia nebulosa*	F	Davenport 1953b, Paris 1955, D. Martin pers. obs.
Polynoe scolopendrina		Polychaete	*Lanice conchilega*	F	Davenport 1953b
Polynoe scolopendrina		Polychaete	*Lysidice ninetta*	F	McIntosh 1900, Davenport 1953b, Paris 1955
Polynoe thouarellicola	Hartmann-Schröder	Cnidarian	*Thouarella* sp.	O	Hartmann-Schröder 1989
Polynoe thouarellicola		Cnidarian	acanthogorgiidae sp.	O	Hartmann-Schröder 1989
Pottsiscalisetosus praelongus	(Marenzeller)	Asteroid	Unidentified	F	Pettibone 1969a
Subadyte pellucida	(Ehlers)	Ophiuroid	*Ophiothrix fragilis*	F	Barel & Kramers 1977, Ockelmann pers. comm.
Subadyte pellucida		Ophiuroid	*Ophiothrix echinata*	F	Devaney 1967
Subadyte pellucida		Ophiuroid	*Ophiothrix alopecurus*	F	Devaney 1967
Subadyte pellucida		Ophiuroid	*Ophiothrix quinquemaculata*	F	Bellan 1964, Guille 1965
Subadyte pellucida		Asteroid	*Solaster papposus*	F	Okuda 1950, Barel & Kramers 1977
Subadyte pellucida		Asteroid	*Astropecten irregularis*	F	Okuda 1950, Barel & Kramers 1977
Subadyte pellucida		Crinoid	*Antedon mediterranea*	F	Barel & Kramers 1977
Subadyte pellucida		Echinoid	*Paracentrotus lividus*	F	Barel & Kramers 1977
Subadyte papillifera	(Horst)	Cnidarian	Unidentified coral	O	Pettibone 1969a
Telolepidasthenia lobetobiensis	Augener & Pettibone	Polychaete	Unidentified chaetopterid (tube)	O	Pettibone 1970
Telolepidasthenia lobetobiensis		Polychaete	*Phyllochaetopterus* sp.	O	Pettibone 1970
Tottonpolynoe symantipatharia	Pettibone	Cnidarian	*Parantipathes* sp.	O	Pettibone 1991b
Tottonpolynoe symantipatharia		Cnidarian	*Sclerisis macquarina*	O	Pettibone 1991b

Table 1 *Continued*

Commensal species	Authority	Host group	Species	Type	References
Uncopolynoe corallicola	Hartmann-Schröder	Cnidarian	Unidentified alcyonarian	O	Hartmann-Schröder 1960
Wilsoniella furcosetosa	(Loshamn)	Polychaete	*Amphitrite edwardsi*	F	Spooner et al. 1957, Pettibone 1993
Sabellaridae					
Phalacrostemma cidariophilum	Marenzeller	Echinoid	*Cidaris cidaris*	O	Barel & Kramers 1977
Sabellidae					
Potamilla symbiotica	Uschacov	Sponge	*Cryptospongia enigmatica*	O	Uschakov 1950
Serpulidae					
Floriprotis sabiuraensis	Uchida	Cnidarian	*Favites abdita*	O	Uchida 1978
Floriprotis sabiuraensis		Cnidarian	*Favia speciosa*	O	Bailey-Brock 1985
Floriprotis sabiuraensis		Cnidarian	*Goniastrea pectinata*	O	Bailey-Brock 1985
Floriprotis sabiuraensis		Cnidarian	*Hydnophora* sp.	O	ten Hove 1989
Floriprotis sabiuraensis		Cnidarian	*Platygyra* sp.	O	ten Hove 1989
Hydroides spongicola	Benedict	Sponge	*Neofibularia nolitangere*	O	ten Hove 1989
Hydroides spongicola		Sponge	*Mycale microsigmatosa*	O	ten Hove 1989
Pseudovermilia madracicola	ten Hove	Cnidarian	*Madracis pharensis*	O	ten Hove 1989
Pseudovermilia madracicola		Cnidarian	*Madracis decactis*	O	ten Hove 1989
Serpula sp. 2	ten Hove	Cnidarian	*Echinopora gemmacea*	O	ten Hove 1994
Spirobranchus gardineri	Pixell	Cnidarian	*Gardenoseris planulata*		ten Hove 1994
Spirobranchus gardineri		Cnidarian	*Porites lutea*		ten Hove 1994
Spirobranchus gardineri		Cnidarian	*Pavona varians*		ten Hove 1994
Spirobranchus giganteus/corniculatus	(Pallas)	Cnidarian	*Agaricia* spp.	O	Hunte et al. 1990
Spirobranchus giganteus/corniculatus		Cnidarian	*Acropora* sp.	O	ten Hove 1994
Spirobranchus giganteus/corniculatus		Cnidarian	*Diploria labyrinthiformis*	O	Hunte et al. 1990
Spirobranchus giganteus/corniculatus		Cnidarian	*Diploria strigosa*	O	Hunte et al. 1990
Spirobranchus giganteus/corniculatus		Cnidarian	*Gardenoseris planulata*	O	ten Hove 1994
Spirobranchus giganteus/corniculatus		Cnidarian	*Hydnophora microconos*	O	ten Hove 1994
Spirobranchus giganteus/corniculatus		Cnidarian	*Madracis* spp.	O	Hunte et al. 1990
Spirobranchus giganteus/corniculatus		Cnidarian	*Millepora alcicornis*	O	Pallas 1766 in ten Hove 1989
Spirobranchus giganteus/corniculatus		Cnidarian	*Millepora complanata*	O	Hunte et al. 1990; Marsden 1992
Spirobranchus giganteus/corniculatus		Cnidarian	*Millepora* sp.	O	ten Hove 1994
Spirobranchus giganteus/corniculatus		Cnidarian	*Montastraea annularis*	O	Hunte et al. 1990

Table 1 *Continued*

Commensal species	Authority	Host group	Species	Type	References
Spirobranchus giganteus/corniculatus		Cnidarian	*Montastrea cavernosa*	O	Hunte et al. 1990
Spirobranchus giganteus/corniculatus		Cnidarian	*Montipora* sp.	O	ten Hove 1994
Spirobranchus giganteus/corniculatus		Cnidarian	*Pavona maldivensis*	O	ten Hove 1994
Spirobranchus giganteus/corniculatus		Cnidarian	*Porites astreoides*	O	Hunte et al. 1990
Spirobranchus giganteus/corniculatus		Cnidarian	*Porites lutea*	O	ten Hove 1994
Spirobranchus giganteus/corniculatus		Cnidarian	*Porites porites*	O	De Vantier et al. 1985
Spirobranchus giganteus/corniculatus		Cnidarian	*Porites* sp.	O	Bailey-Brock 1985, Hunte et al. 1990
Spirobranchus giganteus/corniculatus		Cnidarian	*Siderastraea siderea*	O	Hunte et al. 1990
Spirobranchus giganteus/corniculatus		Cnidarian	*Stylophora* sp.	O	ten Hove 1994
Spirobranchus cf. *nigranucha*	(Fischli)	Cnidarian	*Porites* sp.	O	ten Hove 1989
Spirobranchus polycerus	(Schmarda)	Cnidarian	*Millepora complanata*	O	Marsden 1992
Spirobranchus polycerus		Cnidarian	*Porites astreoides*	O	Marsden 1992
Spirobranchus tetraceros		Cnidarian	*Millepora tuberosa*	O	ten Hove 1970, 1994
Spirobranchus tetraceros		Cnidarian	*Pavona maldivensis*	O	ten Hove 1970, 1994
Sphaerodoridae					
Sphaerodorum ophiurophoretos	Martin & Alvà	Ophiuroid	*Amphipholis squamata*	O	Martin & Alvà 1988, Alvà & Jangoux 1989
Sphaerodorum flavus	Oersted	Ophiuroid	*Amphiura chiajei*	O	Ockelmann pers. comm.
Sphaerodorum flavus		Ophiuroid	*Amphiura filiformis*	O	Ockelmann pers. comm.
Sphaerodoridium guilbaulti	Rullier	Cnidarian	*Paragorgia arborea*	O	Rullier 1974
Commensodorum commensalis	(Lutzen)	Polychaete	*Terebellides stroemi*	O	Lützen 1961
Spintheridae					
Spinther arcticus	(Sars)	Sponge	*Halichondria* sp.	O	George & Hartmann-Schröder 1985
Spinther arcticus		Sponge	*Tedania* sp.	O	George & Hartmann-Schröder 1985
Spinther arcticus		Sponge	Unidentified	O	Sardá 1984, George & Hartmann-Schröder 1985
Spinther arcticus		Cnidarian	Unidentified hydroids	O	George & Hartmann-Schröder 1985
Spinther citrinus	(Stimpson)	Sponge	Unidentified	O	Hartman 1942, Pettibone 1963
Spinther ericinus	Yamamoto & Imajima	Sponge	Unidentified	O	Yamamoto & Imajima 1988
Spinther oniscoides	Johnston	Sponge	*Stylostichon plumosum*	O	Campoy 1982
Spinther oniscoides		Sponge	*Desmacidon* sp.	O	George & Hartmann-Schröder 1985
Spionidae					
Dipolydora carunculata	(Radshevsky)	Sponge	*Halichondria panicea*	F	Radashevsky 1993
Dipolydora carunculata		Sponge	*Dysidea fragilis*	F	Radashevsky 1993

Table 1 *Continued*

Commensal species	Authority	Host group	Species	Type	References
Polydora colonia	Moore	Sponge	*Tedania ignis*	?	Blake 1971, Dauer 1974
Polydora spongicola	Berkeley & Berkeley	Sponge	*Adocia cinerea*	O	Buzhinskaja 1971, Radashevsky 1988
Polydora spongicola		Sponge	*Halichondria panicea*	O	Radashevsky 1993
Polydora spongicola		Sponge	*Ophlitaspongia pennata*	O	Radashevsky 1993
Polydora spongicola		Sponge	*Myxilla incrusatans*	O	Radashevsky 1993
Polydora spongicola		Sponge	*Lyssodendoryx firma*	O	Berkeley & Berkeley 1950
Polydorella prolifera	Augener	Sponge	Unidentified	O	Augener 1914, Blake & Kudenov 1978
Polydorella smurovi	Tzetlin & Britayev	Sponge	Unidentified	O	Tzetlin & Britayev 1985
Polydorella stolonifera	Blake & Kudenov	Sponge	Unidentified	O	Blake & Kudenov 1978
Polydorella dawydoffi	Radashevsky	Sponge	*Petrosia* sp.	O	Radashevsky 1996
Polydorella dawydoffi		Sponge	*Rhaphidiophlus erectus*	O	Radashevsky 1996
Polydorella dawydoffi		Sponge	*Amphimedon* sp.	O	Radashevsky 1996
Polydorella dawydoffi		Sponge	*Callispongia* sp.	O	Radashevsky 1996
Polydorella dawydoffi		Sponge	*Haliclona* sp.	O	Radashevsky 1996
Polydorella dawydoffi		Sponge	*Mycale* sp.	O	Radashevsky 1996
Polydorella dawydoffi		Sponge	*Niphates sp.*	O	Radashevsky 1996
Polydorella dawydoffi		Sponge	*Suberites* sp.	O	Radashevsky 1996
Polydorella dawydoffi		Sponge	*Xetospongia testudinaria*	O	Radashevsky 1996
Polydorella dawydoffi		Sponge	*Polymastia* sp.	O	Radashevsky 1996
Spirorbidae					
Circeis armoricana	(Saint-Joseph)	Decapod	*Pagurus middendorfi*	F	Rzhavsky & Britayev 1988
Circeis armoricana		Decapod	*Pagurus hirsutiusculus*	F	Rzhavsky & Britayev 1988
Circeis paguri	Knight-Jones & Knight-Jones	Decapod	*Eupagurus bernhardus*	O	Al-Ogyly & Knight-Jones 1981
Syllidae					
Amblyosyllis cincinnata	(Verrill)	Sponge	*Haliclona oculata*	O	Riser 1982
Amblyosyllis cincinnata		Sponge	*Halichondria panicea*	O	Riser 1982
Amblyosyllis cincinnata		Sponge	*Isodyctia deichmanae*	O	Riser 1982
Autolytus cornutus	Agassiz	Cnidarian	Unidentified hydroid	?	Pettibone 1963
Bollandia antipathicola	Glasby	Cnidarian	*Antipathes* sp.	O	Glasby 1994
Branchiosyllis exilis	(Gravier)	Ophiuroid	*Ophiocoma echinata*	?	Hendler & Meyer 1982
Branchiosyllis exilis		Ophiuroid	Unidentified	?	Hartmann-Schröder 1978
Brania pusilla	(Dujardin)	Cnidarian	Unidentified coral	?	Gardiner 1976
Exogone sp.		Holothuroid	*Cucumaria planci*	F	Barel & Kramers 1977

Table 1 *Continued*

Commensal species	Authority	Host group	Species	Type	References
Haplosyllides floridana	Augener	Ophiuroid	*Ophiocoma pusilla*	O	Hartmann-Schröder 1978
Haplosyllides floridana		Sponge	*Xetospongia muta*	O	San Martín et al. 1997
Haplosyllides floridana		Sponge	Unidentified	O	Fauvel 1923
Haplosyllides floridana		Cnidarian	Unidentified coral	O	Hartman 1954, Rullier & Amoureux 1979
Haplosyllis agelas	Uebelacker	Sponge	*Agelas dispar*	O	Uebelacker 1982
Haplosyllis anthogorgicola	Utinomi	Cnidarian	*Anthogorgia bocki*	O	Utinomi 1956, Imajima & Hartman 1964
Haplosyllis bisetosa	Hartmann-Schröder	Cnidarian	Unidentified alcyonarian	O	Hartmann-Schröder 1960
Haplosyllis chamaelon	Laubier	Cnidarian	*Paramuricea clavata*	O	Laubier 1960, López et al. 1996
Haplosyllis xeniaecola	Hartmann-Schröder	Cnidarian	*Xenia viridis*	O	Hartmann-Schröder 1993
Pionosyllis magnifica	Uschakov	Decapod	*Paralithodes kamtchatica*	F	T. A. Britayev pers. obs.
Pionosyllis magnifica		Decapod	Unidentified pagurid	F	T. A. Britayev pers. obs.
Proceraea cornuta	(Agassiz)	Cnidarian	Unidentified coral	F	Gardiner 1976
Syllis armillaris	(Müller)	Decapod	Unidentified pagurid	F	T. A. Britayev pers. obs.
Syllis cornuta	Rathke	Decapod	Unidentified Pagurid	F	Fauvel 1923, T. A. Britayev pers. obs.
Syllis cornuta		—	Unidentified inkeepers	F	Fauvel 1923
Syllis gracilis	Grube	Bryozoan	*Pentapora* sp.	F	López et al. 1996
Syllis gracilis		Cnidarian	*Elisella parplexauroides*	F	López et al. 1996
Syllis gracilis		Cnidarian	Unidentified hydroids	F	López et al. 1996
Syllis gracilis		Cnidarian	*Paramuricea clavata*	F	Alós 1988, López et al. 1996
Syllis gracilis		Sponge	Unidentified	F	Amoureux et al. 1980, Campoy 1982
Syllis onkylochaeta	Hartmann-Schröder	Cnidarian	*Xenia* sp.	O	Hartmann-Schröder 1991
Syllis sclerolaema	Hartmann-Schröder	Sponge	Unidentified	F	Ushakov 1955
Trypanosyllis asterobia	Okada	Asteroid	*Louidia quinaria*	O	Imajima 1966

1977) and 87 (Pettibone 1982), with more than 16000 known species (Blake 1994). Commensals are thus present in just over 31% of the known polychaete families and comprise about 1.8% of the known species. For the most part, commensal polychaetes are obligatory symbionts (67%) and belong to families formerly grouped in the artificial Errantia group. However, some others are included in the so-called Sedentary families, namely the Capitellidae, Fauveliopsidae, Flabelligeridae, Sabellariidae, Serpulidae and Spionidae, with symbiotic species being more abundant in this last family (see Tables 1 and 9, p. 313).

According to Clark (1956), almost two-thirds of the polychaetes that were reported as commensals were members of the scaleworm family Polynoidae. With the increasing number of commensal polychaete species described since then, this proportion has been slightly reduced. However, more than half of all currently known commensal species (55%) belong to this family (Fig. 1A). In fact, the polynoids are a very large family, including diverse and numerous species inhabiting different marine habitats. The majority are sluggish and crawling and are found under stones, in crevices and in available tubes or burrows. Most of them are carnivorous, feeding on a great variety of animals. Thus, although the lack of information on most known associations prevented us speculating on the curious dominance of polynoids within commensal polychaetes, the symbiotic habits of this family seem to be more or less easily inferred from their "normal" (i.e. free living) modes of life (Fauchald & Jumars 1979). According to our present evaluation, the family consists of about 700 species, the commensals representing 22.7% of the total. Commensal and free-living scaleworms do not show a marked taxonomical differentiation, the former often belonging to genera that include free-living species as well. The huge genus *Harmothoe*, for example, consists of more than 120 species within which about 15 are commensal, whereas the "typically" free-living genus *Eunoe* includes three commensal species (Table 1). On the other hand, some genera (viz. *Arctonoe*, *Asterophylia*, *Australaugeneria*) and one subfamily, the Arctonoinae (Hanley 1989), could be considered as wholly, or at least mostly, commensal.

The weight of commensal species within the non-polynoid polychaete families is significantly less than within polynoids. The relative number of commensals does not exceed 10% and the number of commensal species is always less that 10, with the exception of syllids, hesionids, spionids and serpulids (Fig. 1A).

Three small polychaete families are entirely symbiotic. The Iphitimidae (*sensu* Fauchald 1970) includes a single genus, *Iphitime*, with one parasitic (see Table 6, p. 286) and seven commensal species, all the latter inhabiting the branchial cavities of decapod crustaceans (crabs and hermit crabs). Also belonging to the Order Eunicida, the family Histriobdellidae consists of two genera and nine species, all of them living epizoically in branchial chambers or carapaces of crustaceans. The single species of *Histriobdella* lives associated with marine lobsters in northern Europe and along the northeastern American coast. The remaining species belong to *Stratiodrilus* and are known as commensals of freshwater crayfish (from Australia, Tasmania, Madagascar, and South and Central America) and marine isopods (from South Africa). The third commensal family, the Nautiliniellidae, has been recently described by Miura & Laubier (1989) and consists of nine species, all of them associated with bivalves (Table 1). The nautiliniellids inhabit deep-water hydrothermal vents and seep-site communities and are closely related to the monotypic "commensal" family Antonbruunidae. *Antonbruunia viridis*, the single species, is also associated with a bivalve. Both families are morphologically very close, suggesting that a further review would probably lead to their union under the older family name, Antonbruuniidae.

Marine host taxa harbouring commensal polychaetes

Commensal polychaetes are associated with as many as 569 species included within the main taxa of marine metazoans (excluding flatworms, nemerteans and nematodes) and even with protozoans (i.e. foraminifers) (Fig. 1B). However, they clearly prefer organisms that provide them with good shelter, such as tube-dwelling or burrowing animals (viz. tubicolous polychaetes, sipunculids, balanoglossids) or relatively large animals possessing "advantageous" protective physiological or morphological characteristics. Among them, organisms having holes, grooves, chambers or channels (e.g. sponges, starfishes), as well as those showing good chemical or physical defences (e.g. sponges, cnidarians, sea urchins) are included. The highest number of host species harbouring polychaetes belong to the echinoderms, and the highest number of commensal polychaetes are also associated with this group. In fact, about 36% of all host species belong to one of the five echinoderm classes and the same percentage of all commensal polychaetes are associated with species of this group (Fig. 1B). The following most dominant groups are the cnidarians (20% of the host species and associated commensal polychaetes) and the polychaetes themselves (16% of the host species and 17% of associated commensal polychaetes), whereas only 7.2% and 13% of the host species are included among such abundant and widely distributed marine animals as molluscs and crustaceans, respectively.

Specificity among commensal polychaetes

All degrees of host-commensal specificity may be found among commensal polychaetes. Although several polychaete species may be associated with many hosts from very different groups, most of them are symbiotic with only one or a few closely related hosts (Table 1). In fact, 59% are strictly "monoxenous associates" (i.e. occurring on one host), this figure becoming 87% if those commensals occurring on two or three different hosts are included. The high degree of specificity of the relationships involving polychaetes may explain these proportions. However, as repeatedly mentioned, the often vague and accidental information on most commensal polychaetes may contribute to exaggeration of the monoxenous pattern. Therefore, only a few species that have been studied in detail could be mentioned as examples of real monoxenous associates. Among them, *Histriobdella homari* inhabiting branchial chambers of two species of Atlantic lobsters (*Homarus vulgaris* and *H. americanus*), *Adyte assimilis* associated with two related species of echinoids (*Echinus esculentus* and *E. acutus*), and *Haplosyllis chamaeleon* associated with the gorgonian *Paramuricea clavata*.

On the other hand, polyxenous polychaetes are usually associated with hosts that frequently belong to the same taxa (i.e. Class, Order or even Family). For example, *Iphitime cuenoti* inhabits the branchial chambers of 11 species of Mediterranean and northern Atlantic crabs, *Paradyte crinoidicola* is associated with 20 species of unstalked crynoids, and *Gastrolepidia clavigera* lives symbiotically with 13 species of tropical holothuroids from two families (Stichopodidae and Holothuridae). Usually, this kind of specificity may be related to strict morphological adaptations of symbionts. Accordingly, *Iphitime cuenoti* has a more or less simplified jaw apparatus (see Fig. 12A, B, p.

272), *Paradyte crinoidicola* is equipped with hooked modified ventral setae (see Fig. 10C, p. 270), and *Gastrolepidia clavigera* has ventral sucker-like lobes (see Fig. 13A, p. 273). A strict preference of commensals for a given host taxon may also be extendable to the family level, such as the nautiliniellids associated with bivalves, the iphitimids with crabs or the serpulids with haermatipic corals (Table 1). More commonly, however, commensals from a given polychaete family are associated with a taxonomically wide range of host groups (e.g. the families Hesionidae, Polynoidae and Syllidae, Table 1).

Polyxenous associations involving host species from very different taxonomic groups are less common among polychaetes, occurring mainly within hesionids and polynoids. The hesionid *Ophiodromus flexuosus* inhabits the ambulacral groves of five starfish species, as well as the tubes of two polychaetes, one balanoglossid and one burrowing holothuroid. *Hololepidella nigropunctata* is associated with 10 ophiuroids, four asteroids, one echinoid, one sponge and one cnidarian. *Arctonoe pulchra* is associated with four asteroids, two holothuroids, one gastropod, one polyplacophoran and 1 polychaete. However, the most surprising polyxenous species is *A. vittata* from the North Pacific ocean. Its host list includes about 30 species from very different taxa: cnidarians, gastropod and polyplacophoran molluscs, polychaetes, asteroids, holothuroids and echinoids! (Table 1). It is interesting to note that the areas of distribution of hosts harbouring *A. vittata* overlapp only partially and, also, that its host preferences differ from one geographical area to another (Davenport 1950, Britayev et al. 1977). For example, its preferred host in Vostok Bay and the southwest coast of Sakhalin Island (Sea of Japan) is the gastropod *Acmaea pallida*, whereas in Avacha Bay (Pacific coast of Kamchatka peninsula) the host is the starfish *Asterias rathbunae* (T. A. Britayev pers. obs.).

These polyxenous associates with a wide range of host species require special attention and may be of great taxonomical interest, as the possibility of sibling or pseudosibling species-complex being hidden under the same species name remains a possibility. Probably one of the most striking cases was exemplified in Pettibone (1993), a recent paper reviewing the formerly "widespread", "polyxenous" scaleworm *Harmothoe lunulata* (delle Chiaje 1841). This species was associated with a huge number of hosts, including asteroids, ophiuroids, holothuroids, cnidarians, polychaetes, sipunculids and balanoglossids. As a result of Pettibone's (1993) careful taxonomic study, the different commensal specimens of this species were transferred to 15 species belonging to three different genera: *Malmgreniella*, *Lepidonopsis* and *Wilsoniella* (plus one additional genus, *Paragattyana*, and several more non-commensal species). The final result was that the number of hosts harbouring each new commensal species was restricted to one or a few closely-related species, from geographically closer areas.

Intra-host distribution patterns of commensal polychaetes

Since the first specific studies by Palmer (1968) and Dimock (1974) on symbiotic polychaete distribution patterns, a great deal of new information has been reported. Nowadays, it seems clear that most adult symbiotic polychaetes with known distributions occur alone on their hosts (Table 2). In other words, their symbiotic populations show regular distributions. This type of distribution not only characterizes most commensal

Table 2 List of commensal polychaetes with known infestation intensities of one symbiont per host.

Species	Species	Species
Arcteobia anticostiensis	*Harmothoe impar*	*Microphthalmus sczelkowii*
Arctonoe vittata	*Harmothoe spinifera*	*Neohololepidella murrayi*
Arctonoe pulchra	*Hesionella mccullochae*	*Nereis fucata*
Bathynoe tuberculata	*Hipponoe gaudichaudi*	*Parahalosydnopsis tubicola*
Benhamipolynoe cairnsi	*Hololepidella nigropunctata*	*Parasyllidea humesi*
Bhawania goodei	*Hololepidella ophiuricola*	*Perinereis cultrifera*
Bhawania pottsiana	*Lepidasthenia digueti*	*Petrecca thyasira*
Cheilonereis cyclurus	*Lepidonotus glaucus*	*Pholadiphila turnerae*
Disconatis contubernalis	*Lepidonotus melanogrammus*	*Pilargis berkeleyae*
Eunice maroovi	*Lumbrineris flabellicola*	*Polyeunoa laevis*
Gattyana cirrosa	*Malmgreniella andreapolis*	*Sphaerodoridium commensalis*
Gorgoniapolynoe bayeri	*Malmgreniella macginitiei*	*Tottonpolynoe symantipatharia*
Halosydna brevisetosa	*Microphthalmus aberrans*	
Harmothoe brevipalpa	*Microphthalmus hamosus*	

polynoids, nereids, pilargids and amphinomids, but also some hesionids, syllids and sphaerodorids. Occasionally, several individuals have been reported on the same host, these worms always being one adult and several juveniles. More often, however, the juvenile individuals of a given species with solitary adults display a random or even an aggregated distribution pattern (Dimock 1974, Britayev & Smurov 1985). The transition from random towards regular patterns of distribution necessarily implies the existence of mechanisms regulating the distribution of adult symbionts on their hosts. As direct observations and specific experiments have shown, several polynoids (e.g. *Arctonoe pulchra*, *A. vittata*, *Halosydna brevisetosa*, *Hololepidella nigropunctata*) and nereids (e.g. *Nereis fucata*) may show strong territorial behaviour, fighting against other conspecific specimens (Devaney 1967, Goerke 1971, Dimock 1974, Britayev 1991). However, we suggest that territorial behaviour may be generalized to most symbiotic polychaetes with regular distributions. Two possibilities arise from the negative intraspecific interactions generated by territorial behaviour: the death of one of the competitors or its relocation to another host, with the final result of the successive fights leading to the formation of regular patterns of distribution. Therefore, both territorial behaviour and intraspecific aggressiveness emerged as major factors controlling the appearance and maintenance of regular distributions within symbiotic polychaetes. However, there are some other possible factors that should not be excluded. For example, the possession of its own host-territory may allow a solitary commensal to exploit by itself the limited resources supplied by its host (i.e. safe place or food sources).

As most polychaetes are heterosexual, the isolation of single specimens will apparently prevent successful reproduction. Therefore, regular distributions must be associated with mechanisms raising the chance of commensal males and females being in close proximity during the reproductive period. Among them, we may mention:

(1) high densities of available hosts allowing successful reproduction of commensals (which do not force the symbiont to develop any specific adaptation),
(2) a synchronized gamete release by the commensals (which will make them independent of the host distribution),

(3) the formation of temporary host aggregations (i.e. linked to its own reproduction or feeding activity) synchronized with the reproductive period of the commensals (which probably implies the recognition of chemical cues linked to host reproduction which will trigger reproduction in commensals).

The occurrence of isolated heterosexual pairs on the same host individual may be considered as a particular solution of reproductive problems within regularly distributed symbionts. This type of distribution, more typical among symbiotic crustaceans (e.g. Huber 1987), has been reported for only five polychaete commensals. These are the syllid *Ambliosyllis cincinnata* (from sponges of the genus *Halichlona, Halichondria* and *Isodyctia*), the pilargid *Antonbruunia viridis* (inhabiting the mantle cavity of the bivalve *Lucinia fosteri*) and the scaleworms *Bathynoe cascadiensis* (associated with the brizingid starfish *Astrocles actinodetus* and *Astrolirus panamensis*), *Harmothoe hyalonemae* (living in atrial cavities of deep-water hexactinellid sponges of the genus *Hyalonema*) and *Gastrolepidia clavigera* (associated with tropical holothuroids) (Hartman & Boss 1965, Riser 1982, Ruff 1991, Martín et al. 1992, Britayev & Zamishliak 1996). Although there is no information on the reproductive cycle of these species, this pattern of distribution seems clearly more favourable to reproductive success. Moreover, there is no indication that the species living in heterosexual pairs are not as aggressive as the species living alone, which strongly supports the existence of sex-specific recognition for the symbiotic species living in pairs. Among polychaetes, sex-specific recognition has been previously reported for aggressive species such as *Neanthes fucata* (Reish 1957).

Several species of commensal polychaetes aggregate on their hosts (Table 3), along with most parasitic boring polychaetes (Table 10, p. 314). This seems to be valid for the symbiotic iphitimids, some histriobdellids, dorvilleids, syllids, serpulids, spirorbids, caobangids and spionids. Since the formation of aggregations among symbiotic polychaetes has not been the subject of specific studies, we can hypothesize that this pattern may result either from a peculiar gregarious habit in the settling larvae (i.e. chemically mediated settlement cues), or from the existence of mechanisms allowing the symbionts to increase their number on the host after an initial infestation by only one or two founders. This second strategy may be accomplished either by short larval development without pelagic phase, viviparity or asexual reproduction.

The aggregated distribution of the serpulid *Spirobranchus giganteus* living on corals (Hunte et al. 1990a), as well as of the spirorbid *Circeis armoricana* and the spionid *Polydora commensalis* on the inner surface of the host hermit crab's shell (Rzhavsky & Britayev 1988, Radashevsky 1989), may result from gregarious behaviour. In fact, aggregated distributions are common among tubicolous polychaetes and has been attributed to the attraction of settling larvae towards the tubes of conspecific adults (Eckelbarger 1978, Jensen & Morse 1984, Pawlik 1988, Marsden et al. 1990). A particular case of aggregated distribution in tubicolous polychaetes occurs in an undescribed species of *Vermiliopsis* from the Caribbean madreporid coral *Stephanocenia michelini*. The worm tubes of this serpulid are surrounded by the coral tissues and, according to Humann (1992), the crowns are regularly distributed on the coral surface (Fig. 4A). We suggest that this special distribution may result from negative adult–larval interactions. In fact, the circulation generated by the crowns may very well prevent the settlement of larvae on the host surface areas controlled by resident worms. Therefore, settling larvae would only be able to start building their tubes at a given distance from all previously existing neighbours, this distance being almost always the same.

Table 3 Commensal polychaetes with known infestation intensities higher than one symbiont per host. (*) indicate that the number of adult worms is one per host, the others being juveniles accompanying the adult; (**) indicate species that are only known from the types. Numbers between brackets indicate unusual maxima.

Species	Intensity	Species	Intensity
Acholoe astericola	1–6 symb./host (*)	*Harmothoe melanicornis*	0.2–0.4 symb./host cm
Adyte assimilis	1–5 symb./host (*)	*Harmothoe spongicola*	1 symb. (**)
Amblyosyllis cincinnata	1–35 symb./host	*Histriobdella homari*	648 symb./host
Arctonoe vittata	1–4 symb./host (*)	*Hydroides spongicola*	several
Arctonoe pulchra	1–7 symb./host (*)	*Iphitime cuenoti*	1–6 (128) symb./host
Branchiosyllis exilis	1–3 symb./host (*)	*Iphitime hartmanae*	1–8 symb./host
Branchipolynoe pettiboneae	4 symb./host (**)	*Iphitime holobranchiata*	1–39 symb./host
Branchipolynoe seepensis	6 symb. (**)	*Mytilidiphila enseiensis*	1–21 symb./host
Branchipolynoe symmytilida	1–3 symb./host	*Nereis grayi*	2 symb. (**)
Brychionoe karenae	1–18 symb./host	*Ophryotrocha geryonicola*	1–12 symb./host
Capitella capitata	few (71) symb./host	*Ophryotrocha mediterranea*	1–20 symb./host
Disconatis accolus	1 symb. (**)	*Ophiodromus pugettensis*	1–20 symb./host
Flabelligera affinis	1–7 symb./host	*Phalacrostemma cidariophilum*	2–3 symb./host
Gaudichaudius cimex	1–3 symb./host	*Polydorella dawydoffi*	130 symb./host cm^2
Gorgoniapolynoe guadalupensis	0.2–0.4 symb./host em	*Polydorella smurovi*	5–10 symb/host cm^2
Gorgoniapolynoe uschacovi	0.2–0.4 symb./host cm	*Sabellaria alcocki*	several symb./host
Haplosyllides floridana	hundreds of thousands	*Spirobranchus giganteus*	0.2–12 symb./host cm^2
Haplosyllis agelas	1–4 symb./host	*Stratiodrilus platensis*	0–30 symb./host
Haplosyllis chamaeleon	>1 symb./host	*Vermiliopsis* sp.	several ten

The dense populations of the small polydorid species of the genus *Polydorella* inhabiting the surface of cork-sponges (Fig. 2C) apparently resulted from their asexual reproductive cycles, which allowed them to reach quickly very high densities of up to 130 ind. cm^{-2} after the colonization of a suitable host sponge by a single new larva (Tzetlin & Britayev 1985, Radashevsky 1996). However, at least one *Polydorella* species (*P. smurovi*) may also reproduce sexually, since cocoons with eggs and embryos at different developmental stages have been found inside the adult tubes. In spite of the absence of direct observations, we can hypothesize that pelagic larvae resulting from sexual reproduction may be responsible for long-distance dispersion. Therefore, the combination of both reproductive strategies may allow them to succeed in finding and quickly colonizing the rare suitable host sponges. A detection of chemical cues from the host sponges by pelagic chemosensitive larvae may be involved in host location, as was suggested for the parasitic syllid *Branchiosyllis oculata* (Pawlik 1983).

In *Histriobdella homari*, the males impregnated the females hypodermically by means of specialized copulatory organs, allowing the spermatozoa to avoid the necessity of swimming to reach the mature eggs, which are directly fertilized in the body cavity of females (Jamieson et al. 1985). Then, the female attaches the eggs both to the ventral surface and egg masses of the host, the young worms then being released as minute adults without planktonic phase (Haswell 1913). Thus, the juveniles directly colonize the host infested by their own parents. This leads to a clustered distribution of the commensals and to high infestation intensities reaching more than 600 worms per host lobster (Simon 1967). The transference from one host to another is probably accomplished by direct migration of adults.

Location of commensal polychaetes on their hosts

Most commensals are characterized by a defined location on the host surface or, less frequently, inside the hosts (Figs 2–6). Since hosts may be considered as commensal refuges, the symbionts are normally located in the most protected sites. For example, commensals of decapods are usually located inside the host's branchial chambers (e.g. most *Iphitime* species, *Ophryotrocha geryonicola*, *O. mediterranea*, *Histriobdella homari*), inside the shells inhabited by hermit crabs (e.g. *Iphitime paguri*, *Nereis fucata*, *Circeis paguri*) or under the tails of egg-bearing female crabs (e.g. *Iphitime hartmanae*). The commensals of tubicolous polychaetes, burrowing holothurians or ghost shrimps (e.g. species of *Harmothoe*, *Lepidasthenia* or *Malmgreniella*) occur inside host tubes or burrows. The numerous associates of starfishes, sea urchins and brittle starfishes (among them, many scaleworms, hesionids and some syllids) prefer the more protected oral surface of their hosts. The commensals of bivalves or limpets often live inside the mantle cavity (e.g. *Parasyllidia humesi*, *Antonbruunia viridis*, *Harmothoe commensalis*, *Arctonoe vittata*), whereas a similar position has been reported for cirriped associates

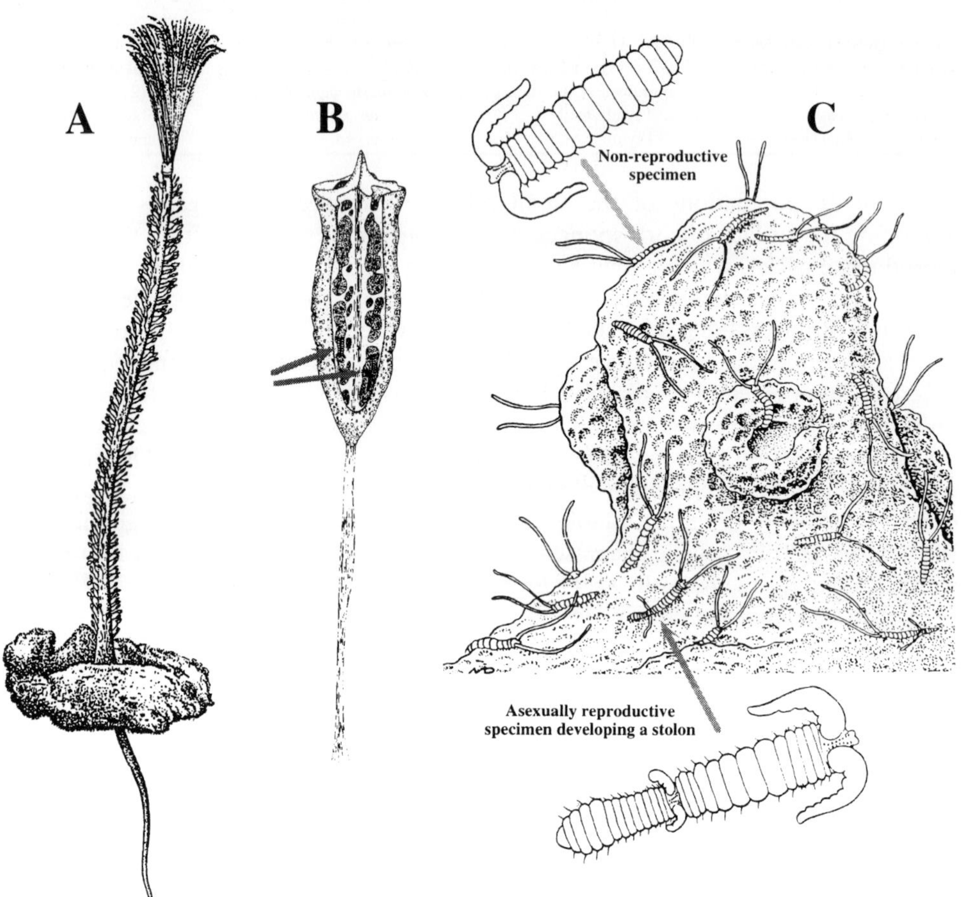

Figure 2 Examples of commensal polychaetes associated to sponges. A, *Potamilla symbiotica* on *Cryptospongia enigmatica* (redrawn from Uschakov 1950). B, *Harmothoe hyalonemae* on *Hyalonema thomsoni* (redrawn from Martín et al. 1992); grey arrows indicate the position of each worm forming a heterosexual pair. C, *Polydorella smurovi* on an unknown red sponge (redrawn from Tzetlin & Britayev 1985).

(e.g. *Hipponoe gaudichaudi*). Although rare, some commensals may live inside the host's body, e.g. the amphinomid *Benthoscolex cubanus* living in the intestine of the deep-water sea urchin *Archeopneustes hystrix* (Emson et al. 1993).

Some commensals have slow movements and, when disturbed, tend to keep themselves firmly attached to their hosts. For example, the syllid *Haplosyllis chamaeleon* living on the branches of the gorgonian *Paramuricea clavata* (Laubier 1960). The worm usually extends its body along the longitudinal axes of the host branches, preferably near zones with a high number of living polyps (mainly the apical parts). When disturbed, the polychaetes introduce their strong hooks into the coenchym and cannot be taken off without damage to the host. However, most commensal polychaetes are agile animals able to change their position according to different stimuli. For example, *Gastrolepidia clavigera* are usually located near anterior or posterior ends of their holothurian host's body, quickly hiding in the oral or cloacal openings when threatened (Britayev & Zamishliak 1996). Most commensal locations are reported for animals extracted from their natural habitats, meaning that observed locations may not be those preferred by the symbionts. Only a few data on the location of commensals in natural conditions are available. The commensal syllid *Branchiosyllis exilis* occurs on the arms or disc of its ophiuroid host *Ophiocoma echinata* (Hendler & Meyer 1982). The polychaete was observed moving along the ventral arm plates, between the alternating rows of arm spines and tube feet. They have also been found partially or entirely hidden in the oral cavity of the ophiuroid. However, immediately after host collection, *Branchiosyllis exilis* occurred more frequently on the arms of the ophiuroid than on the disk. Conversely, *in situ* observations of the scaleworm *Arctonoe vittata* associated with the starfish *Astherias rathbunae* indicated that the worm occurred more frequently on the oral surface of starfish, in the ambulacral grooves or even partially inside the oral cavity (Britayev et al. 1989). Nevertheless, it had been reported as more frequently inhabiting the oral disk (65%) than the arms of the starfish.

It should be pointed out that the locations of *Haplosyllis chamaeleon*, *Branchiosyllis exilis* and *Arctonoe vittata* on their respective hosts are not simply connected with the finding of shelter or physical protection. *Haplosyllis chamaeleon* was observed with its anterior end inside the gastral cavity of the gorgonian polyps. The slow movements and sensibility of the polyps may favour this behaviour, allowing the worm to feed either on the preys inside the polyp's stomach or on the polyps themselves. Although some pigmented parts inside the worm's gut supported the last possibility, no damage was observed in the polyps (Laubier 1960). Thus, it is less likely that *H. chamaeleon* behaves as a commensal or cleaner than as a parasite. *Branchiosyllis exilis* generally lies on the proximal portion of the ventral surface of its host's arms, directly in the path of food boluses carried by the tube feet towards the ophiuroid mouth. As suggested by Hendler & Meyer (1982), such a location allows the commensals to gather food from the alimentary tract of its hosts. Similarly, the preferred location of *Arctonoe vittata* on the host's oral disk seems likely to be connected with food supply. Direct observations on *A. vittata* feeding behaviour in natural conditions demonstrated that the worms gather pieces of their hosts' preys by biting them off both from the everted starfish stomach or directly through the mouth opening (Britayev et al. 1989).

The location on the host may change during the life history of symbionts. The adult specimens of *Hipponoe gaudichaudi* occur inside the branchial chambers of its goose-barnacle host, whereas the juveniles may be located between the mantle and the calcareous plates of the host or attached to the adult symbionts (Fig.

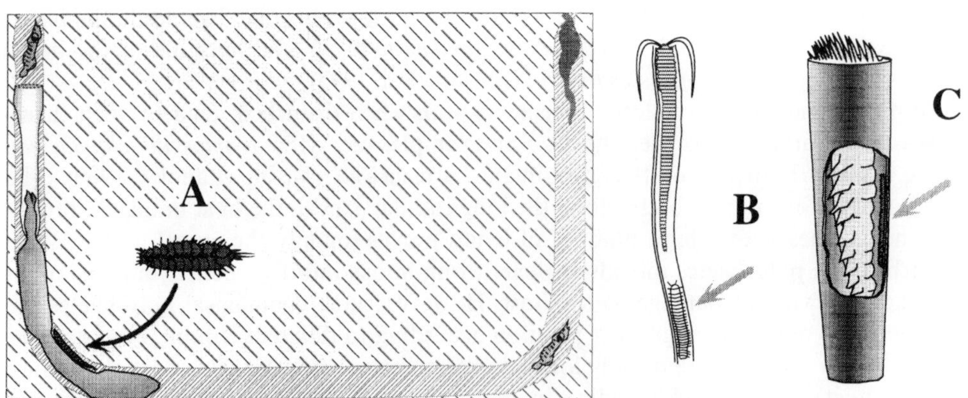

Figure 3 Examples of commensal polychaetes associated to other worms. A, *Hesperonoe adventor* inside the *Urechis caupo* burrow (redrawn from Dales 1957). B, *Harmothoe brevipalpa* inside the *Phyllochaetopterus* sp. tube (redrawn from Cañete et al. 1993). C, *Harmothoe brevipalpa* inside the *Pectinaria chilensis* tube (redrawn from Cañete et al. 1993). The arrows indicate the position of commensal polychaetes.

5D,E) (Britayev & Memmy 1989, Núñez et al. 1991). The characteristic position occupied by *Iphitime cuenoti* is at the extreme anterior dorsal end of the branchial cavity, where the individual worms lie with the ventral surface of body closely applied to the branchiostegal fold separating the viscera from branchiae (Comely & Ansell 1989). When more than one worm occurs in a single host crab, the largest individual generally occupies this position, while other individuals may be found in a similar position in the opposite branchial chamber, at another point on the walls of each chamber, or among the gills. Moreover, in the largest worms, the body is generally coiled. The smallest worms are only found in the interlamellar gill spaces and, when growing, the majority move out and become established in the roof of the branchial chamber (Comely & Ansell 1989).

We suggest that a low availability of optimum larval settlement sites on the host surface could be one of the reasons why the commensals alter their locations during their life cycle. Other probable reasons are: (a) differential preferences of juveniles and adults, (b) inaccessibility to optimum sites for juveniles, as a consequence of the competition with conspecific adults or other symbiotic species. The influence of intraspecific interactions between commensals on their location was demonstrated for the scaleworm *Arctonoe vittata* (Britayev et al. 1989). Its preference for the ventrum of its host starfish's oral disk was apparently inhibited when two worms infested the same host. In fact, the two symbionts were never found together on the oral disk. Whenever one of them occupied the oral disk, the other was on one arm (69% of starfish), the preferred area (i.e. oral disk) being always occupied by the larger worm. Moreover, the location of the two symbionts separately on the arms was also frequent (31% of starfish).

The location of commensal polychaetes must necessarily be influenced by the presence of other symbionts on the same host. However, although it is well known that some commensal polychaetes may share their hosts with symbionts from different taxonomic groups (see Dales 1957 for *Hesperonoe adventor* and the complex network of

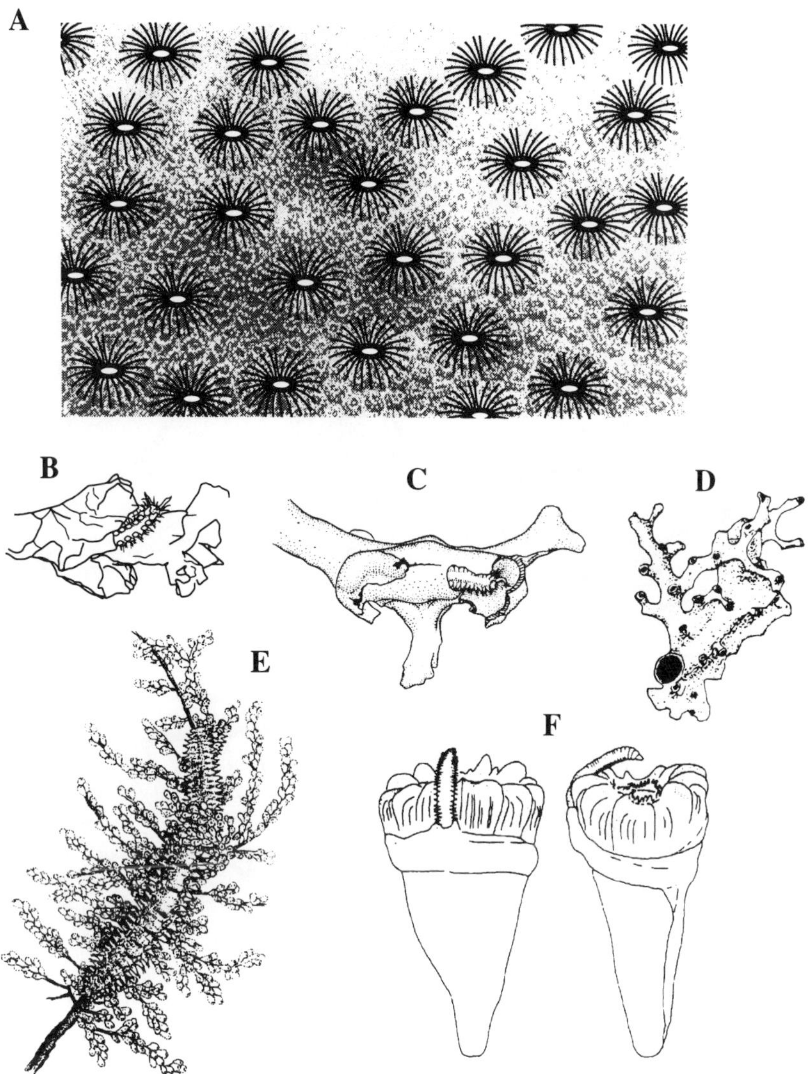

Figure 4 Examples of commensal polychaetes associated to cnidarians. A, *Vermiliopsis* sp. on *Stephanocenia michelini* (redrawn from Humann 1992). B, *Polynoe uschacovi* inside the tube induced in *Callogorgia* sp. (redrawn from Britayev 1981). C, *Gorgoniapolynoe uschakovi* inside a tube induced in *Allopora* sp. (redrawn from Britayev 1981). D, Tube induced by *Eunice floridana* on the branches of *Amphihelia* sp. (redrawn from Paris 1955). E, *Polynoe thouarellicola* on *Thouarella* sp. (redrawn from photograph provided by Dr. G. San Martín). F, *Lumbrineris flabellicola* on *Cariophyllia decapali* (redrawn from Miura & Shirayama 1992).

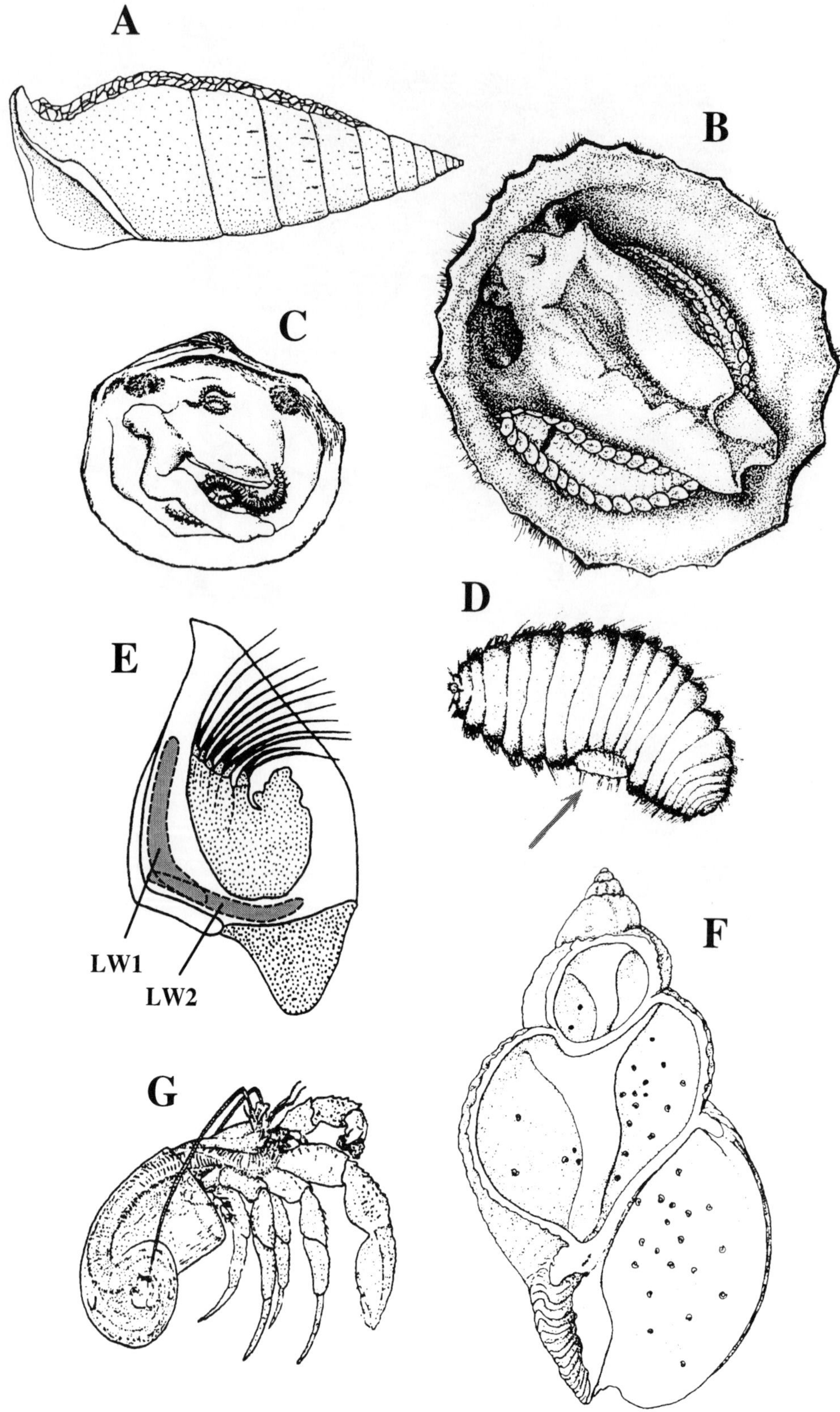
A
B
C
D
E
LW1
LW2
F
G

guests associated with the innkeeper *Urechis caupo*) (Fig. 3A), this discussion has not yet been developed. Changes in the location of commensal species may also vary in accord with the morphology or behaviour of their different host species. The scaleworm *Malmgreniella macginitiei*, for example, has been reported in the burrows of its host shrimp *Callianassa california*, inside the parchment-like tubes of its host polychaete *Axiothella rubrocincta*, or on the arms of its host brittle starfish *Amphiodia urtica* (Pettibone 1993). Another species of the same genus, *Malmgreniella andreapolis*, was found inside the burrows of host holothurians, as well as curved around the disk and mouth of host brittle starfishes.

A special comment should be reserved for those commensal polychaetes that are not "satisfied" by the shelter or the protective abilities of their hosts. These symbionts may either construct their own refuges on the host's surface or stimulate their hosts to build protective structures around them. Among them, the eunicid *Eunice floridana* and the scaleworms *Harmothoe melanicornis*, *Malmgreniella dicirra*, *Gorgoniapolynoe uschakovi* and several species of this genus live inside tunnels or gall-like cavities formed by coenenchymal walls of gorgonian or hydrocoral hosts (Fig. 4B,C,D) whose formation seems to have been brought about by the presence of the commensals (Britayev 1981, Pettibone 1991a,b). The lumbrinerid *Lumbrineris flabellicola* associated with scleractinian corals lives in membranous transparent tubes attached to the side of the host (Zibrowius et al. 1975, Miura & Shirayama 1992) (Fig. 4F) and *Eunice* sp. constructs a fragile tube attached to the length of the upper surface of its gastropod host *Cerithium vertagus* (Gibbs 1969) (Fig. 5A).

Characteristics of the infestations by commensal polychaetes

The prevalences of infestations by commensal polychaetes are highly variable, ranging from very low values (e.g. 0.03% for *Iphitime sartoral* associated with *Portunus spinicarpus*) up to host populations being infested as a whole (e.g. *Liocarcinus corrugatus* infested by *Iphitime cuenoti*, and some host populations infested by *Acholoe astericola* or *Benthoscolex cubanus*) (Table 4). The prevalence has been seldom considered as characteristic at the species level. In *Sphaerodoridium commensalis*, for example, a low prevalence of 3.5% seems to be the normal condition for all known populations of the species (Lützen 1961), whereas in *Acholoe astericola* the characteristic prevalence is as high as 75–100% (Barel & Kramers 1977). However, each population of a given commensal species is usually characterized by different prevalences.

Figure 5 Examples of commensal polychaetes associated to molluscs and crustaceans. A, *Eunice* sp. on *Cerithium vertagus* (redrawn from Gibbs 1969). B, *Arctonoe vittata* in the *Acmaea pallida* mantle cavity. C, *Antonbruunia viridis* in the *Lucinia fosteri* mantle cavity (redrawn from Hartman & Boss 1965). D, An adult *Hipponoe gaudichaudi* and the attached young specimen (redrawn from Núñez et al. 1991). E, The position of two *Hipponoe gaudichaudi* (LW1 and LW2) inside *Lepas anserifera* (redrawn from Britayev & Memmy 1989). F, *Circeis paguri* inside a *Buccinum* shell inhabited by *Eupagurus bernhardus* (redrawn from Al-Ogily & Knight-Jones 1981). G, *Nereis fucata* and its host *Eupagurus bernhardus* (redrawn from a Christmas card by G. Thorson, drawn by P. Winther, in Dales 1957).

Table 4 Commensal polychaetes with known prevalence. Host: number of hosts examined.

Species	Prevalence	Hosts
Acholoe astericola	75–100%	–
Adyte assimilis	10–80%	–
Ancistrosyllis cf *groenlandica*	50%	–
Antonbruunia viridis	80%	>100
Arctonoe vitatta		
Asterias amurensis	0–79.1%	–
Asterias rathbunae	5–59%	–
Acmaea pallida	90–94%	–
Arctonoe pulchra		
Stichopus parvimensis	60–100%	–
Megathura crenulata	25–28%	–
Asterophilia carlae	3.3–13%	–
Benhamipolynoe cairnsi	100%	10
Benthoscolex cubanus	65.5–100%	–
Branchiosyllis exilis	49.5% (29–100%)	142 (1–960)
Capitella capitata	100% (males)–0% (females)	–
Flabelliderma commensalis	6–50%	30–20
Gastrolepidia clavigera	32–80%	–
Gyptis ophiocomae	10%	400
Harmothoe brevipalpa		
Pectinaria	60%	68
Phyllochaetopterus	100%	108
Harmothoe commensalis	22–78%	–
Harmothoe hyalonemae	88.3%	171
Hipponoe gaudichaudi	1.2–1.5%	247–587
Histriobdella homari	100%	–
Hololepidella nigropunctata	20–60%	–
Iphitime cuenoti		
Liocarcinus puber	70%	181
Liocarcinus corrugatus	100%	123
Liocarcinus depurator	5.5–35%	1964–741
Other hosts	<15%	
Iphitime sartoral		–
Libinia spinosa	0.25%	
Portunus spinicarpus	0.03%	
Iphytime paguri	2.6–155%	155–?
Lumbrineris flabellicola	80% (worms)–100% (signals)	20
Mytilidiphila enseiensis	100%	–
Nereis fucata	50–90%	–
Ophryotrocha geryonicola		
Cancer borealis	0.5%	197
Geryon tridens	33%	65
Geryon quinquedens	8.8% (males)–44.4% (females)	277
Ophryotrocha mediterranea	20% (males)–0% (females)	1072
Ophiodromus pugettensis	50%	200
Petrecca thyasira	50%	10
Pilargis berkeleyae	21.7%	–
Sphaerodoridium commensalis	3.5%	1065
Sphaerodorum ophiurophoretos	6.6–11.9%	30–256

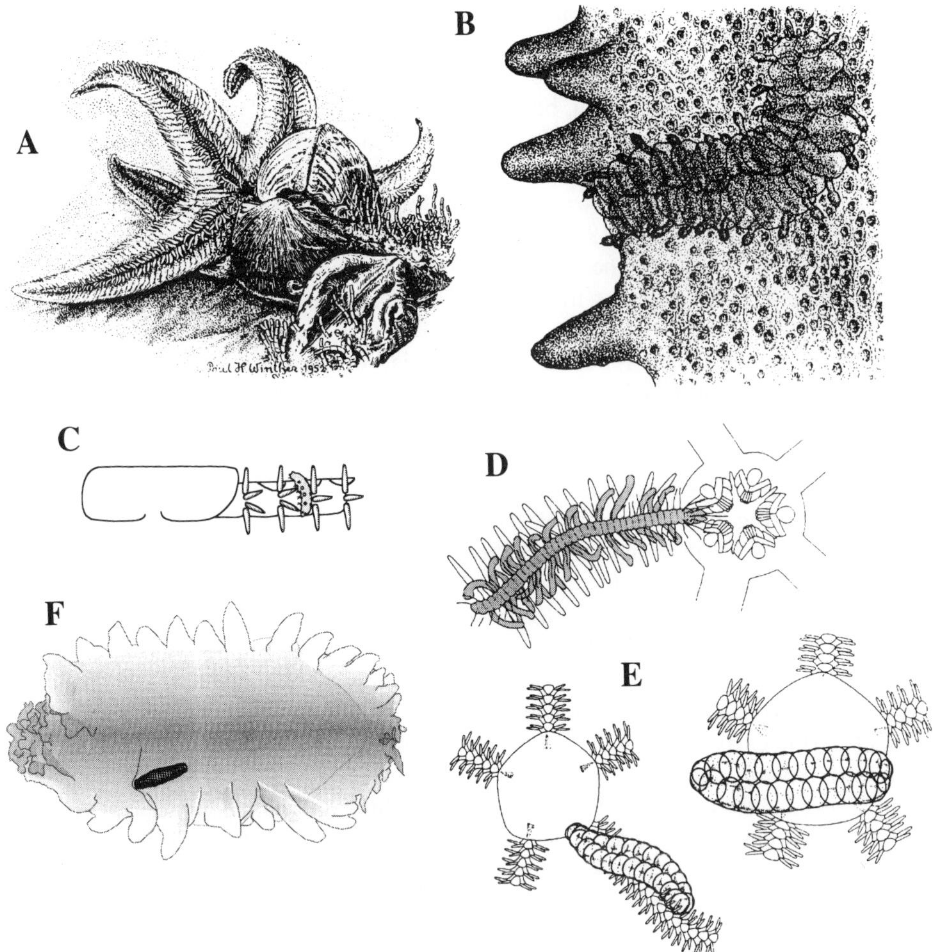

Figure 6 Examples of commensal polychaetes associated to echinoderms. A, *Acholoe astericola* on *Astropecten irregularis* (redrawn from a Christmas card by G. Thorson, drawn by P. Winther, in Davenport 1953a). B, *Gastrolepidia clavigera* on the *Stichopus chloronotus* surface (T. A. Britayev pers. obs.). C, *Sphaerodorum ophiurophoretos* on a *Amphipholis squamata* arm (redrawn from Martin & Alvà 1988). D, *Branchiosyllis exilis* on a *Ophiocoma echinata* arm (redrawn from Alvà & Jangoux 1989 and Hendler & Meyer 1982). E, *Malmgreniella variegata* on *Ophionereis reticulata* arm and oral disc (redrawn from Alvà & Jangoux 1989 and Millot 1953). F, *Eunoe laetmogonensis* on *Laetmogone violacea* (redrawn from Kirkegaard & Billet 1980).

Although bathymetric, spatial and temporal (i.e. seasonal and annual) variability in infestation prevalence and intensity may be expected in commensal populations, the corresponding patterns have seldom been documented.

Changes in infestation prevalence with depth have been reported for the Mediterranean populations of *Iphitime cuenoti* inhabiting *Liocarcinus depurator* (Belloni & Mori 1985, Abelló et al. 1988). Although host abundance increases above 100 m depth, the commensal appears to avoid this depth range, the prevalence of the infestation increasing with depth (Fig. 7A). This seems to be closely related to temperature, being lower and more stable in deep waters (>100 m deep). In fact, the prevalence of the commensal population infesting another Mediterranean host crab, *Macropipus tuberculatus*, was independent of depth, but this host never occurs deeper than 100 m. It is a common fact that species occurring both in the Mediterranean and the Atlantic tend to inhabit cooler waters in the latter (e.g. Laubier 1973). Accordingly, no relationship between depth and prevalence was observed for *Ophryotrocha mediterranea* infesting *Geryon longipes* (Martin et al. 1991). In this case, although the host crab occurs both in the Mediterranean and in the Atlantic, *Ophryotrocha mediterranea* appears to be endemic to the former, whereas in the latter, the ecological niche was occupied by *O. geryonicola*.

Hendler & Meyer (1982) reported spatial variability in infestation prevalence from the different populations of the brittle starfish *Ophiocoma echinata* harbouring the tropical syllid *Branchiosyllis exilis* along the Panamanian coast (Fig. 7B), where the maximum prevalence occurs in late autumn and in winter. Moreover, near Galeta, on the Caribbean coast of Panama, the prevalence of *B. exilis* varied over time between 20% and 90% (Fig 7C). It was highest from November to March (i.e. the dry season) and lowest in the local rainy season. The hesionid *Ophiodromus pugettensis* infesting the starfish *Patiria miniata* at Dana Point (California) reached its highest (79–92%) and lowest (19–32%) prevalence in November–December and summer (prior to the decrease of water temperature), respectively (Lande & Reish 1968) (Fig. 7D). The prevalence of the sphaerodorid *Sphaerodorum ophiurophoretus* infesting the brittle starfish *Amphipholis squamata* at Wimereux (English Channel coast of France) was maximal during early spring (11.9%), decreased in May (6.6%) and totally disappeared in summer (Alvà & Jangoux 1989). The prevalence of *Iphitime cuenoti* in the Scottish waters showed different seasonal patterns depending on the host (Comely & Ansell 1989). In *Liocarcinus puber* and *L. depurator*, the prevalence ranged between 70% and 100%, with no indication of seasonality. In *L. depurator*, however, there was an apparent prevalence fall during April, independently of the catch size. *Carcinus maenas* showed a low level of infestation that apparently varied seasonally, the maximum prevalence occurring in October (i.e. 14%). It was suggested that the decreasing prevalences of *Branchiosyllis exilis* linked to the rainy season resulted either from seasonal changes in salinity (and other environmental parameters) or from seasonal fluctuations in the mortality of the polychaetes because of the recurrent, low-tide emergence of the reef flat (Hendler & Meyer 1982). The marked seasonal pattern of *Ophiodromus pugettensis* seemed to be connected with commensal reproductive dynamics (Lande & Reish 1968). The absence of a clear seasonal pattern in *Iphitime cuenoti* from some of their host crabs was attributed to the persistence of small worms throughout the year, whereas the well defined seasonal peak in the host *Carcinus maenas* was clearly correlated with the peak season for the commensal spawning in October (Comely & Ansell 1989). Conversely, in *Sphaerodorum ophiurophoretos*, there was no further analysis of what caused the seasonal pattern.

With the exception of the impressive fidelity of the year-to-year counts of a regular seasonal trend demonstrated for the infestation prevalence in *Branchiosyllis exilis* (Fig. 7C), the only known data on year-to-year variability are on *Arctonoe vittata* infesting the starfish *Asterias amurensis* (Britayev & Smurov 1985). These authors reported a progressive increase in infestation prevalence from 0% in 1975–76 to 8.4% in 1978 and to 79.1% in 1980.

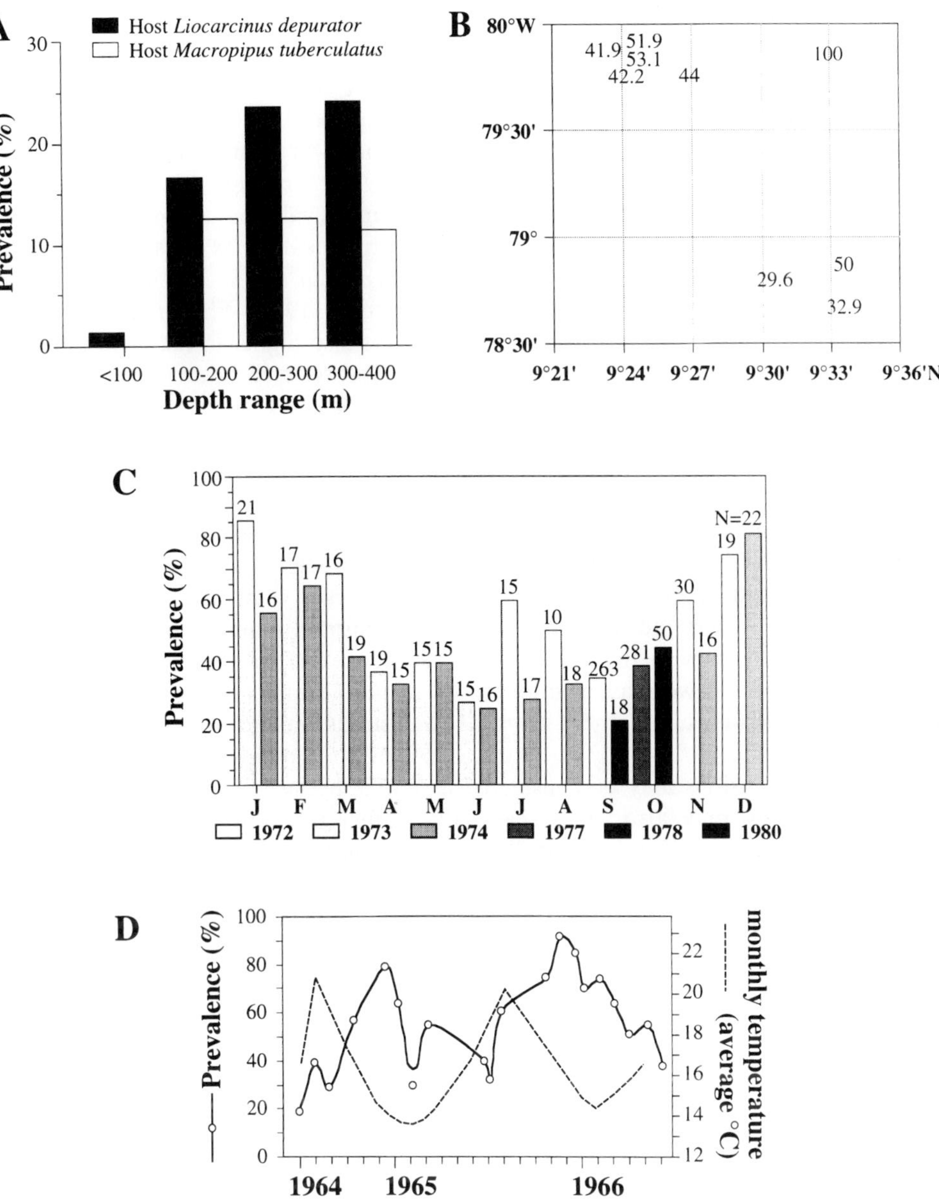

Figure 7 Examples of variability in infestation prevalence of commensal polychaete populations. A, Changes with depth in Mediterranean *Iphitime cuenoti* harboured by *Liocarcinus depurator* and *Macropipus tuberculatus* (redrawn from Abelló et al. 1988). B, Spatial variability of *Branchiosyllis exilis* on *Ophiocoma echinata* along the Panamanian coast (redrawn from Hendler & Meyer 1982). C, Seasonal pattern of *Branchiosyllis exilis* on *Ophiocoma echinata* at Galeta, Caribbean coast of Panama (redrawn from Hendler & Meyer 1982). D, Seasonal pattern *Ophiodromus pugettensis* on *Patiria miniata* at Dana Point, California (redrawn from Lande & Reish 1968).

An additional source of variability in the respective prevalences of commensal populations emerges from the association with different hosts, even if this occurs in the same locality. In these cases, the different prevalences are linked to the greater or lesser suitability of different hosts. As pointed out by Dimock (1974), the infestation prevalence of the scaleworm *Arctonoe pulchra* on the coasts of Santa Barbara (California) was higher in the preferred host, the sea-cucumber *Stichopus parvimensis*, than in the limpet *Megathura crenulata*. Similarly, the infestation prevalences of *Arctonoe vittata* in Vostok Bay progressively increased from the starfish *Aphelasterias japonica* (2–5%) and *Asterias amurensis* (79.1%) to the limpet *Acmaea pallida* (94%), the latter being the preferred host for the scaleworm (Britayev & Smurov 1985). The prevalence of the infestation by *Iphitime cuenoti* in Scottish waters was also higher in the preferred hosts, *Liocarcinus corrugatus* and *L. puber*, than in the non-preferred hosts (such as *L. depurator*, *Carcinus maenas* and *Hyas araneus*) (Comely & Ansell 1989).

Data on intensities of infestations by commensal polychaetes have been more widely reported than prevalences, ranging from 1 to 648 symbionts per host (Tables 2 and 3). However, the 1 : 1 associations are clearly dominant owing to the usually low symbiont densities or to the regular distribution pattern linked to intraspecific aggressive behaviour (see above). Moreover, the list of species characterized by an intensity of one individual per host (Table 2) should probably be larger than it is, as most species included in Table 3 are indeed characterized by a normal intensity of only one large adult worm per host with the occasional presence of one to several more juveniles. Among these species, we may mention those of the genera *Acholoe*, *Adyte*, *Branchiosyllis* and, probably, some others whose growth status was not indicated in the original descriptions.

Infestation intensities may be considered as more species-specific than prevalences. Nevertheless, they may also differ seasonally within the same population, as also occurs with the relative densities (or abundances). However, this variability has been seldom reported. The abundance of the commensal *Ophiodromus pugettensis* was maximum in winter (2.0–2.6 worms per host) and minimum in the middle of summer (0.26–0.44 worms per host) (Lande & Reish 1968) (Fig. 8A). In *Arctonoe pulchra*, intensity and abundance reached maximum values in August for *Stichopus parvimensis* associates (7 and 3.97 worms per host, respectively) and in September for *Megatura crenulata* associates (4 and 1.06 worms per host, respectively) (Dimock 1974). Although it was not clearly demonstrated, these maximum values were due to an increasing number of juvenile commensals. Thus, the seasonal pattern in *Arctonoe pulchra* seems to be related to the reproductive cycle of the species. In *Iphitime cuenoti*, seasonal peaks of infestation intensities apparently occured in the host crabs *Carcinus maenas* (maximum of 5 worms) and *Liocarcinus puber* (maximum of 9 adults and 98 juveniles) in October and May, respectively (Comely & Ansell 1989). Both peaks were clearly related to the reproductive cycle of the symbiont. However, it was difficult to attribute the character of seasonal pattern to the trends shown during a given year, at least in *L. puber*. Comely & Ansell (1989) pointed out that, although they considered it unlikely that any considerable numbers of juveniles were overlooked during their routine examinations of the gills, the peaks of intensity found during one year (1986) were not observed during the next (1987). In fact, this seems to indicate that the seasonal patterns are linked to the reproduction of the symbiont, which may vary from one year to another, as it occurs in virtually all marine organisms (either symbionts or free living).

As occurred with prevalences, we may expect additional changes in intensity of infestation among commensal populations from a given locality that are associated with different hosts. An example of this variability was reported for the scaleworm *Arctonoe pulchra*, whose populations associated with the sea-cucumber *Stichopus parvimensis* and the gastropod *Megathura crenulata* showed infestation intensities of 1 to 7 and 1 to 4 commensals per host, respectively (Dimock 1974). In this case, the differences in intensity are likely to be connected with the amount of living space available to the commensal polychaetes, which is clearly higher in the first host species than in the second. For *Iphitime cuenoti*, the incidence of intensities higher than one commensal per host was higher in the preferred host crabs (Comely & Ansell 1989).

Relationships between host and commensal polychaete characteristics

Whenever reported, the relationship between size structure of commensal and host populations is, if anything, unclear (Martin et al. 1991, 1992, Emson et al. 1993, Rozbaczylo & Ca–ete 1993, Britayev & Zamishliak 1996). This suggests that the life histories of commensal polychaetes are usually independent of their hosts, the host tending to live longer than the symbiont. In many cases, this dynamic implies that the same host population may successively harbour different polychaete populations. More detailed studies focusing on population dynamics of symbiotic polychaetes and their hosts may well reveal the existence of positive correlation between the respective size structures, as occurs among other symbiotic animals (e.g. the pontoniin shrimp *Anchistus custos* associated with the host bivalve *Pinna bicolor* in Britayev & Fahrutdinov 1994).

The size of host may affect the prevalence of infestation, with the largest individuals more often showing the highest prevalence (Fig. 8B, C, D). Examples of this trend are found in the associations between *Circeis paguri* and *Eupagurus bernhardus* (Al-Ogily & Knight-Jones 1981), *Branchiosylis exilis* and *Ophiocoma echinata* (Hendler & Meyer 1982), *Arctonoe vittata* and *Asterias rathbunae* (Britayev et al. 1989), and *Gastrolepidia clavigera* and *Stichopus chloronotus* (Britayev & Zamishliak 1996). Certainly, a similar pattern may be expected for many other associations. Many reasons may explain why large hosts are more favourable to symbionts. They are usually older and so have been exposed to planktonic larvae settlement or adult migration for a longer period than small hosts. Moreover, large hosts provide more room for symbionts. Accordingly, many examples of the existence of low host-size limits have been reported. Among them, starfish *Asterias rathbunae*, with disc radi lower than 35 mm, were not inhabited by *Arctone vittata*, whereas those with radi up to 90 mm harboured only one symbiont and the largest starfish were infested with 1–4 symbionts (Britayev et al. 1989). Also, the scaleworm *Harmothoe commensalis* was not present in shells of the clam *Gari solida* less than 60 mm in length (Rozbaczylo & Cañete 1993). In the case of *Circeis paguri*, small shells and hermit crabs are rarely colonized (Fig. 8C). In fact, the few small shells that are colonized are not very suitable habitats for the symbionts within them, which are rarely or never reach breeding size, presumably because the space available is too small to provide enough food or to satisfy their respiratory needs (Al-Ogily & Knight-Jones 1981).

The associations involving sexually dimorphic hosts merit special mention. One of the clearest examples was of the association between the dorvilleid *Ophryotrocha medi-*

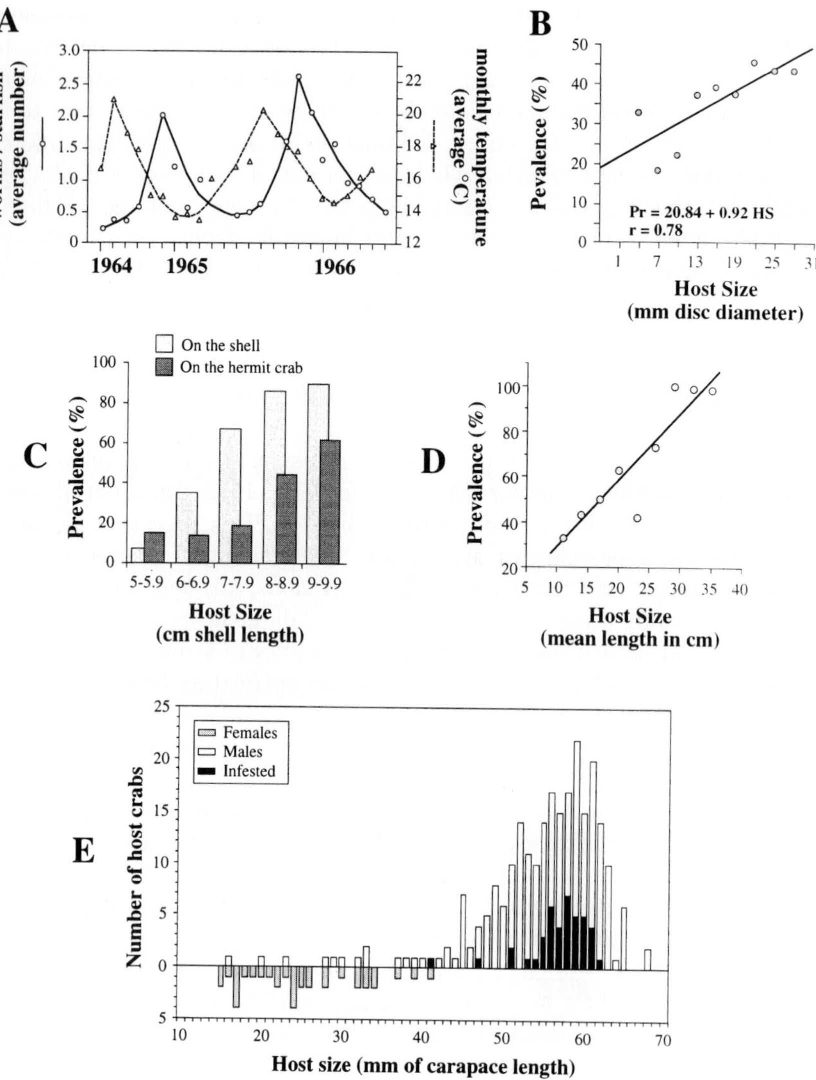

Figure 8 A, Seasonal variability of infestation intensity in *Ophiodromus pugettensis* infesting *Patiria miniata* at Dana Point, California (redrawn from Lande & Reish 1968). B, Relationship between infestation prevalence and host size in *Branchiosyllis exilis* from the *Ophiocoma echinata* oral disk (redrawn from Hendler & Meyer 1982). C, Relationship between infestation prevalence and host size in *Circeis paguri* attached both to the occupied shell and the body of the host *Eupagurus bernhardus* (redrawn from Al-Ogily & Knight-Jones 1981). D, Relationship between infestation prevalence and host size in *Gastrolepidia clavigera* from *Stichopus chloronotus* (redrawn from Britayev & Zamishliak 1996). E, *Ophryotrocha mediterranea* on the *Geryon longipes* branchial chambers (redrawn from Martin et al. 1991).

terranea and the deep-water crab *Geryon longipes* (Martin et al. 1991). The polychaete was not detected in any of the small female crabs examined, while the prevalence among the large male crabs reached up to 19%. Indeed, the carapace length of the smallest infested host male (41 mm) was similar to the maximum length of the female's caparace (Fig. 8E). Thus, it was suggested that the symbionts need a minimum of vital space, which the branchial chambers of juvenile and female crabs do not satisfy. This concurs with the fact that the closely related dorvilleid *Ophryotrocha geryonicola* was detected in females of *Geryon* species that reached larger sizes than males (Gaston & Benner 1981, Pfannenstiel et al. 1982). The opposite situation was reported for *Capitella capitata* living as commensal with the pea-crab *Pinnixia littoralis* (which itself lives as a commensal of the mactrid bivalve *Schizothaeus capax*). The commensals are always associated with the female crabs and never with the males, although both sexes may share the same host bivalve (Clark 1956). The worms lived more often inside thin mucous tubes attached to the crab's carapace, but would also be found burrowing inside a mass of detritus accumulated on the basis formed by the tubes. The most obvious solution was that males were too small to provide a suitable area on which the large *Capitella* could settle and, also, to support the large bolus of detritus. This was unlikely, however, because young female crabs, smaller than full-sized males, did carry the worms. The alternative and more likely explanation, proposed by Clark (1956), was that the males were too active. They were able to leave their own host bivalve, which was never observed of any females crab. By moving, the males prevented the accumulation of detritus on which the worms were presumed to feed and this caused the absence of commensal worms on the male's carapace.

When several hosts are infested by the same commensal species, we may also expect that the relative size of the different host species would affect the respective symbiotic populations. In *Iphitime cuenoti*, for example, the largest commensals are usually reported on those hosts that are also larger (e.g. *Liocarcinus corrugatus* and *L. puber*). Thus, the total size of the commensal may be a function of the size of its crab host. However, this is valid for the small host species of the genus *Macropodia* and *Inachus* but not for the large *Cancer pagurus*, suggesting that other factors may limit the success of the worm (e.g. moulting frequency or stage). The commensal iphitimids seems able to support moulting stress, as worms have been found inhabiting recently moulted soft crabs (Abelló et al. 1988, Comely & Ansell 1989). However, a high moulting frequency may significantly disturb the success of the infestation. Hartnoll (1962) and Comely & Ansell (1989) also suggested that the substrata inhabited by the different hosts or by different individuals of the same host may affect the success of the infestation.

Theoretically, we may expect that the host's population structure will influence the structure of the symbiont population. Nevertheless, the available data to illustrate this statement are very scarce. The relationships between host size and prevalence mentioned above, lead to the speculation that higher values of the infestation indexes will be positively correlated to the presence of large and more numerous hosts. A positive relationship between host population density and infestation characteristics was reported for the starfish *Asterias rathbunae* and its commensal *Arctonoe vittata* (Britayev et al. 1989). In Avatcha Bay, Pacific coast of Kamchatka, the highest infestation prevalence, mean intensity and abundance occurred in the area most densely populated by the host. The high number of intraspecific traumas reflected the frequency of intraspecific interactions resulting from the high density of commensals in the mentioned area (i.e. higher than in the neighbouring ones). This may probably be related to a high mobility of the commensals among hosts, affecting also the structure of the commensal populations.

In most cases, however, the mechanisms to explain how the population density of hosts influences the commensal population remain unclear. We may hypothesize that the accumulation of chemical cues released from dense host populations may attract a high number of settling larvae of the commensal polychaetes more effectively than less dense host populations. Thus, we may expect higher recruitment and more young worms associated with dense host populations. Nevertheless, the influence of other external factors, such as predation pressure, cannot be discarded. In fact, predatory activity will probably decrease the fitness of commensals associated with less abundant host populations more significantly than those harboured by dense host populations.

Adaptations of polychaetes to the commensal mode of life

Cryptic colouring in commensal polychaetes

The colour patterns of commensal polychaetes frequently differ from those of its free-living relatives, even if the latter belong to strictly the same species. For example, the only apparent morphological difference between free-living and commensal specimens of *Lepidonotus glaucus* is the degree of pigmentation (Hanley & Burke 1990). The specimens associated with the *Eunice* hosts were more darkly pigmented, often appearing almost black, when compared with free-living specimens. We may tentatively suggest that the dark colouring of the commensal *Lepidonotus glaucus* is closely related to the dark colour of its host. In fact, one of the most peculiar features of commensal polychaetes (mainly ectocommensal) is their cryptic colouring masking them on the host's surface.

The fact that many commensal animals inhabiting crinoids and holothuroids (including polychaetes) resembled their hosts was pointed out by Potts (1915). Later, the similarity of symbiont worms' colouring to that of their hosts' was stressed by nearly all authors studying commensal associations. Thus, when describing the association of the scaleworm *Malmgreniella variegata* (as *Harmothoe lunulata*) with the ophiuroid *Ophionereis reticulata*, Millot (1953) noted that "worms matched the ophiuran very closely, being white, with a well defined repetitive black pattern borne on elytra". This matching was also reported and well illustrated by Pettibone (1993). Another remarkable similarity in colouring is that of the hesionid *Gyptis ophiocomae* with its ophiuroid host *Ophiocoma scolopendrina* (Storch & Niggemann 1967). Moreover, the effectiveness of the polychaete's camouflage is often heightened by their location on an appropriate site on the host surface. The segments of the syllid *Branchiosyllis exilis* each had a thin white border so that they displayed light striations on close examination. This pattern of pigmentation was reinforced by the location of polychaetes on the black and white banded arms of the ophiuroid host *Ophiocoma echinata*, between colonnades of black and white striped spines (Fig. 6D, p. 259), which made it difficult to distinguish the commensals (Hendler & Meyer 1982).

The colour pattern of commensals inhabiting several host species is usually appropriately similar to the corresponding host colouring. For example, the individuals of the scaleworm *Gastrolepidia clavigera* associated with the uniformly black sea-cucumber, *Stichopus chloronotus*, are usually black, whereas those associated with the yellow to brown *Thelenota ananas* are light brown, and those with the greenish-grey *Stichopus variegatus* are greenish-grey with brown spots (Britayev & Zamishliak 1996). However, some exceptions to this general rule may be observed. Dark brown worms typical to *S. chloronotus* have been found on *S. variegatus*, probably because the scale-worm changed from one host to another (Britayev & Zamishliak 1996). Gibbs (1969) noted similar changes between hosts by this scaleworm species.

In light of the above, we may certainly assume that *Gastrolepidia clavigera* is not able to change its colour in a short time interval, such as octopus or flatfish do. A dual question then arises: does the observed colour pattern result from the host influencing the developing symbiotic polychaetes? or does a commensal population consist of several "sub-populations", genetically pre-adapted to each appropriate host species? With the exception of the experimental demonstration of the trophic origin of the colour pattern in the parasitic syllid *Branchiosyllis oculata* (Pawlik 1983), there are no specific studies on this particular aspect of the arrangement of symbiotic associations. However, the accidental observation of commensal specimens with unusual "hybrid" colouring inclined us towards the first question. Indeed, hybrid colouring may result from the regeneration of posterior sections of the body. In *Arctonoe vittata*, the colour of the posterior part was yellowish-white in accordance with the body colour of its current host, the limpet *Acmaea pallida*, whereas the anterior (i.e. old or original) half of the body was pink-brown, closely resembling the colouring of the worms associated with the starfish *Asterias amurensis* (T. A. Britayev pers. obs.). As relocation from one host to another has been demonstrated (see above), the logical sequence of events may be as follows. A symbiont inhabiting the initial starfish host is disturbed (either as a result of intraspecific aggressive behaviour or predator attack), losing the posterior part of the body (i.e. autotomized or bitten). It is then obliged to relocate to a new host, the limpet in this case, with the subsequent regeneration of the posterior half being in accord with the colour of the new host. It should be pointed out that how these changes in colouring are accomplished has not yet been assessed. In fact, we may hypothesize that the changes may be produced by a substance (or substances) obtained by the polynoids from their hosts (i.e. by partly feeding on them or by incorporating some kind of secretion) but, also, that the substance stimulating the different colourings are obtained both by the hosts and the commensals by feeding on the same kind of food. These two hypotheses clearly had different consequences in the allocation of the involved relationships within parasitisms or commensalisms.

A comparable case of colour mimicry occurs in the relationship between the syllid *Haplosyllis chamaeleon* and its single known host, the gorgonian *Paramuricea clavata* (Laubier 1960). The colouring of the worms varied from pale yellowish to dark red, matching exactly the colours of their host. Although these colours actually occur on the same host colony, there are no evident relationships between the colour of the host branch and that of the attached polychaetes. Moreover, hybrid specimens showing all possible colour combinations may occur. Uniform colour patterns may indicate symbionts growing to their adult size on the same gorgonian branch, then moving to different parts of the colony without any preference for the colour of the branches. Hybrid forms may be the result of these displaced specimens being bitten or autotomized and then regenerating on a differently coloured branch.

Although all these examples support the idea of the influence of the host on the colouring of its commensal polychaetes, the question of how the mimicry is accomplished still remains open. Is there an intrinsic metabolic reaction of the symbiont to the different stimuli provided by the different hosts? or are the symbionts taking pigments directly from the hosts, i.e. feeding on it? If the second answer is correct, then the associations involved should be re-evaluated as being closer to parasitisms. However, without specific descriptions of the behaviour of symbiotic polychaetes and its effects on their hosts, we decided to include them within the commensalisms.

Morphological adaptations of commensal polychaetes

Forty years ago Clark (1956) noted that "Although some worms are almost always, or even exclusively, to be found in more or less intimate association with other animals, very little structural adaptation appears to have taken place to suit them for this mode of life". However, after carefully re-analyzing the significantly high number of commensal species described since this classic paper, it becomes clear that most, or even all, commensals have more or less defined morphological features allowing us to distinguish them from their free-living relatives.

The symbiotic features of commensal polychaetes can be divided into two categories: those significantly adapted to the commensal mode of life, and those repeatedly connected with the commensal mode of life, but lacking any obvious adaptive significance. The first group includes structural adaptations such as: (a) modifications permitting the attachment of the symbionts to their hosts, (b) modifications allowing the symbionts to reproduce without leaving their hosts and so prevent gamete losses, (c) modifications affecting the buccal equipment of symbionts, and (d) modifications enabling the symbionts to mimic their hosts. The best known modifications included in the second group are those of elytra and parapodia in scaleworms, but the eyes may also show interesting modifications. Moreover, it has also been pointed out that commensal polynoids are often larger than their free-living relatives. This modification, however, has been interpreted as a consequence of – instead of an adaptation to – the commensal mode of life. This fact was supposed to be due to a decreasing necessity of exercise in the commensals, and to a food supply being greater and more easily accessible to commensals than to free-living worms (Essenberg 1918).

The observation and correct interpretation of all morphologically evident commensal features (whether they have clearly adaptive significance or not) is important: in particular, whether these features enable symbiotic and free-living polychaetes to be distinguished, despite the specimens being found separately from their hosts. It should be pointed out that this situation is quite common when studying worm collections that are based on trawl or dredge samples, and that often reach the polychaetologists already sorted.

Adaptive morphological modifications Attachment structures are frequently related to the setal arrangement. Hooked setae, apparently modified to facilitate the attachment, have been reported for a wide number of species from many different families (viz. Iphitimidae, Polynoidae, Amphinomidae, Sphaerodoridae, Nautiliniellidae, Spintheridae, Syllidae) (Figs. 9–12). Some of them show remarkable coincidence in shape and, although not demonstrated, probably in functionality, too. In the case of polynoids, some species seem able to replace continually the hooked setae, which become blunt from wear, by new pointed ones growing out from the basis of the parapodia (e.g. *Arctonoe fragilis*). This characteristic is not common to all polynoids but is an exception observed in a few species that habitually attach themselves to other organisms (Essenberg 1918).

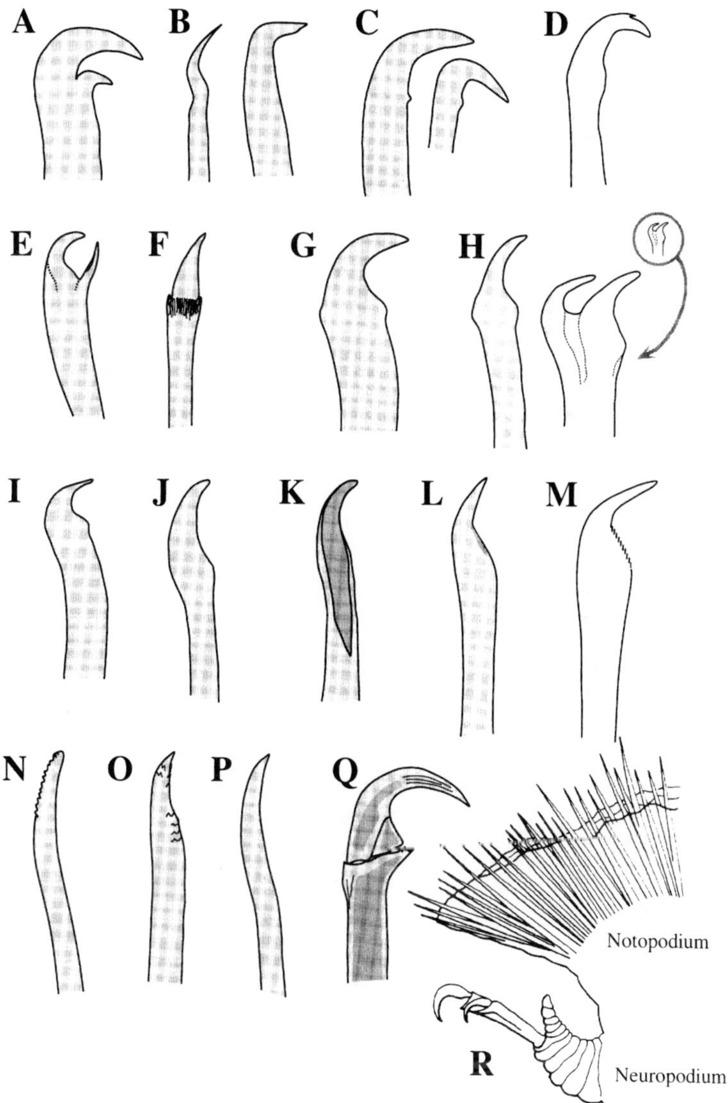

Figure 9 Examples of hooked setae from commensal polychaetes. Amphinomidae: A, *Hipponoe gaudichaudi* (redrawn from Britayev & Memmi 1989). Syllidae: B, *Bollandia antipathicola* (redrawn from Glasby 1994), C, *Syllis onkilochaeta* (redrawn from Hartmann-Scrhröder 1991). Sphaerodoridae: D, *Sphaerodorum ophiurophoretos* (redrawn from Martin & Alvà 1989). Antonbruunidae: E, *Antonbruunia viridis* (redrawn from Hartman & Boss 1965). Nautilinidae: F, *Laubierus mucronatus* (redrawn from Blake 1993), G, *Shinkai longipedata* (redrawn from Miura & Ohta 1991), H, *Natsushima bifurcata* (redrawn from Miura & Laubier 1990), I, *Shinkai sagamiensis* , J, *Petreca thyasira* (redrawn from Blake 1990), K, *Nautiliniella calyptogenicola* (redrawn from Miura & Laubier 1989). Iphitimidae: L, *Iphitime doderleinii* (redrawn from Marenzeller 1902), M, *Iphitime cuenoti* (redrawn from Abelló et al. 1988), M, *Iphitime hartmanae* (redrawn from Kirkegaard 1977), O, *Iphitime paguri* (redrawn from Fage & Legendre 1933), P, *Iphitime holobranchiata* (redrawn from Pilger 1971). Spintheridae: Q, *Spinther citrinus*, R, Detail of the parapodium of *Spinther citrinus* (redrawn from George & Hartmann-Schröder 1985).

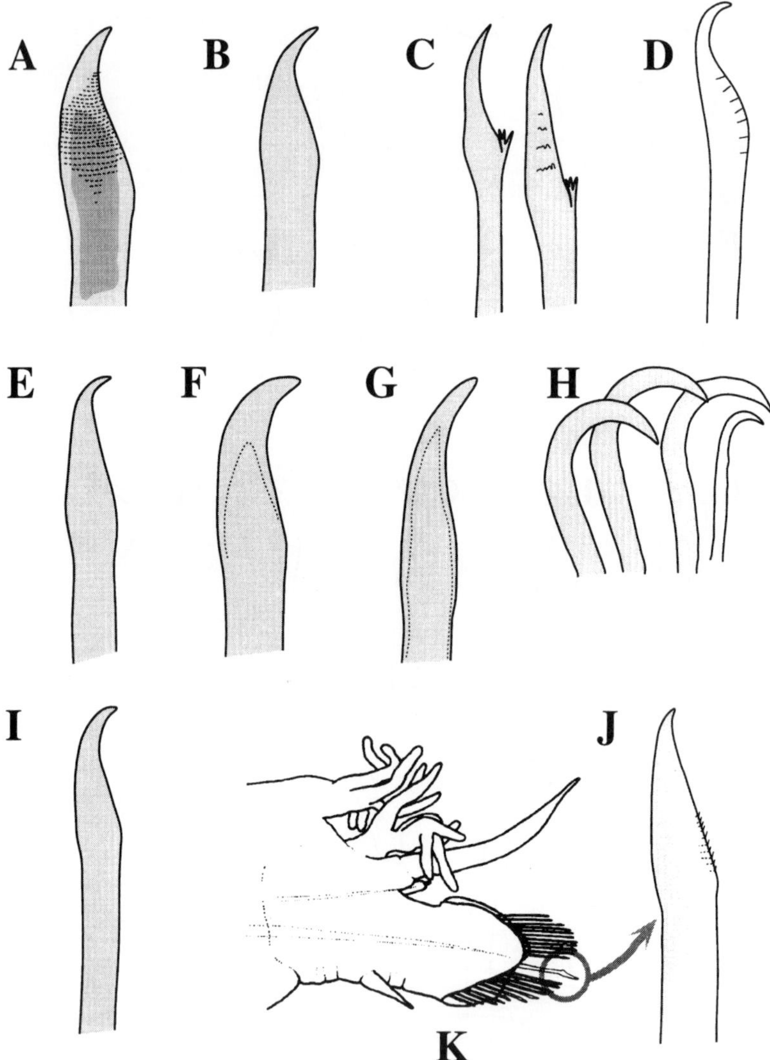

Figure 10 Examples of hooked setae from commensal Polynoidae. A, *Bathynoe cascadiensis* (redrawn from Ruff 1991). B, *Bathynoe tuberculata* (redrawn from Hanley 1989). C, *Paradyte crinoidicola* (redrawn from Hanley 1984). D, *Australaugeneria pottsi* (redrawn from Pettibone 1969a). E, *Arctonoe pulchra*, F, *Arctonoe vittata*, G, *Arctonoe fragilis* (redrawn from Hanley 1989). H, *Uncopolynoe corallicola* (redrawn from Hartmann-Schröder 1960). I, *Branchipolynoe symmytilida* (redrawn from Pettibone 1984a). J, *Branchipolynoe pettiboneae*, K, Detail of the parapodium of *Branchipolynoe pettiboneae* (redrawn from Miura & Hashimoto 1991).

Attachment structures, other than those based on setae, have been reported only twice: for the holothurian-associated scaleworm *Gastrolepidia clavigera* (Fig. 13A) and for the highly modified species of the crustacean-associated family Histriobdellidae. The highly peculiar attaching structures of *G. clavigera* consist of semicircular, serially arranged, shield-like ventral lobes of the segments. These lobes considerably increased the area of contact between the commensal and the smooth, mucus-covered holo-

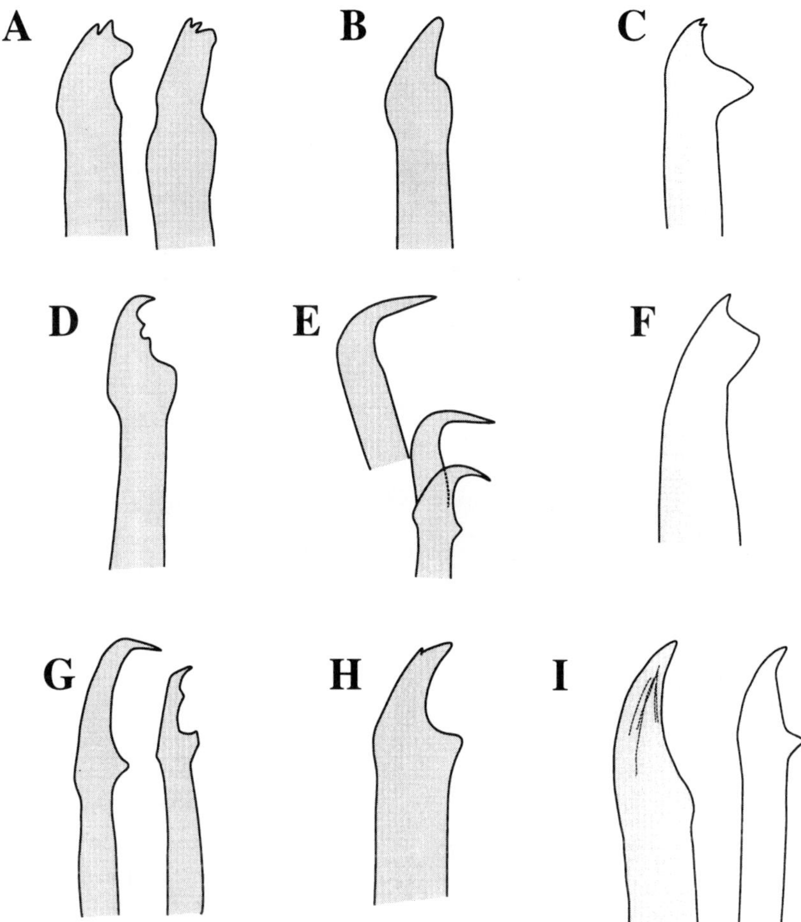

Figure 11 Examples of hooked setae from the known symbiotic species of *Haplosyllis* and related genus. A, *Haplosyllides floridana* (redrawn from San Martín et al. 1997). B, *Haplosyllis dollfusi* (redrawn from San Martín 1984). C, *Haplosyllis spongicola* (D. Martin, personal observation). D, *Haplosyllis agelas* (redrawn from Uebelacker 1982). E, *Haplosyllis xeniaecola* (redrawn from Hartmann-Schröder 1993). F, *Haplosyllis anthogorgicola* (redrawn from Imajima & Hartman 1964). G, *Haplosyllis bisetosa* (redrawn from Hartmann-Schröder 1960). H, *Haplosyllis chamaeleon* (redrawn from Laubier 1960). I, *Trypanosyllis asterobia* (redrawn from Imajima 1966).

thurian body. Moreover, when combined with a characteristic arching of the worm's body, they probably function like suckers (Gibbs 1971). Such an attaching organ, unique among polychaetes, is reminiscent of the small clam *Entovalva semperi* Oshima, which attaches itself to the burrowing sea-cucumber *Protankyra bidentata* by means of a highly modified foot that functions like a suction cup (Dales 1957).

The attachment structure of the histriobdellids clearly differs from that of the scale-worm. The body of these commensal worms is so highly modified that they were initially considered as larval serpulids (Beneden 1853), then as adult leeches (Beneden 1958) and as archiannelids (Hermans 1969). Although some differences may occur at

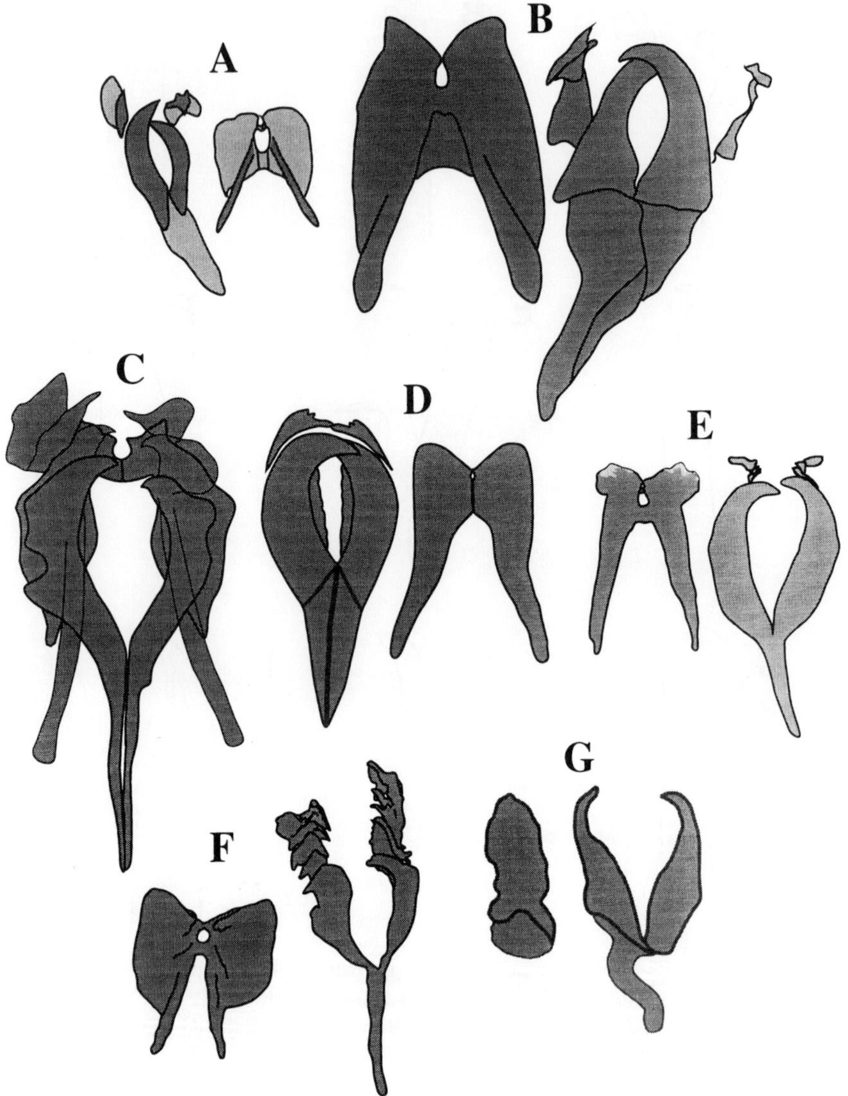

Figure 12 Jaw apparatus of the known species of *Iphitime*. A, *Iphitime cuenoti* "male" (redrawn from Abelló et al. 1988). B, *Iphitime cuenoti* "female" (redrawn from Abelló et al. 1982). C, *Iphitime doderleinii* (redrawn from Marenzeller 1902). D, *Iphitime holobranchiata* (redrawn from Pilger 1971). E, *Iphitime hartmanae* (redrawn from Kirkegaard 1977). F, *Iphitime paguri* (redrawn from Moore & Gorzoula 1973). G, *Veneriserva pygoclava* (redrawn from Rossi 1984).

the species level, the attachment structures are located on the distal end of two widened lateral expansions present on the posterior end of the body (Fig. 13D, E) and consist of several adhesive glands. Modified reproductive structures are also present in this family. Indeed, the males are provided with a copulatory complex (Fig. 13D) composed of a penis with one (*Stratiodrilus* spp.) or two (*Histriobdella*) projecting structures, a highly developed musculature, peniform external structure and several lateral or median expansions or lips. Moreover, the males have also one or two lateral retrac-

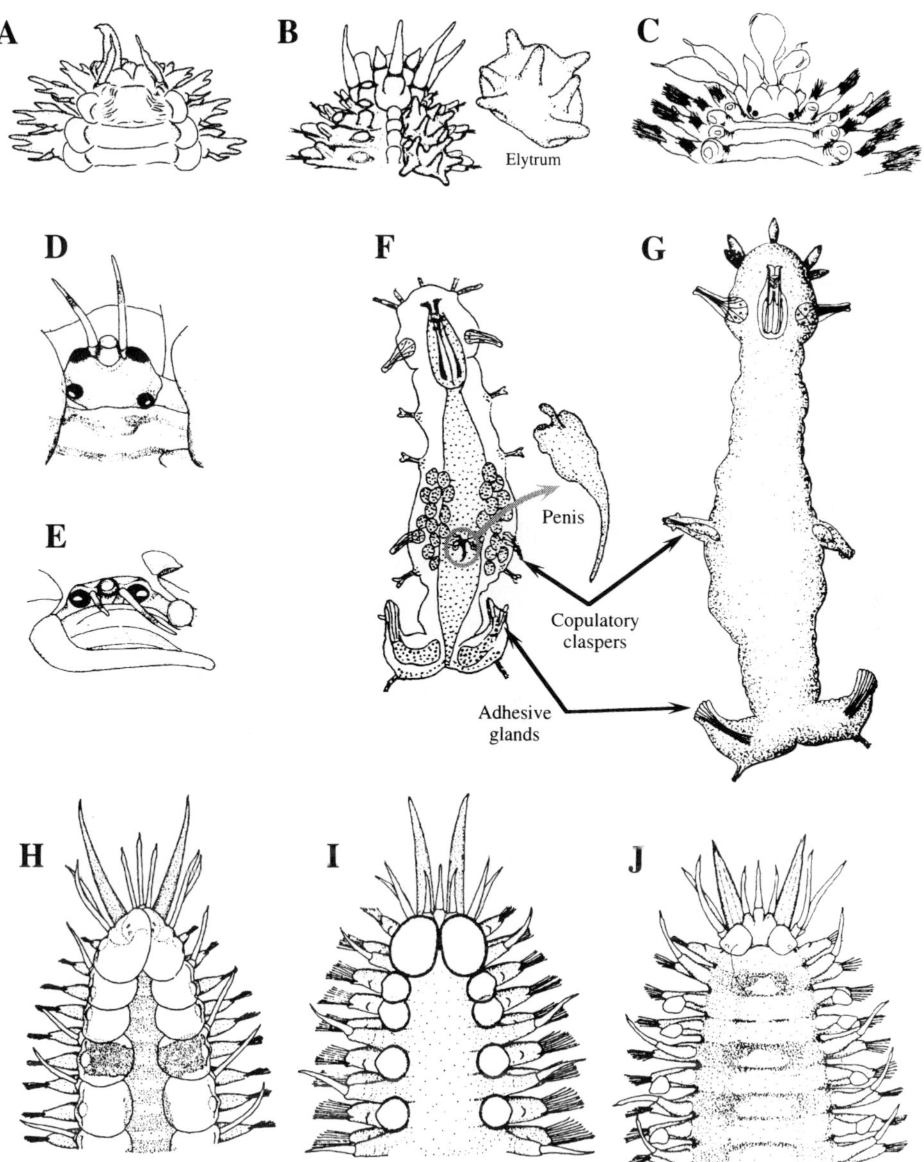

Figure 13 Examples of morphological modifications in commensal polychaetes. Adaptive modifications: A, Ventral view of *Gastrolepidia clavigera* showing the attachment processes (redrawn from Hanley 1989). B, *Bathynoe cascadiensis* showing the large elytra with numerous prominent nodular macrotubercles (redrawn from Ruff 1991). C, *Asterophilia carlae* showing the swollen head appendages (redrawn from Hanley 1989). D, E, Dorsal and frontal views of *Ophthalmonoe pettiboneae* head, showing the modified eye position (redrawn from Petersen & Britayev 1997). F, G, The highly modified males of *Stratiodrilus haswelli* (F, redrawn from Harrison 1928) and *Histriobdella homari* (G, redrawn from George & Hartmann-Schröder 1985). H, I, J, Progressive elytra smallness in *Lepidasthenia* species: *L. mossambica*, *L. microlepis* and *L. stylolepis* (redrawn from Gibbs 1969).

tile copulatory arms or claspers. This complex organ enables the males to inject the sperm directly into the body cavity of females (Lang 1949, Jamieson et al. 1985), assuring the effectiveness of fertilization in the limited space provided by the host.

Modifications of jaw apparatus have not been so often reported for commensals as for parasitic polychaetes (see pp. 297–300). The mandibles of the histriobdellids closely resemble those of the eunicids (Mesnill & Caullery 1922), among which they are currently included (Fauchald 1977), and do not have any apparent adaptive modification. Although it was pointed out that they were not solid but hollow, the relationship of this characteristic with the symbiotic mode of life of these worms has not been defined. In addition, it is often very difficult to describe their exact composition as they vary from individual to individual (Lang 1949). *Stratiodrilus* species have denticulated bucal lips surrounding the mouth opening (Moyano et al. 1993). As these worms feed by grazing on the microflora that grow on the different surfaces of the host's body (Cannon & Jennings 1987), this mouth structure is likely to be related to its feeding mode. In fact, most of the jaws defined as highly specialized do not differ greatly in complexity and function from those of related free-living polychaetes (e.g. commensal *Ophryotrocha* spp. versus other dorvilleids). Moreover, the jaw apparatus may vary in size and appearance according to the size of the worms and the time of usage, such as in *O. geryonicola* (Pfannestiel et al. 1982). The mandibles of the iphitimids also resemble those of the free-living relatives (i.e. eunicids such as *Halla* spp.). However, evidence of tissue damage in their host crabs has never been reported, although all the species have mandibles well equipped for biting and chewing (Fig. 12). In this case, commensal iphitimids are not notably simpler, their jaws being modified only in that the basal pair of maxillae (the only well developed pair) are fused together to form a pair of pincers (Hartnoll 1962). Conversely, the single parasitic iphitimid, *Veneriserva pygoclava*, shows a highly simplified jaw apparatus (see Fig. 12G). Additional variability was initially related to the fact that some species are sexually dimorphic (e.g. *Iphitime cuenoti*), with the males being significantly smaller than females (Hartnoll 1962, Abelló et al. 1988). More recently, however, it was demonstrated that individuals of *I. cuenoti* may have either small or large jaws (Fig. 12A, B), independent of sex, and that those with small jaws were probably young sexually undifferentiated individuals (Comely & Ansell 1989).

Morphological adaptations also allow the commensal polychaetes to mimic the surface structure of their hosts. *Bathynoe cascadiensis*, a scaleworm associated with starfish, has unusually large elytra with numerous prominent nodular macrotubercles (Fig. 13B). These tubercles are likely to be a specific adaptation for living with hosts from the order Forcipulata, a group characterized by having numerous knobs and spines and being heavily armed with large pedicellaria. By mimicking the spines and tube feet of starfish, it has been suggested that the camouflaged worms may protect themselves from being attacked and rejected by their own hosts (Ruff 1991). A spectacular example of mimicry has been demonstrated by the individuals of the scaleworm *Gastrolepidia clavigera* that are associated with the sea-cucumber *Holothuria atra*. The holothurians are black, sometimes with tints of violet, and usually have white coral sand grains adhering to their surface so that they appear to be covered by a silvery-white "coat". The commensal worms are deep violet, almost black, with the elytra dark brown to black with bright white tubercles (Potts 1910, Gibbs 1969). Moreover, the distal part of antennae, palps, and tentacular and dorsal cirri are swollen and white. These white tubercles and appendages on the dorsal side of the worms exactly mimic the holothurian papillae with attached white sand grains (Britayev & Zamishliak 1996). In the case of *Asterophilia carlae*, this uncoloured scaleworm is very conspicuous on the surface of its blue host starfish *Linkia laevigater*. However, the large swellings of the appendages and raised mounds of the posterior edge of each elytron of the polynoid (Fig. 13C), as well as the elliptical outline of its body, perfectly mimic the host in that, often, the starfish has only a portion of each ambulacral groove open and this groove, through which the tube feet protrude, is elliptical in outline (Hanley 1989).

Not obviously adaptive morphological modifications Many commensal polychaetes have structural modifications of their external features that differentiate them from their free-living relatives, but lack an obvious adaptive significance. Often, the finding of some of these characteristics in newly described species allowed its commensal habit to be inferred without reliable observation of the specimens living with any host species (e.g. *Minusculisquama hughesi* and its maldanid hosts in Pettibone 1983). The most typical series of commensal trends are certainly found within the polynoids.

The elytra of many commensals are thin and smooth, lack the ornamentation frequently found on free-living species, and sometimes small (leaving much of the dorsal surface uncovered). A good example of these small elytra occurs in the species of the genus *Lepidasthenia*: in *L. mossambica*, the elytra are very thin, whereas they are smaller in *L. microlepis* and so diminished in overall size as to be mere vestiges at the end of long elytrophores in *L. stylolepis* (Fig. 13H–13J). Additionally, some polynoids (viz. species of the genera *Arctonoe*, *Bathynoe*, *Branchipolynoe*, *Disconatis*, *Gastrolepidia*, *Hololepidella*, *Lepidasthenia*, *Minusculiquama*) have very small notopodia (subbiramous parapodia) with a limited number of purely developed notosetae or even without setae (e.g. *Branchipolynoe pettiboneae* in Fig. 10K). Although some of these modifications may occur in commensal polychaetes associated with bivalves (e.g. *Branchipolynoe symmytilida* in Pettibone 1984a), they appear more frequently as a result of sharing tubes or galleries with a host. Perhaps the protection afforded by the tube, and the clean water circulated through the tube by the host, diminish the need for elytra to form a defensive covering, or the need for large elytra and the presence of notosetae to help keep the dorsal respiratory surface of the body clean. In fact, certain species (e.g. *Disconatis contubernalis*, *Lepidasthenia microlepis*) have the first pair of elytra much larger than subsequent pairs. This may be closely related to the protective role of elytra, particularly if commensal habits include sticking its head out of the tube (Hanley & Burke 1988). A limitation of movement requirements may also be related to the decreasing importance of notosetae in the commensal tube-inhabitant scaleworms.

The scaleworm *Ophthalmonoe pettiboneae*, which lives inside the tubes of the large chaetopterid *Chaetopterus appendiculatus*, also show a very interesting modification. This commensal species has two pairs of eyes, with the anterior one being enlarged, directed anteriorly and with conspicuous lenses (Fig. 13F, G). Both the orientation and structure of the eyes are rather rare among polynoids, but are typical for tube-dwelling acoetids. Therefore, although this eye shape seems clearly to be an adaptation of the scaleworm to inhabiting tubes (Petersen & Britayev 1997), the relationship with its commensal mode of life is not clear.

Adaptive behaviour in commensal polychaetes

Although the behaviour of any commensal polychaete should undoubtedly be adapted to the symbiotic mode of life, very often this adaptive behaviour seems easy to deduce from the normal habits of the free-living polychaetes. Some of the typically symbiotic types of behaviour we expect in commensal polychaetes are: the co-ordination of movements with those of their host when feeding and moving, the finding of a new host (i.e. during recruitment or if they are moved out from their current host, either as a result of intraspecific competition or death of the host), the finding of a reliable way of attaching themselves to their hosts, and the selection of the most protected area in the host body. The last two types of behaviour are closely connected with morphological adaptive modifications, which have already been discussed in other sections. The first two require a more specific discussion. As mentioned in the introduction, commensal relationships are often known as "0+" associations. However, detailed behavioural studies strongly support the view that certain commensal associations are, in fact, "++" or mutualistic relationships, when the benefits are bilateral. This mutualistic behaviour also merits a specific discussion.

It should be pointed out that behavioural information often results from direct field or experimental observations of living associates. Consequently, it tends to be very rare. Nevertheless, some interesting examples have been reported and will be discussed in the following sections.

Co-ordination of host/symbiont movements Remarkable host–symbiont co-ordination in movements and feeding behaviour occurs between the reddish scaleworm *Hesperonoe adventor* and its host echiuroid *Urechis caupo*. The host constructs a large burrow in mud flats and feeds with the aid of a mucous bag, drawing water through the bag by peristaltic movements of the body. The commensal usually lives alongside the host (Fig. 3A, p. 000) and so is obliged to move back and forth along the wall of the burrow to avoid the peristaltic swellings of the host. The commensal feeds on parts of the mucous bag formed by the host, but can reach the bag only when the host itself is eating it (Dales 1957).

The behaviour of *Hesperonoe adventor* closely resembles that of an undetermined species of polynoid (probably belonging to the genus *Malmgreniella* sensu Pettibone 1993) found associated with a tube-dwelling synaptid holothurian on the Catalan coast of the Mediterranean (D. Martin pers. obs.). The scaleworm was observed inside the host burrow, lying below the holothurian with its dorsum touching the ventral side of the host. The symbiont moved in order to avoid the host's peristaltic swellings, exactly as *Hesperonoe adventor* does. In this case, however, there is not a clear trophic relationship between the deposit-feeding host and the probably carnivorous symbiont.

Nature and specificity of host recognition behaviour The most widely studied aspect of host-commensal relationships among polychaetes is probably host recognition behaviour (viz. Davenport 1950, 1953a,b, 1955, 1966a,b, Davenport & Hickok 1951, Bartel & Davenport 1956, Hickok & Davenport 1957, Gerber & Stout 1968, Webster 1968, Dimock & Davenport 1971). In the well known series of experiments using different models of choice-apparatus, most of these authors clearly demonstrated that all the symbiotic polychaetes studied were able to recognize their hosts and to respond to their presence or to some signals (i.e. the so-called host-factor) released from the host into the water (i.e. in absence of the hosts themselves). The known symbiotic polychaetes showing host recognition behaviour in experimental conditions include 14 species and 25 relationships (Table 5). Some commensal polychaetes (viz. some *Arctonoe* spp., *Ophiodromus puggetensis*, *Hesperonoe adventor* and *Hololepidella*

Table 5 Commensal polychaetes with reported host-recognition behavior. Host: AST, asteroid; HOL, holothuroid; OPH, Ophiuroid; GAS, gastropod; POP, polyplacophor.

Symbiont	Host	Species	Reference
Polynoidae			
Arctonoe pulchra	HOL	*Stichopus californicus*	Davenport 1950, Davenport & Hickok 1951, Dimock & Davenport 1971
Arctonoe pulchra	HOL	*Stichopus parvimensis*	Dimock & Davenport 1971
Arctonoe pulchra	GAS	*Megathura crenulata*	Dimock & Davenport 1971
Arctonoe pulchra	AST	*Dermasterias imbricata*	Dimock & Davenport 1971
Arctonoe pulchra	AST	*Louidia foliolata*	Dimock & Davenport 1971
Arctonoe fragilis	GAS	*Diodora aspera*	Davenport 1950
Arctonoe vittata	AST	*Evasterias troschelii*	Gerber & Stout 1968, Palmer 1968
Arctonoe vittata	POP	*Cryptochiton stelleri*	Gerber & Stout 1968, Palmer 1968
Arctonoe vittata	GAS	*Acmaea pallida*	Britayev et al. 1977
Arctonoe vittata	AST	*Asterias amurensis*	Britayev et al. 1977
Arctonoe vittata	AST	*Aphelasterias japonica*	Britayev 1978 (abstract)
Acholoe astericola	AST	*Astropecten irregularis*	Davenport 1953a
Gattyana cirrosa	POL	*Amphitrite johnstoni*	Davenport 1953a
Lepidasthenia argus	POL	*Amphitrite edwardsii*	Davenport 1953b
Halosydna brevisetosa	POL	*Amphitrite robusta*	Davenport & Hickok 1951
Polynoe scolopendrina	POL	*Eupolymnia nebulosa*	Davenport 1953b
Harmothoe spinifera	POL	*Amphitrite gracilis*	Davenport 1953b
Malmgreniella andreapolis	OPH	*Acrocnida brachiata*	Davenport 1953b
(as *Harmothoe lunulata* part.)	HOL	*Leptosynapta inhaerens*	Davenport 1953b
Malmgreniella arenicola	POL	*Amphitrite johnstoni*	Davenport 1953b
Hololepidella nigropunctata	OPH	*Ophiocoma brevipes*	Devaney 1967
Hololepidella nigropunctata	OPH	*Ophiocoma dentata*	Devaney 1967
Lepidasthenia tubicola	POL	*Loimia medusa*	Gaikwad 1993
Hesionidae			
Podarke puggetensis	AST	*Patiria miniata*	Davenport et al. 1960
Podarke puggetensis	AST	*Luidia foliolata*	Davenport et al. 1960

nigropunctata) seemed able to detect their hosts from a distance. Other commensal worms did not show distant host recognition. *Acholoe astericola*, for example, seems only able to recognize its host starfish when adjacent to, or even touching, it. As soon as its head contacts the host, the worm becomes active almost at once, fastening itself to the host either by climbing quickly to the aboral surface or into the ambulacral groove, or by wrapping itself entirely around the tip of the host arm (Davenport 1953a). *Haplosyllis chamaeleon* shows a similar specific response to its gorgonian host (Laubier 1960). When separated from the gorgonian, the polychaetes remain particularly immobile and coiled around themselves. Only after contacting a host branch, do they start to crawl and reach their habitual position on the branches.

The mechanism of host recognition appears highly specific in some commensal polychaetes, at least in experimental conditions. Commensal polychaetes seem able to recognize their host, to distinguish it from taxonomically related species, and to distinguish the original host species from alternative ones. *Arctonoe fragilis* was attracted by water from a tank occupied by its host starfish *Evasterias troschelii*, but not by water from a tank of another starfish of the genus *Pisaster*, with which the worm is not associated. Similarly, *Arctonoe pulchra* was attracted by water coming from the sea-cucumber *Stichopus* (i.e. the host), but not by water from *Cucumaria* (non-associated, but common in the same habitat) (Davenport 1950, Davenport & Hickok 1951).

The specificity of the response is highly variable depending on the different commensal polychaetes. *Acholoe astericola*, for example, may respond positively both to its host and to many other taxonomically related animals (see below). Conversely, *Arctonoe pulchra* showed undifferentiated liking for animals belonging to different taxonomic groups. Indeed, this commensal repeatedly show preference for the arm of the choice-apparatus with water flowing from: starfish (*Dermasterias imbricata*), sea-cucumbers (*Stichopus californicus, S. parvimensis*) and molluscs (*Megathura crenulata*), all of which were in fact their natural hosts. The specimens of *Arctonoe vittata* harboured by *Asterias amurensis* positively respond to chemical signals both from the original host and from its alternative host mollusc *Acmaea pallida*. However, the positive responses of worms associated with *A. pallida* were restricted to the host mollusc (Britayev 1991). The behaviour of *Arctonoe vittata* has clear implications for the population dynamics and life cycle of the species (see Fig. 14). In fact, recruitment may occur on both hosts. However, the symbionts (i.e. juveniles or adults) driven off the host starfish as a result of intraspecific competition will be able to relocate not only on other starfish but also on the host mollusc, whereas those worms initially associated with the mollusc have only one relocation possibility, that is, another host mollusc.

Certain chemical cues released by or from the host to the surrounding water seem to hold the key to host recognition behaviour in polychaetes (Davenport & Hickok 1951). When the asteroid *Evasterias troschelii* was suspended in a bolting-silk bag in an aquarium and then carefully removed, the commensal *Arctonoe fragilis* was able to detect the signal released by the host. When the host starfish was suspended in an aquarium overnight in a dialysis bag, the commensal did not respond to the water outside. However, by splitting the bag with a razor the attractive cue was released and the commensal immediately responded. But it is not at all clear what this cue sub-

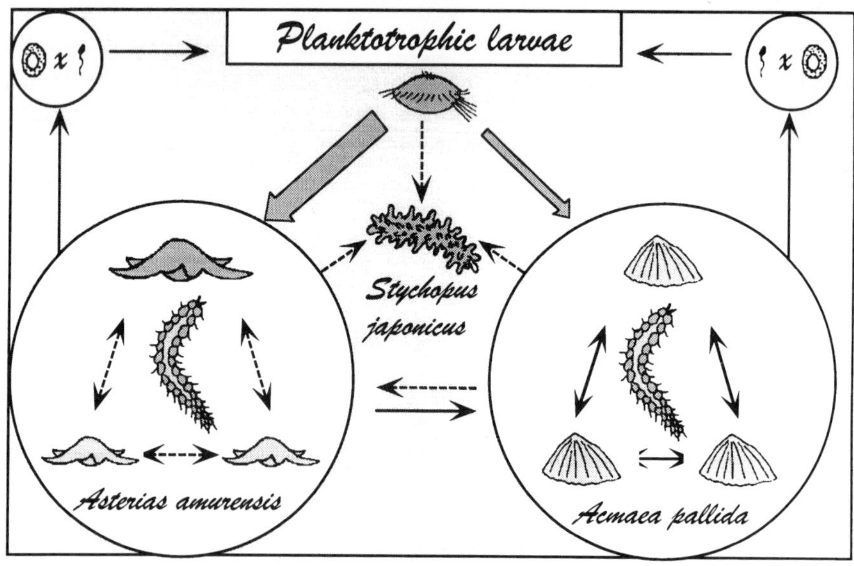

Figure 14 Life cycle of *Arctonoe vittata* (redrawn from Britayev 1991).

stance (or substances) is. It seems to be very labile, as the positive response of *A. fragilis* to the water in absence of the host was only maintained for between 15 min and 30 min. Also, a temperature higher than 64 °C either alters or completely destroys the attractant, eliminating any response (Davenport 1953a).

The fact that, in some cases, commensal responded positively to several taxonomically close hosts suggests a possible biochemical similarity between them, with their cues having the same or very similar chemical composition. The specimens of *Acholoe astericola* originally associated with the starfish *Astropecten irregularis* at Plymouth positively respond not only to their original host but, to some extent, to other asteroids of the order Phanerozonia from the same area (e.g. *Astropecten*, *Louidia*, *Porania*, *Solaster*) and to certain spinulose starfish (*Asterina*, *Palmipes*, *Henricia*), but they did not ordinarily respond to forcipulate starfish (*Asterias*, *Marthasterias*, *Stichastrella*).

The scaleworm *Polynoe scolopendrina* showed a more specific response than *Acholoe astericola* when tested against a wide spectrum of terebellid polychaetes. Although certain reduced responses to two species of *Amphitrite* (not its normal hosts) were detected, the polynoid consistently responded only to its original host (i.e. *Eupolymnia nebulosa*). This commensal has two alternative hosts, the terebellid *Lanice conchilega* (described from deep water on the French coast) and the eunicid *Lysidice ninetta* (occurring also in the Plymouth area). Interestingly, however, the commensal polynoid responded strongly to *L. ninetta* but did not respond to *Lanice conchilega*. Accordingly, it was suggested that, although belonging to the same polychaete family, the attractants released by the hosts *Eupolymnia nebulosa* and *Lanice conchilega* and recognized by the respective commensal populations, did not share the same chemical composition (Davenport 1953b).

Biochemical studies on the host-factor based on disc electrophoresis with polyacrylamide gel revealed that seven protein fractions (five of them mucoproteins) were very likely responsible for the attraction of symbionts. These interesting studies were carried out on a copepod, *Sabelliphilus sarsi*, living in symbiosis with a sabellid polychaete, *Spirographis spallanzani* (Carton 1966a,b, 1967, 1968a,b). Apart from the studies mentioned above, there are no further studies on the nature of the host-factor and on the specificity of host recognition in polychaetes. This is certainly one of the most promising lines of research into symbiosis in polychaetes, particularly as many studies suggest the importance of host recognition behaviour. According to Carthy (1958), the difference between taste (contact chemoreception) and smell (distant chemoreception) in aquatic organisms is a question of threshold, smell receptors having a low, and taste receptors a high, threshold. In support of this, Gerber & Stout (1968) demonstrated that *Arctonoe vittata* (from the host *Diodora aspera*) apparently had two different chemoreceptors, one for distant reception (the antennae) the other for contact reception (the palps). However, these authors also demonstrated that the behaviour of the commensals indicated that the distant chemoreceptors did not function in contact chemoreception, which strongly supported the existence of two different host factors involved in this particular relationship. Gerber & Stout (1968) also pointed out that "further work needs to be done to demonstrate the presence of two possible host factors and their chemical nature and to identify morphologically and physiologically the receptors involved". Although recommended in 1968, this work has still not been done.

Several classic studies on host recognition behaviour have been repeatedly considered conflicting and obscure. The most representative is probably the study on

Harmothoe lunulata and its hosts from Plymouth (Davenport 1953b). The scaleworms associated with the terebellid *Amphitrite johnstoni* unexpectedly demonstrated slight positive responses to the alternative hosts *Arenicola marina*, *Lanice conchilega* and *Golfingia elongata*, but not to species taxonomically related to the original host such as *Amphitrite edwardsi*. In fact, the recent taxonomic review on *Harmothoe lunulata* and related species (Pettibone 1993) enables light to be cast on Davenport's conflicting observation. According to Pettibone (1993), the specimens of "*H. lunulata*" should probably be considered as belonging to different species. The population associated with *Amphitrite johnstoni* was transferred to *Malmgreniella andreapolis*, whereas those associated with the ophiuroid *Acrocnida brachiata* and the holothuroid *Leptosynapta galliennii* were re-allocated to *Malmgreniella arenicola*. Moreover, the three hosts that triggered unexpected positive responses in Davenport's "*H. lunulata*" associated with *Amphitrite johnstoni* (i.e. *A. marina*, *Lanice conchilega* and *Golfingia elongata*) were reported as natural hosts of *Malmgreniella arenicola* by Pettibone (1993). These observations seem to clarify the unexpected commensal responses but pose new questions about the "nature of the host-factor" problem.

It has been demonstrated that the high frequency of commensals migrating from host to host (including other species than the original host) may result from intraspecific competition (Lande & Reish 1968, Palmer 1968, Dimock 1974, Britayev 1991). Therefore, the classic authors, working before these dates, could not fully assess the relevance of the host recognition behaviour they found. In 1966, Davenport wrote that ". . . during adulthood of the commensals, the chaemotaxis serves merely to keep the commensal on the surface of the hosts . . . under natural conditions an annelid . . . would seldom be forced to seek a new host when its own is damaged or dying. Such an event is probably too rare to be considered a factor in the evolution of the powerful chaemotactic response . . .". However, the fact that the spatial pattern formation in symbiotic polychaete populations strongly depends on intraspecific competition and resulting symbiont relocation revealed the significance of host recognition behaviour in life cycles of symbionts.

Host-entering behaviour The section above may suggest that chaemoreception is the dominant mechanism contributing to the maintenance of marine symbiotic partnerships. However, other kinds of stimuli produced by the hosts may trigger characteristic adaptive responses from commensal polychaetes. As pointed out by Gilpin-Brown (1969), the nereidid *Nereis fucata* seemed unable to discriminate between different species of hermit crabs. Thus, the peculiar association of *N. fucata* with hermit crabs seems to display non-chemically regulated behaviour. Surprisingly, this association has been studied by many authors (Rabaud 1939, Thorson 1949, Brightwell 1951, Gilpin-Brown 1969, Goerke 1971, Jensen & Bender 1973, Cram & Evans 1980) and, unusually, it has even been filmed (Goerke 1977a,b).

The most distinctive attribute of *N. fucata* is, probably, its intricate mode of entering the gastropod shell inhabited by the host. Settlement of this species occurs directly on the sea bottom, without any kind of host specificity. Both young recruits and adults that had moved out from their shells owing to the death or moulting of their hosts, are found inside tubes in the bottom sediments. When a hermit crab is placed in the vicinity of such pseudo "free-living" commensals in an experimental aquarium, the worms immediately begin to come out of their tubes. As the worms stretch out of their tubes, the anterior portion of their bodies rise up from the bottom and begin to wave about in a wide, sweeping movement. When the hermit crab enters the search area delimited by these sweeping movements and touches the worm head, the worm starts moving straight to the shell and quickly enters it. A similar reaction was experimentally triggered by bouncing an empty and carefully washed shell on the bottom or even by tapping the edge of the culture vessel with a finger nail. All these experiments demonstrated that the characteristic searching behaviour of *N. fucata* was triggered by the perception of substratum vibrations. Presumably, this behaviour will also be present in natural commensal populations, starting as soon as substratum vibrations are perceived by the worms. This behaviour appears 4 months after settlement and is maintained during the whole life of the commensal. This fact was the main argument used to consider the association of this species and hermit crabs as truly commensal (Gilpin-Brown 1969).

On contacting a host hermit crab, the individuals of *N. fucata* take up a position behind the shell, then crawl over it, usually along its sutures, before eventually reaching the aperture. The worm enters via the upper lip of the aperture and rapidly crawls inside the shell to adopt its habitual position (see Fig. 5G, p. 256). *N. fucata* is able to crawl on the smooth shell surface by secreting mucus from special glands located in its parapodia. These glands have not been reported in free-living nereidids, so constitute a differential character of this symbiotic species. Therefore, *N. fucata*'s characteristic adaptive behaviour is accompanied by a singular physiological adaptation.

Mutualistic behaviour The fact that some commensal associations may probably be transferred to mutualistic interactions may be indicated both by specific behavioural response of hosts to the presence of symbionts and also by the absence of hostile host reaction. A typical example of host response was described in the association between the hesionid *Ophiodromus puggetensis spinapandens* and its echinoid host, the sand dollar *Clypeaster humilis* (Storch & Niggemann 1967). The commensal is usually found crawling among the spines on the oral side of the host. When the head appendages of the symbiont contact the host's spines, these start to move apart, which allows the commensal to go further on the host surface. During its movements, the worm directs its parapodia vertically downwards. Conversely, when different free-living species of polychaetes (viz. nereidids, syllids, chrysopetalids) were experimentally forced to touch the surface of the sea urchin and placed among the host's spines, the spines were motionless. This behaviour prevented any kind of worm movement and was reinforced by the pedicellaria attempting to snatch the intruders. Such very specific host behaviour closely resembled that of giant sea-anemones in their well known association with clown fishes and crabs inhabiting their crowns. In these cases, the symbionts are tolerated by the host, whereas their free-living relatives are damaged or killed by the nematocysts as soon as any contact with sea-anemone tentacles occurs. Symbiont recognition may also occur in the associations between predator hosts and polychaetes, such as those involving hermit crabs and *Nereis fucata*. Adult *N. fucata* stretch from the inner surface of the shell and crawl out to take food from the pagurid's maxilipeds. Either owing to the worm's avoidance of the hermit crab's buccal equipment or to some chemical signal from the symbiont that is effectively recognized by the host, the fact is that the commensals are never attacked by the host. However, other polychaete species may perfectly well be preys of the hermit crab.

Symbiont recognition by hosts would be meaningless if the hosts derived no benefits from the association with their symbionts. Benefits for the host may include defence against predators or cleaning, implying steps from a unidirectional towards a mutualistic relationship. The most evident benefit from the host's association with an aggressive inquiline, like a polynoid (Dimock 1974), is certainly assistance in fending off predators by biting them. This behaviour has been suggested in certain associations involving polynoids (e.g. *Harmothoe hyalonemae* in Martín et al. 1992), but was earlier demonstrated in the association between *Arctonoe vittata* and its host limpet *Diodora aspera* (Dimock & Dimock 1969). In this case, symbiont aggressiveness towards the predatory starfish *Pisaster ochraceus* was not displayed by worms on their own, which strongly suggests that stimuli linked to the host (i.e. chemical or mechanical cues) may trigger the commensal's aggressive behaviour.

Hosts may also benefit from the cleaning activity of symbionts, though this may bear no relation to symbiont aggressive behaviour. Benefits based on cleaning activity have been suggested for the associations between spionids and molluscs (*Polydora glycymerica* and *Glycymeris yessoensis* in Radashevsky 1993) or sponges (*Cliona viridis* and *Polydora rogeri* in Martin 1996). In these cases, the ability of the polydorids to manipulate relatively large particles may favour the filtering activity of the hosts by cleaning the water around the bivalve's siphons or the sponge's inhalant papillae (see the section on boring symbiotic polychaetes). However, the so-called aggressive polynoids may also act as cleaning symbionts either by removing epi- or endobionts or by feeding on mucus or settled detritus. In bathyal muddy benthic environments, the Mediterranean hexactinellid sponges of the genus *Hyalonema* were certainly suitable refuges likely to attract the existing vagile fauna. However, those sponges inhabited by the polynoid *Harmothoe hyalonemae* were never occupied by endobionts other than the commensal worms (Martín et al. 1992). In the association between the polynoid *Arctonoe vittata* and the starfish *Dermasterias imbricata*, the commensal may benefit the host by feeding on sea star mucus and settled detritus (Wagner et al. 1979). In fact, *D. imbricata* lack pedicellariae, structures that are thought to function in part by keeping the aboral surface of other starfish free of epibionts and detritus. As pointed out by Wagner et al. (1979), most starfish hosts reported for *Arctonoe vittata* also lack pedicellariae. All these observations strongly suggest that *A. vittata* may be involved in a mutualistic, rather than a unidirectional, association with its host starfish. Moreover, the series of experiments carried out by Wagner et al. (1979) are the only demonstration of a host, *Dermasterias imbricata*, being attracted towards its commensal polynoid, *Arctonoe vittata*. Although this attraction was clear in the laboratory, its precise role in nature was not clear. Rather than acquire symbionts from a pool of free-living worms (*A. vittata* is considered an obligatory commensal), the host response more likely served to re-establish an existing association after the symbiont had become temporarily separated from its host. In fact, this suggestion coincides with the importance attributed to host-relocation behaviour in the life cycle of symbiont *Arctonoe* species (Dimock 1974, Britayev 1991).

Life-cycles of commensal polychaetes

Little is know on the life cycles of commensal polychaetes. The available data, however, allow us to surmise that, in fact, the life cycle structure of most commensal species does not significantly differ from those of their free-living relatives. The well known generic scheme of polychaetes' life cycle, divided into planktonic larval and benthic adult phases, may perfectly by applied to commensal polychaetes, with a single main difference: the symbiotic habits replace the free-living ones in the benthic phase. The tropical serpulid *Spirobranchus giganteus*, an obligate symbiont that builds its tubes on the surface of living haermatypic corals, instead of boring or excavating the coral skeleton (Marsden 1987, Hunte et al. 1990a,b, Marsden et al. 1990), well illustrates this point. *S. giganteus* is a dioceous species that has external fertilization and a planktonic larval phase of 9–12 days long. These planktonic larvae are attracted by certain coral species but do not respond to certain others. Despite the fact that it has been suggested that commensals' larval settlement should occur on the bottom, with a subsequent juvenile migration towards the respective hosts (Davenport 1966a), it seems more likely that host recognition by settling larvae is a widespread trend among symbiotic polychaetes (parasites included – see the corresponding sections below). Moreover, the planktonic larvae of symbiotic polychaetes may also be attracted by their adult conspecifics, as occurs in *S. giganteus*. In summary, the life cycle of this species consists of a symbiotic adult stage with limited mobility and a free-living planktonic phase that is, in fact, responsible for dispersion.

This scheme may vary depending on the species. For example, in the nereidid *Nereis fucata*, the life cycle is complicated with an additional free-living benthic stage. The gravid heteronereis leave the hosts and spawn freely in the water column from late spring to early summer (Gilpin-Brown 1969, Goerke 1971). Unexpectedly, the planktotrophic larvae settle directly on the soft bottom, rather than on their hosts, and the juveniles live in tubes inside the bottom sediment for several months. During this phase, they feed on detritus and small benthic animals, exactly like their free-living related species. However, 4-month-old juveniles start to develop the ability to recognize the substratum vibrations produced by the hermit crab shell bouncing on the bottom (Gilpin-Brown 1969). The development of the life cycle is completed when a hermit crab reaches the area inhabited by the pseudo free-living juveniles and triggers their characteristic host-entering behaviour (see the corresponding section above). Host entering occurs more intensively in spring, due to the increasing number of "emptied" host shells after the release of gravid heteronereis.

Some commensals tend to simplify their life cycles by reducing, or even eliminating, the free-living pelagic stage. In these cases, the adult symbionts are responsible for dispersal. Known examples of this strategy occur, for example, among the histriobdellids, which possess internal fertilization and whose eggs directly develop inside cocoons attached to the host (e.g. *Histriobdella homari* in Haswell 1913). Conversely, other commensals tend to increase the complexity of their life cycles by having an intermediate host (or hosts). The best known example of such complex life cycles occurs probably in the scaleworm *Arctonoe vittata*. This species is also one of the most widely studied commensal species, whose behaviour, population ecology and reproductive biology has been reported in detail (Davenport 1950, Palmer 1968, Dimock & Dimock 1969, Wagner et al. 1979, Britayev & Smurov 1985, Britayev et al. 1989), allowing us to reconstruct the main traits of its life-cycle (Britayev 1991). In the northwest part of the Sea of Japan (Vostok Bay) this species is associated with several hosts from different taxa (i.e. the starfish *Asterias amurensis*, the limpet *Acmaea pallida* and, more occasionally, with the sea-cucumber *Stichopus japonicus* and the starfish *Aphelasterias japonica*). The spawned eggs are poor in yolk so that the fertilization gives rise to planktotrophic larvae. After a long pelagic phase, settlement occurs on all the above-mentioned hosts, with the main flow of settling larvae being addressed to *Asterias amurensis*, and the low number of successfully settled larvae occurring on *Acmaea pallida* (Fig. 14). However, the latter remain during the whole life on their host limpets, whereas the former may have diverse destinations: the initial starfish individual will be the definitive host for some of them, while some others will be forced to leave it and relocate on other host individuals (i. e. starfish of the same species, or of *Aphelasterias japonica*, but also on limpets or sea-cucumbers). Undoubtedly, intraspecific competition plays the most important role in these relocation processes.

The limpets are probably more "suitable" hosts than the starfish for *Arctonoe vittata*. However, both the high infestation prevalence and the limited available space on the limpet tend to reduce the survival of juveniles. It should be taken into account that the species shows a strong intraspecific aggressive behaviour. However, as a result of the large surface of the host starfish, the frequency of aggressive interactions between symbionts from the same host individual may be significantly reduced. It is thus possible for recently settled juveniles to survive sharing the same host starfish with a large adult. However, if several young specimens settle on the same host, the number of negative interactions between them, as well as that of aggressive actions of the large adult (if present), increase as the young grow (Britayev et al. 1989). The end result is that some symbionts (usually the smallest) are obliged to leave the host and move to another host. In this case, the starfish *Asterias amurensis* may be regarded as an intermediate host, whereas, at least for some symbionts, *Acmaea pallida* acts as the definitive one. In fact, we may speculate that the complex life cycle of *Arctonoe vittata* has an additional interest in that it may probably illustrate how some symbiotic (i.e. parasitic) life cycles including different hosts may have evolved as a result of symbionts' intraspecific competition.

Parasitic polychaetes

Parasites are often specially adapted to facilitate their specialized way of life (Margalef 1980). Thus, a low degree of, or the lack of, specialization has been interpreted as a sign of a recent evolutionary change to parasitic life (Clark 1956). A strict parasitic relationship often implies some degree of co-evolution. In fact, the hosts may even be able to support the presence of their parasites, the infested populations attaining a certain level of equilibrium in normal, non-stressed conditions (Margalef 1980). Endoparasitic relationships tend to be more strict than ectoparasitic relationships. Most ectoparasites are, indeed, haematophagous, dermatophagous or secretophagous. The endoparasites inhabit the tissues or the body cavities of the hosts. As we have included within the commensals those symbiotic polychaetes lacking clear parasitic characteristics, the following sections will refer to the polychaete species with known described morphological or behavioural parasitic trends.

Taxonomical distribution of parasitic polychaete species

Eighty-one polychaete species belonging to 13 different families are reported to be involved in 253 different parasitic relationships (Tables 6, 8 and 9; Fig. 15A). Parasitic

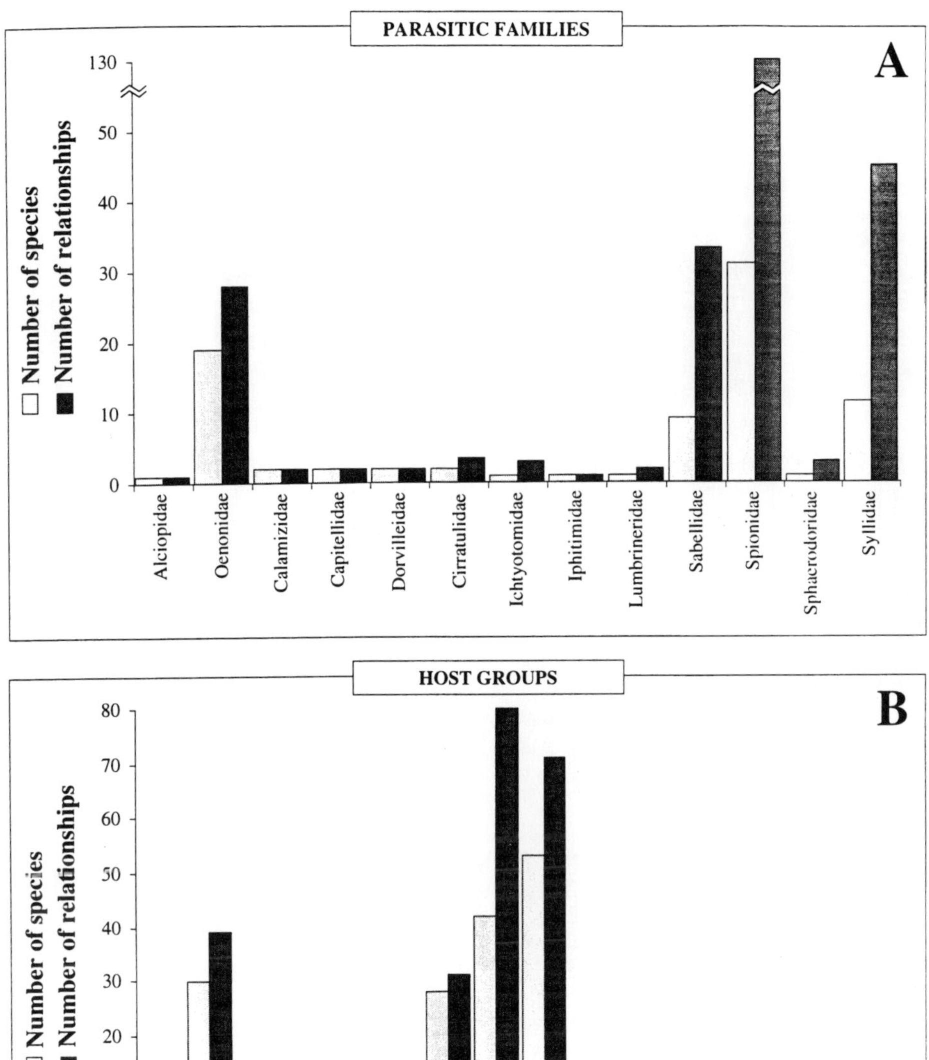

Figure 15 A, Number of parasitic relationships and number of parasitic species from each polychaete family including parasitic forms. B, Number of species from each taxonomic group of host harbouring parasitic polychaetes and number of relationships on which they are involved.

Table 6 List of parasitic polychaete species (boring species excluded). Type: Type of relationship (O, obligatory: F, facultative; P, permanent; T, temporal).

Parasitic species	Authority	Host group	Species	Type	Reference
Alciopidae					
Alciopina parasitica	Claparède & Panceri	Ctenophor	*Cydippe densa*	T	Fauvel 1923, Day 1967
Oenonidae					
Arabella endonata	Emerson	Polychaete	*Diopatra ornata*	P	Emerson 1974
Arabella iricolor	(Montagu)	Polychaete	*Diopatra ornata*	T	Hartman 1968
Drilognathus capensis	Day	Polychaete	*Onuphis holobranchiata*	P	Day 1967
Drilonereis benedicti	Pettibone	Polychaete	*Onuphis magna*	P	Pettibone 1957
Drilonereis caulleryi	Pettibone	Polychaete	*Onuphis* (*Nothria*) *conchylega*	P	Pettibone 1957
Drilonereis forcipes	(Hartman)	Polychaete	*Eunice* sp.	P	Pettibone 1957
Drilonereis forcipes		Polychaete	*Eunice antennata*	P	Pettibone 1957
Drilonereis parasiticus	(Caullery)	Polychaete	Terebellidae p.i.	P	Pettibone 1957
Drilonereis sp.	Pérès	Polychaete	*Protula tubularia*	P	Pérès 1949
Haematocleptes laenae	Hartman & Fauchald	Polychaete	*Leaena minima*	P	Hartman & Fauchald 1971
Haematocleptes terebellidis	Wiren	Polychaete	*Terebellides stroemi*	P	Wiren 1886, Pettibone 1957, George & Hartmann-Schröder 1985
Labrorostratus luteus	Uebelacker	Polychaete	*Haplosyllis spongicola*	P	Uebelacker 1978
Labrorostratus parasiticus	Saint Joseph	Polychaete	*Eusyllis monilicornis*	T	Pettibone 1957
Labrorostratus parasiticus		Polychaete	*Grubeosyllis clavata*	T	Pettibone 1957
Labrorostratus parasiticus		Polychaete	*Odontosyllis ctenostoma*	T	Pettibone 1957
Labrorostratus parasiticus		Polychaete	*Pionsyllis lamelligera*	T	Pettibone 1957
Labrorostratus parasiticus		Polychaete	*Sphaerosyllis pirifera*	T	Wu et al. 1982
Labrorostratus parasiticus		Polychaete	*Syllis prolifera*	T	Pettibone 1957
Labrorostratus prolificus	Amaral	Polychaete	*Perinereis cultrifera*	T	Amaral 1977
Labrorostratus sp.	San Martín & Sardá	Polychaete	*Syllis columbretensis*	P	San Martín & Sardá 1986
Labrorostratus sp.		Polychaete	*Grubeosyllis clavata*	P	San Martín & Sardá 1986
Labrorostratus sp.		Polychaete	*Sphaerosyllis hystrix*	P	San Martín & Sardá 1986
Notocirrus spiniferus	(Moore)	Polychaete	*Diopatra cuprea*	T	Allen 1952, Pettibone 1957
Notocirrus sp. (Juvenile)		Polychaete	*Marphysa sanguinea*	T	Koch 1846, Pettibone 1957
Oligognathus bonelliae	Spengel	Echiuroid	*Bonellia viridis*	P	Pettibone 1957
Oligognathus parasiticus	Cerruti	Polychaete	*Microspio mecznikovianus*	P	Pettibone 1957
Pholadiphilia turnerae	Dean	Bivalve	*Xyloredo ingolfia*	P	Dean 1992
Pholadiphilia turnerae		Bivalve	Xilophangidae p.i.	P	Dean 1992
Calamyzidae					
Calamyzas amphictenicola	Arwidson	Polychaete	*Amphicteis gunneri*	P	Arwidsson 1932, Hartmann-Schröder 1971
Asetocalamyzas laonicola	Tzetlin	Polychaete	*Laonome cirrata*	P	Tzetlin 1985

Table 6 *Continued*

Parasitic species	Authority	Host group	Species	Type	Reference
Capitellidae					
Capitomastus minimus	Langerhans	Cephalopod	*Loligo vulgaris*	T	Clark 1956
Capitella ovinicola	Hartman	Cephalopod	*Loligo opalescens*	T	Clark 1956
Dorvilleidae					
Dorvillea sociabilis	(Webster)	Sponge	*Ircinia campana*	?	Dean 1973
Ophryotrocha puerilis	Claparède & Mecznikov	Holothuroid	*Ocnus planci*	T	Paris 1955, CLark 1956, Jangoux 1987
Ichthyotomidae					
Ichthyotomus sanguinarius	Eisig	Fish	*Myrus vulgaris*	P	Paris 1955, Fauchald & Jumars 1979
Ichthyotomus sanguinarius		Fish	*Conger vulgaris*	P	Paris 1955, Fauchald & Jumars 1979
Ichthyotomus sanguinarius		Fish	*Sphagebranchus inberbis*	P	Clark 1956
Iphitimidae					
Veneriserva pygoclava	Rossi	Polychaete	*Aphrodita longipalpa*	P	Rossi 1984
Lumbrineridae					
Ophiuricola cynips	Ludwig	Ophiuroid	*Ophiura irrorata*	P	Ludwig 1905, Clark 1956 Alvà & Jangoux 1989
Ophiuricola cynips		Ophiuroid	*Ophioglypha tumulosa*	P	Jangoux 1987
Sphaerodoridae					
Sphaerodorum flavus	Oersted	Ophiuroid	*Ophiura* sp.	F	K. W. Ockelmann pers. obs.
Sphaerodorum flavus		Ophiuroid	*Amphiura filiformis*	F	K. W. Ockelmann pers. obs.
Sphaerodorum flavus		Ophiuroid	*Amphiura chiajei*	F	K. W. Ockelmann pers. obs.
Syllidae					
Autolytus penetrans	Wright & Woodwick	Cnidarian	*Allopara californica*	O	Wright & Woodwick 1977
Branchiosyllis exilis (Juvenile)	(Gravier)	Sponge	*Speciospongia vespera*	F	Clark 1956
Branchiosyllis oculata	Elhers	Sponge	*Aplysilla longispina*	O	Pawlik 1983
Branchiosyllis oculata		Sponge	*Aplysina fistularis*	O	Pawlik 1983
Branchiosyllis oculata		Sponge	*Chondrilla nucula*	O	Pawlik 1983
Branchiosyllis oculata		Sponge	*Cinachyra alloclada*	O	Pawlik 1983
Branchiosyllis oculata		Sponge	*Ircinia felix*	O	Pawlik 1983
Branchiosyllis oculata		Sponge	*Pseudoceratina crassa*	O	Pawlik 1983
Branchiosyllis oculata		Sponge	*Speciospongia othella*	O	Pawlik 1983
Branchiosyllis oculata		Sponge	*Speciospongia vespera*	O	Dauer 1974
Branchiosyllis oculata		Sponge	*Tedania ignis*	O	Pawlik 1983
Branchiosyllis oculata		Sponge	*Tethya actinia*	O	Pawlik 1983
Haplosyllis spongicola	(Grube)	Sponge	*Aaptos* cf *aaptos*	O	T. A. Britayev pers. obs.
Haplosyllis spongicola		Sponge	*Adocia neens*	O	Dauer 1974
Haplosyllis spongicola		Sponge	*Aplysina cauliformis*	O	Tsurumi & Reiswig 1997

Table 6 *Continued*

Parasitic species	Authority	Host group	Species	Type	Reference
Haplosyllis spongicola		Sponge	*Clatrina coriacea*	O	Clark 1956
Haplosyllis spongicola		Sponge	*Gellioides digitalis*	O	Uebelacker 1978
Haplosyllis spongicola		Sponge	*Geodia gibberos*	O	Dauer 1974
Haplosyllis spongicola		Sponge	*Halichondria* sp.	O	Spooner et al. 1957
Haplosyllis spongicola		Sponge	*Halichondria agglomerans*	O	Uriz 1983
Haplosyllis spongicola		Sponge	*Haliclona* sp.	O	T. A. Britayev pers. obs.
Haplosyllis spongicola		Sponge	*Ircinia campana*	O	Dauer 1974
Haplosyllis spongicola		Sponge	*Ircinia fasciculata*	O	Alós et al. 1982
Haplosyllis spongicola		Sponge	*Ircinia ramosa*	O	Dauer 1974
Haplosyllis spongicola		Sponge	*Petrosia ficiformis*	O	Alós et al. 1982
Haplosyllis spongicola		Sponge	*Phoriospongia* (*Chondropsis*) sp.	O	Magnino & Gaino in press
Haplosyllis spongicola		Sponge	*Reniera rosea*	O	Clark 1956
Haplosyllis spongicola		Sponge	*Rhizaxinella pyrifera*	O	Uriz 1983
Haplosyllis spongicola		Sponge	*Speciospongia vesparia*	O	Dauer 1974
Haplosyllis spongicola		Sponge	*Theonella swinhoei*	O	Magnino & Gaino in press
Haplosyllis spongicola		Sponge	*Verongula reiswigi*	O	Reiswig 1973
Haplosyllis spongicola		Sponge	*Xytopsene sigmatum*	O	Dauer 1974
Haplosyllis spongicola		Sponge	Young specimen	O	Bacescu 1971
Haplosyllis spongicola		Sponge	Unidentified	O	Lee & Rho 1994
Haplosyllis spongicola		Polychaete	*Eunice* sp.	T	Treadwell 1909, Potts 1912
Myrianida pinningera	(Montagu)	Tunicate	*Phallusia mammilata*	T	Spooner et al. 1957
Myrianida pinningera		Tunicate	*Ascidiella aspersa*	T	Spooner et al. 1957
Parasitosyllis claparede	Potts	Nemertini	Unidentified	P	Potts 1912
Parasitosyllis claparede		Polychaete	Unidentified	P	Potts 1912
Procerastea fasciata	Hartmann-Schröder	Cnidarian	*Eudendrium carneum*	F	E. Cruz-Rivera pers. comm.
Procerastea hallenziana	Malaquin	Cnidarian	*Syncoryne eximia*	F	Allen 1915, 1921, Spooner et al. 1957, Alós 1989
Procerastea halleziana		Cnidarian	*Tubularia indivisa*	F	Caullery 1925, Spooner et al. 1957
Procerastea hydrozoicola	Hartmann-Schröder	Cnidarian	*Pseudosolanderia* sp.	O	Hartmann-Schröder 1992
Procerastea parasimpliseta	Hartmann-Schröder	Cnidarian	*Pseudosolanderia* sp.	O	Hartmann-Schröder 1992
Proceraea sp.	Britayev, San Martín & Sheiko	Cnidarian	*Abietinaria turgida*	O	Britayev et al. unpubl.
Typosyllis extenuata		Sponge	*Speciospongia vesparia*	F	Pearse 1932, Dauer 1974

polychaetes are thus present in about 15% of known polychaete families and about 0.5% of known species, which is far from being the highest figure for the presence of commensals (see above, p. 221 and 246).

Probably one of the most striking facts arising from the comparison between classic papers, such as those by Paris (1955) and Clark (1956), and the present review is that some of the typical examples of parasitic polychaetes have now been excluded from this category, and included in the commensals. Clearly this is due to increased knowledge of the characteristics of the involved associates. *Histriobdella homari* and *Ophryotrocha geryonicola*, together with species of the genus *Iphitime*, were reported as ectoparasites of the branchial chambers of crustaceans by the above classic authors. More recent studies, however, demonstrated that *Histriobdella homari* behave more as a cleaning symbiont than as a parasite (Jennings & Gelder 1976), whereas the relationships between crabs and *Iphitime* spp. (Hartnoll 1962, Pilger 1971, Abelló et al. 1988, Comely & Ansell 1989) and *Ophryotrocha* spp. (Gaston & Benner 1981, Martin et al. 1991) may be regarded as commensalisms. In fact, these worms feed on material captured or attracted by the host, and their presence did not negatively affect the gills of the host.

The highest number of polychaete species considered, to some degree, as parasites belong to the family Spionidae (38%), which are, in fact, boring species. This family also includes the highest number of relationships (52%), which suggests that most of the parasitic spionid species have a low degree of specificity. In other words, the same species may frequently bore (and infest) many different hosts. However, the symbiotic associations involving boring polychaete species (spionids, but also the so-called parasitic species of the family Sabellidae) show very peculiar characteristics and their main trends will be analyzed in separate sections.

Most of the polychaetes that have been reported, in principle, as real parasites are obligatory symbionts (45%), whereas only 30% are permanent parasites (Table 6). Parasitic polychaetes mainly belong to the family Oenonidae (about 24%). These worms are internal parasites and, with the exception of *Oligognathus bonelliae* (parasitic on the echiuroid *Bonellia viridis*) and *Pholadiphilia turnerae* (parasitic on the bivalve *Xyloredo ingolfia*), all their hosts are other polychaetes (Table 6). The parasitic oenonids apparently show a high degree of specificity: of the 19 known species, 16 are known to infest a single host. This trend may be explained, however, by their extreme rarity, which gives rise to the fact that most species are only known from a single original record.

The syllids are the second most well represented family within the "non-boring" parasites (14%). Members of this family have been mainly reported from sponge hosts, the most representative parasitic syllid being the cosmopolitan species *Haplosyllis spongicola*. The remaining polychaete families include only one or two parasitic species. Among them, the highly specialized families Calamizidae (two species) and Ichtyotomidae (one species) are entirely parasitic.

Marine taxa harbouring parasitic polychaetes

The parasitic polychaetes infest 186 different host species included in 16 groups of organisms (Table 6, 8 and 9; Fig. 15B). With the exception of three fishes and one plant, all remaining host species are benthic invertebrates. Gastropod and bivalve molluscs showed the highest percentage of species harbouring parasitic polychaetes (29% and 23%, respectively). Thus, they are involved in most of the parasitic relationships (28% and 31%, respectively). With the exception of the oenonid *Pholadiphilia turnerae*, all parasitic polychaetes infesting gastropods and bivalves belong either to the caobangids or to the spionids. Similarly, the host cirripeds and bryozoans, a few of the host sponges and cnidarians, and the host seagrass are also infested by spionids.

The next most common host groups (each one representing about 16% of the host species) are the sponges and the polychaetes themselves. The former are mainly parasitized by syllids, particularly *Haplosyllis spongicola* and *Branchiosyllis oculata*. The latter are infested by the parasitic oenonids (with the exception of *Pholadiphilia turnerae*), two calamizids, one iphitimid and one syllid. The host polychaete species belong to the families Aphroditidae, Eunicidae, Onuphidae, Terebellidae, Spionidae, Ampharetidae, Syllidae, Nereidae and Sabellidae.

Specificity among parasitic polychaetes

It is very difficult to analyze the degree of specificity among parasitic polychaetes. The analysis of the available data seems to point to a clear monoxenous pattern. More than half of the parasitic polychaete species (about 69%) infest a single host species, while the figure rises to 91% if those infesting two or three different host species are included. However, this monoxenous pattern is clearly biased by the virtual absence, in most cases, of data other than the original citation. As new data are reported, the host spectrum tends to increase. This may be the case of the oenonid *Labrorostratus parasiticus*. This species, with six known host polychaete species (up to nine if we assume that the specimens described by San Martín & Sardá (1986) as *Labrorostratus* sp. are indeed *L. parasiticus*), is certainly the most frequently reported and the one with most recent citations. *L. parasiticus* is also the only oenonid known from more than one host polychaete. According to our interpretation, the latter may be a direct consequence of the former. A certain degree of specificity in the parasitic habits of *L. parasiticus* can be assumed, as all known hosts belong to the family Syllidae (Table 6).

Polyxenous relationships are mainly present within the syllids parasitizing sponges (i.e. *Branchiosyllis oculata* and *Haplosyllis spongicola*). However, the apparently polyxenous behaviour of *H. spongicola* (it has been reported from as many as 21 different host sponges) must be interpreted with care. The external morphology of this species is very well adapted to its life style and appears to be relatively simple. However, two different morphologies have been reported (viz. different arrangements of cirri, different shapes of setae) and are considered either as belonging to different subspecies, *H. spongicola spongicola* and *H. spongicola tentaculata* (Imajima 1966), as different growth stages of the same species (San Martín 1984) or, even, as different species, *H. spongicola* and *H. tentaculata* (Lee & Rho 1994). The existence of different colourings has been related to the colour of the host sponge (San Martín 1984, Magnino & Gaino in press). In light of the above, we suggest that the possible existence of a hidden sibling or pseudo-sibling species complex is still an open question for this species.

After the above-mentioned syllids, the number of host species parasitized by the same polychaete species sharply decreases: two species, *Ichthyotomus sanguinarius* and

Sphaerodorum flavus, parasitize three host species, whereas only *Pholadiphila turnerae* parasitizes two host species. It must be pointed out, however, that these parasitic species closely resemble the behaviour of *Labrorostratus parasiticus* in that the taxonomic status of their respective hosts is always very close. All hosts are: sponges for *Haplosyllis spongicola*, fishes (species of conger) for *Ichthyotomus sanguinarius*, ophiuroids for *Sphaerodorum flavus*, and pholadid bivalves for *Pholadiphila turnerae*.

Infestation characteristics and intra-host distribution patterns of parasitic polychaetes

The known prevalences of infestation by non-boring parasitic polychaetes are extremely low: 0.2% within 28604 hosts examined for *Labrorostratus luteus* (Uebelacker 1978), 0.05% within more than 6000 hosts examined for *Labrorostratus* sp. (San Martín & Sardá 1986), and a maximum of 4% for *Arabella endonata* but with only 25 hosts examined (Emerson 1974). Probably, the infestation prevalence for *A. endonata* could not be accurately predicted from the small host sample examined and a decrease in prevalence could be expected with an increase in the amount of hosts examined. However, Emerson (1974) pointed out that several hundred of the host polychaete *Diopatra ornata* were examined from a population located in the vicinity of the type locality and none of the specimens were infested.

Infestation intensities have been more often reported than prevalences (Table 7). A regular pattern of distribution seems to characterize most of the known parasitic relationships involving polychaetes, with each host harbouring only one single adult parasite. This statement can not be easily confirmed because of a lack of information on the incidence of parasitic relationships involving polychaetes. However, the presence of one parasite in each host in the most frequently reported species, *Labrorostratus parasiticus*, supports the hypothesis of a regular distribution for all related species. The presence of two parasites on the same host specimen was reported for *Haplosyllis spongicola* on a eunicid host (as *H. cephalata* in Treadwell 1909), for *Ichthyotomus sanguinarius* (one male and one female, see Fig. 20C, p. 301) on an eel host (Eisig 1906, Fauvel 1923) and for *Calamyzas amphictenicola* (also one male and one female) on its ampharetid host (Franzén 1982).

Although less frequently, an aggregated distribution has also been reported for parasitic polychaetes (Table 7). The aggregates always seem to be connected to the reproduction of the involved species, with the exception of the parasitic syllids *Branchiosyllis oculata* and *Haplosyllis spongicola*. The highest known number of parasites (more than 20000) inside a single host has been reported for *H. spongicola* (Bacescu 1971). However, *Branchiosyllis oculata*, *Proceraea* sp. and some of the endoparasitic oenonids also show numerous individuals inside a single host (Table 7). In *Branchiosyllis oculata*, the high densities were surprisingly supported by the host sponges with no apparent effect on them (Pawlik 1983). In *Proceraea* sp., the aggregated distribution occurs only within the juveniles, the mature adult phase being probably free-living (Britayev et al. unpubl.). In the oenonids *Notocirrus spiniferus* (Allen 1952) and *Arabella endonata* (Emerson 1974) the individuals were juveniles showing many different developmental stages. Amaral (1977) reported that two adult individuals of *Labrorostratus prolificus* showed traces of asexual reproduction (stolons) coexisting with their numerous juvenile stages. Conversely, there was not additional information about the several individuals of *Veneriservia pygoclava* found inside its aphroditid host.

Location of parasitic polychaetes on their hosts

Most parasitic polychaetes are endoparasites (Fig. 16), either inhabiting the tissues of their hosts (e.g. *Alciopina parasitica, Dorvillea sociabilis, Branchiosyllis exilis, Autolytus penetrans, Myrianida pinnigera, Typosyllis extenuata*), the peri-intestinal blood sinus (e.g. *Haematocleptes terebellidis*), the coelomic cavities (e.g. *Ophryotrocha puerilis, Veneriservia pygoclava, Oligognathus bonelliae*, most oenonids parasitizing polychaetes), the aquiferous system (e.g. *Haplosyllis spongicola*), the hydrothecae (e.g. *Proceraea* sp.) or special galls formed by the host because of the presence of the parasite (e.g. *Ophiuricola cynips*). With the exception of *Autolytuspenetrans* and the last three species, no special morphological, structural or behavioural reactions to the presence of the parasites in their hosts has been described.

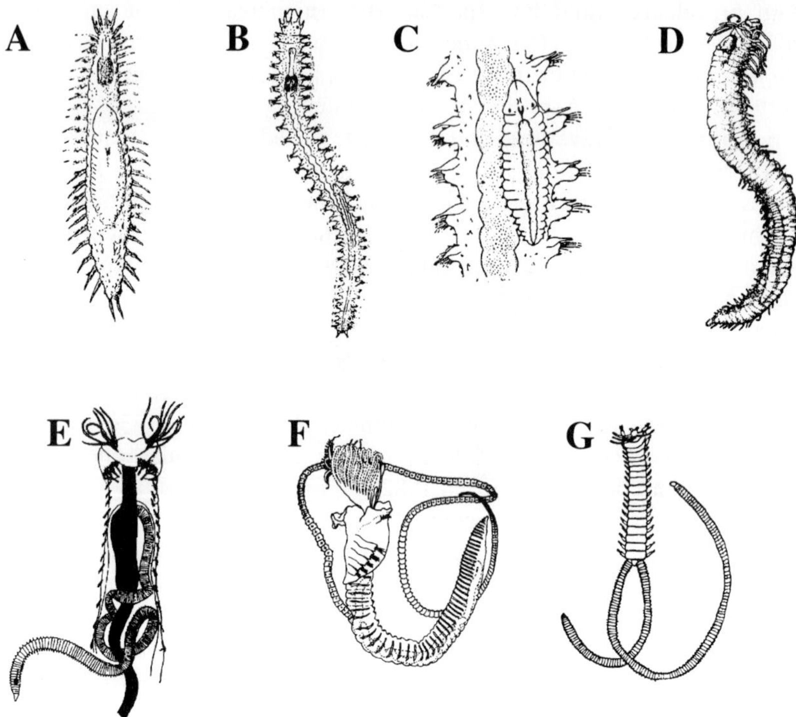

Figure 16 Examples of endoparasitic polychaetes infesting their hosts. A, *Labrorostratus* sp. on *Grubeosyllis clavata* (redrawn from San Martín & Sardá 1986). B, C, *Labrorostratus parasiticus* on *Sphaerosyllis pirifera* (redrawn from Wu et al. 1982). D, *Labrorostratus luteus* on *Haplosyllis spongicola* (redrawn from Uebelacker 1978). E, *Drilonereis parasiticus* on a terebellid (redrawn from Caullery 1914). F, *Drilonereis* sp. on *Protula tubularia* (redrawn from Pérès 1949). G, *Drilonereis caulleryi* on *Onuphis conchylega* (redrawn from Pettibone 1957).

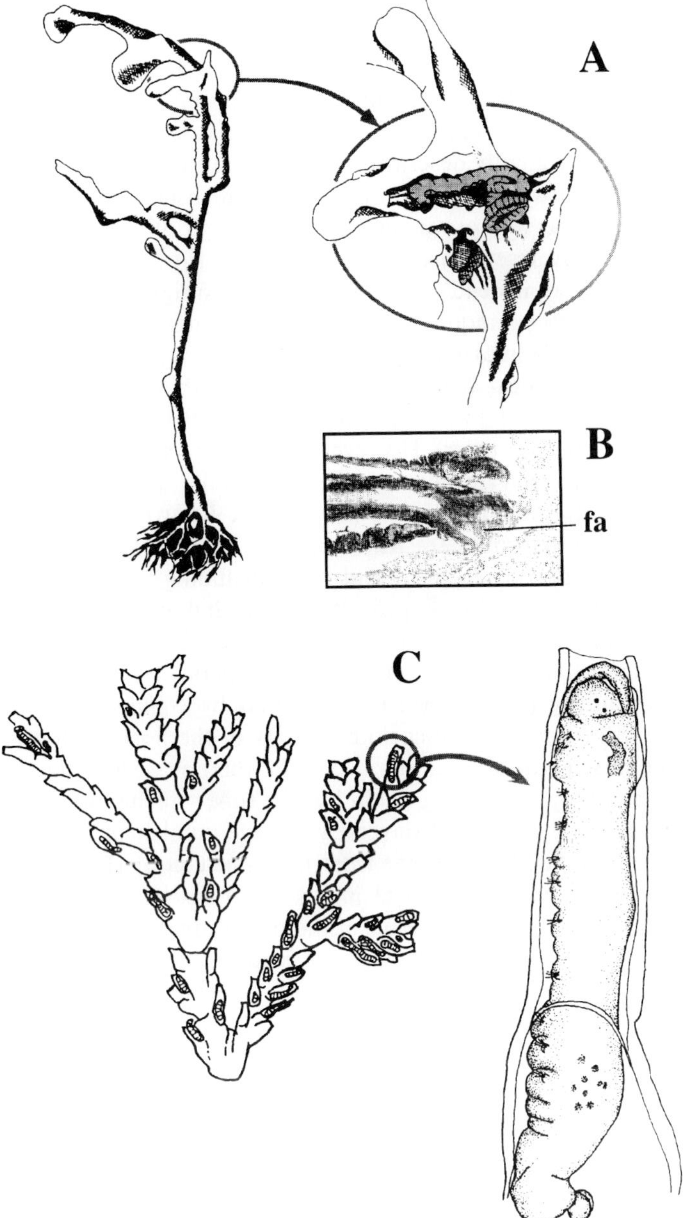

Figure 17 A, Two specimens of *Haplosyllis spongicola* inside a special chamber formed by the host sponge *Rhizaxinella pyrifera* (redrawn from Uriz 1983). B, *Haplosyllis spongicola* inside a canal of the *Aplysina cauliformis* aquiferous system, penetrating the canal wall and grazing on the sponge tissues (redrawn from Tsurumi & Reiswig 1997); fa: feeding activity. C, Juvenile *Proceraea* sp. infesting an *Abietinaria turgida* colony, showing a detail of one worm inside a modified hydrothecae (redrawn from Britayev et al. unpubl.).

Wright & Woodwick (1977) described the blisters formed by the small syllid *Autolytus penetrans* on the hydrocoral *Allopora californica*. This worm apparently penetrated the surface of the host resulting in a hyperplasia of cellular and calcareous material that produce a characteristic vermiform mound. The blisters may be located almost anywhere on the living hydrocoral with no specific association with the coral polyps. This species was not originally described as parasite. In fact, there were no indications of any damage to the host – other than the blisters themsleves – linked to the presence of the symbiont, nor was any profit suggested. Although this species does not actually bore, its blisters closely resemble those of the polydorids. In this case, however, the blister lining had cellular, instead of muddy, make up (Wright & Woodwick 1977).

The induction of special modifications in host sponges by *Haplosyllis spongicola* apparently depends on the morphology of its hosts. If the sponge canals are larger than the parasite, it may be located inside these canals without inducing special modifications in the host (Bacescu 1971, Pansini & Daglio 1980–1981, Tsurumi & Reiswig 1977, Magnino & Gaino in press). However, it may directly suck the sponge-bacteria tissues by inserting the pharynx through the canal walls, as was observed in the host reef bacteriosponge *Aplysina cauliformis* (Tsurumi & Reiswig 1977) (Fig. 17B). Where the canals are smaller than the parasite, *Haplosyllis spongicola* directly penetrates into the sponge tissues, living inside them and often including malformations in the host sponge. The formation of a thick protective epithelial layer surrounding the zone inhabited by the worms has been mentioned for the sponges *Rhizaxinella pyrifera* (Uriz 1978) (Fig. 17A) and *Aaptos* cf *aaptos* (T. A. Britayev pers. obs.)

According to Ludwig (1905), the galls produced by *Ophiuricola cynips* are rather large and partly protrude into the coelomic cavity of the host ophiuroid *Ophioglypha tumulosa*. In the host hydroid *Abietinaria turgida*, there is a single large mature adult *Proceraea* sp. per colony, inhabiting a thin hyaline tube attached to the axis of the host branches (Britayev et al. unpubl.). Conversely, a large number of juveniles of this parasite occurs inside modified hydrothecae from the same colony (Fig. 17C, Table 7). The modified hydrotheca consist of a basal jug-like part (as in uninhabited ones) and a distal cylindrical part, which is the result of the polyp's response to the presence of the worm. Moreover, the presence of the parasite leads to morphological changes in the polyp, which lengthens its body during the initial stages of infestation (when the hydrothecae are shared with the parasite). At a further stage polyps are totally absent from the infested hydrothecae. Finally, the worms leave the hydrothecae and build mucous tubes attached to the stem or branches of the colony to continue their development until the adult stage when they are able to leave the colony.

The oenonid *Pholadiphila turnerae* can be considered as an ectoparasit, as it inhabits the mantle (infrabranchial) cavity of the host bivalve (Dean 1992). The ectoparasitic polychaetes, however, tend to live more or less intimately attached to different parts of the host's body (see Fig. 20, p. 301). They may simply live on the host's surface (e.g. *Sphaerodorum flavus*), inside tubes attached to the host surface (e.g. the species of the genus *Procerastea*), attached to an external appendix (e.g. a parapodial cirrus in *Haplosyllis spongicola* as *H. cephalata*) or penetrating the host tissues (e.g. the body wall in *Asetocalamyzas laonicola* and *Parasitosyllis claparede*, the branchiae in *Calamyzas amphictenicola*, the fins in *Ichthyotomus sanguinarius*). *Branchiosyllis oculata* has been mentioned as living inside the canals of *Speciospongia vespera* (Pearse 1932, Westinga & Hoetjes 1981), but has been more often found living (and feeding) on the external surface of the remaining sponge host species (see Table 6). It was found often embed-

Table 7 Parasitic polychaetes with known infestation intensities. The asterisk indicates the mentioning of male and female individuals living near together on the same host.

Species	Intensity
Arabella endonata	>4 parasites/host
Asetocalamyzas laonicola	1 parasites/host
Branchiosyllis oculata	4–34 parasites/host
Calamyzas amphictenicola	>2* parasites/host
Drilognathus capensis	1 parasites/host
Drilonereis benedicti	1 parasites/host
Drilonereis caulleryi	1 parasites/host
Drilonereis forcipes	1 parasites/host
Drilonereis parasiticus	1 parasites/host
Drilonereis sp.	1 parasites/host
Haematocleptes leaenae	1 parasites/host
Haplosyllis spongicola	
Young sponge	21 000 parasites/host
Verongula reiswigi	50–100 parasites/host cm^3
Theonella swinhoei	68 parasites/host cm^3
Phorospongia sp.	56–240 parasites/host cm^3
Haplosyllis spongicola (as *H. cephalata*)	2 parasites/host
Ichthyotomus sanguinarius	1–2* parasites/host
Labrorostratus luteus	1 parasites/host
Labrorostratus parasiticus	1 parasites/host
Labrorostratus prolificus	33 parasites/host
Labrorostratus sp.	1 parasites/host
Notocirrus sp.	numerous parasites/host
Notocirrus spiniferus	53 parasites/host
Pholadiphila turnerae	1 parasites/host
Proceraea sp.	8–47 parasites/host colony
Veneriservia pygoclava	4–8 parasites/host

ded in a trench in the sponge ectosome (excavated by the setae), with their long dorsal cirri extending outward onto the sponge surface (Pawlik 1983). Moreover, *Branchiosyllis oculata* is morphologically adapted for dwelling on sponge surface and even takes on the colouring of some of its hosts. Thus, it can be considered a specialized ectoparasite.

Relationships between host and parasitic polychaete characteristics

It is commonly accepted, and is implicit in the definition of a parasitic relationship, that the parasitic organisms must be smaller than their hosts. Most parasitic associations involving polychaetes are so. However, some exceptions can be found within the oenonids. Some parasitic species are not smaller than their free-living relatives (e.g. *Notocirrus* spp., *Drilonereis* spp.), some other species are similarly sized or even larger than their hosts (Fig. 16). In these cases, the species involved would probably be closer

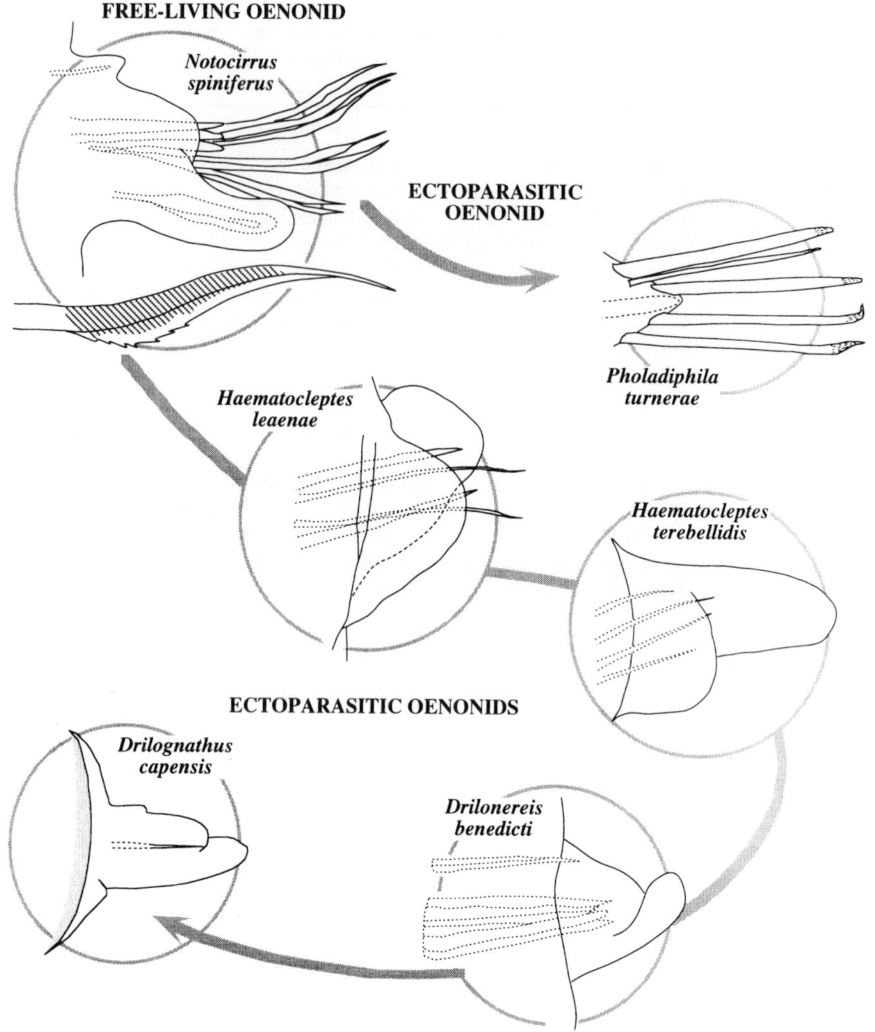

Figure 18 Scheme of the modifications of the typical setal pattern of Oenonidae as a consequence of the progressive adaptation to the parasitic mode of life (see Table 6 for the respective authorities).

to a predator than to a real parasite. In the case of *Pholadiphila turnerae*, the fact that the mere presence of the large worm in the branchial cavity causes significant stress to the host, as it hinders water flow, was the basis for considering the relationship parasitic (Dean 1992).

The feeding activity of *Haplosyllis spongicola* apparently caused a decrease in the growth rate of host sponge *Verongula reiswigi* (Reiswig 1973). On the other hand, the *Haplosyllis spongicola* population inhabiting another closely related host sponge, *Aplysina cauliformis*, was reproductively active at all times of the year, and was thus qualified as a "good" parasite (*sensu* Ricklefs 1990) by Tsurumi & Reiswig (1997). The same authors mentioned that there was no correlation between the timing of sexual reproduction in *Haplosyllis spongicola* and *Aplysina cauliformis*. However, because of its continuous grazing and removal of host biomass, the syllid certainly must have an impact, although of unknown significance, upon both sexual and asexual reproduction of the host sponge (Tsurumi & Reiswig 1997). In *Proceraea* sp., growth from late larva to juvenile is accompanied by mouth and gut development but also by the disappearance of polyps from the infested hydrotheca. Moreover, juveniles from empty hydrothecae were observed inserting their heads into thecae openings from neighbouring polyps. Not surprisingly, pigment granules characteristic of the hydroid were observed in the polychaete intestine. All these observations strongly support the theory that polychaetes feed on hydroid tissues, and so are real parasites (Britayev et al. unpubl.). As a consequence of the parasite's feeding activity that causes a significant reduction in the number of active polyps per colony, we suggest that considerable stress, affecting the overall well being of the infested host, should be expected. A very close feeding habit was described for *Procerastea halleziana* living on the hydroid host *Tubularia indivisa*. The worm lives in hyaline tubes attached to the host branches and was observed to suck the contents of the gastro-vascular cavity of its host after piercing the bases of the hydroid polyps exteriorly (Caullery 1925). On the basis of this observation (and lacking specific information), we have considered all remaining species of *Procerastea* from hydroid hosts to be parasites (see Table 6).

Adaptations of polychaetes to the parasitic mode of life

Morphological adaptations in parasitic polychaetes

As a general rule, endoparasites show a reduction or specialization of locomotive appendages and mouth parts, and the are usually smaller than their free-living relatives (Clark 1956). However, endoparasitic polychaetes often undergo nothing but the most trivial changes. Unfortunately, several factors prevent the development of satisfactory hypotheses: first, the rarity of these parasites and secondly, the fact that we know practically nothing of their life histories.

Reductions in the setal pattern of the endoparasitic oenonids (Fig. 18) are seen in the species *Drilonereis benedicti*, which has no emergent setae, and in the two species of *Haematocleptes*, which possess several slender, distally pointed setae rather than the typical limbate setae of the oenonids. In *Drilognathus capensis*, the setae are completely absent and only a single stout pointed aciculum, often projecting through the skin, is present on each parapodium. Moreover, study of the jaw apparatus of all endoparasitic oenonid species (Fig. 19), shows that they tend to have smaller jaws. This tendency can be related to some extent to a more specialized parasitic mode of life. The mandibles are well developed in the free-living oenonids, as well as in the parasitic *Notocirrus spiniferus*, *Arabella iricolor*, *A. endonata* and *Drilognathus capensis*. Conversely, they are only small plates in *Labrorostratus*, *Haematocleptes* and *Oligognathus*, and are completely lacking in the parasitic species of *Drilonereis*. The free-living oenonids, like the parasitic *Notocirrus* and *Arabella*, possess five pairs of well developed maxillae, whereas the parasitic *Drilonereis* possess four. In the remaining parasitic genera, the maxillae consist of three pairs of unidentate hooks in *Oligognathus* while in *Drilognathus*, *Labrorostratus* and *Haematocleptes* they are two pairs of small cuticular pieces. Free-living oenonids and both free-living and parasitic species of *Notocirrus* and *Drilonereis* possess a pair of long, slender maxillary carriers with a third carrier attached

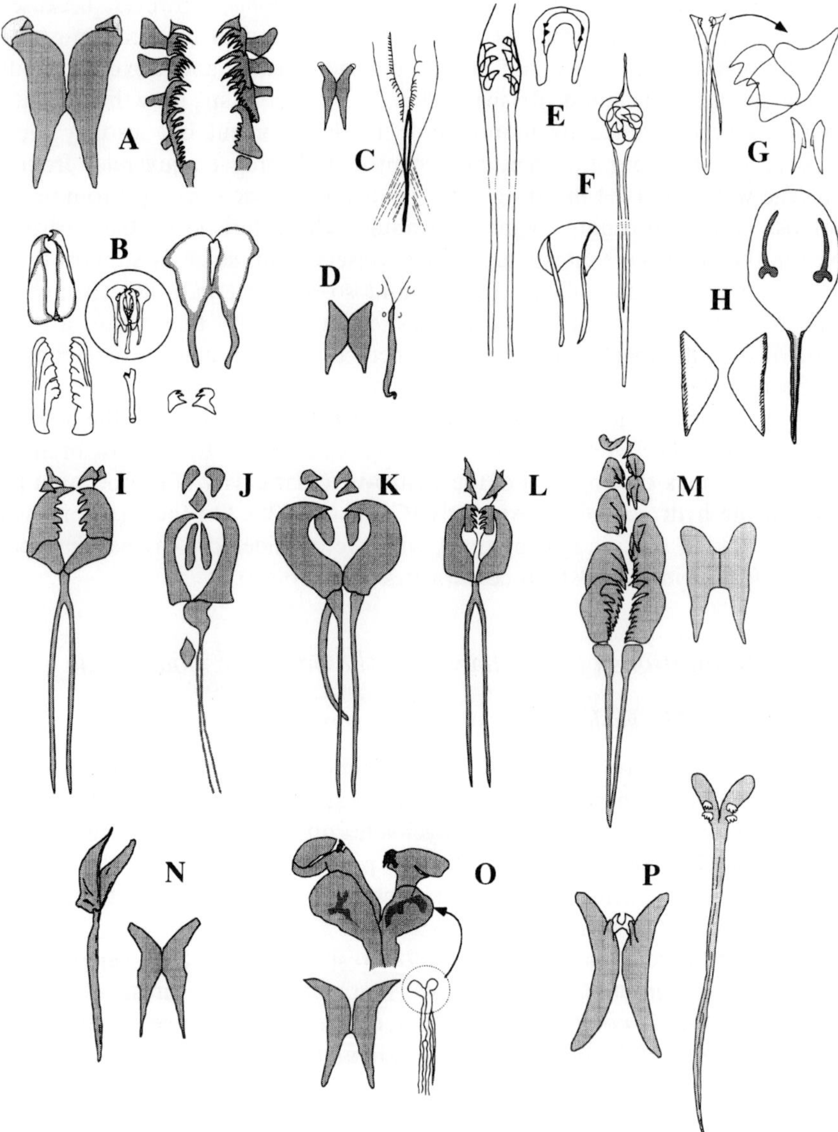

Figure 19 Jaw apparatus of parasitic Oenonidae, including a representative example of a typical free-living species. A, Free-living adult *Notocirrus spiniferus* (redrawn from Pettibone 1957). B, *Pholadiphila turnerae* (redrawn from Dean 1992). C, Parasitic juvenile *Notocirrus spiniferus* (redrawn from Pettibone 1957). D, *Drilognathus capensis* (redrawn from Day 1967). E, *Oligognathus bonelliae* (redrawn from Pettibone 1957). F, *Oligognathus parasiticus* (redrawn from Pettibone 1957). G, *Haematocleptes leaenae* (redrawn from Hartman & Fauchald 1971). H, *Haematocleptes terebellides* (redrawn from George & Hartmann-Schröder 1985). I, *Drilonereis caulleryi* (redrawn from Pettibone 1957). J, *Drilonereis parasiticus* (redrawn from Pettibone 1957). K, *Drilonereis forcipes* (redrawn from Pettibone 1957). L, *Drilonereis benedicti* (redrawn from Pettibone 1957). M, *Arabella endonata* (redrawn from Emerson 1974). N, *Labrorostratus prolificus* (redrawn from Amaral 1977). O, *Labrorostratus luteus* (redrawn from Uebelacker 1978). P, *Labrorostratus parasiticus* (redrawn from Pettibone 1957).

ventrally. The paired maxillary carriers show a different degree of fusion in *Labrorostratus* and *Haematocleptes* than in *Drilognathus*, *Oligognathus* and the parasitic species of *Arabella*, with the third carrier retained in the former two genera and lost in the latter three. The maxillae and maxillary carriers of the single known species of *Drilognathus*, *D. capensis*, were considered vestigial (Day 1960, 1967). However, they closely resembled those of the young *Notocirrus spiniferus* found living as parasite in *Diopatra cuprea*, which were considered incompletely developed (Pettibone 1957). Whether the jaw apparatus of *Drilognathus capensis* corresponds to a young developing specimen or to an adult with highly modified jaws is impossible to assess. However, the shape of the jaw apparatus and the size of the worm (3 mm long for only 60 segments) suggest that the single known specimen was not adult.

Apart from the structural changes in the jaw apparatus, which may be more or less subjectively related to a progressive adaptation to the endoparasitic mode of life, little is known on the possible sources of food for the endoparasitic oenonids and, thus, on the real significance of the above-mentioned smallness in jaws. Emerson (1974) mentioned that the *Diopatra ornata* females harbouring *Arabella endonata* contained significantly fewer oocytes than the uninfested females. This author suggested that the gametes may represent the major food source for the later growth stages of the parasite, when it has a fully functional jaw apparatus and digestive tract, and that smaller or juvenile parasites lacking jaws could survive by either direct consumption of the coelomic fluid or absorption of dissolved nutrients from it. Moreover, gamete consumption by adults may also provide additional space within the coelomic cavity of the host for the parasites at the earlier stages.

Surprisingly, it is within the ectoparasitic polychaetes that we find the highest degree of structural and behavioural adaptations to their peculiar mode of life. An exception, however, are those temporary ectoparasites that are not at all modified, such as *Myrianida pinnigera*, which feeds on the body fluids of ascidians, or the capitellids *Capitomastus lividus* and *Capitella ovincola* which parasitize the egg-masses of *Loligo* (Table 6, p. 286).

The mouth parts of *Pholadiphila turnerae* are small, like those of the endoparasitic oenonids (Fig. 19), with well developed mandibles and three pairs of maxillar plates, which suggests reliance on the host for nutrition (Dean 1992). Although their maxillae are less modified than those of several endoparasitic oenonids, the slender, short (but not broad) maxillary carrier with indications of tripartite end (which strongly suggests fusion of the original priognat carrier) are believed to be the consequence of a parasitic lifestyle (Dean 1992). Possible food sources within the pholad mantle cavity may be the gill filaments, the mucous coat on these filaments or perhaps partly digested material within the pholad's wood-storing caecum. The setal arrangement of *P. turnerae* is the same as in all oenonid species (free living or endoparasitic). However, its acuminate setae and strong spines may be an adaptation to its parasitic mode of life, affording strong purchase against the body of its mollusc host. These setae are more highly developed than those seen in other parasitic oenonids, being more robust, more numerous and extending out much further from the parapodia, which may be a consequence of the ectoparasitic habit of the species (Fig. 18).

The ectoparasitic sphaerodorids of the species *Sphaerodorum flavus* feed on the epithelium of their host ophiuroids and, as its regeneration is not possible, they may even kill the hosts if the wounds are too big (K. W. Ockelmann, pers. comm.). This association, together with that between *Pholadiphila turnerae* and the pholadid mollusc and those of the temporary ectoparasites, are probably the simplest relationships developed by ectoparasitic polychaetes. Otherwise, there is a progressively increasing gradient in intimacy in host–parasite relationships.

Haplosyllis cephalata, which has been thought identical to *H. spongicola* (San Martín 1984), was found attached to a parapodial cirrus of the host eunicid. The cirrus had been swallowed and occupied most of the pharyngeal cavity of the parasite. However, no special modifications of the mouth, pharynx or pharyngeal cavity seem to be present in this species. *Asetocalamyzas laonicola* penetrates the body wall of the host terebellid using its large everted pharynx, which was distended after penetrating the host body (Tzetlin 1985) allowing the parasite to keep firmly attached to the host (Fig. 20A). A more specialized pharynx can be found in *Parasitosyllis claparede* (Potts 1912). The pharynx projects from the pharynx-sheath and resembles an elongated vase with a short stem, the rim of which is embedded in the host body while the stem communicates posteriorly with the proventriculum (Fig. 20B). Except in this posterior region, the pharynx is enormously thickened and composed of at least four layers of chitin. Anteriorly, these chitinous layers run far into the host, bending round at right angles as they enter the host body. As they penetrate, these layers become thinner and thinner so that it is soon extremely difficult to distinguish between the tissues of parasite and host. The lumen of the pharynx is narrow anteriorly and there is a cushion muscle posteriorly. A narrow duct can be observed apparently communicating the pharynx with the proventriculum, which would serve for the passage of fluids absorbed from the host. The other known species of calamizid, *Calamyzas amphictenicola*, possesses a buccal sucker and a specialized eversible pharynx with stylet-shapped sucker tube to penetrate the branchiae of its host ampharetid (Arwidsson 1932, Paris 1955, Fauvel 1959, Hartmann-Schröder 1971).

Certainly, the highest degree of specialization occurs in the single known ichthyotomid species *Ichthyotomus sanguinarius* (Eisig 1906). Its peristomium, with the subterminal mouth, forms an oral cone that is capable of being protracted beyond the prostomium and retracted to form a cup-shaped oral sucker by which it fixes on the skin of the host fins (Fig. 20C). The ventral cirri have spinning glands opening at their tips, serving also to attach to the host. The muscular pharynx has a pair of articulating, scissorlike jaws, each formed by a distal spoonlike stylet provided with recurved teeth, a middle articulating joint and a bifurcated basal stem for muscle attachment (Fig. 20D). When the oral cone is protracted, the stylets project from the opening, pierce the skin of the eel and attach firmly. In such a way, the worm cuts the blood capillaries of the host and uses its muscular pharynx to suck the host blood. The blood passes into dorsal and ventral haemophilic pharyngeal glands and into the gut, which is enlarged with diverticula extending into the parapodia. The modifications of this species are not only morphological but physiological. Effectively, like the mosquito, it has developed an anticoagulant to facilitate its feeding mode. Moreover, the circulatory apparatus and branchiae are absent. According to Eisig (1906), the fact that the eel's blood remains always liquid and constantly flows towards the gut diverticula may enable the tissues of the parasite to breathe.

Behavioural adaptations in parasitic polychaetes

Probably the only study reporting behavioural aspects of a parasitic polychaete is about *Branchiosyllis oculata* on different sponge hosts (Pawlik 1983) (Table 6). This

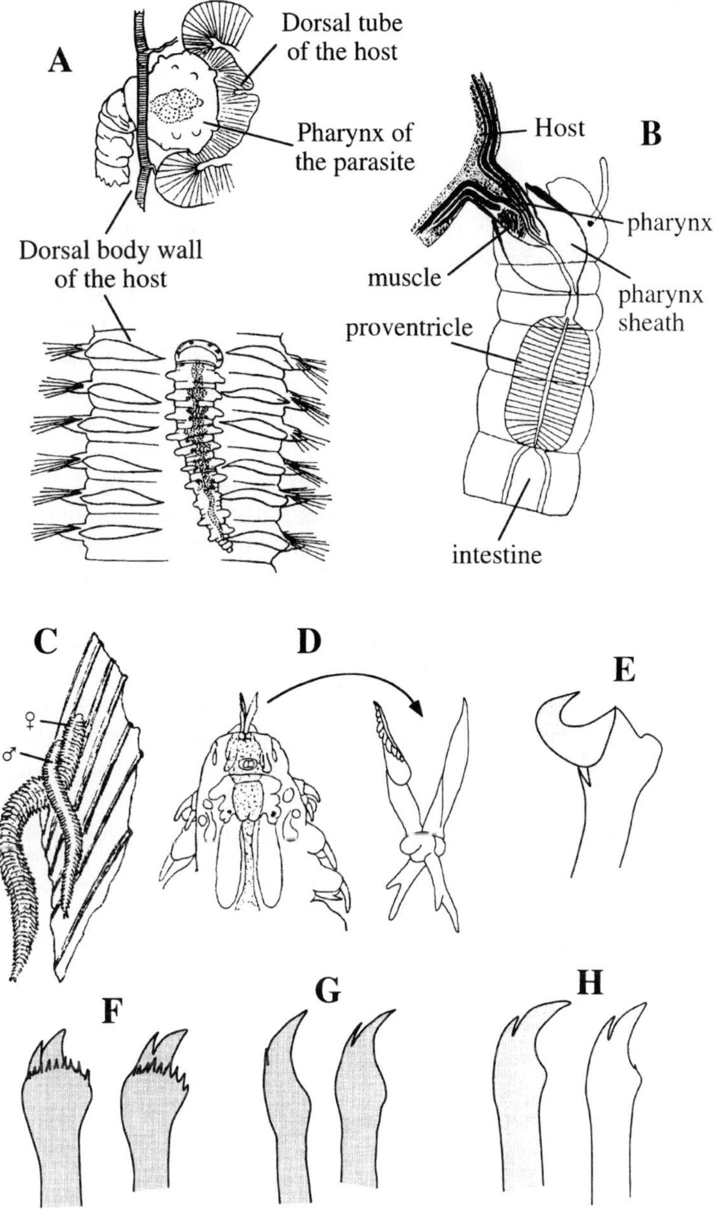

Figure 20 Examples of ectoparasitic polychaetes infesting their hosts. A, *Asetocalamyzas laonicola* attached to *Laonome cirrata* (redrawn from Tzetlin 1985). B, *Parasitosyllis claparede* attached to an unidentified polychaete (redrawn from Potts 1912). C, *Ichthyotomus sanguinarius* attached to a host eel (redrawn from Eisig 1906 in Fauvel 1923). D, Detail of the anterior end and scissor-like mandibles of *Ichthyotomus sanguinarius* (redrawn from Eisig 1906 in Paris 1955). E, Claw-like hook of *Branchiosyllis oculata* (redrawn from Pawlik 1983). F, Hooked setae of *Procerastea parasimpliseta* (redrawn from Hartmann-Schröder 1992). G, Hooked setae of *Procerastea halleziana* (redrawn from Alós 1989). H, Hooked setae of *Procerastea hydrozoicola* (redrawn from Hartmann-Schröder 1992).

author reported a series of data which define this species as a specialized parasite apparently well adapted to living on the surface of its hosts. By means of a combined effect of parapodial muscular contractions and the claw-like hooks of the setae (Fig. 20E), the worm may either keep itself firmly attached to the host tissues or advance across the sponge's surface. The long dorsal cirri appear to have a mechanosensory function. Touching the worm, cirri produced a rapid reaction: a writhing backward motion (to dislodge setae) followed by a rapid forward advance. Hooked setae are also present in other parasitic syllids such as *Haplosyllis spongicola* (see Fig. 11C, p. 271) and the species of *Procerastea* (Fig. 20F, G, H). Although *Haplosyllis spongicola* may use its setae as *Branchiosyllis oculata* does (i.e. to move on and inside its host sponges), the species of *Procerastea* more probably use their setae to move back and forth along their tubes and to keep themselves attached to the branches of their host hydroids when feeding.

The colour of the gut contents and body of *Branchiosyllis oculata* on the sponges *Cinachyra alloclada*, *Speciospongia othella* and *Tedania ignis* was yellow, brownish-black and red, respectively, each the same colour as its sponge host. Worms from the remaining host species had uncoloured bodies with the gut contents varying from brown to grey. When transferring uncoloured worms to the red *T. ignis*, they were observed to have taken on red gut-contents a few hours after transplantation. This was considered a clear demonstration of the fact that the polychaete feeds on sponge tissue. In addition, transplanted autotomized worms regenerated their posterior setigers with pigmentation identical to that of *Branchiosyllis oculata* found on the surrogate sponges, while their anterior setigers did not show any detectable change in colour. This provided evidence that the polychaetes incorporate sponge pigments into their body tissues, although only the pigments of some species were incorporated. The worms having the same colour as the sponges they inhabit may have a distinct antipredatory advantage as the uncoloured worms were much more readily eaten by fishes under laboratory conditions.

Life cycle and reproductive adaptations in parasitic polychaetes

It seems evident that a parasitic mode of life should lead to modifications in the life cycle of parasites as they adapt to their particular mode of life. However, little is known on the developmental and reproductive cycles of parasitic polychaetes. In *Ichtyothomus sanguinarius* and *Calamyzas amphictenicola*, male and female individuals have been reported on the same host. The former has a hypertrophied genital apparatus (Eisig 1906), whereas the later has aflagelate spermatozoa (Franzén 1982). There is not a clear relationship between the hypertrophy of the genital apparatus and the parasitic mode of life or the presence of heterosexual pairs. Conversely, the aflagelate spermatozoa have extremely low motility. In this case, the relationship of low motility with male and female individuals always being found close together seems clear, as the spermatozoa do not need to swim to reach the eggs.

Fully mature specimens carrying eggs or sperm have not been reported for the solitary endoparasitic oenonids such as *Labrorostratus parasiticus*, while some specimens were reported as free living among algae. On the basis of this slender evidence Caullery & Mesnil (1916) suggested that infestation took place when the parasites were extremely young, that adults left the host as they approached sexual maturity and that they spawned outside it. If this theory is general among solitary endoparasitic oenonids, it would explain the retention by some of them of characteristics of free-living species of the family (Clark 1956). According to this author, however, what is more inexplicable and puzzling than the lack of morphological specialization, is how the endoparasitic habit was ever evolved in these worms, particularly if there is no substantial reduction in size.

Occasionally, however, the size of endoparasitic oenonids is much smaller than that of their hosts. In these cases, the existence of endoparasitic juvenile stages, very abundant in a single host, together with the presence of asexually reproductive adults inside the host body (Amaral 1977) suggest two different strategies of dispersion. In the first case, we may assume an adult free-living reproductive phase. Then, the initial infestation would be accomplished by larvae (probably pelagic), whereas the host infestation would progress as a result of successive cycles of asexual multiplication of the parasites inside it. The final step in the infestation would be the release of adults that would reach maturity as free-living forms. In the second case, we may assume an absence of the sexual reproductive phase. Although the infestation inside the host could progress the same as in the first strategy, the infestation of new hosts would be undertaken by direct migration of the stolons released from the first infested host after entirely consuming the available host resources. This second strategy involves a more contagious distribution of both host and parasite populations. In theory, the two strategies could coexist in the same species. The asexual phase would lead to the infestation of hosts from the same population while the sexual phase would allow the colonization of new host populations. Unfortunately, the low incidence of these parasites and poor knowledge of their life histories do not permit satisfactory general extrapolations.

In the case of the parasitic syllid *Proceraea* sp., a first approach to the initial infestation phases suggested that the infestation may occur through planktonic larvae reaching the colony and subsequently penetrating the hydrothecae. However, three different reasons lead Britayev et al. (unpubl.) to discard this strategy: (a) the penetration of the well defended thecae by the delicate larvae of the worm seems very dificult, (b) that the larvae have no ciliary cover but a small apical tuff strongly suggests the absence of a pelagic larval phase, and (c) there is only one parasite per host hydrothecae. Therefore, these authors suggested that the female *Proceraea* sp. (*Sacconereis*) take care of their offspring by directly laying their eggs inside the hydrothecae. Thus, both a reliable shelter and fresh food are provided for the initial developmental stages of the parasite. This behaviour closely resembles that of the ichneumonid insects that lay their eggs inside the body of other insects and keep them alive to supply their larvae with fresh food. This, in turn, leads to the observed clustered distribution of commensals among the host colonies and also favours the presence of a high number of commensals infesting each host colony. Accordingly, swimming fertilized females are probably responsible for commensal dispersion from one host colony to another.

In *Branchiosyllis oculata*, the fact that adult worms removed from the host sponges had little capacity to relocate their original host or to locate a new sponge suggested that sponge selection and colonization was carried out solely by larvae (possible swimming or crawling) (Pawlik 1983). However, all adults were morphologically identical and were able to survive after being experimentally relocated to different host sponges. Thus, it appeared unlikely that recruitable larvae were more specifically adapted to particular sponge species. The same author also pointed out that the differential densities on some host species (particularly high in *Tedania ignis*, intermediate in the coloured species and lower in the colourless species) may result from the combination of two factors: (a) differential recruitment as a result of simple differences of density or accessibility of the host species, or chemical cues and corresponding chemosensitive larvae; (b) differential mortality linked either to worms' colour mimicry, to physical (spicules) or chemical defences of hosts (whether these defences are transferable or not to the worms) or to the inaccessibility of the host to the potential predators. Differences in infestation intensity within the polychromatic host *Tethya actinia* may indicate that whatever controls host pigmentation may also have a direct effect on *Branchiosyllis oculata* larval recruitment and/or adult survival.

From the two examples reported above, it seems clear enough that an accurate study of a given parasitic relationship (from an ecological, behavioural or even morphological point of view) provides pointers towards a reliable hypothesis on the life cycle of the symbiont. However, unfortunately, these studies are few and far between.

Symbiotic boring polychaetes

Usually, boring polychaetes did not show special adaptations other than those allowing them to bore into other organisms with more or less hard structures (mainly of a calcareous nature). The association between them and the bored organisms often appeared to be fortuitous, as the worms might be found boring either into living organisms or non-living substrata. Forty-eight species of polychaetes, belonging to the families Spionidae Caobangidae, S, Sabellidae and Cirratulidae, have been reported as borers of other living marine organisms (Tables 8 and 9). Some interactions between these boring worms and their hosts, however, seemed to be negative, which lead to the species being labelled as "parasitic" borers. They are pests that drain the energy by causing the host to work harder to keep them away from the interior of the shell, but they are often uninterested in the hosts as such. The host suffers, but is not attacked directly: it is just stressed by mud-eating worms that are "looking for a home". This may reflect an old-fashion view of a "parasitic" association. Hence, although we decided to mention them in this review, we realized that some of the species cannot be considered as real symbionts and, thus, we have included these boring polychates in the present specific section. In other cases, there were either no negative interactions described or the nature of the interactions were assessed by behavioural observations and some benefits were suggested or even demonstrated, enabling classification of the worms as "commensal" borers. There are a few quantitative studies on infestations by boring polychaetes: the known data on intensities and prevalences are re-summarized in Table 10. The two types of interactions involving boring polychaetes (i.e. parasitic and commensalistic) will be reviewed in the following three sections.

Parasitic boring polychaetes

The most common parasitic boring polychaetes are, certainly, the spionids (Table 8). Among them, the species of *Polydora* and related genera are the only one able to bore (Blake 1969, 1980, 1996, Blake & Evans 1973). However, little is known concerning the initial steps of their establishment in the calcareous habitats (but see Hopkins 1958), and the physical (linked to the modified setae of the 5th setiger) or chemical (a not-yet identified acidic substance) mechanisms of burrow expansion are still a matter of controversy (Haigler 1969, Blake & Evans 1973, Zottoli & Carriker 1974, Sato-Okoshi & Okoshi 1993).

We also know little about how the polydorids interact with their hosts. Most polydorids that live in molluscs do not harm their hosts. The worms may make different kinds of burrows (viz. U-shaped, branched, cylindrical) (Fig. 21), which may be lined with a variety of parchment-like secretions or with mud or sand grains. When reaching the inner walls of the shells, however, this boring activity often leads the mollusc hosts to react by producing "mud-blisters", probably the most damaging effect of polydorids. The mud-blisters are masses of mud accumulated on the inner surface of the shell by the recently settled worms (Fig. 21D, E). The hosts react by first secreting over the mud a roof of conchiolin and then a layer of nacreous material. The mud-filled chambers formed are occupied by the worms which communicate with the exterior via pairs of tubes either at or close to the periphery of the shell. Although it is difficult to evaluate the expenditure of metabolic energy linked to this extraordinary secretion of protective layers (and, thus, to determine to what extent the life of the mollusc might be affected), the mollusc host may suffer a certain degree of distress (Kent 1979). Additionally, mud-blisters may be so numerous that they lead to the host's death. In fact, all this boring activity weakens the shells (Kent 1981) and so makes them less resistant to diseases, other parasites and predators.

The presence of blisters was also reported on the hydrocoral *Allopora californica.* In this case, however, the blisters were formed by the syllid *Autolytus penetrans,* which apparently penetrated the surface of the host resulting in a hyperplasia of cellular and calcareous material (Wright & Woodwick 1977). However, *A. penetrans* does not actually bore. Although its blisters closely resemble those of the polydorids, they were lined with cells instead of mud. In spite of blisters, there were no indications of any other damage to the coral linked to the presence of the syllid. Therefore. *A. penetrans* was not originally described as parasite. However, two points allowed us to include this relationship among parasitisms (see Table 6, p. 287, and p. 294): (a) no apparent profits were reported for the host and (b) the behaviour of the symbiont closely resembles that of a cysticole parasite.

Most boring polydorids have been considered as parasitic organisms from an anthropogenic rather than ecological point of view. As soon as they infest species of molluscs that are cultivated or harvested as a fishery resource (viz. mussels, pearl oysters, edible oysters, scallops), they are immediately viewed as pests (e.g. Haswell 1885, Houlbert & Galaine 1916, Lunz 1940, 1941, Korringa 1951, Landers 1967, Blake & Evans 1973, Tkachuk 1988). However, the polydorids may, although rarely, cause the demise of the cultivated host populations thus affecting commercial profit. A decrease of mollusc growth-rates linked to the presence of polydorid species has seldom been reported (but see Imajima & Sato 1984, Mori et al. 1985). However, they may damage the appearance of the shell, making it less attractive to potential consumers and so lowering the market value of the mollusc. In most cases, the shell damage caused by boring species may be linked to their gregarious habits, leading to the presence of large numbers of individuals on the same host (Table 10).

The boring polydorids may also inhabit other organisms with calcareous structures. In some cases, the interactions with their hosts (e.g. cirriped crustaceans, serpulid polychaetes) closely resembles the interactions established with molluscs (Table 8). They simply bore into the calcareous parts of these hosts and no further information

Table 8 List of symbiotic species of boring polychaetes of the family Spionidae. Type: Type of relationship (O, obligatory; F, facultative).

Symbiotic borers	Authority	Host	Species	Type	Reference
Boccardia accus	(Rainer)	Bivalve	*Atrina pectinata zetlandica*	F	Read 1975
Boccardia accus		Bivalve	*Chione stuchtburyi*	F	Read 1975
Boccardia accus		Bivalve	*Ostrea* sp.	F	Read 1975
Boccardia accus		Bivalve	*Paphies subtriangularum*	F	Read 1975
Boccardia accus		Bivalve	*Perna canaliculus*	F	Read 1975
Boccardia accus		Gastropod	*Haliotis iris*	F	Read 1975
Boccardia androgyna	Read	Sponge	*Ancorina alata*	O	Read 1975
Boccardia androgyna		Sponge	*Ancorina* spp.	O	Read 1975
Boccardia androgyna		Sponge	*Anchinoe* spp.	O	Read 1975
Boccardia androgyna		Sponge	*Halichondria* spp.	O	Read 1975
Boccardia androgyna		Sponge	*Tethya* spp.	O	Read 1975
Boccardia anophthalma	(Rioja)	Gastropod	*Muricanthus nigricus*	O	Blake 1980
Boccardia berkeleyorum	Blake & Woodwick	Gastropod	*Pododesmus macrochisma*	O	Blake 1980
Boccardia chilensis	Blake & Woodwick	Bivalve	*Atrina pectinata zetlandica*	F	Read 1975
Boccardia chilensis		Bivalve	*Chione stuchtburyi*	F	Read 1975
Boccardia chilensis		Gastropod	*Haliotis iris*	F	Read 1975
Boccardia knoxi	Rainer	Bivalve	*Atrina pectinata zetlandica*	F	Read 1975
Boccardia knoxi		Gastropod	*Haliotis iris*	F	Read 1975
Boccardia lamellata	Rainer	Bivalve	*Atrina pectinata zetlandica*	F	Read 1975
Boccardia outakouica	Rainer	Bivalve	*Atrina pectinata zetlandica*	F	Read 1975
Boccardia outakouica		Bivalve	*Ostrea luritaria*	F	Rainer 1973
Boccardia tricuspa	(Hartman)	Bivalve	*Concholepas concholepas*	F	Carrasco 1974
Boccardia tricuspa		Gastropod	*Haliotis* sp.	F	Blake 1980
Boccardia tricuspa		Gastropod	*Muricanthus nigricus*	F	Blake 1980
Boccardiella hamata	(Webster)	Bivalve	*Crassostrea gigas*	F	Radashevsky 1993
Boccardiella hamata		Bivalve	*Mizuchopecten yessoensis*	F	Radashevsky 1993
Boccardiella hamata		Bivalve	Unidentified	F	Webster 1879a,b, Hartman 1951, Blake & Evans 1973

Table 8 *Continued*

Symbiotic borers	Authority	Host	Species	Type	Reference
Dipolydora alborectalis	(Radashevsky)	Bivalve	*Crassostrea gigas*	F	Radashevsky 1993
Dipolydora alborectalis		Bivalve	*Crenomytilus grayanus*	F	Radashevsky 1993
Dipolydora alborectalis		Bivalve	*Mizuchopecten yessoensis*	F	Radashevsky 1993
Dipolydora alborectalis		Bivalve	*Modiolus kurilensis*	F	Radashevsky 1993
Dipolydora alborectalis		Bivalve	*Pododesmus macrochisma*	F	Radashevsky 1993
Dipolydora alborectalis		Bivalve	*Swiftopecten swifti*	F	Radashevsky 1993
Dipolydora alborectalis		Cirriped	*Balanus rostratus*	F	Radashevsky 1993
Dipolydora alborectalis		Gastropod	*Acmaea pallida*	F	Radashevsky 1993
Dipolydora alborectalis		Gastropod	*Crepidula derjugini*	F	Radashevsky 1993
Dipolydora alborectalis		Gastropod	*Littorina squalida*	F	Radashevsky 1993
Dipolydora alborectalis		Gastropod	*Nucella heyseana*	F	Radashevsky 1993
Dipolydora alborectalis		Gastropod	*Tegula rustica*	F	Radashevsky 1993
Dipolydora armata	(Langerhans)	Cnidarian	*Leptastrea purpurea*	F	Okuda 1937
Dipolydora armata		Bryozoan	*Holoporella sardonica*	F	Laubier 1958
Dipolydora armata		Bryozoan	*Porella concinna*	F	Laubier 1958
Dipolydora armata		Bryozoan	*Schizoporella armata*	F	Laubier 1958
Dipolydora barbilla	(Blake)	Gastropod	*Muricanthus nigricus*	F	Blake 1981
Dipolydora bidentata	(Zachs)	Bivalve	*Crassostrea gigas*	F	Radashevsky 1993
Dipolydora bidentata		Bivalve	*Crenomytilus grayanus*	F	Radashevsky 1993
Dipolydora bidentata		Bivalve	*Mizuchopecten yessoensis*	F	Radashevsky 1993
Dipolydora bidentata		Bivalve	*Modiolus kurilensis*	F	Radashevsky 1993
Dipolydora bidentata		Bivalve	*Pododesmus macrochisma*	F	Radashevsky 1993
Dipolydora bidentata		Bivalve	*Swiftopecten swifti*	F	Radashevsky 1993
Dipolydora bidentata		Gastropod	*Acmaea pallida*	F	Radashevsky 1993
Dipolydora bidentata		Gastropod	*Buccinum middendorffii*	F	Radashevsky 1993
Dipolydora bidentata		Gastropod	*Margarites* sp.	F	Radashevsky 1993
Dipolydora bidentata		Gastropod	*Muricanthus nigricus*	F	Blake 1981

Table 8 *Continued*

Symbiotic borers	Authority	Host	Species	Type	Reference
Dipolydora bidentata		Gastropod	*Nucella hyseana*	F	Radashevsky 1993
Dipolydora bidentata		Gastropod	*Tegula rustica*	F	Radashevsky 1993
Dipolydora carunculata	(Radashevsky)	Bivalve	*Crassostrea gigas*	F	Radashevsky 1993
Dipolydora carunculata		Bivalve	*Crenomytilus grayanus*	F	Radashevsky 1993
Dipolydora carunculata		Bivalve	*Mizuchopecten yessoensis*	F	Radashevsky 1993
Dipolydora carunculata		Gastropod	*Acmaea pallida*	F	Radashevsky 1993
Dipolydora carunculata		Gastropod	*Nucella heyseana*	F	Radashevsky 1993
Dipolydora commensalis	(Andrews)	Decapod	*Clibanarius vittatus*	O	Dauer 1991
Dipolydora commensalis		Decapod	*Pagurus annulipes*	O	Dauer 1991
Dipolydora commensalis		Decapod	*Pagurus branchiomastus*	O	Radashevsky 1989
Dipolydora commensalis		Decapod	*Pagurus capillatus*	O	Radashevsky 1989
Dipolydora commensalis		Decapod	*Pagurus granosimanus*	O	Hatfield 1965
Dipolydora commensalis		Decapod	*Pagurus longicarpus*	O	Andrews 1891a,b, Hatfield 1965, Dauer 1991
Dipolydora commensalis		Decapod	*Pagurus middendorfi*	O	Radashevsky 1989
Dipolydora commensalis		Decapod	*Pagurus pollicaris*	O	Dauer 1991
Dipolydora commensalis		Decapod	*Pagurus samuelis*	O	Blake 1969
Dipolydora concharum	(Verrill)	Bivalve	*Mercenaria mercenaria*	F	Blake 1969
Dipolydora concharum		Bivalve	*Mercenaria stimpsoni*	O	Radashevsky 1993
Dipolydora concharum		Bivalve	*Mizuchopecten yessoensis*	O	Mori et al. 1985, Radashevsky 1993
Dipolydora concharum		Bivalve	*Pandora wardiana*	O	Radashevsky 1993
Dipolydora concharum		Bivalve	*Placopecten magellanicus*	F	Blake 1969
Dipolydora concharum		Gastropod	*Buccinum verkruzeni*	O	Radashevsky 1993
Dipolydora concharum		Gastropod	*Crepidula derjugini*	O	Radashevsky 1993
Dipolydora concharum		Gastropod	*Cyprina islandica*	O	Blake & Evans 1973
Dipolydora concharum		Gastropod	*Neptunea bulbacea*	O	Radashevsky 1993
Dipolydora concharum		Gastropod	*Neptunea constricta*	O	Radashevsky 1993
Dipolydora concharum		Gastropod	*Neptunea lyrata*	O	Radashevsky 1993

Table 8 *Continued*

Symbiotic borers	Authority	Host	Species	Type	Reference
Dipolydora concharum		Gastropod	*Neptunea polycostata*	O	Radashevsky 1993
Dipolydora concharum		Gastropod	*Turritella fortilirata*	O	Radashevsky 1993
Dipolydora elegantissima	(Blake & Woodwick)	Bivalve	*Tivela stultorum*	O	Blake & Woodwick 1971, Blake 1980
Dipolydora rogeri	(Martin)	Sponge	*Cliona viridis*	O	Martin 1996
Dipolydora socialis	(Schmarda)	Bivalve	*Placopecten magellanicus*	F	Blake & Evans 1973
Dipolydora trilobata	(Radashevsky)	Bivalve	*Crassostrea gigas*	F	Radashevsky 1993
Dipolydora trilobata		Bivalve	*Crenomytilus grayanus*	F	Radashevsky 1993
Dipolydora trilobata		Bivalve	*Modiolus kurilensis*	F	Radashevsky 1993
Dipolydora trilobata		Bivalve	*Pododesmus macroshisma*	F	Blake & Woodwick 1971
Dipolydora trilobata		Cirriped	*Balanus rostratus*	F	Radashevsky 1993
Dipolydora trilobata		Cirriped	*Semibalanus cariosus*	F	Radashevsky 1993
Dipolydora trilobata		Gastropod	*Acmaea pallida*	F	Radashevsky 1993
Dipolydora trilobata		Gastropod	*Buccinum middendorffii*	F	Radashevsky 1993
Dipolydora trilobata		Gastropod	*Margarites* sp.	F	Radashevsky 1993
Dipolydora trilobata		Gastropod	*Nucella heyseana*	F	Radashevsky 1993
Dipolydora trilobata		Polychaete	*Cruzigera zygophora*	F	Radashevsky 1993
Polydora alloporis	Light	Cnidarian	*Allopora californica*	O	Light 1970a
Polydora bioccipitalis	Blake & Woodwick	Bivalve	*Mesodesma donacium*	F	Blake & Woodwick 1972, Blake 1983
Polydora brevipalpa	Zachs	Bivalve	*Mizuchopecten yessoensis*	F	Imajima & Sato 1984, Mori et al. 1985, Radashevsky 1993
Polydora ciliata	(Johnston)	Bivalve	*Mytilus edulis*	F	Michaelis 1978, Ramberg & Schram 1982
Polydora ciliata		Bivalve	*Mizuchopecten yessoensis*	F	Mori et al. 1985
Polydora ciliata		Bivalve	*Mya arenaria*	F	Blake & Evans 1973, Ramberg & Schram 1982
Polydora ciliata		Bivalve	*Ostrea caruncullata*	F	Blake & Evans 1973
Polydora ciliata		Bivalve	*Ostrea edulis*	F	Korringa 1951, Blake & Evans 1973, Ramberg & Schram 1982
Polydora ciliata		Gastropod	*Littorina littorea*	F	Blake 1980, Ramberg & Schram 1982
Polydora cornuta	Bosc	Phanerogam	*Zostera marina*	F	Michaelis 1978

Table 8 *Continued*

Symbiotic borers	Authority	Host	Species	Type	Reference
Polydora curiosa	Radashevsky	Gastropod	*Acmaea pallida*	F	Radashevsky 1993
Polydora glycymerica	Radashevsky	Bivalve	*Glycymeris yessoensis*	O	Radashevsky 1993
Polydora glycymerica		Bivalve	*Anadara broughtoni*	O	Radashevsky 1993
Polydora haswelli	Blake & Kudenov	Bivalve	*Crassostrea commercialis*	O	Blake & Evans 1973, Blake & Kudenov 1978
Polydora hoplura	Claperède	Bivalve	*Atrina pectinata zetlandica*	F	Read 1975
Polydora hoplura		Bivalve	*Crassostrea gigas*	F	Blake & Kudenov 1978
Polydora hoplura		Bivalve	*Ostrea edulis*	F	Korringa 1951, Blake 1980
Polydora hoplura		Bryozoan	*Schizoporella auriculata*	F	Laubier 1959
Polydora latispinosa	Blake & Kudenov	Bivalve	*Pectein alba*	O	Blake & Kudenov 1978
Polydora latispinosa		Bivalve	Oysters	O	Blake & Kudenov 1978
Polydora vulgaris	Mohammad	Bivalve	*Pinctada margaritifera*	F	Mohammad 1972, Blake 1980
Polydora websteri	Hartman	Bivalve	*Aequipecten gibbus*	F	Blake & Evans 1973
Polydora websteri		Bivalve	*Crassostrea commercialis*	F	Blake & Evans 1973
Polydora websteri		Bivalve	*Crassostrea gigas*	F	Blake & Evans 1973
Polydora websteri		Bivalve	*Crassostrea virginica*	F	Blake & Evans 1973
Polydora websteri		Bivalve	*Hinnites multirugosus*	F	Blake & Evans 1973
Polydora websteri		Bivalve	*Mercenaria mercenaria*	F	Blake & Evans 1973
Polydora websteri		Bivalve	*Mesodesma deauratum*	F	Blake & Evans 1973
Polydora websteri		Bivalve	*Modiolus modiolus*	F	Blake & Evans 1973
Polydora websteri		Bivalve	*Mytilus edulis*	F	Blake & Evans 1973
Polydora websteri		Bivalve	*Ostrea lurida*	F	Blake & Evans 1973
Polydora websteri		Bivalve	*Ostrea permollis*	F	Blake & Evans 1973
Polydora websteri		Bivalve	*Ostrea* sp.	T	Blake 1969
Polydora websteri		Bivalve	*Patinopecten caurinus*	F	Blake & Evans 1973
Polydora websteri		Bivalve	*Pecten irradians*	F	Blake & Evans 1973
Polydora websteri		Bivalve	*Placopecten magellanicus*	F	Blake 1971, Blake & Evans 1973
Polydora websteri		Gastropod	*Crepidula fornicata*	F	Blake 1971
Polydora websteri		Gastropod	*Littorina littorea*	F	Blake 1971
Polydora websteri		Gastropod	*Muricanthus nigricus*	F	Blake 1980
Polydora websteri		Gastropod	*Thais lapillus*	F	Blake 1971
Polydora wobberi	Light	Cnidarian	*Lophogorgia* sp.	O	Light 1970b
Polydora woodwicki	Blake & Kudenov	Gastropod	*Haliotis ruber*	F	Blake & Kudenov 1978

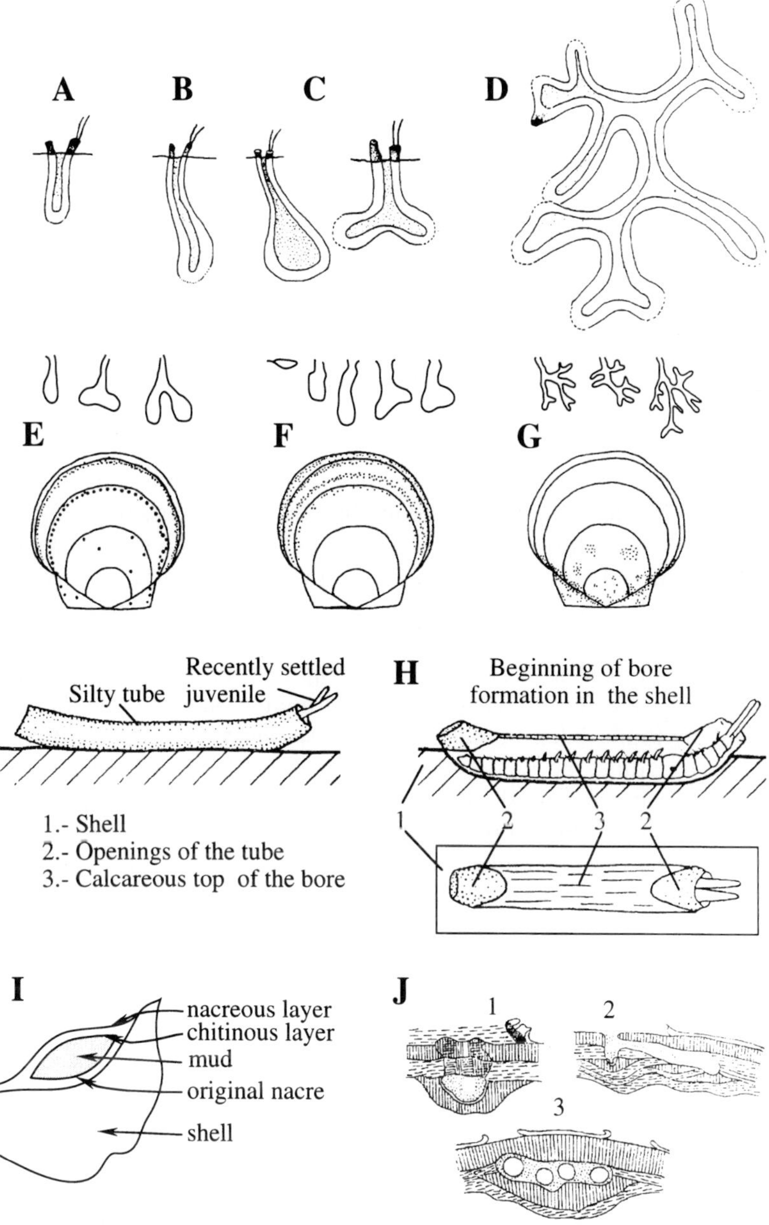

Figure 21 Bore holes of *Polydora ciliata* (A), *Polydora websteri* on *Mercenaria mercenaria* (B) and on *Placopecten magellanicus* (C) and *Dipolydora concharum* on *Placopecten magellanicus* (D) redrawn from Blake & Evans (1973). Bore holes of *Polydora variegata* (E), *Polydora ciliata* (F) and *Dipolydora concharum* (G) on *Patinopecten yessoensis* redrawn from Mori et al. (1985). H, Settlement and tube formation of *Dipolydora commensalis* (redrawn from Radashevsky 1989). I, Scheme of a generic mud-blister of polydorid worms (redrawn from Blake & Evans 1973). J, Sequence of formation of a *Polydora ciliata* mud-blister on *Mytilus edulis*: 1. Settlement, 2. Initial bore hole, 3. Mud-blister (redrawn from Tkachuk 1988).

about the interactions between host and polydorids is available (Radashevsky 1993). Similarly, *Dipolydora armata* was reported as symbiont with the madreporid *Leptastrea purpurea* by Okuda (1937), but there was no additional information. *Polydora alloporis* inhabits deep burrows in the coenosteum of the hydrocoral *Allopora californica*, forming large colonies without apparently damaging the host (Light 1970a). Each *Polydora alloporis* burrow has two distinctive openings and the worms may be seen in living material with the palpi projecting from one of the openings and the pygidium just visible below the lip of the other.

Probably the most remarkable associations with non-mollusc hosts are those between *Dipolydora armata* and *Polydora hoplura* and the bryozoans *Holoporella sardonica, Porella concinna, Schizoporella armata* and *S. auriculata* (Laubier 1958, 1959). The bryozoans are infested throughout their basal layer (usually dead) fixed to the substratum, the worms entering the colony using the existing crevices or breaks. In *Holoporella sardonica, Dipolydora armata* excavates a dense complex of galleries, some of them reaching the outer surface of the colony. At the beginning, the openings are only made of silt and mucus. Then, the bryozoan builds a thin calcareous layer (about 2 mm high and 0.25 mm in diameter) surrounding the worm tubes (Fig. 22A). In *Porella concina*, the initial infestation steps are the same, but there is no trace of galleries inside the colony. The special structures surrounding the polychaete tubes are, in this case, composed of host zooids (Fig. 22B). The tube heights are similar but wider (about 1 mm wide). At the end of these formations, a thin calcareous zone (probably a growing zone) resembles that in *Holoporella sardonica* (Fig. 22A,B). In *Schizoporella armata* the two types of calcareous formations (with and without zooids) may be found. *Polydora hoplura* was only found infesting *Schizoporella auriculata*. In this case, the calcareous formations surrounding the worm tubes are made of zooids, reaching about 6 mm high. The formations are conical with 4–5 mm at the base and 0.5–1 mm at the top. A thin calcareous layer of no more than 0.1 mm forms the border of the cones. The infestation of the bryozoan colonies by the polydorids occurs in two phases. The first one (initial penetration of the colonies) does not differ from the normal excavating habits of the species. The second phase is characterized by a stimulation of the bryozoan to build the calcareous formations. These two points strongly suggest that these

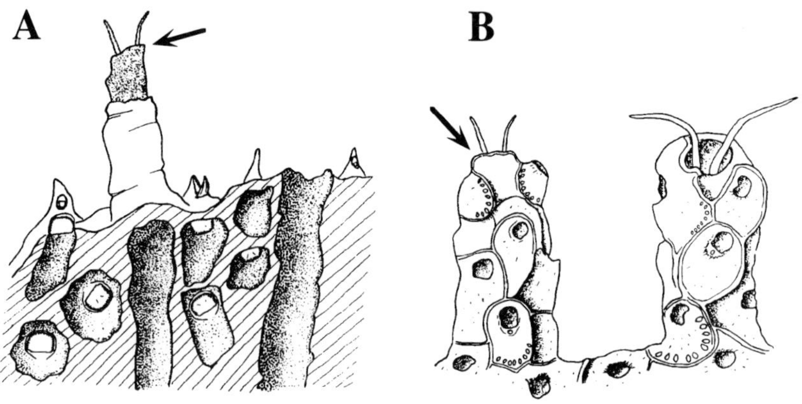

Figure 22 *Dipolydora armata* inhabiting bryozoans (redrawn from Laubier 1958). A, *Holoporella sardonica*; B, *Porella concinna*. The arrows indicate the position of tube openings.

Table 9 List of symbiotic species of boring polychaetes of the families Sabellidae and Cirratulidae. Type: Type of relationship (O, obligatory; F, facultative).

Symbiotic borers	Authority	Host	Species	Type	Reference
Sabellidae					
Caobangia abboti	Jones	Gastropod	*Antimelania asperata*	O	Jones 1974
Caobangia abboti		Gastropod	*Balanocochlis glans*	O	Jones 1974
Caobangia abboti		Gastropod	*Brotia dactylus*	O	Jones 1974
Caobangia abboti		Gastropod	*Brotia pageli*	O	Jones 1974
Caobangia abboti		Gastropod	*Brotia pontificalis*	O	Jones 1974
Caobangia abboti		Gastropod	*Brotia pontificalis agrestis*	O	Jones 1974
Caobangia abboti		Gastropod	*Brotia pontificalis pageli*	O	Jones 1974
Caobangia abboti		Gastropod	*Stenomelania* sp. 1	O	Jones 1974
Caobangia abboti		Gastropod	*Stenomelania* sp. 2	O	Jones 1974
Caobangia billeti	Giard	Gastropod	*Melania aubryana*	O	Jones 1974
Caobangia brandti	Jones	Gastropod	*Brotia pseudosulcospira pseudosulcospira*	O	Jones 1974
Caobangia brandti		Gastropod	*Brotia pseudosulcospira armata*	O	Jones 1974
Caobangia brandti		Gastropod	*Brotia binodosa binodosa*	O	Jones 1974
Caobangia brandti		Gastropod	*Brotia binodosa subgloriosa*	O	Jones 1974
Caobangia brandti		Gastropod	*Anulotaia* cf *mekongensis*	O	Jones 1974
Caobangia brandti		Gastropod	*Cipangopaludina chinensis*	O	Jones 1974
Caobangia brandti		Gastropod	*Mekongia jullieni*	O	Jones 1974
Caobangia brandti		Bivalve	*Hyriopsis delaportei*	O	Jones 1974
Caobangia ceylonica	Jones	Gastropod	*Ganga abbreviata*	O	Jones 1974
Caobangia ceylonica		Gastropod	*Paludomus hanleyi*	O	Jones 1974
Caobangia ceylonica		Gastropod	*Tanalia aculeata*	O	Jones 1974
Caobangia indica	Jones	Gastropod	*Antimelania menkiana*	O	Jones 1974
Caobangia indica		Gastropod	*Hemimintra stephanus*	O	Jones 1974
Caobangia morrisoni	Jones	Gastropod	*Brotia* sp. 1	O	Jones 1974
Caobangia morrisoni		Gastropod	*Brotia* sp 2	O	Jones 1974
Coabangia morrisoni		Gastropod	*Brotia testudinaria*	O	Jones 1974
Caobangia smithi	Jones	Gastropod	*Brotia costula siamensis*	O	Jones 1974
Caobangia smithi		Gastropod	*Brotia* cf *baccata*	O	Jones 1974
Caobangia smithi		Gastropod	*Brotia gloriosa*	O	Jones 1974
Caobangia smithi		Gastropod	*Brotia* sp. 3	O	Jones 1974
Caobangia smithi		Gastropod	*Paludomus conica*	O	Jones 1974
Pseudofabricia sp.	Kudenov	Gastropod	*Haliotis* sp.	F	Oakes & Fields 1996
Pseudopotamilla reniformis	(Müller)	Bivalve	*Placopecten magellanicus*	F	Blake 1969
Cirratulidae					
Dodecaceria concharum	Oersted	Bivalve	*Arctica islandica*	F	M. E. Petersen pers. comm.
Dodecaceria fimbriata	(Verrill)	Bivalve	*Placopecten magellanicus*	F	Blake 1969
Dodecaceria fimbriata		Bivalve	*Mercenaria mercenaria*	F	Blake 1969
Dodecaceria fimbriata		Bivalve	*Crepidula fornicata*	F	Blake 1969
Dodecaceria sp.		Bivalve	*Ostrea edulis*		D. Martin pers. obs.

relationships may be considered specific parasitisms (closer to cysticole behaviour). In addition, we may speculate on a possible cleaning behaviour of the worm, which could benefit the host and so lead to reallocate these relationships between the commensalisms (see *Polydora glycymerica* and *Dipolydora rogeri* in the next section).

Parasitic polydorids were also reported from non-calcareous organisms (Table 9). *Polydora cornuta* forms cysts in the eelgrass *Zostera marina* (Michaelis 1978). Any sponge with a substantial, firm body appears to be a suitable habitat for the hermaphroditic species *Boccardia androgyna* (Read 1975). The worm inhabits U-shaped mud-lined tubes within the sponges and a large mud-filled cavity may be formed if the sponge species has soft tissue with little spongine material. Finally, *Polydora wobberi* causes structural damage to the colony of the gorgonian *Lophogorgia* sp. (Light 1970b). The worm lives in U-shaped burrows along short branches that arise from the bases of the coelenterate colonies. The burrows open to the exterior at the tips of short stubby branches of the gorgonian that are about 20–30 cm in length (the longer uninfested branches may reach 70 cm in length). The final result is that the woody central stems of the gorgonian are entirely missing in infested branches. However, it seems not to seriously affect the overall well being of the host.

A situation similar to that of the shell-boring polydorids was described for the sabellid *Pseudopotamilla reniformis* boring in *Placopecten magellanicus* (Blake 1969) and *Pseudofabricia* sp. in *Haliotis rufescens* (Oakes & Fields 1996). *Pseudopotamilla reniformis* inhabits large ridges on the inside of the shell, which appear to be the response of the scallop to occasional penetration of the inside of the shell by the worm. The inside of the ridges is filled by mud and worm tubes. The large size of *P. reniformis* and the possibility of high infestation intensities (Table 10) qualifies this worm as a poten-

Table 10 Summary of known data on infestations by boring polychaetes. Numbers between brackets indicate the amount of hosts examined.

Parasitic shell borers	Intensity	Prevalence
Boccardia accus	several	—
Boccardiella hamata	10 ind./host shell	—
Caobangia brandti	28 ind./host shell	—
Dipolydora alborectalis	1500 ind./host valve	—
Dipolydora bidentata	some ten ind./host shell	—
Polydora brevipalpa	some ten ind./host shell	—
Dipolydora caruncolata	1–2 ind./host shell	—
Dipolydora concharum	dense aggregates	—
Polydora ciliata	1->4 ind./host shell	—
Polydora glycymerica	8 ind./host shell	—
Dipolydora trilobata	several hundreds/host shell	—
Polydora vulgaris	—	4.68%–41.2%
Pseudopotamilla reniformis	30 ind./host shell	—
Parasitic borers of organisms other than molluscs		
Dipolydora armata	6–50 ind. per cm^3	—
Polydora hoplura	1 ind. per cm^2	—
Polydora wobberi	10–16 ind./host colony	
Commensal borers		
Dipolydora commensalis	1–10–17 ind./host	36.6% (508)–70.9% (158)
Polydora glycymerica	8 ind./host	—
Dipolydora rogeri	10–18 ind./host cm^3	100% (3)

tial menace to the scallop. The very small specimens of *Pseudofabricia* sp. form cylindrical burrows at the outer part of the abalone shell's edge. As the result of their activity, the shell becomes very weak and porous, which prevents the abalone from enlarging its shell horizontally. Even a heavy infestation by *Pseudofabricia* sp. is generally not life-threatening for the host. It is, however, a very serious problem for the commercial exploitation of abalones, significantly decreasing their growth rate and hence causing additional costs and reducing profits.

The species of the freshwater caobangid polychaetes (reviewed by Jones 1974), which are currently included within the family Sabellidae (Fitzhugh 1989, 1991, Rouse & Fitzhugh 1994, Fauchald & Rouse (1997) are exclusive inhabitants of mollusc shells (Table 9). Little is known of the life-history of the *Caobangia* species since most of them are known only from dried specimens in museums. The worms bore into shells of both gastropods and bivalve molluscs in Asian rivers. The adult burrows are restricted to the collumelar area of the snails (i.e. the thicker calcareous region of the shell), although encapsulated metamorphosing larvae were encountered all over the shell. The bore holes, mainly located in the apex of the snail shell, weaken this portion, which causes a truncated profile (Figs 23A, 23B). The burrows, lined with a thin secreted membrane, never connect with one another and are teardrop-shaped. In the case of boring bivalve shells, the worms were distributed throughout the available surface.

Cirratulids of the genus *Dodecaceria* are well known boring polychaetes (Table 9). Although some of them may be pests in bivalves of commercial value such as *Placopecten magellanicus*, they are not particularly specific (or, perhaps, not specific at all). They bore into virtually any kind of calcareous substratum (e.g. live or dead mollusc or barnacle shells, encrusting coralline algae) without showing any apparent preference for a given species. Moreover, some species of this genus cannot be considered as strict borers. *Dodecaceria fewkesi*, for example, forms its own free tubes, whereas *D. con-*

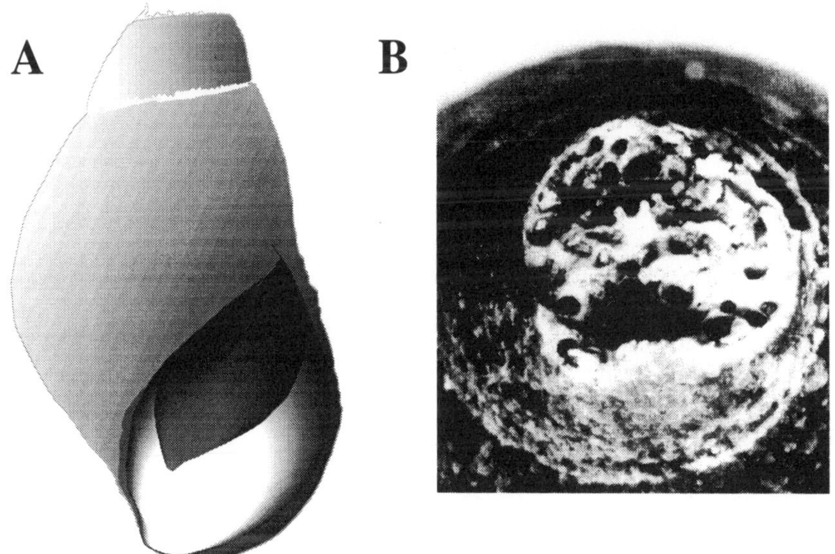

Figure 23 *Caobangia brandti* bore holes in *Brotia pseudosulcospira* (redrawn from Jones 1974). A, Scheme of the shell with truncated apex. B, Detail of the bore holes on the truncated apex.

charum also occurs as a cryptofaunal nestler (i.e. it is commonly found between the internal ribs of dead barnacle shells in *Laminaria* holdfasts and occurs in protected cracks and crevices in shells). In fact, Evans (1969) found North American populations of *Dodecaceria concharum* to be secondary borers (i.e. occupying and adapting vacant burrows of primary borers, but unable to make a burrow of their own), and Gibson & Clark (1976: 659) found that "in the laboratory ... adults or asexually produced individuals of *D. caulleryi* (=*D. concharum*) fail to make burrows in pieces of smooth rock or smooth *Lithothamnion*. Small holes made in the surface with a needle were sometimes temporarily occupied but a definitive burrow was never made in them". Although *Dodecaceria concharum* may seem able to build a tube of sorts, it appears that it more often secretes just a smooth calcium lining in already built burrows. In the laboratory, worms in a dish with detritus will make a kind of tube and then secrete a calcium layer sticking it together and corresponding to the lining of the tube (Gibson & Clark 1976, M. E. Petersen pers. comm.). In fact, the impact of borers or nestlers with some boring ability such as the species of *Dodecaceria*, would probably be higher on molluscs with exposed shells (e.g. scallops or oysters) than on those that are normally buried in the substratum (e.g. *Arctica islandica*). In Danish waters, *Dodecaceria concharum* occurs in shells of *Arctica islandica* previously excavated by the sponges of the genus *Cliona*. Usually, these are shells of dead molluscs, but *Dodecaceria concharum* also occurs in live ones if they are not completely buried in the sediment and if they have been attacked by the sponge first (M. E. Petersen & O. Tendal pers. comm.). However, this association has not been attributed to a worm–sponge symbiosis, but to the fact that older shells partly decalcified by the sponge may be easier to excavate than fresh ones. The behaviour of some *Dodecaceria* species as "secondary borers" may also help to explain their occurrence on sponge-excavated calcareous substrata. In the Mediterranean, specimens of *Dodecaceria* sp. have also been observed inhabiting calcareous substrata (i.e. encrusting coralline algae, oysters) previously excavated by *Cliona viridis* (D. Martin & D. Rosell pers. obs.).

Commensal boring polychaetes

One of the best known commensal borers is the spionid *Dipolydora commensalis*, which bores into snail shells occupied by numerous species of hermit-crabs (Table 9). Studies on its boring activity revealed that larvae mainly settle on the columella and on the inner surface of the last convolution of the shells, near the outer edge (Radashevsky 1989). Although it was initially suggested that growing worms may relocate themselves from the outer edge to the columella, those that do not settle on this last site will actually perish (V. Radashevsky pers. comm.). As a consequence, all tube openings of *D. commensalis* are located along the columellar side (near the inner lip) of the opening of the shell occupied by the host hermit crabs. The worms excavate shallow channels along the columella, and some tubes can be found in the umbilicus of the shell. The tubes of large gravid females may even reach the top of the columella. The tubes are coiled within the sediment that fills the apex of the infested shell. The worm tubes are covered by a thin arched calcareous "roof" covering the galleries on the inner surface of the hermit host's shell (Radashevsky 1989). The worm constructs the roofs above the excavated depression from redeposited calcareous material

(V. Radashevsky pers. comm.). Each host shell is usually inhabited by one large female and several smaller males and juveniles, whose tube openings are characteristically non-visible externally. In addition, simultaneous hermaphroditism has also been reported for *D. commensalis* and, in the case of death of the female, the remaining largest individual becomes a female (Radashevsky 1989). The worms have been observed feeding either by removing suspended particles from the branchial (respiratory) current of the hermit crabs or by removing particles attached to the setae on the legs of the hermit crabs (Dauer 1991). When feeding as suspension feeders, worms held the frontal surface of their palps directly in the branchial current of the crab, and constantly contacted the legs of its host with the abfrontal and lateral surfaces of the palps (which are very short).

Polydora glycymerica may be also considered as a specialized commensal borer (Radashevsky 1993). The inner walls of the long and wide U-shaped burrow are lined with silt, its two openings being located at the posterior end of the shell in the region of the mollusc siphons. This position is closely related to the feeding mode: projecting out of the burrow, the worms expose their palps and catch suspended particles from the water current produced by the mollusc. Similar behaviour may be assumed for *Boccardia accus*, whose burrows are also U-shaped and usually follow the curve of the host's shell growth lines. The borings are not lined with sand apart from a sand-grain partition at the short, U-shaped external chimneys. As the host cockle lies buried just beneath the sediment surface, the worm extends out through the external chimneys around the bivalve siphons (Read 1975). This is the easiest way to contact with the water column just over the sediment, but also the closeness of the siphons may allow the worm to feed on the water current produced by the host.

A particularly curious relationship occurs between *Dipolydora rogeri* and the boring sponge *Cliona viridis* (Martin 1996). Both organisms bore into calcareous algae, but the sponge may develop at different growth stages, progressively over growing the calcareous excavated substratum (Rosell & Uriz 1991). At the papillate stage of growth, only the papillae of the sponge protrude from the calcareous substratum and the tops of the polychaete tubes were observed protruding from the surface of the algae. The external part of the polydorid tube was formed by fine sediment grains embedded in an organic matrix, whereas the linear internal parts were only formed by the organic matrix. Internal tubes passed through the calcareous algae and the sponge body (choanosome) until an appropriate portion of an aquiferous canal of the sponge was reached. Then, the tubes disappeared, being replaced by the sponge canals (Fig. 24A,B). In encrusting and massive sponge specimens with almost no trace of visible calcareous substratum within the sponge surface, the anterior ends of the polychaete tubes were also observed directly protruding from the sponge surface. The polychaetes tend to place the openings of their tubes near the inhalant papillae of the sponge. Thus, their feeding activity can be favoured by the inhalant flow originated by the filtering activity of the host. Moreover, the sponge offers physical protection to the worm when the calcareous algae have completely disappeared (i.e. massive stage). The ability of the polychaetes to manipulate relatively large particles (either to feed or to build its tubes) may favour the filtering activity of the sponge by cleaning the area around the inhalant papillae, thus preventing the collapse of their orifices. Moreover, the polychaetes do not disturb the normal progress of the host through successive growth stages and the aquiferous canal occupied by the polydorid does not collapse. If the canal belongs to the exhalant system, water flow would contribute to the discharge of worm faeces,

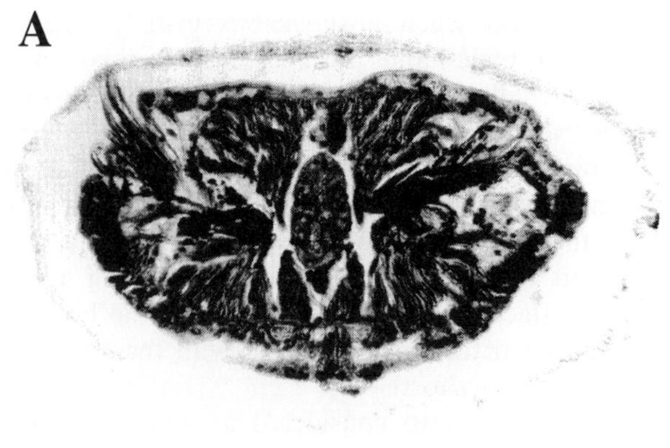

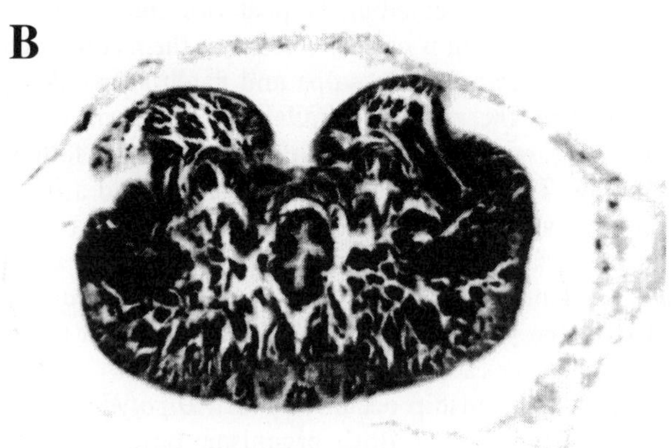

Figure 24 Sections of an anterior part of the *Dipolydora rogeri* body inside the calcareous algae excavated by the host sponge *Cliona viridis* (redrawn from Martin 1996). A, Section through a segment from the postbranchial region; B, section through branchial segment.

while faeces could be used as food by the sponge if the canal belongs to the inhalant system. In light of the above, the association between *Dipolydora rogeri* and *Cliona viridis* was tentatively considered as a mutualism (Martin 1996).

Specificity in the symbiotic relationships involving boring polychaetes

The analysis of the available data (Tables 8, p. 306 and 9, p. 313) on the specificity of the relationships between boring polychaetes and their hosts seems to display a generic monoxenous pattern. About 37% of the polychaetes infest a single host species, this figure becoming 63% if borers infesting two or three host species are included. It must be pointed out, however, that some species are reported on 10 to 19 different hosts and are, thus, clearly polyxenous. On the other hand, studies enabling the exact degree of specificity to be assessed are virtually lacking, although stimulating (but inconclusive) hypotheses may be found throughout the literature. It should be taken into account that more than half (53%) parasitic boring polychaetes are facultative symbionts, boring also into non-living substrata (Tables 8 and 9). Often, the appraisal of their interactions is tentative, as in the case of defining caobangids a parasites. There are no studies on the possible costs to the hosts arising from the infestation of these worms (other than the weakening of the shell); additionally, only one species of *Caobangia* was reported on a single host (probably the single specimen known), while the remaining were reported on two to nine different hosts (Table 9). In fact, in the rivers, snails and bivalves are the most common and, often, the only available sources of calcareous substrata for these worms. However, the caobangids have also been found boring into empty shells of dead molluscs. Although there are no real observations, it was assumed that they might bore into non-living substrata such as calcareous limestone outcrops that, it was though, could serve as populational reservoirs allowing the infestation of new molluscs.

Among "parasitic" *Polydora* species, there are a few observations suggesting a certain degree of specificity in the relationships with their hosts. Ramberg & Schram (1982) observed that the shells of the gastropod *Littorina littorea* smaller than 10 mm were never infested by *Polydora ciliata.* As the gastropod host did not become mature until it reached this size, it was suggested that larvae of *P. ciliata* may be guided to the snail by some substance that the snails secrete into the water. However, this was not supported by data. The existence of chemical cues in the interaction between the polydorids infesting bryozoans was suggested by Laubier (1959) on the basis of two facts: (a) the worm seems to control the height of the calcareous formations induced in the host colony, this height being always the same (on average); (b) even in colonies where the polydorids cannot be currently found, the calcareous formations remain open, the bryozoan apparently being unable to close them. Finally, the special configuration of the prostomium (markedly projected forward), the presence of interbranchial septa and the reduced pygidium, led Light (1970b) to suggest that *P. wobberi* could be a highly specialized parasitic borer.

Within the commensal boring polychaetes, little is known about the specificity of the relationships. *Dipolydora commensalis*, for example, is known to be associated with as many as nine different species. This may habitually be related to a lack of specificity. However, the worms seem to be highly specialized, as all these host species are hermit crabs (Table 9). Moreover, adult *D. commensalis* have never been found free-living and, although it was demonstrated that the species can be experimentally induced to burrow in shells that can never be occupied by hermit crabs (i.e. bivalve shells) this has never been reported in natural conditions. Conversely, the association of *Polydora glycymerica* and the burrowing bivalve *Glycymeris yessoensis* may be considered as a rare example of a high degree of specificity within commensal boring polydorids. Despite the intensive research carried out on this species, it was always found in the shells of *G. yessoensis*, with the exception of a very few localized cases where it was boring into the shell of another clam, *Anadara broughtoni* (Radashevsky 1989).

Food availability, sheltering or avoidance of stagnation were suggested as factors cementing the relationship between *Dipolydora commensalis* and the host hermit crabs (Hatfield 1965). When the host leaves an infested shell, the worms then have nutrient granules in the stomach wall to ensure their survival. Although these absences are necessarily temporary, they imply that the shell could be re-occupied by a different host individual or even by a different hermit crab species. Since the polydorids are, in principle, sedentary, and their feeding mode always incurs the added risk of being attacked by the host, we may assume that the crabs are able to recognize the worms as associates, such as in other polychaetes commensal with hermit crabs (e.g. *Nereis fucata*, Gilpin-Brown 1969). There are no studies on this particular aspect of the relationship, but some very interesting questions arise from the possibility of a chemically mediated recognition. Are the possible hosts "genetically" adapted to recognize the presence of the symbionts? Have the symbionts developed behavioural adaptations to solve the problem of host exchange (e.g. a delay in the start of feeding activity at the beginning of each new host occupancy, till the recognition of the commensal and the "permission" to feed)? Is the behaviour of the symbiont the same, independently of the re-occupation of the shell by the same or a different host species? Additional questions are posed by the fact that shells covered by the hydroid *Hydractinia echinata* were more heavily infested by *Dipolydora commensalis* (Andrews 1891a,b). May the hydroid be a third associated symbiont? How could the presence of the hydroid favour that of the worm? All these questions remain open, making the symbiotic relationships involving *D. commensalis* a very interesting subject for further research.

The relationship between *D. rogeri* and its host sponge *Cliona viridis* was considered endemic to the NW Mediterranean, as it was only known from the type locality (Martin 1996). *Cliona viridis* is a widely distributed species (Rosell & Uriz 1991, Rosell 1996) but the presence of its associated worm had not been previously reported. On the other hand, *Dipolydora armata*, the species most closely resembling *D. rogeri*, is a cosmopolitan polydorid that was reported to bore in calcareous algae from the same area in the NW Mediterranean. These algae were, however, non-excavated by *Cliona viridis* (Martin 1987). The behaviour of *Polydora* or *Dipolydora rogeri* inhabiting calcareous algae excavated by papillate specimens of *Cliona viridis* appears not to differ greatly from that of other coralline-boring species of *Dipolydora*. The main difference, and indeed that characterizing the association and supporting its specificity, is that the symbiotic association is still maintained when the sponge progresses to the encrusting and massive stages, with almost no trace of visible calcareous matrix. The association between *D. rogeri* and *Cliona viridis* has only been considered from a descriptive point of view, based on dead specimens. Further morphological and experimental studies are required before its exact nature may be assessed. Morphological aspects that might be addressed are, for example, both the cellular organization and the aquiferous system structure of the sponge around the worm tubes. Experiments could also usefully analyze the substratum preferences (excavated calcareous algae versus others) during larval settlement of polychaetes, or assess the possible differences in growth rates (from papillate to massive stages) of infested and non-infested sponges.

The way forward

Over the last forty years, the information on symbiotic polychaetes has grown enormously. This review has attempted to give an overview of the current state of knowledge. What is clear from this review is that the relationships involving symbiotic polychaetes are often remarkably complex and this may be true no matter the approach we use to analyze them (i.e. taxonomical, biological, physiological, ecological, behavioural). Our understanding of the interactions between the symbiotic polychaetes and their hosts is still pitifully inadequate. In this final section, therefore, we shall conclude by suggesting avenues of further research which, in the next few years, could help to redress this shortfall.

One of the most troublesome facts dealing with many symbiotic polychaetes is their rarity. We cannot change this and, unfortunately, this still leaves the acquisition of new information on many species (either already known or not yet described) to chance. This may be the case of the parasitic oenonids, which are known, for the most part, only from the original descriptions. In fact, there are many interesting hypotheses arising from the comparison of all trends that characterize each known relationship involving the parasitic oenonids. These hypotheses are necessarily open and cannot be convincingly refuted or validated if no new data on the relationships are reported. However, additional information, other than a single mention, may provide relevant data. The host-specificity (i.e. it always infests syllids) characterizing *Labrorostratus parasiticus* may be a good example.

The rarity of parasites also makes difficult the evaluation of their role in natural communities. We may expect, however, that changes in the environmental conditions affecting a given community would lead to an increase in the number of parasites, so that they would significantly affect the natural "economy" of biocoenosis. In fact, symbionts have been often regarded as "curiosities of the nature", but the few known studies on the biology and behavioural trends of some target species clearly demonstrated their significance in the economy of marine communities (e.g. cleaners such as *Histriobdella homari* or *Arctonoe vittata*, defenders such as *Arctonoe vittata* or *Spirobranchus giganteus*). A different concept of "economy" comes from an anthropomorphic point of view. In this case, the species boring on shells of commercially exploited organisms may play a relevant role. Although these species are common in "healthy" cultures, the possibility of unpredictable outbreaks of their number, followed by damage to commercial farms, might not be excluded (e.g. the undescribed boring sabellid in Oakes & Fields 1996). In addition, the impact of borers on their host populations may significantly restrict the number of the latter, subsequently leading to community perturbations.

Most symbiotic polychaetes are known from dead specimens and their quality of symbionts has been either proposed after directly finding them on their respective hosts or inferred from some widely accepted "symbiotic attributes". As a consequence, we often know little about the nature of most associations. Although, in practice, it is difficult to obtain unequivocal demonstrations of all factors affecting a given interaction, the few known examples of symbiotic associations whose nature has been studied in depth revealed a more complex network of relationships than expected, these associations being apparently more bi-directional than unidirectional (e.g. Dimock & Dimock 1969, Wagner et al. 1979, Britayev 1991, Radashevsky 1993, Martin 1996). A significant step forward would be, therefore, to look at the nature of the associations on the basis of direct observations of living organisms (i.e. *in situ* or under experimentally controlled situations). This approach may provide information, for example, on host–symbiont colour matching. In fact, the source of this colour matching is itself a matter of controversy. Whether a symbiont acquires its host's colour by feeding on the host or as a response to any stimuli coming from the host will substantially contribute to decision on the exact nature of a given association. A closely related matter of study, which is also relevant to assessing the nature of a given association, may be the type of trophic relationship established between a host and its symbiont(s). We also point out that, often, a symbiont is not alone on its host. Thus, a very interesting direction for further research will be to study the interactions (i.e. competition, predator–prey) between different symbionts (i.e. from the same or from different species) co-inhabiting on the same host individual (i.e. the association as a "microcosm"). A major problem may be that, often, it is difficult to obtain enough living partners to undertake experimental essays, but also, that the experimentally observed trends must be completed with the inference of their functional role in natural symbiotic populations.

Today, biologists have a number of new and interesting tools at their disposal and new and exciting techniques are being developed continually (in particular, we are all aware of the potential of electron microscopy or molecular biology). It is certain that such tools and techniques will be employed to great benefit in the area of polychaete symbiotic relationships. These approaches should not, however, be pursued to the exclusion of more established techniques, which still have a valuable contribution to make. Faunistic and systematic studies based on classical morphological characters are still fundamental in localizing and describing new associations. The elegant review on symbiotic polynoids by Pettibone (1993) may be a good example of this. Moreover, it should be taken into account that remote areas, such as bathyal bottoms (e.g. Kirkegaard & Billet 1980, Martín et al. 1992, Emson et al. 1993), cold seep-sites (e.g. Miura 1988, Miura & Laubier 1990) or hydrothermal vents (e.g. Pettibone 1984a,b, Miura & Hashimoto 1991, Miura & Ohta 1991) are becoming more and more accessible to scientists. The limiting conditions of these environments are probably connected with the diversity of new symbiotic associations found there. The functionality of these associations, as well as the development of both host and symbiont life cycles, are interesting fields still open to further research.

Although the polychaetes are probably one of the most widely studied groups of marine benthic organisms, the basic information on descriptive ecology (e.g. spatial distribution, bathymetric distribution, seasonality, population structure and dynamics) is still lacking for most known symbiotic polychaetes and, consequently, for the ecological unit formed by the host–symbiont partnership. Symbiotic associations involving polychaetes may serve as useful models to study biological interactions among marine organisms. The associations could be analyzed as elementary "biocoenosis", which include a reduced number of species and relationships. In that way, the biotic interactions (intra- and interspecific) could be studied and analyzed with less expense. Moreover, symbiotic associations are "natural" sample units to study distribution patterns (Palmer 1968). As all (or, at least, most) commensals are found on the hosts, each host is the basic habitat unit for a given commensal. Therefore, changes or variations in the commensal's distribution are likely to be of ecological interest. The symbionts may serve as a convenient model of distribution, it then being possible to generalize on the basis of the main trends described from the association.

Finally, despite several examples where chemically mediated host–symbiont relationships are know to exist because behavioural responses of both partners have been demonstrated, there are none where any details of their chemical nature are understood. In this respect, polychaete physiology lags behind that of almost all other invertebrate groups. The habitat restrictions imposed on a symbiont by its intimate association with its host is probably at the basis of most suggested roles of chemical cues in symbiotic associations. The first classical role is the host-recognition behaviour by adult symbiotic polychaetes, which is involved in the specificity of the response of the symbionts to their hosts. The few studies on this subject are now considered as classic papers but, in fact, the origin of the host-factor, its chemical nature, the way of reaching the symbiont and how the symbionts detect them, are basically unknown for most symbiotic associations involving polychaetes. Among other roles, chemical cues may be also involved in, (a) location of host during larval settlement, (b) sex-specific recognition when heterosexual pairs of symbionts inhabit the same host individual, (c) hosts' reproduction triggering symbionts' reproduction, and (d) symbiont recognition by hosts. Molecular tools such as the biochemical and chromatographic techniques, which are employed in purification and structural analysis of chemicals, have advanced significantly, so the identification of the substance (or substances) involved in the maintaining and regulation of symbiotic associations are today easier than ever. Chemical characterization of cues may be a way to clarify the different patterns of symbiont specificity (i.e. if the host shared or differed in the nature of the attractants). But we cannot forget the possibility of hidden sibling species-complex, being involved in non-specific relationships. Once the existence of this problem has been demonstrated the way is then open for analysis using a variety of modern techniques. Fine morphological studies using electron microscopy, molecular techniques such as protein electrophoresis and RNA/DNA sequencing may provide answers to questions on sibling species involved in symbiotic associations.

Certainly, we are not providing a complete list of shortcomings connected with the study of symbiotic associations involving polychaetes in this necessarily brief, conclusive section. A good final epilogue would probably be that, no matter the scientific field used to approach the symbiotic polychaetes, there are so many open questions that the efforts devoted will be always rewarded.

Acknowledgements

The authors wish to thank M. E. Petersen, M. H. Pettibone, J. R Hanley, J. Gil, M. J. Uriz, H. A. ten Hove, K. Fitzhugh, V. A. Sveshnikov, V. I. Radashevsky, A. V. Rzhavsky and G. San Martín, both for kindly providing advice on the taxa on which they are specialists or for help with bibliographical work. We also wish to thank G. Read and the Annelida electronic forum, through which we were able to discuss symbiotic polychaetes. D. Rosell carefully read the manuscript, highly contributing to improve its readability. E. Solohina kindly drew the originals of Figs 4E, 5B and 6B. M. Reiswig kindly provided the original of Fig. 17B. E. Cuñado and X. Cartes were able to locate (and to obtain offprints of) most papers we asked they for. The first author benefited from a research contract from the High Council of Scientific Research of Spain (CSIC) and has been partly financed during 1996 by a grant from the Associated European Laboratory of Marine Sciences. The study was also supported by Russian Foundation of Basic Researches (grant no 96-04-49077) and Russian Foundation "Biodiversity". This review is a contribution to the co-operation agreement between the Centre d'Estudis Avançats de Blanes from the CSIC of Spain and A. N, Severtzov of Animal Morphology and Ecology from the Russian Academy of Sciences.

References

Abelló, P. 1985. *Iphitime cuenoti* (Polychaeta: Iphitimidae), commensale des crabes en Mediterranée. *Rapports et Procès-Verbaux des Réunions Commission Internationale pour l'Exploration Scientifique de la Mer Mediterranée* **29**, 5 only.

Abelló, P., Sardá, R. & Masalles, D. 1988. Infestation of some Mediterranean brachyuran crabs by the polychaete *Iphitime cuenoti*. *Cahiers de Biologie Marine* **29**, 149–62.

Abelló, P., Villanueva, R. & Gili, J. M. 1990. Epibiosis in deep–sea crab populations as indicator of biological and behavioural characteristics of the host. *Journal of the Marine biological Association of the United Kingdom* **70**, 687–95.

Achari, G. P. K. 1974[1977]. On the polychaete *Gattyana deludens* Fauvel associated with the hermit crab *Diogenes diogenes* Herbst and *D. custos* Fabricius. *Journal of the Marine Biological Association of India* **16**, 839–43.

Addicot, J. F. 1984. Mutualistic interactions in population and community processes. In *A new ecology. Novel approaches to interactive systems*, P. W. Priece, et al. (eds). New York: Willey–Interscience, 438–53.

Ajeska, R. A. & Nybakken, J. 1976. On *Halosydna brevisetosa* commensal with *Melibe leonina*. *Veliger* **19**, 19–26.

Allen, E. J. 1915. Polychaeta of Plymouth and the South Devon coast including a list of the Archiannelida. *Journal of the Marine Biological Association of the United Kingdom* **10**, 592–646.

Allen, E. J. 1921. Regeneration and reproduction of the syllid *Procerastea*. *Philosophical Transactions of the Royal Society of London* **211**, 131–77.

Allen, M. J. 1952. An example of parasitism among polychaetes. *Nature* **169**, 197 only.

Al–Ogily, S. M. & Knight–Jones, E. W. 1981. *Circeis paguri* the spirorbid polychaete associated with the hermit crab *Eupagurus bernhardus*. *Journal of the Marine Biological Association of the United Kingdom* **61**, 821–6.

Alós, C. 1988. *Anélidos poliquetos de Cabo de Creus (Alt Empordà)*. Doctoral thesis, Universitat de Barcelona.

Alós, C. 1989. Adiciones à la fauna de Anélidos Poliquetos de la península Ibérica: familia Syllidae. *Cahiers de Biologie Marine* **30**, 329–37.

Alós, C., Campoy, A. & Pereira, F. 1982. Contribución al conocimiento de los Anélidos Poliquetos endobiontes de Esponjas. *Actas Simposium Ibérico de Estudio del Bentos Marino* **3**, 139–59.

Alvà, V. & Jangoux, M. 1989. Sur les symbioses entre polychètes et ophiures avec la description d'un nouveau cas de commensalisme. In *Echinodermes: actuels et fossiles. Actes du VIe Séminaire International sur les Echinodermes*, M. B. Régis et al. (eds). Aix–Marseille III: CERAM; Faculté des Sciences et Techniques de Saint–Jérôme; Foundation Océanographique Ricard, 185–92.

Amaral, A. C. Z. 1977. Um poliqueto endoparasita, *Labrorostratus prolificus* sp. nov. em Nereídeo. *Boletim do Instituto Oceanográfico, Sao Paulo* **26**, 285–92.

Amoureux, L., Josef, G. & O'Connor, B. 1980. Annélides Polychètes de l'éponge *Fasciospongia cavernosa* Schmidt. *Cahiers de Biologie Marine* **21**, 387–92.

Andrews, E. A. 1891a. A commensal annelid. *American Naturalist* **24**, 25–35.

Andrews, E. A. 1891b. Report upon the Annelida Polychaeta of Beaufort, North Carolina. *Proceedings of the United States National Museum* **14**, 277–302.

Arvanitidis, C. & Koukouras, A. 1994. Polychaete fauna associated with the coral *Cladocora caespitosa* (L.) in the eastern Mediterranean. *Mémoires du Muséum National d'Histoire Naturelle Série A (zoologie)* **162**, 347–53.

Arwidsson, I. 1932. *Calamyzas amphictenicola*, ein ektoparasitischer Verwandter der Sylliden. *Zoologiska Bidrag fran Uppsala* **14**, 153–218.

Augener, H. 1914. Polychaeta II, Sedentaria. *Fauna Südwest–Australia* **5**, 1–170.

Augener, H. 1922. Results of Dr E. Mjoberg's Swedish scientific expeditions to Australia 1910–1333. Polychaeten. *Kungliga Svenska Vétenskapsakademiens Handlings* **63**, 1–49.

Bacescu, M. 1971. Les Spongiaires: un des plus intéressants biotopes benthiques marins. *Rapports et Procés-Verbaux des Réunions Commission Internationale pour l'Exploration de la Scientifique Mer Mediterranée* **20**, 239–41.

Bailey–Brock, J. H. 1985. Polychaetes from Fijian coral reefs. *Pacific Science* **39**, 195–220.

Barel, C. D. N. & Kramers, P. G. N. 1977. A survey of the echinoderm associates of the North-East Atlantic area. *Zoologische Verhandelingen, Leiden*, **156**, 1–159.

Bartel, A. H. & Davenport, D. 1956. A technique for the investigation of chemical responses in aquatic animals. *British Journal of Animal Behaviour* **4**, 117–19.

Beauchamp, P. de & Zachs, I. 1913. Esqisse d'une monographie bionomique de la plage de Terrénès. *Mémoires de la Societé de Zoologie Francaise* **26**, 197–237.

Bellan, G. 1959. Repartition biogeographique et bionomique de quelques annélides polychètes de la Méditerranée et du proche océan. *Recueil des Travaux de la Station Marine d'Endoume* Bull. 17, Fasc. 29, 127–72.

Bellan, G. 1964. Contribution a l'étude systématique, bionomique et écologique des Annélides Polychètes de la Méditeranée. *Recueil des Travaux de la Station Marine d'Endoume* Bull. 33, Fasc. 49, 1–371.

Belloni, S. & Mori, M. 1985. Presenza di *Iphitime cuenoti* (Annelida: Iphitimidae) in Mar Ligure. *Nova Thalassia* **7**, 402 only.

Ben–Eliahu, M. N. 1975a. Polychaete cryptofauna from rims of similar intertidal vermetid reefs on the Mediterranean coast of Israel and in the Gulf of Elat: Sabellidae (Polychaeta Sedentaria). *Israel Journal of Zoology* **24**, 54–70.

Ben–Eliahu, M. N. 1975b. Polychaete cryptofauna from rims of similar intertidal vermetid reefs on the Mediterranean coast of Israel and in the Gulf of Elat: Nereidae (Polychaeta Errantia). *Israel Journal of Zoology* **24**, 177–91.

Ben–Eliahu, M. N. 1976. Polychaete cryptofauna from rims of similar intertidal vermetid reefs on the Mediterranean coast of Israel and in the Gulf of Elat: Serpulidae (Polychaeta Sedentaria). *Israel Journal of Zoology* **25**, 103–19.

Beneden, P. J. van 1853. Note sur une larve d'annelide d'une forme toute particuliere, rapportee avex doutes aux serpules. *Bulletin de l'Academie Royale de Bruxelles* **20**, 69–72.

Beneden, P. J. van 1858. Histoire naturelle d'un animal nouveau designe sous le nom d'*Histriobdella. Bulletin de l'Academie Royale de Bruxelles* **5**, 270–302.

Beneden, P. J. van 1869. Le commensalisme dans le Regne Animal. *Proceedings of the Royal Irish Academy* **28**, 621–36.

Berkeley, E. 1924. On a new case of commensalism between echinoderm and annelid. *Canadian Field Naturalist* **38**, 193 only.

Berkeley, E. & Berkeley, C. 1941. On a collection of Polychaeta from Southern California. *Bulletin of the Southern California Academy of Sciences* **40**, 16–60.

Berkeley, E. & Berkeley, C. 1942. North Pacific Polychaeta, chiefly from the west coast of Vancouver Island, Alaska and Bering Sea. *Canadian Journal of Science* **20**, 183–208.

Berkeley, E. & Barkeley, C. 1948. *Canadian Pacific fauna.* Toronto: Fisheries Research Board of Canada.

Berkeley, E. & Berkeley, C. 1950. Notes on Polychaeta from the coast of western Canada. IV. Polychaeta Sedentaria. *Annals and Magazine of Natural History* 12th Series **3**, 50–69.

Blake, J. A. 1969. Systematics and ecology of shell–boring polychaetes from New England. *American Zoologist* **9**, 813–20.

Blake, J. A. 1971. Revision of the genus *Polydora* from the East coast of North America. *Smithsonian Contributions to Zoology* **75**, 1–32.

Blake, J. A. 1980. *Polydora* and *Boccardia* species (Polychaeta: Spionidae) from western Mexico, chiefly from calcareous habitats. *Proceedings of the Biological Society of Washington* **93**, 947–62.

Blake, J. A. 1983. Polychaetes of the family Spionidae from South America, Antarctica, and adjacent seas and islands. In *Biology of the Antarctic Seas XIV. Antarctic Research Series. Vol. 39* (*3*), L. Kornicker (ed.). Washington DC: American Geophysical Union, 205–88.

Blake, J. A. 1990. A new genus and species of Polychaeta commensal with a deep–sea thyasirid clam. *Proceedings of the Biological Society of Washington* **103**, 681–6.

Blake, J. A. 1993. New genera and species of deep–sea polychaetes of the family Nautiliniellidae from the Gulf of Mexico and the Eastern Pacific. *Proceedings of the Biological Society of Washington* **106**, 147–57.

Blake, J. A. 1994. Introduction to the Annelida. In *Taxonomic atlas of the benthic fauna of the Santa Maria Basin and Western Santa Barbara Channel. The Annelida, Part 1*, J. A. Blake et al. (eds). Santa Barbara, California: Santa Barbara Museum of Natural History, 39–113.

Blake, J. A. 1996. Family Spionidae Grube, 1850. Including a review of the genera and species from california and a revision of the genus *Polydora* Bosc, 1802. In *Taxonomic atlas of the benthic fauna of the Santa Maria Basin and Western Santa Barbara Channel. The Annelida, Part 3*, J. A. Blake et al. (eds). Santa Barbara, California: Santa Barbara Museum of Natural History, 81–224.

Blake, J. A. & Evans, J. W. 1973. *Polydora* and related genera as borers in mollusc shells and other calcareous substrates. *Veliger* **15**, 235–49.

Blake, J. A. & Kudenov, J. D. 1978. The Spionidae (Polychaeta) from southeastern Australia and adjacent areas with a revision of the genera. *Memoirs of the National Museum of Victoria* **39**, 171–280.

Blake, J. A. & Woodwick, K. H. 1971. New species of *Polydora* (Polychaeta: Spionidae) from the coast of

California. *Bulletin of the Southern California Academy of Sciences* **70**, 72–9.

Boucher, D. H., James, S. & Keeler, K.H. 1982. The ecology of mutualism. *Annual Review of Ecology and Systematics* **13**, 315–47.

Brettey, J. & Campbell, A. 1985. Occurrence of *Histriobdella homari* (Annelida: Polychaeta) on the American lobster in the Canadian maritimes. *Canadian Journal of Zoology* **63**, 392–5.

Brightwell, L. R. 1951. Some experiments with the common hermit crab (*Eupagurus bernhardus* Linn.) and transparent univalve shells. *Proceedings of the Zoological Society of London* **121**, 279–83.

Britayev, T. A. 1978. The role of chemical cues in *Arctonoe vittata* behaviour. In *All Union Symposium on the behaviour of aquatic invertebrates.* F. D. Morduhay-Boltovskoy (ed.). Borok: Institute of Biology of Inland Waters. (In Russian).

Britayev, T. A. 1981. Two new species of commensal polynoids (Polychaeta: Polynoidae) and bibliography on polychaetes, symbionts of Coelenterata. *Zoologicheski Zhurnal* **60**, 817–24. (In Russian).

Britayev, T. A. 1989. The symbiotic polychaetes: morphology, ecology and distribution. In *Symbiosis among marine animals,* V. A. Sveshnikov (ed.). Moscow: A. N. Severtzov Institute, Russian Academy of Sciences, 60–74. (In Russian).

Britayev, T. A. 1991. Life cycle of the symbiotic scaleworm *Arctonoe vittata* (Polychaeta: Polynoidae). *Ophelia* (Suppl.) **5**, 305–12.

Britayev, T. A. 1993. *Pilargis berkeleyae* (Polychaeta, Pilargidae) as a commensal of a sedentary polychaete *Chaetopterus cautus* (Chaetopteridae). *Zoologicheski Zhurnal* **72**, 147–51. (In Russian).

Britayev, T. A. & Fahrutdinov, R. R. 1994. Pontoniin shrimps associated with molluscs at the coast of South Vietnam. In *Hydrobionts of South Vietnam,* D. S. Pavlov & J. N. Sbikin (eds). Moscow: Nauka, 128–45. (In Russian).

Britayev, T. A., Ivashchenko, N. I. & Litvinov, E. G. 1977. Peculiarities of the commensal complex in the polychaete *Arctonoe vittata. Biologiya Morya* **1978**, 76–8. (In Russian).

Britayev, T. A. & Memmy, M. P. 1989. Symbiotic association between the pelagic polychaete *Hipponoe gaudichaudi* (Polychaeta, Amphinomidae) and goose barnacles (Cirripedia, Lepadomorpha). *Zoologicheski Zhurnal* **68**, 30–35. (In Russian).

Britayev, T. A. & Smurov, A. V. 1985. The structure of a population of symbionts and related biological features, *Arctonoe vittata* (Polychaeta, Polynoidae) taken as an example. *Zhurnal Obschey Biologii* **46**, 355–66. (In Russian).

Britayev, T. A., Smurov, A. V., Adrianov, A. V., Bazhin, A. G. & Rhzavsky, A. V. 1989. Ecology of symbiotic polychaete *Arctonoe vittata* according to the peculiarity of starfish *Asterias ratbunae* ecology. In *Symbiosis among marine animals,* V. A. Sveshnikov (ed.). Moscow: A. N. Severtzov Institute, Russian Academy of Sciences, 102–27. (In Russian).

Britayev, T. A. & Zamishliak, E. A. 1996. Association of the commensal scaleworm *Gastrolepidia clavigera* (Polychaeta: Polynoidae) with holothurians near the coast of South Vietnam. *Ophelia* **45**, 175–90.

Buzinskaja, G. N. 1967. On the ecology of the polychaetous annelids of the Possjet Bay (the Sea of Japan). *Isseldovaniya Fauni Moorei* **5**, 78–124. (In Russian).

Buzinskaja, G. N. 1971. New and interesting species of polychaete worms of the Possjet Bay of the Sea of Japan. *Isseldovaniya Fauni Moorei* **8**, 123–45. (In Russian).

Buzinskaja, G. N., Obut, A. M. & Potin, V. V. 1980. Erantiate polychaetes (Polychaeta, Errantia) of the coral reefs and islands of the Indian and Pacific oceans (based on the material of the "KALLISTO" expedition of 1974, 1975). In *Biology of Coral Reefs.* Moscow: USSR Academy of Sciences; Institute of Marine Biology, Far East Science Centre, 225–56. (In Russian).

Campoy, A. 1982. Fauna de España. Fauna de Anélidos Poliquetos de la Península Ibérica. *Publicaciones de la Universidad de Navarra. Serie Zoológica* **7**, 1–400.

Cañete, J. I., Gallardo, V. A., Enríquez, S. & Baltazar, M. 1993. *Harmothoe brevipalpa* Begström, 1916 (Polynoidae), poliqueto asociado a los tubos de dos especies de poliquetos frente a la Bahía de Concepción, Chile. *Boletín de la Sociedad de Biología de Concepción, Chile* **64**, 43–6.

Cannon, L. R. G. & Jennings, J. B. 1987. Occurrence and nutritional relationships of four ectosymbiotes of the freshwater crayfishes *Cherax dispar* Riek and *Cherax punctatus* Clark (Crustacea: Decapoda) in Queensland. *Australian Journal of Marine and Freshwater Research* **38**, 419–27.

Carrasco, F. D. 1974. Spionidae (Polychaeta) provenientes de la Bahía de Concepción y lugares adyacentes. *Boletín de la Sociedad de Biología de Concepción, Chile* **48**, 185–201.

Carthy, J. D. 1958. *An introduction to the behaviour in invertebrates.* London: George Allen & Unwin.

Carton, Y. 1966a. Etude du comportement et de la specificite parasitaire de *Sabelliphilus sarsi* Claparède, copepode parasite de *Spirographis spallanzani* Viviani. *Bulletin du Museum National d'Histoire Naturelle*

Paris **37**, 807–17.

Carton, Y. 1966b. Specificite parasitaire de *Sabelliphilus sarsi*, parasite de *Spirographis spallanzani*. 1. Intérêt de son étude dans la taxonomie de divers Sabellidae. *Archives de Zoologie Expérimentale et Générale* **107**, 428–42.

Carton, Y. 1967. Specificite parasitaire de *Sabelliphilus sarsi*, parasite de *Spirographis spallanzani*. 2. Reactions histologiques des hôtes lors des contaminations expèrimentales. *Archives de Zoologie Expérimentale et Générale* **108**, 387–411.

Carton, Y. 1968a. Specificite parasitaire de *Sabelliphilus sarsi*, parasite de *Spirographis spallanzani*. 3. Mise en evidence d'une attraction biochimique du Copepode par l'Annelide. *Archives de Zoologie Expérimentale et Générale* **109**, 123–44.

Carton, Y. 1968b. Specificite parasitaire de *Sabelliphilus sarsi*, parasite de *Spirographis spallanzani*. 4. Etude biochimique des hotes et de leurs secretions. *Archives de Zoologie Expérimentale et Générale* **109**, 269–86.

Caullery, M. 1914. *Labidognathus parasiticus* n. g. n. sp. Cas nouveau d'endoparasitisme évolutif chez les Euniciens. *Compte Rendu des Séances de la Société de Biologie, Paris* **77**, 490–93.

Caullery, M. 1925. Schizogenese et schizogamie de *Procerastea halleziana* Malaquin. Parasitism de ce Syllidien sur les Tubulaires. *Bulletin de la Société Zoologique de France* **50**, 204–8.

Caullery, M. & Mesnil, F. 1916. Notes biologiques sur les mares à *Lithothamnion* de la Hague. 1. Presentation d'un *Labrorostratus parasiticus* S. J., parasite interne d'*Odonthosyllis ctenostoma* Clap. *Bulletin de la Société Zoologique de France* **40**, 160–61.

Cazaux, C. 1968. Etude morphologique du developpement larvaire d'annelides polychetes (Basin d'Arcachon). I. Aphroditidae, Chrysopetalidae. *Archives de Zoologie Expérimentale et Générale* **109**, 477–543.

Clark, R. B. 1956. *Capitella capitata* as a commensal, with a bibliography of parasitism and commensalism in the polychetes. *Annals and Magazine of Natural History* 12th Series, **9**, 433–48.

Comely, A. & Ansell, A. D. 1988. Invertebrate associates with the sea-urchin *Echinus sculentus* L. from the Scottish west coast. *Ophelia* **28**, 111–37.

Comely, C. A. & Ansell, A. D. 1989. The occurrence of the eunicid polychaetes *Iphitime cuenoti* Fauvel and *I. paguri* Fage & Legendre in crabs from the Scottish west coast. *Ophelia* **31**, 59–76.

Cordero, E. H. 1927. Un nuevo Arquianélido, *Stratiodrilus platensis*, n. sp., que habita sobre *Aegla laevis* (Latr.). Nota preliminar. *Physis, Revista de la Sociedad Argentina de ciencias naturales* **8**, 574–8.

Cram, A. & Evans, S. M. 1980. Shell entry behaviour in the commensal ragworm *Nereis fucata*. *Marine Behaviour and Physiology* **7**, 57–64.

Cuenot, L. 1912. Contribution a la faune du Bassin d'Arcachon. Echinodermes. *Bulletin de la Station Biologique d'Arcachon* **14**, 17–1227.

Dales, R. P. 1957. Interrelations of organisms. A. Commensalism. In *Treatise on marine ecology and paleoecology*, J. W. Hedgpeth (ed.). *Memoirs of the Geological Society of America* No. 67, 391–412.

Dauer, D. M. 1974. Polychaete fauna associated with Gulf of Mexico sponges. *Florida Scientist* **36**, 192–6.

Dauer, D. M. 1991. Functional morphology and feeding behavior of *Polydora commensalis* (Polychaeta: Spionidae). *Ophelia* (Suppl.) **5**, 607–14.

Davenport, D. 1950. Studies in the physiology of commensalism. I. The polynoid genus *Arctonoë*. *Biological Bulletin* **98**, 81–93.

Davenport, D. 1953a. Studies in the physiology of commensalism. III. The polynoid genera *Acholoe, Gattyana*, and *Lepidastenia*. *Journal of the Marine Biological Association of the United Kingdom* **32**, 161–73.

Davenport, D. 1953b. Studies in the physiology of commensalism. IV. The polynoid genera *Polynoe, Lepidasthenia* and *Harmothoe*. *Journal of the Marine Biological Association of the United Kingdom* **32**, 273–88.

Davenport, D. 1954. Notes on the early stages of the commensal polynoid *Acholoe astericola* (delle Chiaje). *Journal of the Marine Biological Association of the United Kingdom* **33**, 123–7.

Davenport, D. 1955. Specificity and behaviour in symbiosis. *Quarterly Review of Biology* **30**, 29–46.

Davenport, D. 1966a. Echinoderms and the control of behavior in associations. In *Physiology of Echinodermata*, R. A. Boolootian (ed.). New York: Insterscience Publishers, 145–56.

Davenport, D. 1966b. The experimental analysis of behaviour in symbioses. In *Symbiosis. I. Associations of microorganisms, plants and marine organisms*, S. M. Henry (ed.). New York: Academic Press, 381–429.

Davenport, D., Camougis, G. & Hickok, J. F. 1960. Analysis of the behaviour of commensals in host-factor. I. A hesionid polychaete and pinnotherid crab. *Animal Behaviour* **8**, 209–18.

Davenport, D. & Hickok, J. F. 1951. Studies in the physiology of commensalism. 2. The polynoid genera *Arctonoe* and *Halosydna*. *Biological Bulletin* **100**, 71–83.

Davenport, D. & Hickok, J. F. 1957. Notes on the early stages of the facultative commensal *Podarke puget-*

tensis Johnson (Polychaeta, Hesionidae). *Annals and Magazine of Natural History* 12th Series **10**, 625–31.

Day, J. H. 1960. The polychaete fauna of South Africa. Part 5. Errant species dredged off Cape coasts. *Annals of the South African Museum* **45**, 261–373.

Day, J. H. 1962. Polychaeta from several localities in the western Indian Ocean. *Proceedings of the Zoological Society of London* **139**, 627–56.

Day, J. H. 1967. A monograph on the polychaetes of southern Africa. Part 1. Errantia. London: British Museum (Natural History).

Day, J. H. 1973. Polychaeta collected by U. D. Gaikwad at Ratnagiri, south of Bombay. *Zoological Journal of the Linnean Society* **52**, 337–61.

Dean, H. K. 1992. A new arabellid polychaete living in the mantle cavity of deep-sea wood boring bivalves (family Pholadidae). *Proceedings of the Biological Society of Washington* **105**, 224–32.

Desportes, I., Laubier, L. & Theodorides, J. 1977. Présence d'*Ophryotrocha geryonicola* (Esmark) (Polychète, Dorvilleidae) en Méditerranée Occcidentale. *Vie et Milieu* **27**, 131–3.

Devaney, D. M. 1967. An ectocommensal polynoid associated with Indo-pacific echinoderms, primarily ophiuroids. *Occasional Papers of the Bernice P. Bishop Museum* **23**, 287–304.

Devaney, D. M. 1974. Shallow water echinoderms from British Honduras, with a description of a new species of *Ophiocoma* (Ophiuroidea). *Bulletin of Marine Science* **24**, 122–64.

DeVantier, L. M., Reichelt, R. E. & Bradbury, R. H. 1985. Does *Spirobranchus giganteus* protect host *Porites* from predation by *Acanthaster planci*: predator pressure as a mechanism of coevolution? *Marine Biology* **32**, 307–10.

Dimock Jr, R. V. 1974. Intraspecific aggression and the distribution of a symbiotic polychaete on its host. In *Symbiosis in the sea*, W. B. Vernberg (ed.). Columbia: University of South Carolina Press, 29–44.

Dimock Jr, R. V. & Davenport, D. 1971. Behavioral specificity and the induction of host recognition in a symbiotic polychaete. *Biological Bulletin* **141**, 472–84.

Dimock Jr, R. V. & Dimock, J. G. 1969. A possible "defense" response in a commensal polychaete. *Veliger* **12**, 65–8.

Ditlevsen, H. 1917. Annelids. I. *Ingolf Report* **4**(4), 1–71, Plates 1–6.

Eckelbarger, K. J. 1978. Metamorphosis and settlement in the Sabellariidae. In *Settlement and metamorphosis of marine invertebrate larvae*, F. S. Chia & M. E. Rice (eds). New York: Elsevier, 127–44.

Eisig, H. 1906. *Ichthyotomus sanguinarius*, eine auf Aalen schmarotzende Annelide. *Fauna und Flora des Golfes von Neapel* **28**, 1–300.

Emerson, R. R. 1974. A new species of polychaetous annelid (Arabellidae) parasitic in *Diopatra ornata* (Onuphidae) from Southern California. *Bulletin of the Southern California Academy of Sciences* **73**, 1–4.

Emson, R. H., Young, C. M. & Paterson, G. L. J. 1993. A fire worm with a sheltered life: studies of *Benthoscolex cubanus* Hartman (Amphinomidae), an internal associate of the bathyal sea-urchin *Archeopneustes hystrix* (A. Agassiz, 1880). *Journal of Natural History* **27**, 1010–28.

Essenberg, C. 1918. The factors controlling the distribution of the Polynoidae of the Pacific coast of North America. *University of California Publications in Zoology* **18**, 171–238.

Estcourt, I. N. 1967. Burrowing polychaete worms from a New Zealand estuary. *Transactions of the Royal Society of New Zealand* **9**, 65–78.

Euders, H. E. 1905. Notes on the commensals found in the tube of *Chaetopterus pergamentaceus*. *American Naturalist* **34**, 1–3.

Evans, J. W. 1969. Borers in the shell of the sea scallop *Placopecten magellanicus*. *American Zoologist* **9**, 775–82.

Fage, L. 1936. Sur l'association d'un annélide polychète *Lumbriconereis flabellicola* n. sp. et d'un madrépore *Flabellum pavoninum distinctum* E. et H. *Comptes Rendus du XIIe Congrés International de Zoologie* **1**, 941–5.

Fage, L. & Legendre, R. 1925. Sur une Annélide Polychète (*Iphitime cuenoti* Fauv.) commensale des crabes. *Bulletin de la Societe Zoologique de France* **50**, 219–25.

Fage, L. & Legendre, R. 1933. Les anélides polychètes du genre *Iphitime*. A propos d'une espèce nouvelle commensale des pagures, *Iphitime paguri*. *Bulletin de la Societé Zoologique de France* **58**, 299–305.

Fauchald, K. 1970. Polychaetous annelids of the families Eunicidae, Lumbrineridae, Iphitimidae, Arabellidae, Lysaretidae and Dorvilleidae from western Mexico. *Allan Hancock Monographs in Marine Biology* No. 5, 1–335.

Fauchald, K. 1977. Polychaete worms, definitions and keys to the orders, families and genera. *Natural History Museum of Los Angeles County Science Service* **28**, 1–190.

Fauchald, K. 1992. A review of the genus *Eunice* (Polychaeta: Eunicidae) based upon type material. *Smith-*

sonian Contributions to Zoology No. 523, 1–422.

Fauchald, K. & Jumars, P. A. 1979. The diet of worms: a study of polychaete feeding guilds. *Oceanography and Marine Biology: an Annual Review* **17**, 193–284.

Fauchald, K. & Rouse, G. W. 1977. Polychaete systematics: past and present. *Zoologica Scripta* **26**, 71–138.

Fauvel, P. 1913. Quatrième note préliminaire sur les Polychètes provenant des campagnes de l'"*Hirondelle*" et de la "*Princesse-Alice*", ou déposées dans le Musée Océanografique de Monaco. *Bulletin de l'Institut Océanographique* (*Foundation Albert 1er, Prince de Monaco*) **270**, 1–80.

Fauvel, P. 1914. Un Eunicien énigmatique *Iphitime cuenoti* n. sp. *Archives de Zoologie Expérimentale et Générale* **53**, 34–7.

Fauvel, P. 1923. Faune de France. Polychètes Errantes. *Faune de France* **5**, 1–488.

Fauvel, P. 1959. Classe des Annélides Polychètes. Annelida Polychaeta (Grube, 1851). In *Traité de zoologie. Anatomie, systématique, biologie*, P. P. Grassé (ed.). Paris: Masson et Cie, 13–196.

Fitzhugh, K. 1989. A systematic revision of the Sabellidae-Caobangidae-Sabellongidae complex (Annelida: Polychaeta). *Bulletin of the American Museum of Natural History* **192**, 1–104.

Fitzhugh, K. 1991. Polychaete phylogenetics and the growth of scientific knowledge. *Ophelia* (Suppl.) **5**, 55–62.

Fox, R. S. & Ruppert, E. E. 1985. Shallow-water marine benthic macroinvertebrates of South Carolina, species identification, community composition and symbiotic associations. *The Belle W. Baruch Library in Marine Sciences* No. 14, 1–130.

Franzén, Å. 1982. Ultrastructure of the biflagellate spermatozoon of *Tomopteris helgolandica* Greef, 1879 (Annelida, Polychaeta). *Gamete Research* **6**, 29–37.

Frith, D. W. 1976. Animals associated with sponges at North Hayling, Hampshire. *Zoological Journal of the Linnean Society* **58**, 353–62.

Führ, I. M. 1971. A new histriobdellid on a marine isopod from South Africa. *South African Journal of Science* **67**, 325–6.

Gaikwad, U. D. 1993. Some ecological observations on commensal behaviour of polychaetes, *Loimia medusa* and *Lepidasthenia tubicola*. *Journal of Ecobiology* **5**, 143–5.

Gallardo, V. A. 1967. Polychaeta from the bay of Nha Trag, South Viet Nam. *Scientific Researches on the Marine Invertebrates of South China Sea and the Gulf of Thailand, 1959–1961. NAGA Report* **4**, 35–279.

Ganapati, P. & Radhakrishna, Y. 1962. Inquilinism between a new hesionid polychaete and a holothurian *Molpadia* sp. *Current Science* **31**, 382–3.

Gardiner, S. L. 1976. Errant polychaete annelids from North Carolina. *Journal of the Elisha Mitchell Scientific Society* **91**, 77–220.

Gaston, G. R. & Benner, D. A. 1981. On Dorvilleidae and Iphitimidae (Annelida: Polychaeta) with a redescription of *Eteonopsis geryonicola* and a new host record. *Proceedings of the Biological Society of Washington* **94**, 76–87.

George, J. D. & Hartmann-Schroeder, G. 1985. *Polychaetes: British Amphinomida, Spintherida and Eunicida. Synopses of the British fauna (New Series). Vol. 32.* London: Linean Society of London and Estuarine and Brackish-Water Sciences Association.

Gerber, H. S. & Stout, J. F. 1968. Sensory basis of the symbiotic relationship of *Arctonoe vittata* (Grube) (Polychaeta, Polynoidae) to the keyhole limpet *Diodora aspera*. *Physiological Zoology* **41**, 169–79.

Gibbs, P. E. 1969. Aspects of polychaete ecology with particular reference to commensalism. *Philosophical Transactions of the Royal Society of London* **255**, 443–58.

Gibbs, P. E. 1971. The polychaete fauna of the Solomon Islands. *Bulletin of the British Museum of Natural History* (*Zoology*) **21**, 101–211.

Gibson, P. H. & Clark, R. B. 1976. Reproduction of *Dodecaceria caulleryi* (Polychaeta: Cirratulidae). *Journal of the Marine Biological Association of the United Kingdon* **56**, 649–74.

Gilpin-Brown, J. B. 1969. Host-adoption in the commensal polychaete *Nereis fucata*. *Journal of the Marine Biological Association of the United Kingdom* **49**, 121–7.

Glasby, C. J. 1994. A new genus and species of polychaete *Bollandia antipathicola* (Nereidoidea, Syllidae), from black coral. *Proceedings of the Biological Society of Washington* **107**, 615–21.

Goerke, H. 1971. *Nereis fucata* (Polychaeta, Nereidae) als kommensale von *Eupagurus bernhardus* (Crustacea, Decapoda) Entwicklung einer Population und verhalten der Art. *Veröffentlichungen des Instituts für Meeresforschung in Bremerhaven* **13**, 79–81.

Goerke, H. 1977a. *Nereis fucata* (Nereidae) – Kommensalismus: Eindringen in das Gehause von *Eupagurus bernhardus*. Film E 1891 des IWF, Gottingen 1977. *Publikationen zu Wissenschaftlichen Filmen* **58**, 3–12.

Goerke, H. 1977b. *Nereis fucata* (Nereidae) – Kommensalismus: Nahrungserwerb. *Publikationen zu Wissenschaftlichen Filmen* **59**, 3–11.

Gravier, C. 1900. Sur le commensalisme de l'*Eunice harasii* Audouin et M.-Edwards et de l'*Ostrea edulis* L. *Bulletin du Museum National d'Histoire Naturelle, Paris* **6**, 415–17.

Gravier, C. 1905a. Sur un Polynoidien (*Lepidasthenia digueti* n. sp.) commensal d'un *Balanoglossus*. *Comptes Rendus de l'Academie des Sciences de Paris* **140**, 875–8.

Gravier, C. 1905b. Sur les genres *Lepidasthenia* Malmgren et *Lepidametria* Webster. *Bulletin du Museum National d'Histoire Naturelle, Paris* **11**, 181–4.

Gravier, C. 1918. Note sur une Actinie (*Thoractis*) et un Annélide Polychète (*Hermadion fauveli*) commensaux d'une éponge silicieuse (*Sarcostegia oculata*). *Bulletin de l'Institut Océanographique* No. 334, 1–20.

Guille, A. 1965. Contribution a l'étude de la systématique et l'écologie d'*Ophiotrix quinquemaculata* d. Ch. *Vie et Milieu* **15**, 243–308.

Haigler, S. A. 1969. Boring mechanism of *Polydora websteri* inhabiting *Crassostrea virginica*. *American Zoologist* **9**, 821–8.

Hamond, R. 1966. The Polychaeta of the coast of Norfolk. *Cahiers de Biologie Marine* **7**, 383–436.

Hanley, J. R. 1984. A new host and locality records of the commensal *Adyte crinoidicola* (Polychaeta, Polynoidae). *The Beagle, Records of the Northern Territory Museum of Arts and Sciences* **1**, 87–92.

Hanley, J. R. 1989. Revision of the scaleworm genera *Arctonoe* Chamberlin and *Gastrolepidia* Schmarda (Polychaeta, Polynoidae) with the erection of a new subfamily, Arctonoinae. *The Beagle, Records of the Northern Territory Museum of Arts and Sciences* **6**, 1–34.

Hanley, J. R. 1992. Checklist of scaleworms (Polychaeta: Polynoidae) from Hong Kong. In *The marine flora and fauna of Hong Kong and southern China. III. Proceedings of the Fourth International Marine Biological Workshop: the Marine Flora and Fauna of Hong Kong and Southern China, Hong Kong, 11–29 April 1989*, B. Morton (ed.). Hong Kong: Hong Kong University Press, 361–9.

Hanley, J. R. 1993. Scaleworms (Polychaeta: Polynoidae) of Rottnest Island, Western Australia. In *Proceedings of the Fifth International Marine Biological Workshop: the Marine Flora and Fauna of Rottnest Island, Western Australia*, F. E. Wells, et al. (eds). Perth: Western Australian Museum, 305–20.

Hanley, J. R. & Burke, M. 1988. A new genus and species of commensal scaleworm (Polychaeta, Polynoidae). *The Beagle, Records of the Northern Territory Museum of Arts and Sciences* **5**, 5–15.

Hanley, J. R. & Burke, M. 1989. A new genus and species of commensal scaleworm (Polychaeta, Polynoidae) from Broome, Western Australia. *The Beagle, Records of the Northern Territory Museum of Arts and Sciences* **6**, 97–103.

Hanley, J. R. & Burke, M. 1990. Scaleworms (Polychaeta: Polynoidae) of Albany, Western Australia. In *Proceedings of the Third International Marine Biological Workshop: the Marine Fauna of Albany, Western Australia*, F. E. Wells et al. (eds). Perth: Western Australia Museum, 203–36.

Hanley, J. R. & Burke, M. 1991a. A new genus and species of scaleworm (Polychaeta, Polynoidae) from the Cascade Plateau, Tasman Sea. *The Beagle, Records of the Northern Territory Museum of Arts and Sciences* **8**, 97–102.

Hanley, J. R. & Burke, M. 1991b. Polychaeta Polynoidae: scaleworms of the Chesterfield Islands and Fairway Reefs, Coral Sea. In *Résultats des Campaignes* MUSORSTOM, *Vol. 8*, A. Crosnier (ed.). *Mémoires du Museum National d'Histoire Naturelle* Série A **151**, 9–82.

Harrington, N. R. 1897. On nereids commensal with hermit crabs. *Transactions of the New York Academy of Science* **16**, 214–21.

Harrison, L. 1928. On the genus *Stratiodrilus* (Archiannelida: Histriobdellidae) with a description of a new species from Madagascar. *Records of the Australian Museum* **16**, 116–21.

Hartman, O. 1941. New species of polychaetous annelids from Southern California. *Allan Hancock Pacific Expedition* **7**, 159–71.

Hartman, O. 1942. The polychaetous Annelida. Report on the scientific results of the Atlantis Expedition to the West Indies under the joint auspices of the University of Havana and Harvard University. *Memorias de la Sociedad Cubana de Historia Natural* **16**, 89–104.

Hartman, O. 1948. The polychaetous annelids of Alaska. *Pacific Science* **2**, 1–58.

Hartman, O. 1951. The littoral marine annelids of the Gulf of Mexico. *Publications of the Institute of Marine Sciences* **2**, 7–124.

Hartman, O. 1952. *Iphitime* and *Ceratocephala* (Polychaetous annelids) from California. *Bulletin of the Southern California Academy of Sciences* **51**, 9–20.

Hartman, O. 1954. Marine annelids from the northern Marshall Islands. *United States Geological Survey Professional Papers* **260Q**, 615–44.

Hartman, O. 1966. Polychaetous annelids of the Hawaiian Islands. *Occasional Papers of the Bernice P. Bishop Museum* **23**, 163–252.

Hartman, O. 1968. *Atlas of the errantiate polychaetous annelids from California*. Los Angeles: Allan Hancock Foundation, University of Southern California.

Hartman, O. & Boss, K. J. 1965. *Antonbruunia viridis*, a new inquiline annelid with dwarf males, inhabiting a new species of pelecypod, *Lucinia fosteri* in the Mozambique channel. *Annals and Magazine of Natural History* 13th Series, **8**, 177–86.

Hartman, O. & Fauchald, K. 1971. Deep-water benthic polychaetous annelids off New England to Bermuda and other North Atlantic Areas. Part II. *Allan Hancock Monographs in Marine Biology* **6**, 1–327.

Hartmann-Schröder, G. 1960. Polychaeten aus dem Roten Meer. *Kieler Meeresforschungen* **16**, 69–125.

Hartmann-Schröder, G. 1971. Annelida, Borstenwürmer, Polychaeta. In *Tierwelt Deutschlands und der angrenzenden Meeresteile nach ihren Maermalen und nach ihrer Lebensweise*, M. Dahl & F. Peus (eds). Jena: Gustav Fischer Verlag, **58**, 1–594.

Hartmann-Schröder, G. 1978. Einige Sylliden-Arten (Polychaeta) von Hawaii und aus dem Karibischen Meer. *Mitteilungen aus dem Hamburgischen Zoologischen Museum und Institut* **75**, 49–61.

Hartmann-Schröder, G. 1981. *Hololepidella alba* n. sp. (Polynoidae: Polychaeta), a commensal with the crown-of-thorns starfish from Mactan, Cebu. *Philippine Scientist* **18**, 10–14.

Hartmann-Schröder, G. 1983. Zur Kenntnis einiger Foraminiferengehäuse bewohnender Polychaeten aus dem Nordostatlantik. *Mitteilungen aus dem Hamburgischen zoologischen Museum und Institut* **80**, 169–76.

Hartmann-Schröder, G. 1984. Zwie neue kommensalische Polychaeten der Gattung *Hololepidella* Willey (Polynoidae) von der Philippinen. *Mitteilungen aus dem Hamburgischen Zoologischen Museum und Institut* **81**, 63–70.

Hartmann-Schröder, G. 1985. *Polynoe caeciliae* Fauvel (Polynoidae), ein mit Korallen assoziierter Polychaet. *Mitteilungen aus dem Hamburgischen Zoologischen Museum und Institut* **82**, 31–5.

Hartmann-Schröder, G. 1989. *Polynoe thouarellicola* n. sp. aus der Antarktis, assoziiert mit Hornkorallen, und Wiederbeschreibung von *Polynoe antarctica* Kinberg, 1858 (Polychaeta, Polynoidae). *Zoologischer Anzeiger* **3/4**, 205–21.

Hartmann-Schröder, G. 1991. *Syllis onkylochaeta* sp.n., ein korallenfressender Polychaet (Syllidae) aus dem Korallenaquarium des Löbbecke-Museums. *Helgoländer Meeresuntersuchungen* **45**, 59–63.

Hartmann-Schröder, G. 1992. Drei neue Polychaeten-arten der familien Polynoidae und Syllidae von Neu-Kaledonien, assoziiert mit einer verkalten Hydrozoe. *Helgoländer Meeresuntersuchungen* **46**, 93–101.

Hartmann-Schröder, G. 1993. *Haplosyllys xeniaecola*, ein neuer Polychaet (Syllidae) von der Molukken (Indonesien). *Helgoländer Meeresuntersuchungen* **47**, 305–10.

Hartnoll, R. G. 1962. *Iphitime cuenoti* Fauvel (Eunicidae), a polychaete new to British waters. *Annals and Magazine of Natural History* 13th Series, **5**, 93–6.

Haswell, W. A. 1885. On a destructive parasite of the rock-oyster (*Polydora ciliata* and *P. polybranchia* n. sp.). *Proceedings of the Linnean Society of New South Wales* **10**, 273–5.

Haswell, W. A. 1900. On a new Histriobdellid. *Quarterly Journal of Microscopical Science* **43**, 299–335.

Haswell, W. A. 1913. Notes on the Histriobdellidae. *Quarterly Journal of Microscopical Science* **59**, 197–226.

Hatfield, P. A. 1965. *Polydora commensalis* Andrews – larval development and observations on adults. *Biological Bulletin* **128**, 356–68.

Hendler, G. & Meyer, D. L. 1982. An association of a polychaete, *Branchiosyllis exilis*, with an ophiuroid, *Ophiocoma echinata*, in Panama. *Bulletin of Marine Science* **32**, 736–44.

Henry, S. M. (ed.) 1966. *Symbiosis. I. Associations of microorganisms, plants and marine organisms*. New York: Academic Press.

Hermans, C. O. 1969. The systematic position of the Archiannelida. *Systematics and Zoology* **18**, 85–102.

Hickok, J. F. & Davenport, D. 1957. Further studies in the behavior of commensal polychaetes. *Biological Bulletin* **113**, 397–406.

Hilbig, B. 1994. 9. Family Hesionidae Sars, 1862. In *Taxonomic atlas of the benthic fauna of the Santa Barbara Basin and western Santa Barbara Channel. Vol. 4. The Annelida, Part 1. Oligochaeta and Polychaeta: Phyllodocida (Phyllodocidae to Paralacydoniidae)*, J. A. Blake & B. Hilbig (eds). Santa Barbara: Santa Barbara Museum of Natural History, 243–69.

Hopkins, S. H. 1958. The planktonic larvae of *Polydora websteri* Hartman (Annelida, Polychaeta) and their settling on oysters. *Bulletin of Marine Science of the Gulf and Caribbean* **8**, 268–77.

Hornell, J. 1891. Report on the polychaetous annelids of the L.M.B.C. district. *Proceedings and Transactions of the Liverpool Biological Society* **5**, 223–68.

Horst, R. 1915. On new and little-known species of Polynoinae from the Netherland's East-Indies. *Zool-*

ogische Mededeelingen Leiden **1**, 2–20.

Houlbert, C. & Galaine, C. 1916. Sur le chambrage des huiters et l'infection possible desc chambers parle fait d'une annelide tubicole parasite de la coquille (*Sclerocheilus*). *Comptes Rendus de l'Academie des Sciences de Paris* **162**, 54–6.

Hove, H. A. ten 1970. Serpulinae (Polychaeta) from the Caribbean. 1. The genus *Spirobranchus*. *Studies on the Fauna of Curaçao and Caribbean Islands* **32**, 14–19.

Hove, H. A. ten 1989. Serpulinae (Polychaeta) from the Caribbean. IV *Pseudovermilia madracicola* sp.n., a symbiont of corals. *Natuurwetenschappelijke Studiekring voor Suriname en de Nederlandse Antillen* **123**, 135–44.

Hove, H. A. ten 1994. Serpulidae (Annelida: Polychaeta) from the Seychelles and Amirante Islands. In *Oceanic reefs of the Seychelles. Cruise reports Netherlands Indian Ocean Programm II*, J. van der Land (ed.). Leiden: National Natural Museum of Leiden, 107–16.

Huber, M. E. 1987. Aggressive behavior of *Trapezia intermedia* Miers and *T. digitalis* Latreille (Brachiura: Xantidae). *Journal of Crustacean Biology* **7**, 238–48.

Humann, P. 1992. *Reef creature identification. Florida, Caribbean, Bahamas.* Orlando: Vaughan Press.

Hunte, W., Colin, B. E. & Marsden, J. R. 1990a. Habitat selection in the tropical polychaete *Spirobranchus giganteus*. 1. Distribution on corals. *Marine Biology* **104**, 87–92.

Hunte, W., Marsden, J. R. & Conlin, B. E. 1990b. Habitat selection in the tropical polychaete *Spirobranchus giganteus*. III. Effects of coral species on body size and body proportions. *Marine Biology* **104**, 101–7.

Imajima, M. 1966. The Syllidae (polychaetous annelids) from Japan. IV. Syllinae. *Publications of the Seto Marine Biological Laboratory* **14**, 219–52.

Imajima, M. & Hartman, O. 1964. The polychaetous annelids from Japan. I. *Allan Hancock Foundation Special Publication* **26**, 1–237.

Imajima, M. & Sato, W. 1984. A new species of *Polydora* (Polychaeta: Spionidae) collected from Abashiri Bay, Hokkaido. *Bulletin of the National Science Museum of Tokyo, Series A* **10**, 57–62.

Izuka, A. 1912. The errantiate Polychaeta of Japan. *Journal of the College of Science, Imperial University of Tokyo* **30**, 1–262.

Jamieson, B. G. M., Afzelius, B. A. & Franzén, Å. 1985. Ultrastructure of the acentriolar, aflagellate spermatozoa and the eggs of *Histriobdella homari* and *Stratiodrilus novaehollandiae* (Histriobdellidae, Polychaeta). *Journal of Submicroscopical Cytology* **17**, 363–72.

Jangoux, M. 1987. Diseases of Echinodermata. III. Agents metazoans (Annelida to Pisces). *Diseases in Aquatic Organisms* **3**, 59–83.

Jennings, J. B. & Gelder, S. R. 1976. Observations on the feeding mechanism, diet and digestive physiology of *Histriobdella homari* van Beneden 1858: an aberrant polychaete symbiotic with North American and European lobsters. *Biological Bulletin* **151**, 489–517.

Jensen, K. & Bender, K. 1973. Invertebrates associated with snail shells inhabited by *Pagurus bernhardus* (L.) (Decapoda). *Ophelia* **10**, 185–92.

Jensen, R. A. & Morse, D. E. 1984. Intraspecific facilitation of larval recruitment: gregarious settlement of the polychaete *Phragmatopoma californica* (Fewkes). *Journal of Experimental Marine Biology and Ecology* **54**, 677–84.

Jones, M. L. 1974. On the Caobangidae, a new family of Polychaeta, with redescription of *Caobangia billeti* Giard. *Smithsonian Contributions to Zoology* **175**, 1–55.

Jones, S. 1964. Notes on animal associations. The starfish *Pentaceros hedemanni* and the hesionid polychaete *Podarke angustifrons*. *Journal of the Marine Biological Association of India* **6**, 249–50.

Kent, R. M. L. 1979. The influence of heavy infestations of *Polydora ciliata* on the flesh content of *Mytilus edulis*. *Journal of the Marine Biological Association of the United Kingdom* **59**, 289–97.

Kent, R. M. L. 1981. The effect of *Polydora ciliata* on the strength of *Mytilus edulis*. *Journal du Conseil, Conseil Permanent International pour l'Exploration de la Mer* **39**, 252–5.

Kirkegaard, J. B. 1977. A new species of *Iphitime* (Polychaeta: Iphitimidae) living under the tail of *Hyas* (Crustacea: Decapoda) in the Oslo Fjord. In *Essays on polychaetous annelids in memory of Dr Olga Hartman*, D. J. Reish & K. Fauchald (eds). Los Angeles: Allan Hancock Foundation, University of Southern California Press, 199–209.

Kirkegaard, J. B. & Billett, D. 1980. *Eunoe laetmogonensis*, a new species of polynoid worm, commensal with the bathyal holothurian *Laetmogone violacea*, in the north-east Atlantic. *Steenstrupia* **6**, 101–9.

Kirkegaard, J. B. & Santhakumaran, L. N. 1967. On a new species of annelid associate of marine wood borers; *Cirriformia limnoricola* n. sp. (Polychaeta). *Videnskabelige Meddelelser fra Dansk naturhistorisk Forening* **130**, 213–16.

Koch, H. 1846. Einige Worte zur Entwicklungsgeschichte von Eunice, mit einem Nachwortevon A. Koelliker. *Denkschriften der Allgemeimen Schweizenschen Gesellschaftfürdie Gesammten Naturwissenschaften* **8**, 1–31.

Korringa, P. 1951. The shell of *Ostrea edulis* as a habitat: observations on the epifauna of oysters living in the Oosterchelde, Holland, with some notes on polychaete worms occuring there in other habitats. *Archives Néerlandaises de Zoologie* **10**, 32–152.

Koukouras, A., Russo, A., Voultsiadou-Koukoura, E., Arvanitidis, C. & Stefanidou, D. 1996. Macrofauna associated with sponge species of different morphology. *P.S.Z.N.I.: Marine Ecology* **17**, 569–82.

Koukouras, A., Russo, A., Voultsiadou-Koukoura, E., Dounas, C. & Chintiroglou, C. 1992. Relationship of sponge macrofauna with the morphology of their hosts in the north Aegean Sea. *Internationale Revue der Gesammten Hydrobiologie* **77**, 609–19.

Kudenov, J. D. 1975. The occurrence of the polynoid *Harmothoe* cf *lunulata* from the tube of the maldanid *Axiothella rubrocincta* (Polychaeta). *Research Notes of the Pacific Marine Station of California* **34**, 42–3.

Lande, R. & Reish, D. J. 1968. Seasonal occurrence of the commensal polychaetous annelid *Ophryodromus pugettiensis* on the starfish *Patiria miniata*. *Bulletin of the Southern California Academy of Sciences* **67**, 104–11.

Landers, W. S. 1967. Infestation of the hard clam, *Mercenaria mercenaria*, by the boring polychaete worm, *Polydora ciliata*. *Proceedings of the National Shellfisheries Association* **57**, 63–6.

Lang, K. 1949. A contribution to the morphology of *Stratiodrilus platensis* Cordero (Histriobdellidae). *Arkiv für Zoologi* **42A**, 1–30.

Laubier, L. 1958. Contribution a la faunistique de coraligène. I. Quelques particularités biologiques de *Polydora armata* Langerhans. *Vie et Milieu* **9**, 412–15.

Laubier, L. 1959. Contribution a la faunistique de coraligène. III. Deux spionidiens inquilins sur des bryozoaires chilostomides. *Vie et Milieu* **10**, 347–9.

Laubier, L. 1960. Une nouvelle sous-espèce de Syllidien: *Haplosyllis depressa* Augener ssp. nov. *chamaeleon*, ectoparasite sur l'octocoralliaire *Muricea chamaeleon* von Koch. *Vie et Milieu* **11**, 75–87.

Laubier, L. 1973. Découverte d'une annélide polychète de l'Atlantique boréal dans l'étage bathyal de Méditerranée occidentale. *Vie et Milieu* **23**, 255–61.

Lee, J.-W. & Rho, B.-J. 1994. Systematic studies on Syllidae (Annelida, Polychaeta) from the South Sea and the East Sea in Korea. *The Korean Journal of Systematic Zoology* **10**, 131–44.

Light, W. J. 1970a. *Polydora alloporis*, new species, a commensal spionid (Annelida, Polychaeta) from a hydrocoral off Central California. *Proceedings of the California Academy of Sciences, 4th Series* **37**, 459–72.

Light, W. J. 1970b. A new spionid (Annelida: Polychaeta) from the Gulf of California. *Bulletin of the Southern California Academy of Sciences* **69**, 74–9.

López, E., San Martín, G. & Jiménez, M. 1996. Syllinae (Syllidae, Annelida, Polychaeta) from Chafarinas Islands (Alborán Sea, W. Mediterranean). *Miscellània Zoològica* **19**, 105–18.

Ludwig, H. 1905. Ein endoparasitischer Chaetopod in einer Tiefsee Ophiure. *Zoologischer Anzeiger* **29**, 397–9.

Lunz, G. R. 1940. The annelid worm *Polydora* as an oyster pest. *Science* **92**, 310 only.

Lunz, G. R. 1941. *Polydora*, a pest in South Carolina oysters. *Journal of the Elisha Mitchell Scientific Society* **57**, 273–83.

Lützen, J. 1961. Sur une nouvelle espèce de Polychète, *Sphaerodoridium commensalis* n. gen., n. spec. (Polychaeta Errantia, famille des Sphaerodoridae) vivant en commensal de *Terebellides stroemi* Sars. *Cahiers de Biologie Marine* **2**, 409–16.

MacNae, W. & Kalk, M. 1962. The fauna and flora of sand flats at Inhaca Island, Mocambique. *Journal of Animal Ecology* **31**, 93–128.

Magnino, G. & Gaino, E. in press. *Haplosyllis spongicola* (Polychaeta, Syllidae) associated with two species of sponges from East Africa (Tanzania, Indian Ocean). *P.S.Z.N.I.: Marine Ecology.*

Marenzeller, E. 1874. Zur Kenntnis der adriatischen Anneliden. *Sitzungsberichte der Kaiserliche Akademie der Wissenschaften Wien, Mathematische-Naturwissenschaftliche Klasse* **69**, 407–78.

Marenzeller, E. 1895. *Phalacrostemma cidariophilum*, eine neue Gattung und Art der Hermelliden. *Anzeiger, Akademie der Wissenschaften, Wien* (*Mathematische Naturwissenschaften Klasse*) **32**, 191–2.

Marenzeller, E. 1902. Südjapanische Anneliden III. *Denskschriften der Akademie der Wissenschaften Wien, Mathematische-Naturwissenschaftliche Klasse* **72**, 563–82.

Marenzeller, E. 1904. Steinkorallen. *Wissenschaftliche Ergebnisse der Deutschen Tiefsee-Expedition auf dem Dampfer "Valdivia" 1898–1899* **7**, 261–318.

Margalef, R. 1980. *Ecología*. Barcelona: Ediciones Omega S. A.

Margolis, L., Esch, G. W., Holmes, J. C., Kuris, A. M. & Schad, G. A. 1982. The use of ecological terms in parasitology (report of a committee of the American Society of Parasitologists). *Journal of Parasitology* **68**, 131–3.

Marsden, J. R. 1987. Coral preference behaviour by planktotrophic larvae of *Spirobranchus giganteus corniculatus* (Serpulidae: Polychaeta). *Coral Reefs* **6**, 71–4.

Marsden, J. R. 1992. Reproductive isolation in two forms of the serpulid polychaete, *Spirobranchus polycerus* (Schmarda) in Barbados. *Bulletin of Marine Science* **51**, 14–19.

Marsden, J. R., Conlin, B. E. & Hunte, W. 1990. Habitat selection in the tropical polychaete *Spirobranchus giganteus*. II. Larval preferences for corals. *Marine Biology* **104**, 93–99.

Martin, D. 1987. La comunidad de Anélidos Poliquetos de las concreciones de algas calcáreas del litoral Catalán. Caracterización de las especies. *Publicaciones del Departamento de Zoología de Barcelona* **13**, 45–54.

Martin, D. 1996. A new species of *Polydora* (Polychaeta, Spionidae) associated with the excavating sponge *Cliona viridis* (Porifera, Hadromerida) in the north-western Mediterranean Sea. *Ophelia* **45**, 159–74.

Martin, D., Abelló, P. & Cartes, J. 1991. A new species of *Ophryotrocha* (Polychaeta: Dorvilleidae) commensal in *Geryon longipes* (Crustacea: Brachyura) from the western Mediterranean Sea. *Journal of Natural History* **25**, 1–14.

Martin, D. & Alvà, V. 1988. *Sphaerodorum ophiurophoretos* n. sp., une nouvelle éspece de Sphaerodoridae (Annelida, Polychaeta) commensal sur *Amphipholis squamata* (Echinodermata, Ophiuridae). *Bulletin de l'Institut Royal des Sciences Naturelles de Belgique* **58**, 45–9.

Martín, D., Rosell, D. & Uriz, M. J. 1992. *Harmothöe hyalonemae* sp. nov. (Polychaeta, Polynoidae), an exclusive inhabitant of different Atlanto-Mediterranean species of *Hyalonema* (Porifera, Hexactinellida). *Ophelia* **35**, 169–85.

McGinitie, G. E. & McGinitie, N. 1949. *Natural history of marine animals*. New York: McGraw-Hill.

McIntosh, W. C. 1900. *A monograph of the British Annelida. II. Polychaeta, Amphinomidae to Sigalionidae*. London: Ray Society.

Mesnil, F. & Caullery, M. 1922. L'appareil maxillaire d'*Histriobdella homari*; affinités des hitriobdellides avec les euniciens. *Comptes Rendus de l'Academie des Sciences de Paris* **174**, 913–17.

Mettam, C. 1980. On the feeding habits of *Aphrodita aculeata* and commensal polynoids, a short note. *Journal of the Marine Biological Association of the United Kingdom* **60**, 833–4.

Michaelis, H. 1978. Zur Morphologie und Oekologie von *Polydora ciliata* und *Polydora ligni* (Polychaeta, Spionidae). *Helgoländer Meeresuntersuchungen* **31**, 102–16.

Millot, N. 1953. A remarkable association between *Ophionereis reticulata* (Say) and *Harmothoe lunulata* (Delle Chiage). *Bulletin of Marine Science* **3**, 96–9.

Miura, T. 1988. A new species of the genus *Protomystides* (Annelida, Polychaeta) associated with a vestimentiferan worm from the Hatsushima cold-seep site. *Proceedings of the Japanese Society of Systematics and Zoology* **38**, 10–14.

Miura, T. & Hashimoto, J. 1991. Two new branchiate scale-worms (Polynoidae: Polychaeta) from the hydrothermal vent of the Okinawa Trough and the volcanic seamount off Chichijima Island. *Proceedings of the Biological Society of Washington* **104**, 166–74.

Miura, T. & Hashimoto, J. 1993. *Mytilidiphila*, a new genus of nautiliniellid polychaetes living in the mantle cavity of deep-sea mytilid bivalves collected from the Okinawa Trough. *Zoological Science* **10**, 169–74.

Miura, T. & Laubier, L. 1989. *Nautilina calyptogenica*, a new genus and species of parasitic polychaete on a vesicomyid bivalve from the Japan Trench, representative of a new family Nautilinidae. *Zoological Science* **6**, 387–90.

Miura, T. & Laubier, L. 1990. Nautiliniellid polychaetes collected from the Hatsushima cold-seep site in Sagami Bay, with descriptions of new genera and species. *Zoological Science* **7**, 319–25.

Miura, T. & Ohta, S. 1991. Two polychaete species from the deep-sea hydrothermal vent in the Middle Okinawa Trough. *Zoological Science* **8**, 383–7.

Miura, T. & Shirayama, Y. 1992. *Lumbineris flabellicola* (Fage, 1936), a lumbrinerid polychaete associated with a Japanese haermatypic coral. *Benthos Research* **43**, 23–7.

Mohammad, M.-B. M. 1972. Infestation of the pearl oyster *Pinctada margaritifera* (Linné) by a new species of *Polydora* in Kuwait, Arabian Sea. *Hydrobiologia* **39**, 463–77.

Monro, C. C. A. 1924. On the Polychaeta collected by HMS "Alert", 1881–1882. Families Polynoidae, Sigalionidae, and Eunicidae. *Zoological Journal of the Linnean Society* **36**, 37–64.

Monticelli, F. S. 1892. Notizia preliminare intorno ad alcuni inquilini degli Holothuroïdea del Golfo di Napoli. *Monitore di Zoologia Italiani* **3**, 248–56.

Moore, J. P. 1905. New species of polychaetes from the North Pacific, chiefly from Alaskan waters. *Proceedings of the Academy of Natural Sciences of Philadelphia* **57**, 525–54.

Moore, J. P. 1908. Some polychaetous annelids of the north Pacific of North America. *Proceedings of the Academy of Natural Sciences of Philadelphia* **60**, 321–64.

Moore, J. P. 1910. The polychaetous annelids dredged by the USS "Albatross" off the coast of southern California in 1904. II. Polynoidae, Aphroditidae and Sigalionidae. *Proceedings of the Academy of Natural Sciences of Philadelphia* **62**, 328–402.

Moore, P. G. & Gorzula, S. J. 1973. A note on the commensal eunicean *Iphitime paguri* Fage and Legendre (Polychaeta) in the Firth of Clyde. *Journal of Natural History* **7**, 161–4.

Morgado, E. H. & Amaral, A. C. Z. 1981a. Anelídeos poliquetos associados a um briozoário. III. Polynoidae. *Boletim do Instituto Oceanográfico, Sao Paulo* **30**, 91–6.

Morgado, E. H. & Amaral, A. C. Z. 1981b. Anelídeos poliquetos associados a um briozoário I. Eunicidae, Lumbrineridae, Lysaretidae e Dorvilleidae. *Iheringia* **60**, 33–54.

Morgado, E. H. & Amaral, A. C. Z. 1981c. Anelídeos poliquetos associados a um briozoário. II. Palmyridae. *Boletim do Instituto Oceanográfico, Sao Paulo* **30**, 87–9.

Morgado, E. H. & Amaral, A. C. Z. 1984. Anelídeos poliquetos associados ao briozoário *Schizoporella unicornis* (Johnston). IV. Phyllodocidae e Hesionidae. *Revista Brasileira de Biologia* **2**, 49–54.

Morgado, E. H. & Amaral, A. C. Z. 1985. Anelídeos poliquetos associados ao briozoário *Schizoporella unicornis* (Johnston). V. Syllidae. *Revista Brasileira de Biologia* **3**, 219–27.

Mori, K., Sato, W., Nomura, T. & Imajima, M. 1985. Infestation of the Japanese scallop *Pecten (Patinopecten) yessoensis* by the boring polychaetes, *Polydora* on the Okhotsk Sea coasts of Hokkaido, especially Abashiri waters. *Bulletin of the Japanese Society of Scientific Fisheries* **51**, 27–41.

Mori, M. & Belloni, S. 1985. Distribution, abundance and infestation of *Ophryotrocha geryonicola* (Annelida: Dorvilleidae) in *Geryon longipes* (Crustacea: Decapoda: Geryonidae) of Ligurian bathyal bottoms. *Oebalia* **11**, 277–87.

Moyano, H. I. G., Carrasco, F. D. & Gacitua, S. 1993. On Chilean species of the genus *Stratiodrilus* Haswell, 1900 (Polychaeta, Histriobdellidae). *Boletín de la Sociedad de Biología de Concepción, Chile* **64**, 147–57.

Newell, G. E. 1954. The marine fauna of Whitstable. *Annals and Magazine of Natural History 12th Series* **7**, 321–50.

Núñez, J., Brito, M. C. & Ocaña, O. 1991. Anélidos poliquetos de Canarias: familia Amphinomidae. *Cahiers de Biologie Marine* **32**, 469–76.

Oakes, F. R. & Fields, R. C. 1996. Infestation of *Haliotis refescens* shells by a sabellid polychaete. *Aquaculture* **140**, 139–43.

Okuda, S. 1934. On a tubicolous polychaete living in commensal with a pycnogonid. *Annotationes Zoologicae Japonenses* **14**, 437–9.

Okuda, S. 1936a. Description of two polychaetous annelids found in burrows of an apodous holothurian. *Annotationes Zoological, Japonenses* **15**, 410–15.

Okuda, S. 1936b. Japanese commensal polynoids. *Annotationes Zoologicae Japonenses* **15**, 561–71.

Okuda, S. 1937. Spioniform polychaetes from Japan. *Journal of the Faculty of Science, Hokkaido Imperial University, Series 6, Zoology* **5**, 217–54.

Okuda, S. 1950. Notes on some commensal polychaetes from Japan. *Annotationes Zoologicae Japonenses* **24**, 49–53.

Orton, J. H. 1923. Some new commensals from the Plymouth District. *Nature* **112**, 861 only.

Orton, J. H. & Smith, C. 1935. Experiments with *Amphitrite* and its commensals. *Annals and Magazine of Natural History 10th Series* **16**, 644–5.

Overestreet, R. M. 1983. Metazoan symbionts of crustaceans. In *The biology of Crustacea*, A. J. Provenzano Jr (ed.). New York: Academic Press, 155–250.

Paiva, P. C. & Nonato, E. F. 1991. On the genus *Iphitime* (Polychaeta: Iphitimidae) and description of *Iphitime sartorae* sp. nov. a commensal of brachyuran crabs. *Ophelia* **34**, 209–15.

Palmer, J. B. 1968. *An analysis of the distribution of a commensal polynoid on it hosts.* PhD thesis, University of Oregon.

Pansini, M. & Daglio, S. 1980–1981. Osservazioni sull'inquilinismo di policheti erranti in alcune demospongie del litorale ligure. *Bollettino del Museo del Istituto de Biologia dell'Universita di Genova* **48–49**, 55–60.

Paris, J. 1955. Commensalisme et parasitisme chez les annélides polychètes. *Vie et Milieu* **6**, 525–36.

Pascual, M., Núñez, J. & San Martín, G. 1996. *Exogone* (Polychaeta: Syllidae: Exoginae) endobiontics of sponges from the Canary and Madeira Islands with description of two new species. *Ophelia* **45**, 67–80.

Pawlik, J. R. 1983. A sponge-eating worm from Bermuda: *Branchiosyllis oculata* (Polychaeta, Syllidae). *P.S.Z.N.I.: Marine Ecology* **4**, 65–79.

Pawlik, J. R. 1988. Larval setlement and metamorphosis of sabellariid polychaetes, with special reference to *Phragmatopoma lapidosa*, a reef-building species, and *Sabellaria floridensis*, a non-gregarious species. *Bulletin of Marine Science* **43**, 41–60.

Pearse, A. S. 1934a. Observations on the parasites and commensals found associated with crustaceans and fishes at Dry Tortugas, Florida. *Papers of Tortugas Laboratory of Carnegie Institution Washington* **28**, 103–15.

Pearse, A. S. 1934b. Inhabitants of certain sponges at Dry Tortugas. *Papers of Tortugas Laboratory of Carnegie Institution Washington* **28**, 117–24.

Pérès, J.-M. 1949. Sur un cas nouveau de parasitisme chez les Polychètes. *Bulletin de l'Institut Océanographique* No. 945 1–4.

Petersen, M. E. & Britayev, T. A. 1997. A new genus and species of polynoid scaleworm commensal with *Chaetopterus appendiculatus* Grube from the Banda Sea (Annelida: Polychaeta), with a review of commensals of Chaetopteridae. *Bulletin of Marine Science* **60**, 261–76.

Pettibone, M. H. 1953. *Some scale-bearing polychaetes of Puget Sound and adjacent waters.* Seattle: University of Washington Press.

Pettibone, M. H. 1954. Marine polychaete worms from Point Barrow, Alaska, with additional records from the North Atlantic and North Pacific. *Proceedings of the United States National Museum* **103**, 203–356.

Pettibone, M. H. 1955. New species of polychaete worms of the family Polynoidae from east coast of North America. *Journal of the Washington Academy of Sciences* **45**, 119–26.

Pettibone, M. H. 1956. Some polychaete worms of the families Hesionidae, Syllidae, and Nereidae from the coast of North America, West Indies, and Gulf of Mexico. *Journal of the Washington Academy of Sciences* **46**, 281–94.

Pettibone, M. H. 1957. Endoparasitic polychaetous annelids of the family Arabellidae with descriptions of new species. *Biological Bulletin* **113**, 170–87.

Pettibone, M. H. 1961. New species of polychaete worms from the Atlantic Ocean, with a revision of the Dorvilleidae. *Proceedings of the Biological Society of Washington* **74**, 167–86.

Pettibone, M. H. 1963. *Marine polychaete worms of the New England region. 1. Families Aphroditidae through Trochochaetidae.* Washington, DC: Smithsonian Institution.

Pettibone, M. H. 1969a. Review of some species referred to *Scalisetosus* McIntosh (Polychaeta, Polynoidae). *Proceedings of the Biological Society of Washington* **82**, 1–30.

Pettibone, M. H. 1969b. Remarks on the North Pacific *Harmothoe tenebricosa* Moore (Polychaeta, Polynoidae) and its association with asteroids (Echinodermata, Asteroidea). *Proceedings of the Biological Society of Washington* **82**, 31–42.

Pettibone, M. H. 1969c. The genera *Polyeunoa* McIntosh, *Hololepidella* Willey, and three new genera (Polychaeta, Polynoidae). *Proceedings of the Biological Society of Washington* **82**, 43–62.

Pettibone, M. H. 1969d. *Australaugenira pottsi*, new name for *Polynoe longicirrus* Potts, from the Maldive Islands (Polychaeta: Polynoidae). *Proceedings of the Biological Society of Washington* **82**, 519–24.

Pettibone, M. H. 1969e. Revision of the aphroditoid polychaetes of the family Eulepthidae Chamberlin (= Eulepidinae Darboux; = Pareulepidae Hartman). *Smithsonian Contributions to Zoology* **41**, 1–44.

Pettibone, M. H. 1970. Polychaeta Errantia of the Siboga Expedition. Part IV. Some additional polychaetes of the Polynoidae, Hesionidae, Nereidae, Goniadidae, Eunicidae, and Onuphidae, selected as new species by the late Dr Hermann Augener with remarks on other related species. In *Siboga-Expeditie Uitkomsten op Zoologisch, Bonatisch, Oceanographisch en Geologisch gebied verzameld in Nederlandsch Oost-Indië 1899–1900*, M. Weber et al. (eds). Leiden: E. J. Brill, 199–270.

Pettibone, M. H. 1975. Review of the genus *Hermenia* with a description of a new species (Polychaeta: Polynoidae: Lepidonotinae). *Proceedings of the Biological Society of Washington* **88**, 233–48.

Pettibone, M. H. 1977. Review of *Halosydnopsis* and related genera (Polychaeta: Polynoidae: Lepidonotinae). In *Essays on polychaetous annelids in memory of Dr Olga Hartman*, D. J. Reish & K. Fauchald (eds). Los Angeles: University of Southern California Press, 39–62.

Pettibone, M. H. 1982. Annelida. Polychaeta. In *Synopsis and classification of living organisms*, Vol. 2, S. P. Parker (ed.). New York: McGraw-Hill, 3–43.

Pettibone, M. H. 1983. *Minusculisquama hughesi*, a new genus and species of scaleworm (Polychaeta: Polynoidae) from Eastern Canada. *Proceedings of the Biological Society of Washington* **96**, 400–406.

Pettibone, M. H. 1984a. A new scaleworm commensal with deep-sea mussels on the Galapagos hydrothermal vent (Polychaeta: Polynoidae). *Proceedings of the Biological Society of Washington* **97**, 226–39.

Pettibone, M. H. 1984b. Two new species of *Lepidonotopodium* (Polychaeta: Polynoidae: Lepidonotopodinae) from hydrothermal vents off the Galapagos and East Pacific Rise at 21 °N. *Proceedings of the Biological Society of Washington* **97**, 849–63.

Pettibone, M. H. 1986a. A new scaleworm commensal with deep-sea mussels in the seep-sites at the Florida Escarpment in the eastern Gulf of Mexico (Polychaeta: Polynoidae: Branchipolynoidae). *Proceedings of the Biological Society of Washington* **99**, 444–51.

Pettibone, M. H. 1986b. Review of the Iphioninae (Polychaeta: Polynoidae) and revision of *Iphione cimex* Quatrefages, *Gattyana deludens* Fauvel, and *Harmothoe iphionelloides* Johnson (Harmothoinae). *Smithsonian Contributions to Zoology* **428**, 1–43.

Pettibone, M. H. 1989a. A new species of *Benhamipolynoe* (Polychaeta: Polynoidae: Lepidastheniinae) from Australia, associated with the unattached stylasterid coral *Conopora adeta*. *Proceedings of the Biological Society of Washington* **102**, 300-304.

Pettibone, M. H. 1989b. Revision of the aphroditoid polychaetes of the family Acoetidae Kinberg (= Polyodontidae Augener) and reestablishment of *Acoetes* Audouin and Milne-Edwards, 1832, and *Euarche* Ehlers, 1887. *Smithsonian Contributions to Zoology* **464**, 1–138.

Pettibone, M. H. 1990. New species and new records of scaled polychaetes (Polychaeta: Polynoidae) from the axial seamount caldera of the Juan de Fuca Ridge in the Northeast Pacific and the East Pacific Ocean off northern California. *Proceedings of the Biological Society of Washington* **103**, 825–38.

Pettibone, M. H. 1991a. Polynoids commensal with gorgonian and stylasterid corals, with a new genus, new combinations, and new species (Polychaeta: Polynoidae: Polynoinae). *Proceedings of the Biological Society of Washington* **104**, 688–713.

Pettibone, M. H. 1991b. Polynoid polychaetes commensal with antipatharian corals. *Proceedings of the Biological Society of Washington* **104**, 714–26.

Pettibone, M. H. 1993. Scaled polychaetes (Polynoidae) associated with ophiuroids and other invertebrates and review of species referred to *Malmgrenia* McIntosh and replaced by *Malmgeniella* Hartman, with descriptions of new taxa. *Smithsonian Contributions to Zoology* **538**, 1–92.

Pettibone, M. H. 1996. Revision of the scaleworm genera *Acholoe* Claparède, *Arctonoella* Buzhinskaja, and *Intoshella* Darboux (Polychaeta, Polynoidae) with the erection of the new subfamily Acholoinae. *Proceedings of the Biological Society of Washington* **109**, 629–44.

Pfannenstiel, H. D., Grothe, C. & Kegel, B. 1982. Studies on *Ophryotrocha geryonicola* (Polychaeta, Dorvilleidae). *Helgoländer Meeresuntersuchungen* **35**, 119–25.

Pilger, J. 1971. A new species of *Iphitime* (Polychaeta) from *Cancer antennarius* (Crustacea: Decapoda). *Bulletin of the Southern California Academy of Sciences* **70**, 84–7.

Potts, F. A. 1910. Polychaeta of the Indian Ocean. 2. The Palmyridae, Aphroditidae, Polynoidae, Acoetidae and Sigalionidae. *Transactions of the Linnean Society of Zoology, Series 2*, **16**, 325–53.

Potts, F. A. 1912. A new type of parasitism in the Polychaeta. *Proceedings of the Cambridge Philosophical Society* **16**, 409–13.

Potts, F. A. 1915. The fauna associated with the crinoids of a tropical coral reef; with species reference to its colour variations. *Papers of the Department of Marine Biology of the Carnegie Institution of Washington* **8**, 71–96.

Rabaud, E. 1939. *Nereis fucata* Sav. et la notion de commensalisme. *Bulletin de Biologie de la France et de Belgique* **73**, 293–302.

Radashevsky, V. I. 1988. Morphology, ecology, reproduction and larval development of *Polydora uschakovi* (Polychaeta, Spionidae) in the Peter the Great Bay of the Sea of Japan. *Zoologicheski Zhurnal* **67**, 870–78. (In Russian).

Radashevsky, V. I. 1989. Ecology, sex determination, reproduction and larval development of commensal Polychaetes *Polydora commensalis* and *Polydora glycymerica* in the Japanese Sea. In *Symbiosis in marine animals*, V. A. Sveshnickov (ed.). Moscow: A. N. Severtzov Institute, Russian Academy of Sciences, 137–64. (In Russian).

Radashevsky, V. I. 1993. Revision of the genus *Polydora* and related genera from the North West Pacific (Polychaeta, Spionidae). *Publications of the Seto Marine Biological Laboratory* **36**, 1–60.

Radashevsky, V. I. 1996. Morphology, ecology and asexual reproduction of a new *Polydorella* species (Polychaeta: Spionidae) from the South China Sea. *Bulletin of Marine Science* **58**, 684–93.

Rainer, S. F. 1973. *Polydora* and related genera (Polychaeta: Spionidae) from Otago waters. *Journal of the Royal Society of New Zealand* **3**, 345–564.

Ramberg, J. P. & Schram, T. A. 1982. A systematic review of the Oslofjord species of *Polydora* Bosc and *Pseudopolydora* Czerniavsky, with some new biological and ecological data (Polychaeta: Spionidae).

Sarsia **68**, 233–47.

Read, G. B. 1975. Systematics and biology of polydorid species (Polychaeta: Spionidae) from Wellington Harbour. *Journal of the Royal Society of New Zealand* **5**, 395–419.

Reise, K. & Ax, P. 1979. A meiofaunal "thiobios" limited to the anaerobic sulfide system of marine sand does not exist. *Marine Biology* **54**, 225–37.

Reish, D. J. 1957. The life history of the polychaetous annelid *Neanthes caudata* (delle Chiaje), including a summary of development in the family Nereidae. *Pacific Science* **11**, 216–28.

Reiswig, H. M. 1973. Population dynamics of Jamaican Demospongiae. *Bulletin of Marine Science* **23**, 191–226.

Rhode, K. 1981. The nature of parasitism. In *Australian Ecology Series: ecology of marine parasites*, H. Heatwole (ed.). St Lucia: University of Queensland Press, 4–5.

Ricketts, E. F., Calvin, J., Hedgpeth, J. W. & Phillips, D. W. 1985. *Between Pacific tides*. Stanford: Stanford University Press.

Ricklefs, R. E. 1990. *Ecology*. New York: W. H. Freeman.

Riser, N. W. 1982. Observations on some poorly known syllid polychaetes from the Gulf of Maine. *Canadian Journal of Zoology* **60**, 105–11.

Rosell, D. 1996. *Systematics, biology and ecology of Mediterranean excavating sponges*. Doctoral thesis, Universitat de Barcelona, Barcelona.

Rosell, D. & Uriz, M. J. 1991. *Cliona viridis* (Schmidt, 1862) and *Cliona nigricans* (Schmidt, 1862) (Porifera, Hadromerida): evidence which shows they are the same species. *Ophelia* **33**, 45–53.

Rossi, M. M. 1984. A new genus and species of iphitimid parasitic in an aphoditid (Polychaeta), with an emendation of the family Iphitimidae. *Bulletin of the Southern California Academy of Sciences* **83**, 163–6.

Rouse, G. W. & Fitzhugh, K. 1994. Broadcasting fables: is external fertilization really primitive? Sex size, and larvae in sabellid polychaetes. *Zoologica Scripta* **23**, 271–312.

Rozbaczylo, N. & Cañete, J. I. 1993. A new species of scale-worm, *Harmothoe commensalis* (Polychaeta: Polynoidae), from mantle cavities of two Chilean clams. *Proceedings of the Biological Society of Washington* **106**, 666–72.

Ruff, R. E. 1991. A new species of *Bathynoe* (Polychaeta: Polynoidae) from the northeast Pacific Ocean commensal with two species of deep-water asteroids. *Ophelia* **5**, 219–30.

Ruff, R. E. 1995. 3. Family Polynoidae Malmgren, 1867. In *Taxonomic atlas of the benthic fauna of the Santa Barbara Basin and western Santa Barbara Channel. Vol. 5. The Annelida, Part 2. Polychaeta: Phyllodocida (Syllidae and scale-bearing families) Amphinomida and Eunicida*, J. A. Blake et al. (eds). Santa Barbara: Santa Barbara Museum of Natural History, 105–66.

Rullier, F. 1972. Annélides polychètes de Nouvelle-Calédonie recueillies par Y. Plessis et B. Salvat. *Expédition Franaise sur les Récifs Coralliens de la Nouvelle-Calédonie* **6**, 1–169.

Rullier, F. 1974. Quelques annelides polychetes de Cuba recueillies dans des eponges. *Travaux du Muséum d'Histoire Naturelle Grigore Antipa* **14**, 9–77.

Rullier, F. & Amoureux, L. 1979. Campagne de la Calypso au large des côtes Atlantiques de l'Amérique du Sud (1961–2). I. 33. Annélides Polychètes. *Annales do Instituto de Oceanografia* **55**, 145–206.

Rzhavsky, A. V. & Britayev, T. A. 1988. Specific features of colonies of *Circeis armoricana* (Spirorbidae, Polychaeta) on hermit crabs on the East Kamchatka coast. *Zoologicheski Zhurnal* **61**, 17–22. (In Russian).

Saint-Joseph, A. de 1888. Les Annélides polychétes des côtes de Dinard. 2. *Annals of Scientia Naturales* **5**, 141–338.

San Martín, G. 1984. *Estudio biogeográfico, faunístico y sistemático de los poliquetos de la familia sílidos (Syllidae, Polychaeta) en Baleares*. Doctoral thesis, Universidad Complutense.

San Martín, G., Ibarzábal, D., Jiménez, M. & López, E. 1997. Redescription of *Haplosyllides floridana* Augener, 1924 (Polychaeta, Syllidae, Syllinae), with notes on morphological variability and comments on the generic status. *Bulletin of Marine Science* **60**, 364–70.

San Martín, G. & Sardá, R. 1986. Sobre la presencia de un Arabélido (Polychaeta: Arabellidae) parásito de Sílidos (Polychaeta: Syllidae), del género *Labrorostratus* en las costas españolas. *Boletín de la Real Sociedad Española de Historia Natural (Sección Biología)* **82**, 141–6.

Sardá, R. 1984. *Estudio sobre los poliquetos de las zonas mediolitoral e infralitoral de la región del Estrecho de Gibraltar*. Doctoral thesis, Universitat de Barcelona, Barcelona.

Sato-Okoshi, W. & Okoshi, K. 1993. Microarchitecture of scallop and oyster shells infested with boring *Polydora*. *Nippon Suisan Gakkaishi* **59**, 1243–7.

Simon, J. L. 1967. Behavioral aspects of *Histriobdella homari*, an annelid commensal of the American lobster. *Biological Bulletin* **133**, 450 only.

Simon, J. L. 1968. Incidence and behavior of *Histriobdella homari* (Annelida: Polychaeta), a commensal of the American lobster. *Bioscience* **18**, 35–6.

Southern, R. 1914. Archiannelida and Polychaeta, in Clare Island survey. Part. 47. *Proceedings of the Royal Irish Academy, Dublin* **31**, 1–160.

Southward, E. C. 1956. On some Polychaeta of the Isle of Man. *Annals and Magazine of Natural History 12th Series*, **9**, 257–79.

Spies, R. B. 1977. Reproduction and larval development of *Flabelliderma commensalis* (Moore). In *Essays on polychaetous annelids in memory of Dr Olga Hartman*, D. J. Reish & K. Fauchald (eds). Los Angeles: The Allan Hancock Foundation, University of Southern California, 323–45.

Spooner, G. M., Wilson, D. P. & Trebble, N. 1957. Phylum Annelida. In *Plymouth Marine Fauna*. 3rd edn. Plymouth: Marine Biological Association UK.

Steinbeck, J. & Ricketts, E. F. 1941. *Sea of Cortez. A leisurely journal of travel and research. With a scientific appendix comprising materials for a source book on the marine animals of the Panamic faunal province*. New York: Viking Press.

Stewart, W. C. 1970. *A study of the nature of the attractant emitted by the asteroid host of the commensal polychaete* Ophiodromus pugettiensis. Dissertation, University of California.

Stock, J. H. 1986. Cases of hyperassociation in the Copepoda (Herphyllobiidae and Nereicolidae). *Systematics and Parasitology* **8**, 71–81.

Storch, V. & Niggemann, R. 1967. Auf Echinodermen lebende Polychaeten. *Kieler Meeresforschungen* **23**, 156–64.

Storch, V. & Rosito, R. M. 1981. Polychaetes from inter-specific associations found off Cebu. *Philippine Scientist* **18**, 1–9.

Storm, V. 1881. Bidrag til Kundskab om Throndhjemsfjordens Fauna III. *Kongelige Norske videnskabers selskabs skrifter* 1881, 73–96.

Thesunov, A. V., Britayev, T. A., Larionov, V. V., Khodkina, I. V. & Tzetlin, A. B. 1989. Notes on commensals of some crinoids from Maldive coral reefs. In *Symbiosis among marine animals*, V. A. Sveshnikov (ed.), Moscow: A. N. Severtzov Institute, Russian Academy of Sciences, 166–92. (In Russian).

Thorson, G. 1949. Om eremit kraeftan och dens kommensal *Nereis fucata*. *Svensk Faunistisk Revy* **11**, 93–8.

Tkachuk, L. P. 1988. Vermins of the Black Sea mussel plantations. *Ekologica Morya* **30**, 60–4. (In Russian).

Treadwell, A. L. 1909. *Haplosyllis cephalata* as an ectoparasite. *Bulletin of the American Museum of Natural History* **26**, 359–60.

Treadwell, A. L. 1921. Leodicidae of the West Indian region. Papers of the *Department of Marine Biology of the Carnegie Institution of Wasington* **15**, 1–131.

Tsurumi, M. & Reiswig, M. 1997. Sexual versus asexual reproduction in an oviparous rope-form sponge, *Aplysina cauliformis* (Porifera, Verongida). *Invertebrate Reproduction and Development* **32**, 1–9.

Tzetlin, A. B. 1985. *Asetocalamyzas laonicola* gen. et sp. n., a new ectoparasitic polychaete from the White Sea. *Zoologicheski Zhurnal* **64**, 269–98. (In Russian).

Tzetlin, A. B. & Britayev, T. A. 1985. A new species of Spionidae (Polychaeta) with asexual reproduction associated with sponges. *Zoologica Scripta* **14**, 177–81.

Uchida, H. 1975. Ectocommensal polynoids (Annelida, Polychaeta) associated with the echinoderms. *Bulletin of the Marine Park Reservation Stations* **1**, 19–30.

Uchida, H. 1978. Serpulid tube worms (Polychaeta, Sedentaria) from Japan with the systematic review of the group. *Bulletin of the Marine Park Reservation Stations* **2**, 1–98.

Uebelacker, J. M. 1978. A new parasitic polychaetous annelid (Arabellidae) from the Bahamas. *Journal of Parasitology* **64**, 151–4.

Uebelacker, J. M. 1982. *Haplosyllis agelas*, a new polychaetous annelid (Syllidae) from the Bahamas. *Bulletin of Marine Science* **32**, 856–61.

Uriz, M. J. 1983. Monografia. I. Contribución a la fauna de esponjas (Demospongia) de Catalunya. *Anales de la Sección de Ciencias del Colegio Universitario de Gerona* **7**, 1–220.

Uschakov, P. V. 1950. Polychaetes from the Sea of Okhotsk. *Issledovaniya dal'nevostochnykh Morei SSSR* **2**, 140–236. (In Russian).

Uschakov, P. V. 1955. Polychaetous annelids of the Far Eastern Seas of the USSR. *Akademiya Nauk SSSR* **56**, 1–433.

Uschakov, P. V. 1962. Polychaetous annelids of the family Phyllodocidae and Aphroditidae from the Antarctic and Subantarctic. *Isseldovaniya Fauni Moorei* **1**, 129–89. (In Russian).

Uschakov, P. V. 1982. *Fauna of the USSR. 2(1): Polychaetes. Polychaetes of the suborder Aphroditiformia of the Arctic Ocean and the northwestern part of the Pacific, Families Aphroditidae and Polynoidae.* Moscow:

NAUKA, Academy of Sciences of the USSR, Zoological Institute. (In Russian).

Uschakov, P. V. & Wu, B. L. 1959. The polychaetous annelids of the families Phyllodocidae and Aphroditidae from the Yellow Sea. *Archives de l'Institut d'Oceanologie Sinica* **1**, 1–40. (In Russian).

Utinomi, H. 1956. On the so-called "Umi-Utiwa" a peculiar flabellate gorgonacean, with notes on a syllidean polychaete commensal. *Publications of the Seto Marine Biological Laboratory* **5**, 243–50.

Uzmann, J. R. 1967. *Histriobdella homari* (Annelida: Polychaeta) in the American lobster *Homarus americanus. Journal of Parasitology* **53**, 210–11.

Verrill, A. E. 1882. New England Annelida. Part I. Historical sketch, with annotated lists of the species hitherto recorded. *Transactions of the Connecticut Academy of Arts and Sciences* **4**, 285–324.

Vila, P. I. & Bahamonde, N. 1985. Two new species of *Stratiodrilus, S. aeglaphilus* and *S. pugnaxi* (Annelida: Histriobdellidae) from Chile. *Proceedings of the Biological Society of Washington* **98**, 347–50.

Voultsiadou-Koukoura, H. E., Koukouras, A. & Eleftheriou, A. 1987. Macrofauna associated with the sponge *Verongia aerophoba* in the North Aegean Sea. *Estuarine, Coastal and Shelf Science* **24**, 265–78.

Wagner, R. H., Phillips, D. W., Standing, J. D. & Hand, C. 1979. Commensalism or mutualism: attraction of a sea star towards a symbiotic polychaete. *Journal of Experimental Marine Biology and Ecology* **39**, 205–10.

Webster, H. E. 1879a. The Annelida Chaetopoda of New Jersey. *Annual Reports of the New York State Museum of Natural History* **32**, 101–28.

Webster, H. E. 1879b. The Annelida Chaetopoda of the Virginian coast. *Transactions Albany Institute New York* **9**, 202–72, plates 1–13.

Webster, S. K. 1968. An investigation of the commensals of *Cryptochiton stellaris* (Middendorf, 1846) in the Monterey Peninsula area, California. *Veliger* **11**, 121–5.

Wesenberg-Lund, E. 1938. *Ophryotrocha geryonicola* (Bidenkap) (= *Eteonopsis geryonicola* Bidenkap) refound and redescribed. *Göteborgs Königlia Vetenskaps och Vitter Hets-Samhälles Handlingar, Series B*, **6**, 1–13.

Wesenberg-Lund, E. 1941. Notes on the Polychaeta I. 1. *Harmothoe bathydomus* Hj. Ditlevsen refound. *Videnskabelige Meddelelser fra Dansk Naturhistorisk Forening i Köbenhavn* **105**, 31–2.

Westheide, W. 1982. *Microphthalmus hamosus* sp. n. (Polychaeta, Hesionidae) – an example of evolution leading from the interstitial fauna to a macrofaunal interspecific relationship. *Zoologica Scripta* **11**, 189–93.

Westinga, E. & Hoetjes, P. C. 1981. The intrasponge fauna of *Speciospongia vesparia* (Porifera, Demospongiae) at Curaçao and Bonaire. *Marine Biology* **62**, 139–50.

Willey, A. 1905. Report on the Polychaeta collected by Professor Herdman, at Ceylon, in 1902. *Report to the Government of Ceylon on the Pearl Oyster Fisheries of the Gulf of Manaar, with supplementary reports upon the Marine Biology of Ceylon, by Other Naturalists. Part IV Supplementary Report* **30**, 212–324.

Wiren, A. 1886. *Haematocleptes terebellidis* nouvelle annélide parasite de la famille des euniciens. *Bihang Till Kungliga Svenska Vetenskaps-Akademiens Handlingar* **11**, 1–10.

Wright, J. D. & Woodwick, K. H. 1977. A new species of *Autolytus* (Polychaeta: Syllidae) commensal on a Californian hydrocoral. *Bulletin of the Southern California Academy of Sciences* **76**, 42–8.

Wu, B. L., Sun, R. P. & Yang, D. J. 1982. On the occurrence of endoparasitic polychaetous annelids in Chinese waters. *Oceanologia et Limnologia Sinica* **13**, 201–4. (In Chinese).

Wu, B. L., Sun, R. P. & Yang, D. J. 1985. *Nereidae (polychaetous annelids) of the Chinese Coasts.* Berlin: Springer Verlag/China Ocean Press Beijing.

Yamamoto, R. & Imajima, M. 1988. A new species of the genus *Spinther* (Polychaeta, Spintheridae) from Japan. *Bulletin of the National Science Museum of Tokyo, Series A*, **11**, 129–35.

Zibrowius, H. 1981. Associations of Hydrocorallia Stylasterina with gall-inhabiting Copepoda Siphonostomatoidea from the south-west Pacific. Part I. On the stylasterine hosts, including two new species, *Stylaster papuensis and Crypthelia cryptonema. Bidragen tot de Dierkunde* **51**, 268–86.

Zibrowius, H., Southward, E. C. & Day, J. H. 1975. New observations on a little-known species of *Lumbrineris* (Polychaeta) living on various cnidarians, with notes on its recent and fossil scleractinian hosts. *Journal of the Marine Biological Association of the United Kingdom* **55**, 83–108.

Zottoli, R. A. & Carriker, M. R. 1974. Burrow morphology, tube formation and microarchitecture of shell dissolution by the spionid polychaete *Polydora websteri. Marine Biology* **27**, 307–16.

Oceanography and Marine Biology: an Annual Review 1998, **36**, 341–71

UCL Press

CEPHALOPOD EGGS AND EGG MASSES*

SIGURD v. BOLETZKY
Observatoire Océanologique de Banyuls – Laboratoire Arago
C.N.R.S. – UMR 7628, 66651 Banyuls-sur-Mer, France

Abstract Cephalopod eggs and egg masses turn up in samples taken during oceanographic cruises (including subsamples consisting of the stomach contents of marine predators); they are encountered by SCUBA divers and crews of manned submersibles, and stranded egg masses can be found on ocean beaches around the world. If it is comparatively easy to identify such material as "squid eggs", it is much more difficult to recognize the group or species to which the eggs belong. There are various reasons why the identification of eggs and egg masses is often difficult or impossible, especially for the non-specialist. The lack of standardized illustrated keys showing both the embryonic stages and the corresponding aspects of egg capsules for well known species is a major impediment to identification or further developments needed to optimize the chances of finding hitherto unknown material.

To fill the remaining gaps in our basic knowledge of cephalopod development (e.g. *Spirula*, *Vampyroteuthis*), ostensibly rare material may be sought in the rich depositories of samples taken during past oceanographic cruises. The chances of finding such material in good condition, with appropriate accompanying provenance records, increase with any effort made to improve present and future sampling strategies and preservation routines. Although it is understandable that "gluey stuff" found in plankton nets or dredges brought up by oceanographic vessels is unattractive and therefore tends to be hosed overboard, it should be realized that scientific treasures may thus be lost.

This review surveys the available literature and makes a plea for increased efforts to optimize the chances of finding the rare material by working carefully through the "common stuff"; the latter then provides other information of great scientific importance. To mention one aspect that is often overlooked: egg masses may be the only tangible proof of reproductive activity of a species in a given area, and that proof may be needed year after year in biological surveys even of common species.

Introduction

Cephalopod eggs and egg masses have been described many times, but no comprehensive presentation of the subject exists to date. What is more annoying is that even common inshore species for which eggs have been described in detail are still difficult to identify from spawn samples alone. It is not surprising, therefore, that much less is known about eggs and egg masses of pelagic and deep-sea cephalopods.

How far can this situation be improved? A rather crucial problem that continues to face us is that female sexual products in cephalopods cannot be described as concisely as the male products, that is, spermatozoids and spermatophores (and resulting spermatangia) (see Roper & Voss 1983). This is due to the fact that both eggs and egg masses undergo thorough modifications over days, weeks or months before the

* In memory of **Pierre Tardent** (1927–97).

developing young are ready to hatch. The structurally simple ovum, once fertilized (then called the zygote), transforms itself into an increasingly complicated organism, while the capsule enclosing the growing embryo is slowly modified in structure and generally also in size (e.g. Robson 1926, 1932b). However, the difficulties thus arising for identification should not obscure the biologically important fact that all these changes occurring in the embryo and in the capsule material are expressions of the genetic make-up of a species (epigenetic control of certain processes notwithstanding). Thus, they are as interesting and characteristic features of the phenotype as the adult organs and behaviour, deserving treatment in general text-books (Tardent 1993).

So long as specific identification of capsule material alone is impossible, it remains inevitable that both the embryo and the egg case be considered in an attempt to recognize a species from egg masses collected at sea. But even where these two sources of information are available, problems of intraspecific variability in both embryogenesis and spawning may persist.

The aim of this review is twofold, namely to speak to the field collector with no cephalopod knowledge (who can use a photocopy of the Appendix as an aid for gross identification), and to draw the attention of cephalopod specialists to the remaining gaps in our knowledge of eggs and egg masses. The available literature is briefly surveyed for a discussion about how to improve that knowledge. Joint efforts of scientific crews aboard research vessels, SCUBA divers engaged in field work, users of manned submersibles, and aquarium biologists will be necessary to further the studies of ecologists, oceanographers, and fisheries biologists interested in cephalopods as high level consumers and as prey for large vertebrates. It will procure also new material for the work of zoologists studying the general biology of cephalopods. Identified eggs and egg masses indeed provide abundant information on the reproductive activity of species in time and space (reflecting seasonal cycles, for example) and relative to hydrological conditions, on synecological relations among species that may be competitors, predators or prey. Finally, embryos of "living fossils" like *Spirula* and *Vampyroteuthis* may elucidate some major riddles of evolutionary morphology in coleoid cephalopods.

Historical Résumé

One of the earliest scientific descriptions with figures of cephalopod eggs and egg masses is the thesis of Bohadsch (1752). This author described "*Sepia* eggs" from loliginid squid egg capsules, the embryos figured resembling indeed early sepiid stages. It should be remembered, however, that even an octopus was a "*S. octopodia*" according to the original Linnean designation (Robson 1929). On the other hand, actual misidentifications appear quite often in scientific literature. Thus Jatta (1896) figured the eggs of *S. orbignyana* (Plates 7, 8) taken from the inside of sponges but labelled them as *S. elegans*, whereas his *S. elegans* eggs (Plates 2, 7) are labelled *S. orbignyana*. In spite of these errors and some other problems of species identification, the colour lithographs prepared by C. Merculiano for Jatta's monograph are among the finest illustrations of cephalopod eggs and egg masses. Even today misidentifications are easily overlooked by authors or editors of excellent books; for example, sepiid eggs were described as sepiolid eggs in the photo plates accompanying the work of Nesis (1982: 55). In particular, popular books and articles are prone to such misinformation, e.g. confusion of loliginid egg "fingers" (see Appendix) and octopus egg strings (Möhres 1964: 123).

With the improvement of microscopes in the nineteenth century, observation of eggs and embryos became increasingly detailed. The embryos figured by Blainville (1837) and by Férussac & d'Orbigny (1835–48) and Owen (1835) were rather simple sketches, whereas the embryological descriptions by Kölliker (1844) already provide much finer details. With such information at hand, species identifications from embryonic material become increasingly reliable, especially when size indications are precise and special features such as chromatophore patterns of hatchlings are recognizable (e.g. Jatta 1896, Naef 1923, 1928, Yung 1930).

Observations of spawning in common cuttlefish (*Sepia officinalis*) and lesser octopus (*Eledone cirrhosa*) by Joubin (1888) and (in *E. cirrhosa*) Gravely (1908), in loliginid squid (*Loligo pealei*) by Drew (1911) set the stage for combined field and aquarium studies of cephalopod reproduction culminating in the first review that considers the possibility of culturing cephalopods in captivity (Grimpe 1928). More recently, Lane (1960) collected black and white photographs of cephalopod spawning, eggs and egg masses. Apart from the information available through several articles in Boyle (1983), reviews with figures of the most representative types of eggs and egg masses were published by Jaeckel (1958), Mangold-Wirz (1963), Arnold & Williams-Arnold (1977), Fioroni (1978), Arnold (1984), Boletzky (1974a, 1986), Mangold (1989), and Guerra (1992). The most recent review of spawning (Nesis 1996) provides figures of eggs and egg masses for about a dozen species and lists information on egg features for about 40 families.

Group-typical capsule forms

Until a few decades ago, one would have been content with the following key to identify the greater cephalopod groups of the subclass Coleoidea (= Endocochleata, Dibranchiata) based on their spawned eggs as described by Naef (1923, 1928).

- Singly encapsulated eggs (laid in more or less dense clusters):
  decapod order Sepioidea.
- Capsules or jelly masses containing several to many eggs:
  decapod order Teuthoidea.
- Eggs devoid of capsules other than the chorion, the latter being drawn out into a stalk for fixation:
  "brooding" octopod (sub)order Incirrata.

This simple key was accompanied by three "special cases":

- the single, encapsulated eggs of *Nautilus*, subclass Nautiloidea (Willey 1902)
- and for the coleoids: the single, encapsulated eggs of "cirromorph" octopods, (sub)order Cirrata (Verrill 1885), along with the single, unencapsulated eggs of *Vampyroteuthis*, order Vampyromorpha (Pickford 1949a,b).

Two sorts of new information have changed the picture in recent years. On the one hand, more diverse spawning modes and egg structures have become known for various coleoid groups (while total lack of knowledge persists with many others). On the other hand, some of the great taxonomic groups mentioned above are called in question by phylogenetic systematics, especially when based on molecular methods. Although the old taxon Decapoda (=Decabrachia) stands unchallenged, the hitherto recognized subgroups "Sepioidea" and "Teuthoidea" are no longer satisfactory taxa (it thus appears that "sepiolid squids" is a correct designation in relation to "teuthoid squids"). To give an example, the Idiosepiidae seemed to be closely related to the Sepiadariidae and Sepiolidae, but now turn out as a likely close ally of some oegopsid squids like Ommastrephidae and Enoploteuthidae (Bonnaud et al. 1997). These new aspects are of great importance to cephalopod systematics, but they cannot be translated at once into a widely used jargon to characterize cephalopod groups.

What can be kept in use for the present purpose is the basic split between the ectocochleate nautiloids and the endocochleate coleoids (Table 1). Within the "modern" Coleoidea, the Decabrachia or Decapoda (name preoccupied by decapod crustaceans) are one of two natural (very likely monophyletic) groups; which taxonomic rank (e.g. superorder) it is given is of little importance. Decabrachia contain two or three ill-defined "Orders". The Sepioidea include two families with a calcified, chambered shell (the monotypic Spirulidae and the speciose Sepiidae). The Sepiolioidea have been separated from the Sepioidea to encompass two "bobtail" families with strongly reduced or lacking shell remnants (Sepiadariidae, Sepiolidae) and the pygmy cuttlefish (Idiosepiidae), whereas the Sepiolida include only the Sepiolidae and the Idiosepiidae. The Teuthoidea include two myopsid squid families (the Loliginidae, and the monogeneric Pickfordiateuthidae), the comb-fin squid (Chtenopterygidae, apparently related to the myopsid squids), and a great variety of oegopsid squid families – in fact more than 20 according to present taxonomic lists.

The other major coleoid group (Table 1), counterpart of the Decabrachia, is a combination of the Octobrachia and the monotypic Vampyromorpha (containing the single species *Vampyroteuthis infernalis*, once considered a cirrate octobrachian); this combination is called Vampyropoda. Setting aside the Vampyromorpha, we are left with the group Octobrachia that contains two subgroups, the Cirroctopoda and the Octopoda (Cirroctopoda being a new name for "Octopoda cirrata", Octopoda being the original name of what had become "Octopoda incirrata" in earlier systematic revisions). What is important to remember is that the incirrate (finless) Octopoda are a specialized group apparently derived from a cirroctopodan (finned) ancestor that was closer to the present than the common ancestor of the vampyromorphs and the octobrachians (Boletzky 1992a).

Table 1 Outline of recent cephalopod classification (for explanation of the taxa Octopoda, Cirroctopoda and Vampyropoda (see Young 1989, 1997, Boletzky 1992a).

Class Cephalopoda	
Subclass Nautiloidea (one extant, monogeneric family: Nautilidae)	
Subclass Coleoidea (=Endocochleata, or Dibranchiata)	
Superorder	Decabrachia (=Decapoda)
Order	Sepioidea
Order	Sepiolioidea (>Sepiolida)
Order	Teuthoidea
	Suborder Myopsida
	Suborder Oegopsida
Superorder	Vampyropoda
Order	Vampyromorpha
Order	Octobrachia (=Octopoda Grimpe, non Leach)
	Suborder Cirroctopoda (="Cirrata")
	Suborder Octopoda (="Incirrata")

With Table 1 in mind, we can now refine the old key mentioned at the beginning of this section.

Pearly Nautilus

Nautilus (*Nautilus* spp.) are unique in laying very large eggs (20–30 mm) in individual capsules the outer case of which is partly doubled (a bell-shaped basal rim of nidamental jelly being drawn over the main capsule body during egg-laying), with a series of longitudinal folds around canals (Hamada et al. 1980, Arnold & Carlson 1986; Arnold et al. 1993b).

Cuttlefish and sepiolid squids

Cuttlefishes and sepiolids ("bobtail squids") lay medium sized eggs (≈3–10 mm) one after the other (Fig. 1), each ovum being wrapped in a spirally coiled, gelatinous envelope or capsule (Fig. 5a); these eggs are generally fixed to a substratum in loose or dense batches (Bott 1938, Okutani 1978). Sepiolids often pile up their eggs in several layers (Thorson 1946, Arnold et al. 1972); the translucent outer jelly capsule becomes leathery (Fig. 4) or turns opaque and rigid (*Rossia* spp.; Racovitza 1894, Steenstrup 1900, Boletzky & Boletzky 1973, Anderson & Shimek 1994). Cuttlefishes of the family Sepiidae often attach their eggs by first pulling the jelly coat into a pair of filaments and then sticking these filaments around a slender support structure (Fig. 1). The resulting fixation ring (Fig. 3) becomes fairly solid. In some species, the egg capsule

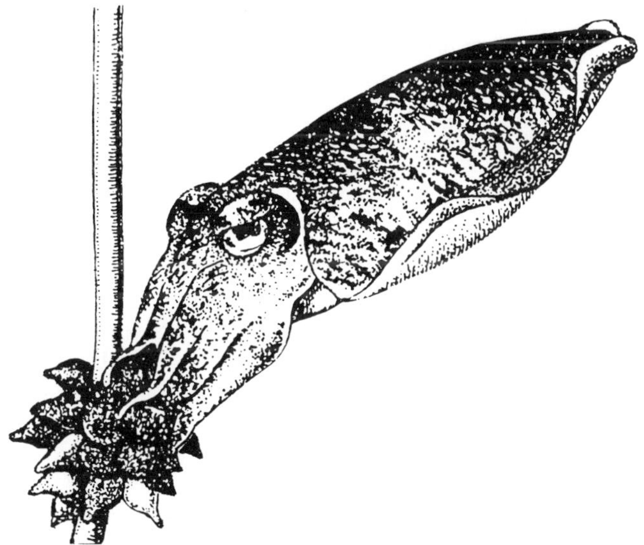

Figure 1 Female *Sepia officinalis* depositing an egg in a batch of eggs already attached to a piece of tubing in an aquarium. For the aspect of the ovum inside the gelatinous capsule material (which is stained with ink), see Figure 2. (Drawn from a photograph in Bott, 1938; about one-half natural size).

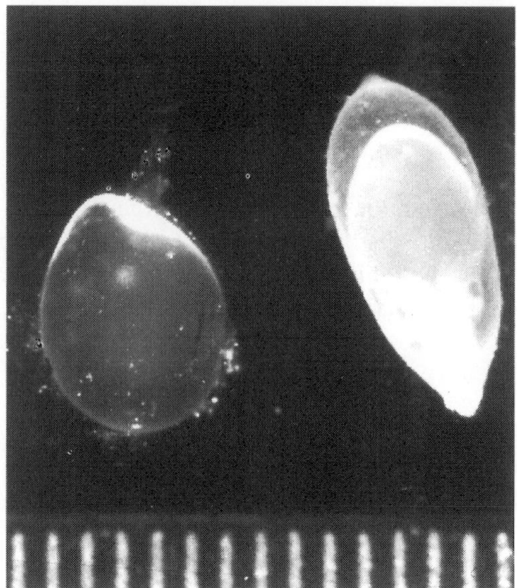

Figure 2 "Unprotected eggs" of *Sepia officinalis* (left) and *S. orbignyana* (right). The egg at left is shown without the gelatinous nidamental envelopes (a few remains of the oviducal jelly still adhere to the chorion); this situation occurs in "untidy" spawning when encapsulation (a process taking place in the mantle cavity of the female) is disturbed, so that the egg comes to lie at the surface of the capsule where it is exposed to fouling organisms. The egg at right is complete with (unstained) nidamental envelopes and therefore could develop normally, if it were attached to a support; out of the "nursing" sponge which normally provides protection throughout embryonic development, it is rapidly bogged down in a muddy substratum (scale with millimetre divisions).

apparently sticks to the substratum without a fixating ring (*Metasepia tullbergi*, Natsukari 1979). Some sepiid species (*Sepia orbignyana*, Naef 1923; *Sepia* sp., which is either *S. misakiensis* or *S. tokyoensis*, Okada 1927) but also certain sepiolids (*Rossia* sp., Aldrich & Lu 1968; *R. pacifica* and *R. palpebrosa*, K. N. Nesis, pers. comm.) may introduce their eggs into sponges (generally Demospongiae), which may thus act as nursing organisms (see Fig. 2). Sepiadariids lay single eggs somewhat similar to those of *Rossia*, but individually attached to rocks (*Sepioloidea lineolata*, Dew 1959b). Nothing is known about spawning in *Spirula*; Nesis (1996) suggests that eggs are probably laid close to or on the bottom.

Pygmy cuttlefish

Pygmy cuttlefish (*Idiosepius* spp.) lay eggs with individual jelly capsules devoid of a distinct outer coat (Natsukari 1970, English 1981). One should remember that the

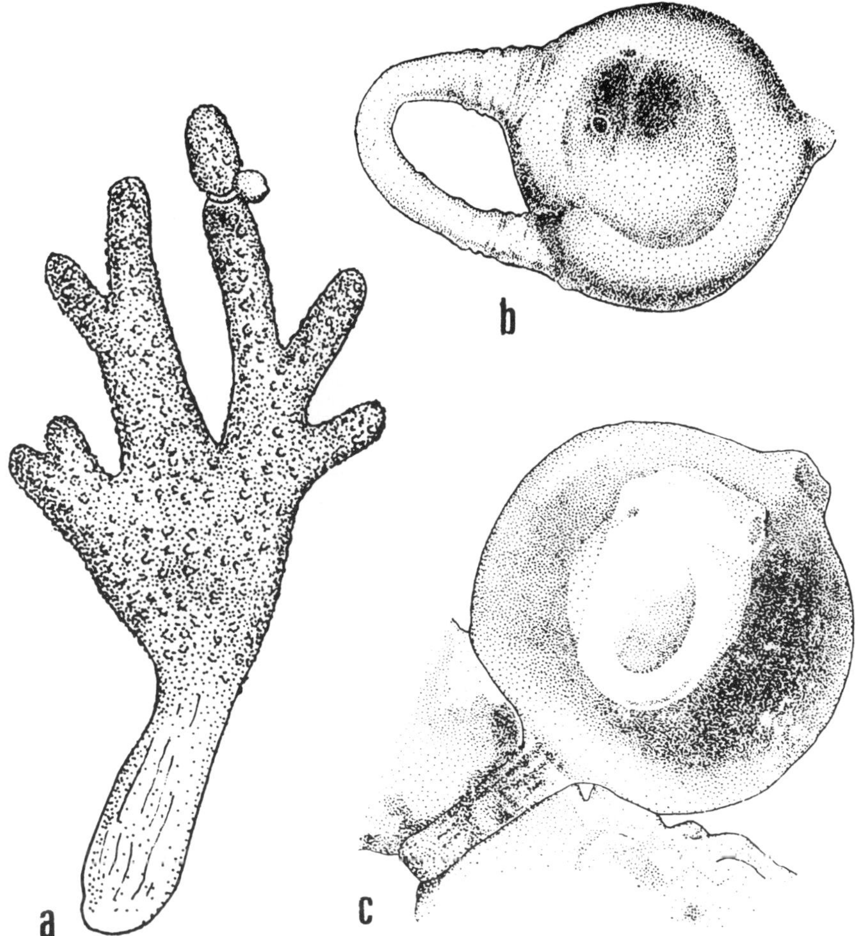

Figure 3 An egg of *Sepia elegans* attached to one of the branches of a colony of the octocoral *Alcyonium palmatum* (with contracted polyps) (a, about natural size). An egg taken from its support to show the fixating ring (b, greatly enlarged); this ring originally was formed from two filaments into which the female draws out the gelatinous envelope at the moment of attaching the egg to a support structure (see Fig. 1). A late embryonic stage (c) seen through the expanded, nearly transparent envelopes, which are generally unpigmented (same maginfication as b). (From Bouligand, 1961).

ovum measures about 1 mm and is thus very large for a miniature adult measuring only about 10–15 mm in mantle length. When approaching the question of a possible "teuthoid" relationship of *Idiosepius*, one should consider that the production of a capsule containing more than one ovum (as is typical for teuthoid squids) might be technically impossible in this dwarfish animal.

Teuthoid squids

Teuthoid squids generally lay their eggs in batches wrapped by a common jelly mass or capsule that never turns leathery or rigid; it may be either amorphous or structured

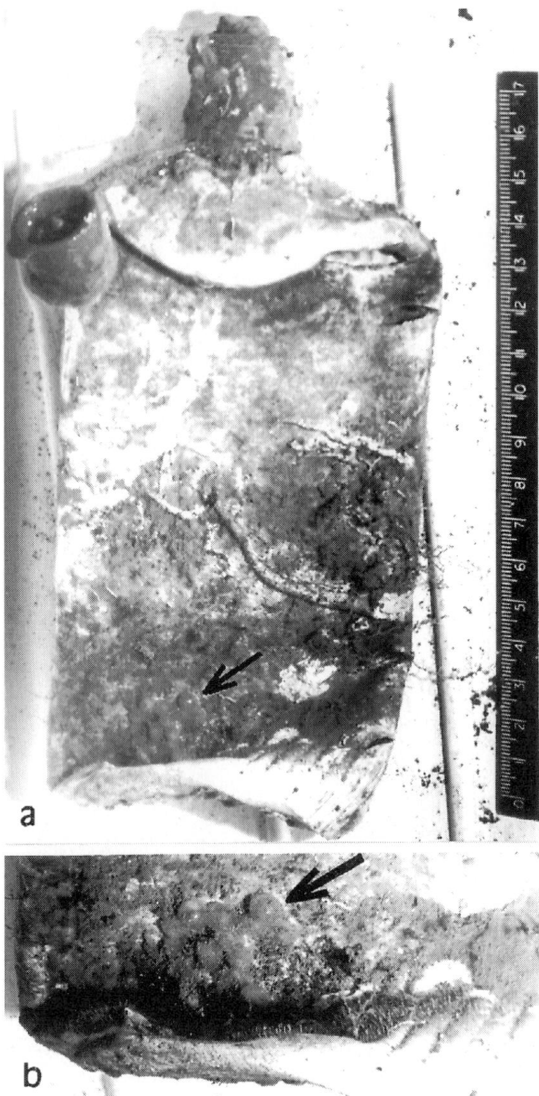

Figure 4 A batch of eggs (arrow) laid by an unidentified species of sepiolid squid on the lower surface of a compressed plastic container lying on a muddy bottom (a). At higher magnification (b), the eggs can be seen to be attached laterally to one another.

in a way similar to the spiral coiling of cuttlefish or sepiolid egg envelopes (Fig. 5a). In the myopsids, the original size of a common capsule depends partly on the size (about 1–5 mm) and partly on the number of ova it contains (Choe & Oshima 1961, Choe 1966). The minimum number of ova per capsule is two in "Type 2" of *Sepioteuthis* sp. (cf. *lessoniana*) (Segawa et al. 1993; see also Segawa 1987 and Izuka et al. 1996), whereas normally in this genus, the relatively large ova are in numbers of three or

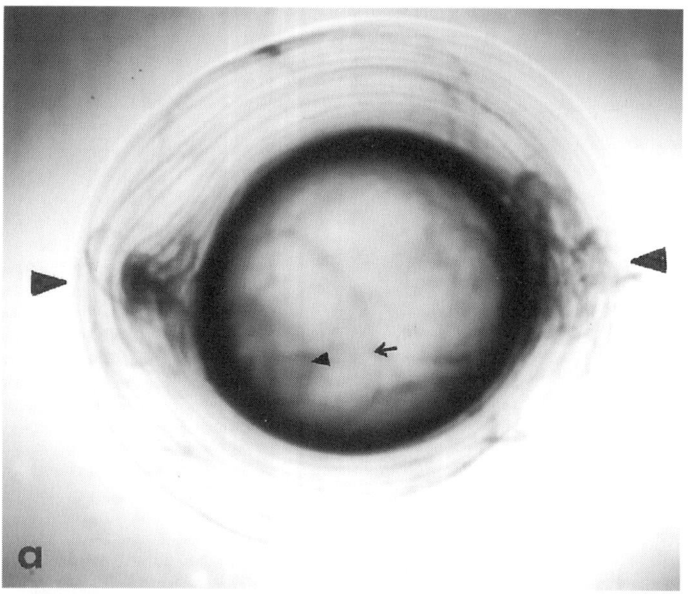

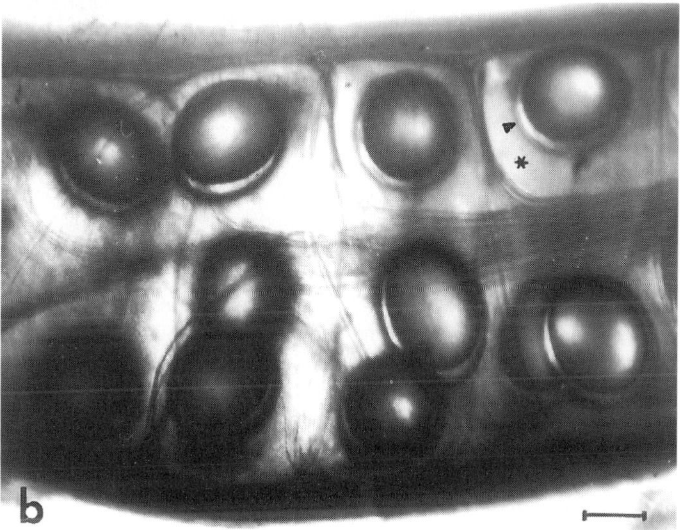

Figure 5 Decabrachian eggs laid singly (a) or in series in a common capsule (b) shown at same magnification (see scale bar at bottom). The single egg (a) is from the large sepiolid squid, *Rossia macrosoma*; the opaque, rigid egg case has been removed to expose the spiral nidamental jelly, which is coiled around the axis marked by large arrow heads. The small arrow points at the margin of the thin epibolic gastrula (growing over the yolk mass) with the prospective embryonic body in the centre (small arrow head). The serial eggs (b) are from the myopsid squid, *Loligo vulgaris*, at an early gastrula stage (arrow head points at chorion slightly lifted off the yolk surface). The axis of spiral coiling of nidamental envelopes is in the same orientation as in a; but instead of a single ovum, an egg "string" (formed by an amorphous oviducal jelly*) containing many ova is tightly wrapped and thus forced into a helix (cf. Fig. 6). Scale bar 1 mm, applies to both a and b.

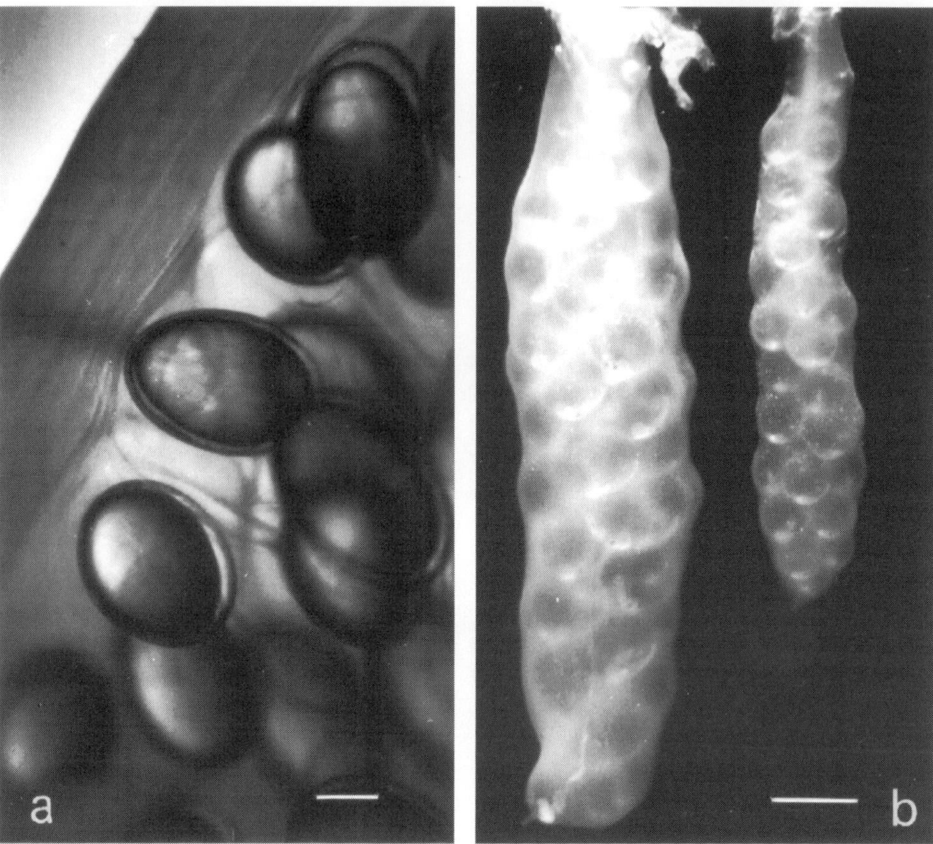

Figure 6 Detail of an egg capsule of *Loligo forbesii* (a) at similar developmental stage as eggs shown in Fig. 5; note larger size and more elongate shape of individual eggs, and greater number of eggs per turn than in 5b (the central axis lies outside the frame) (Scale bar 1 mm). Egg capsules at late embryonic stages (b) with greatly enlarged volume of perivitelline fluid in a perfectly spherical chorion: *Loligo forbesii* (left) and *Loligo vulgaris* (right). Scale bar 10 mm.

more per capsule (e.g. *Sepioteuthis* sp., Wülker 1913; *S. sepioidea*, Arnold 1965b, LaRoe 1971, Moynihan & Rodaniche 1982), with considerable variations, however (e.g. 2–6 in *S.* bilineata, Larcombe & Russell 1971). Other loliginids lay egg capsules or egg "fingers" generally containing dozens of ova (Fig. 6). These capsules are either deposited on a horizontal substratum and "stand" or lie in the water as "mops" (e.g. Drew 1911, McGowan 1954, Fields 1965, Roper 1965, Hall 1970, Natsukari 1976, Vecchione 1988, Hanlon et al. 1992, Sauer & Smale 1993, Sauer et al. 1997), or they are made to stick to the ceiling of rocky overhangs or to branched sessile organisms (or anchor lines) and hang down in the water (e.g. *Loligo vulgaris*: Jecklin 1934).

Among the oegopsid squids of pelagic waters, *Thysanoteuthis rhombus* is remarkable in producing huge floating gelatinous egg capsules (100–130 cm in length, 15–20 cm across) containing a double helix of small eggs (Clarke 1966, Sabirov et al. 1987, Nigmatullin et al. 1995, Guerra & Rocha, 1997; Fig. 7), This arrangement is understandable from the simultaneous functioning of paired oviducts in the spawning female (Nigmatullin et al. 1991). Other reports describing a single helix of eggs are apparently

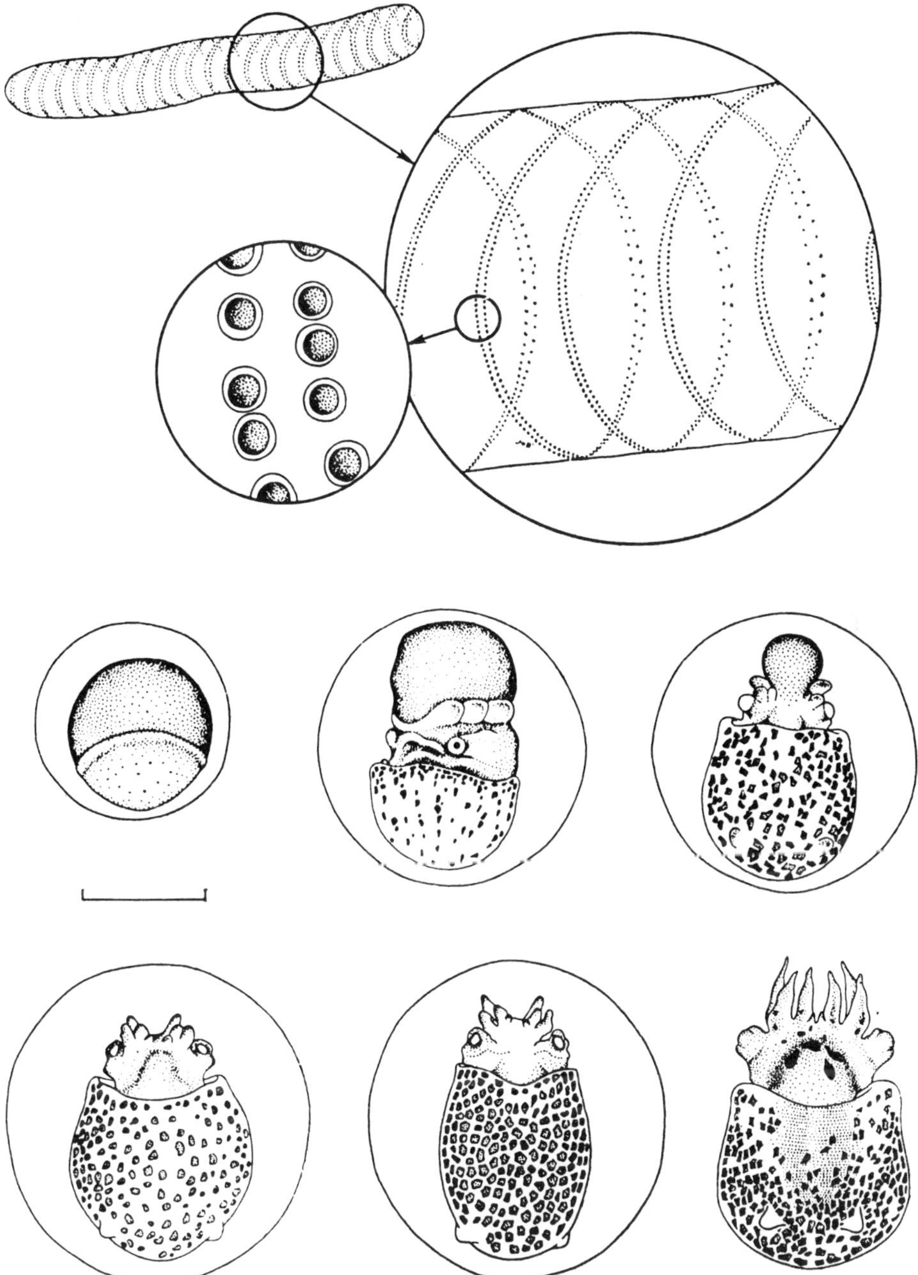

Figure 7 Above, diagrammatic representation of the large, floating egg capsule of the pelagic diamondback squid *Thysanoteuthis rhombus*, with details of egg strings in double helix. Below, selected embryonic stages in increasingly expanded chorion, and hatchling (lower right) at same magnification. Scale bar 1 mm. (From Sabirov et al. 1987).

due to observations of mechanically disturbed egg masses in which the original arrangement of the eggs is no longer recognizable (Misaki & Okutani 1976, Suzuki et al. 1979). In contrast, in ommastrephid squids the paired source of eggs is no longer recognizable in the amorphous egg mass, which forms a spherical jelly "balloon" measuring about 80–100 cm in diameter in *Illex illecebrosus* (Durward et al. 1980, O'Dor & Balch 1985) and *Todarodes pacificus* (Hamabe 1961, 1963, Bower & Sakurai 1996). Naef (1928: 172) discussed various records of the early literature (e.g. Quoy & Gaimard 1830) in relation to his unidentified "oegopsid x" and "ommastrephid y" eggs. For pelagic squids, other than those that have been observed during spawning of single eggs (e.g. *Watasenia scintillans*, T. Okutani pers. comm.), it is difficult to reconstruct the spawning mode from fractionated egg masses or isolated eggs recovered from plankton samples (Hayashi et al. 1987), although in some cases (e.g. enoploteuthid squids, *Brachioteuthis* sp.) spawning of single eggs appears very likely (Young & Harman 1985, Young et al. 1985a,b).

The greatest surprise of recent years has been the observation that female squids of the family Gonatidae carry egg masses in their arm crown (Okutani et al. 1995, B. A. Seibel pers. comm.; see also Björke et al. 1997). F. G. Hochberg (pers. comm.) considers that all squid species with eggs over 3 mm may brood their eggs, which goes hand in hand with tentacle loss. Okiyama (1993) views tentacle loss in both sexes as related to mating. The observed female behaviour is the first instance of distinct post-spawning egg-care in teuthoid squids. Loliginid males paying attention to freshly laid egg masses have been observed earlier (Tardent 1962), and spawning female cuttlefish may become very aggressive and attack an experimenter taking samples from an egg mass (S. v. Boletzky pers. obs.), but such signs of ostensibly protective behaviour are not comparable to a continued egg-care behaviour, which was believed to be a specialty of the (incirrate) octopods (see below, p. 354).

Vampyromorpha

Vampyroteuthis infernalis lays nearly spherical eggs about 3.5 mm in diameter, apparently without (preservable) jelly coats, but embryonic stages have not been observed (Pickford 1949a, 1950a). From the youngest planktonic stage ever described (Pickford 1949b), one can draw a "search image" for the identification of advanced embryos in eggs taken with plankton nets at depths ranging from about 3000 m to 6000 m: typical cephalopod embryos, probably with a distinct outer yolk sac until shortly before hatching, and with 10 arms or arm rudiments, none of which is greatly enlarged relative to the others, but with the second pair of arms (dorsolateral arms) somewhat reduced in size and devoid of special structures (cirri) on their adoral side.

Finned octopods

The finned Cirroctopoda lay large, individually encapsulated eggs measuring about 10–30 mm in length, which they apparently stick to a solid substratum. The rigid outer capsule may be smooth or sculptured; it either encloses the embryo tightly or provides space for strong expansion of the chorion (Fig. 8). Such eggs are found in dredge samples from depths generally greater than 1000 m (Boletzky 1982, 1985, Hochberg et

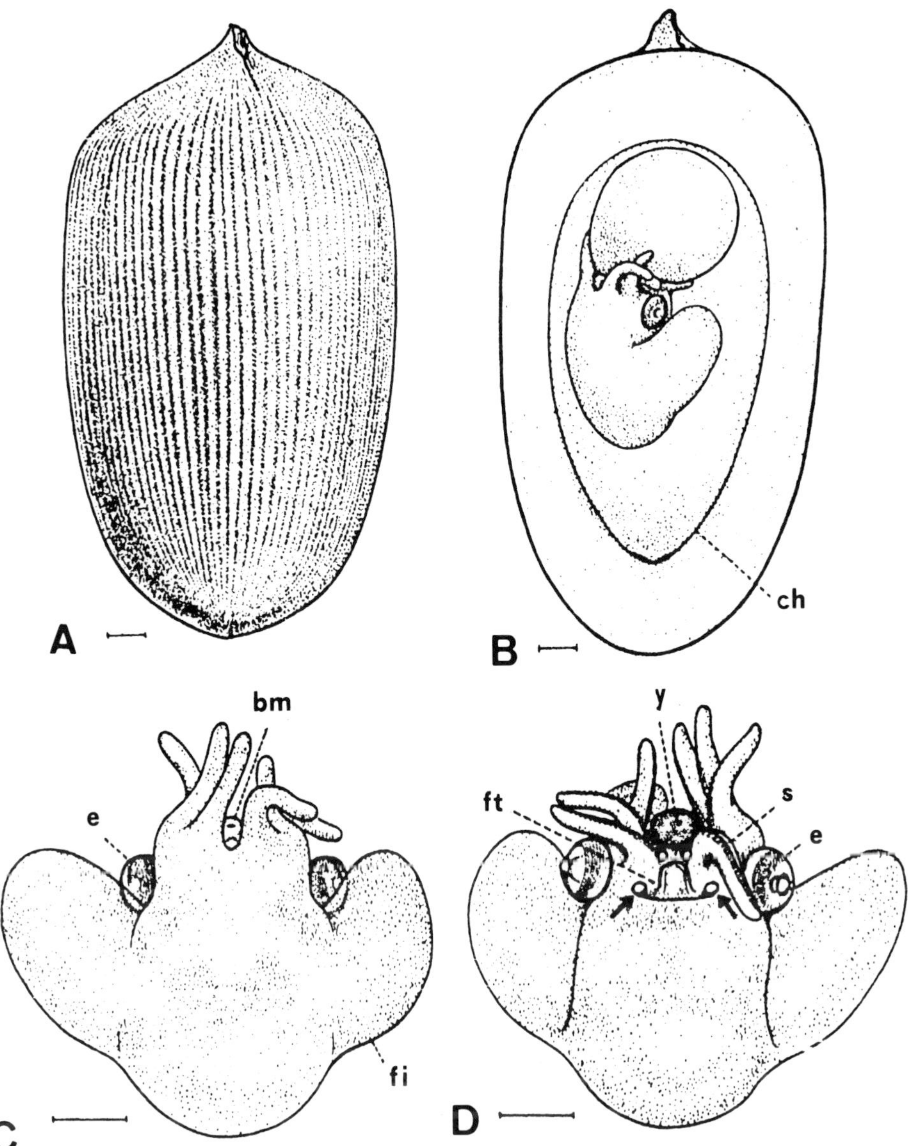

Figure 8 Cirroctopod egg from the subantarctic waters in the area of the Kerguelen Islands. A, outer aspect of the rigid egg case with regular longitudinal ribs. B, egg case opened to expose the much smaller chorion (ch) surrounded by gelatinous material. The embryo is not tightly enclosed in the chorion (in contrast to other cirroctopod embryos observed at similar stages). C, dorsal view of the embryo proper (without the yolk sac) showing the buccal mass (bm) between the bases of the dorsal arms, the eyeballs (e) not yet covered by the corneal skin, the large fins (fi). D, ventral view of the embryo showing the yolk (y) severed, the arms with one row of rudimentary suckers (s), the eyeballs (e) with the eye lens, the funnel tube (ft) emerging from the mantle aperture between the olfactory vesicles (arrows). Scale bars 1 mm. (From Boletzky 1982).

al. 1992). Very large gastropod eggs may look similar to certain cirroctopod eggs, so positive identification depends on the observation of a recognizable cephalopod embryo inside the capsule.

Finless octopods

The finless or incirrate Octopoda lay eggs without gelatinous envelopes around the egg chorion. The eggs of this group of octopods are guarded by the female until the young hatch. During the final stages of oogenesis, the follicular tissue surrounding the chorion

Figure 9 An advanced immature oocyte with follicular folds (left) and a nearly mature oocyte (right), both from the ovary of an *Octopus vulgaris*. The follicular tissue still covers the entire chorion surface (including the micropyle mp), and ends in a follicle stalk (fs); a chorion stalk is not yet differentiated (for a mature egg, see middle of Fig. 10).

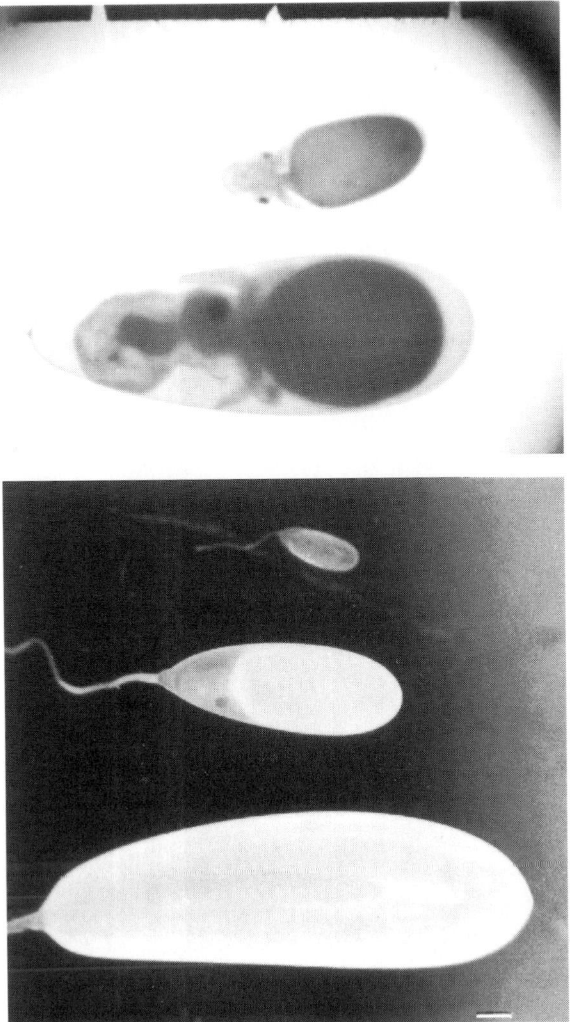

Figure 10 Eggs from five octopodid species shown at same magnification. The upper two eggs are from shallow water octopus species of the western Atlantic (Florida): at the top a pygmy octopus, *Octopus* sp. X (formerly considered *O. joubini*, see Forsythe & Toll 1991); second from top the Caribbean reef octopus, *O. briareus*, both at advanced developmental stages, the outer yolk sac lying at the side of the micropyle and the embryo proper close to the chorion stalk (as a result of the so-called first reversal). The lower three eggs are from species of the eastern Atlantic and Mediterranean (see also Fig. 11): the common octopus, *O. vulgaris*, the chorion measuring only 2 mm (without the stalk) at an early embryonic stage; the horned octopus (or northern lesser octopus) *Eledone cirrhosa*, similar in size to egg of *Octopus* sp. X; the musky octopus (or Mediterranean lesser octopus) *Eledone moschata* at a very early embryonic stage (top scale with indentations at 5 mm intervals; bottom scale bar 1 mm).

(previously secreted around the ovum) forms a stalk that forces the proximal part of the chorion to thin out (Fig. 9). Whether the number and arrangement of follicular folds visible during these oogenetic stages (Orelli 1960) provide generic or species-specific characters remains to be verified (F. G. Hochberg pers. comm.). The fully

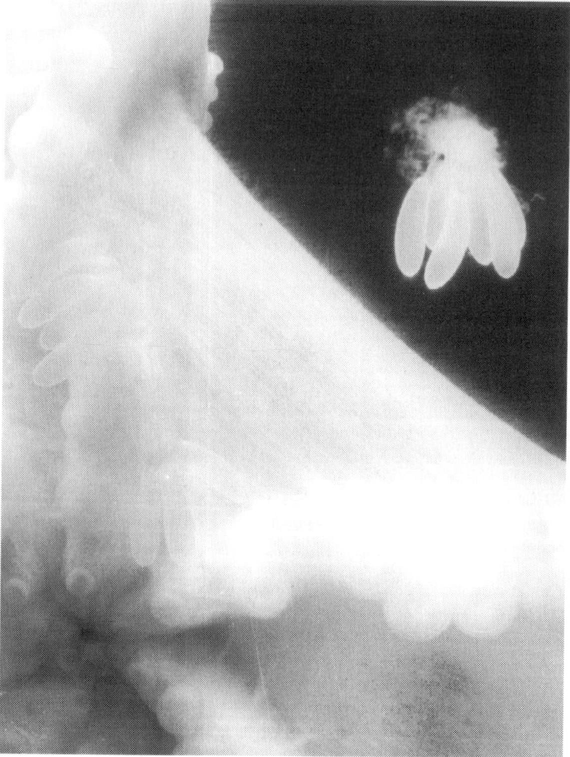

Figure 11 Egg strings (festoons) or clusters of three octopodids from the eastern Atlantic and Mediterranean shown at same magnification: short strings from *Eledone cirrhosa* (top left), long strings from *Octopus vulgaris* (top right), clusters in *Eledone moschata* (bottom; "brooding" female taken through the aquarium window). Scale bar 10 mm.

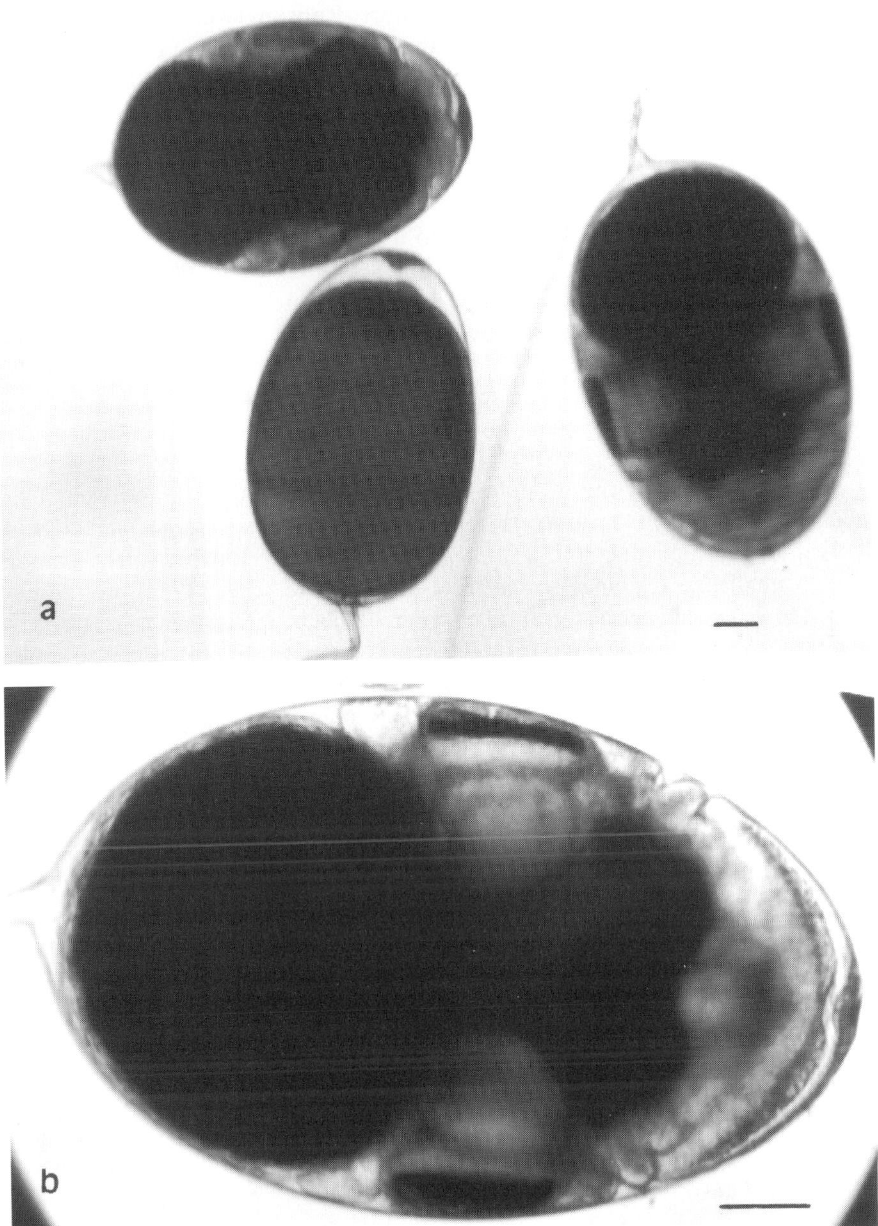

Figure 12 *Argonauta argo*: random sample of eggs taken from the brood shell of a female (a). Note size difference between late gastrula stage (a, bottom left), mid-organogenetic stage (a, top left) and advanced organogenetic stages (a, right and b). Note also the position of the embryo in relation to the chorion stalk (b: at left) and micropyle hump (b: at right) (no reversion of embryo in this species). Scale bar 0.1 mm.

formed chorion stalk may lack a distinct demarcation from the egg chamber, as in *Octopus maorum* (Batham 1957); it may be short (Robson 1932a, Boletzky 1981, Ito 1983, Stranks 1988, Lu & Stranks 1991, O'Shea & Kubodera 1996) or long, depending on the species. Indeed its final length is species-specific (Pickford 1950b) as is the size and shape of the chorionic chamber containing the ovum (Fig. 10). The chorion chambers in spawned eggs of benthic octopods range in size from 2 mm to almost 40 mm (Hochberg et al. 1992).

In the bottom-living Octopodidae, another feature that varies between species is the way in which the chorion stalks are glued to a substratum (with material secreted by the oviducal glands), either individually (Arakawa 1962), or in small clusters with a common fixation disk, or as festoons made from many interwoven stalks that are glued together (Fig. 11). Such festoons, which can be branching (Norman 1992a,b), are either glued to a hard surface in a den (Fisher 1923, Rees 1956, Vevers 1961, Brough 1965, Boletzky 1969, 1981, 1984, Cousteau & Diolé 1973, Overath & Boletzky 1974, Gabe 1975, Joll 1976, Mangold-Wirz et al. 1976, Van Heukelem 1977, Ambrose 1981, Haaker 1985, Cosgrove 1993), or in a bivalve or other mollusc shell (Robson 1929), or they are carried around by the brooding female in the ventral arms and web. The last mode has been observed with relatively large eggs (*Hapalochlaena maculosa*: Dew 1959a, Tranter & Augustine 1973; *Graneledone* sp. F. G. Hochberg pers. comm.) as well as with small eggs (*Octopus aegina*: Eibl-Eibesfeldt & Scheer 1962; probably other species of the *O. aegina* species group: Norman 1992a,b; *Octopus burryi*: Forsythe & Hanlon 1985; *Scaeurgus* sp.: W. F. Van Heukelem pers. comm., see also Boletzky 1984). But these are a minority among the bottom-living octopodids. When such festoons are found in the stomach contents of predators, it may be difficult to identify the species and to reconstruct the spawning or carrying mode (Nesis & Nigmatullin 1978).

To carry the eggs represents the normal mode of egg-care in the half-dozen pelagic octopod families. The most elaborate forms of egg-care involve modifications of the arm crown as in the bolitaenids (Young 1972) or in *Haliphron* (Young 1995); production, by the dorsal arms, of special egg-carriers as in *Tremoctopus* (Naef 1923, 1928, Hamabe 1973) or a "brood shell" as in *Argonauta* (Kölliker 1844, Naef 1923, 1928); or ovovivipary as in *Ocythoe* (Naef 1923, 1928). Figure 12 shows embryos taken from the brood shell of a female *Argonauta argo*; they are at very different developmental stages, a situation typical for the virtually continuous, multiple spawning of this species. Apart from their very small size, the eggs of *Argonauta* spp. can be distinguished from any other octopod egg by the simple fact that the embryos do not undergo a reversal in the chorion chamber (Portmann 1933, 1937, Nesis 1977, Boletzky & Fioroni 1990); the mantle remains at the side of the thickened micropyle area of the chorion throughout embryonic development. Eggs of other pelagic octopods are still poorly known (Robson 1932a, Thore 1949, Hochberg et al. 1992).

Capsule form changes related to embryonic development

As briefly mentioned in the Introduction, a major difficulty in identifying cephalopod eggs and egg masses of unknown origin is the variety of "aspects" they offer depending on the developmental stage of the embryos they contain. The most obvious changes occur in the embryos themselves, leading from a spherical, ovoid, or sub-cylindrical zygote through a series of morphogenetic stages to the stage of the hatchling, which already is a miniature cephalopod with virtually all the organs of an adult (Boletzky 1974b). At late embryonic stages it is generally easy to distinguish decabrachian from octobrachian embryos by counting the arm rudiments. However, certain squids (e.g. Ommastrephidae: Naef 1923, 1928, Okiyama 1965, Okiyama & Kasahara 1975, O'Dor et al. 1982, Laptikhovsky & Murzov 1990, Watanabe et al. 1996) do not form the full complement of arms before hatching; moreover the characteristic fusion of the tentacles forming the proboscis of the so-called rhynchoteuthion stage, so typical of ommastrephids, occurs only late in embryonic development. Similarly, the ventral arms of the pelagic bobtail *Heteroteuthis dispar* remain very inconspicuous so that an advanced embryonic stage appears as "octobrachian" (Boletzky 1978). In the pygmy cuttlefish, *Idiosepius pygmaeus*, the hatchling is really eight-armed in that the tentacle rudiments are arrested in an early morphogenetic stage (Natsukari 1970) and grow out only during juvenile development (i.e. the opposite of what happens in tentacle-less squids such as *Gonatopsis*, in which tentacles are present at hatching and during juvenile life only).

The envelopes or capsules enclosing the developing embryos undergo their own changes, in more or less close relation to the metabolism of the embryo. The first task they have is to offer a track to the spermatozoa added during spawning (Ikeda et al. 1993). There is now evidence for the presence of sperm from more than one male in an egg capsule (Shaw & Boyle 1997). It is apparently the accumulation of large molecules in the perivitellin fluid that generates water uptake through the semi-permeable chorion (De Leersnyder & Lemaire 1972, Gomi et al. 1986, Boletzky 1987). The ensuing volume increase is generally stronger in decabrachian eggs than in the eggs of octopods. This is probably due to the presence of oviducal jelly on the outside of the chorion, since lack of this material impedes normal chorion swelling (Arnold & O'Dor 1990). Perhaps the limited volume increase of octopus eggs is also the result of reduced pore size in the chorion membrane, but in contrast to what is often said, octopus eggs do increase in volume during embryonic development. Wells & Wells (1977) calculated that the increase in volume of the chorion capsule during development is 80% in *Octopus vulgaris*, 125% in *O. briareus*, and more than 150% in *O. joubini* (now *Octopus* sp. X; see Forsythe & Toll 1991) and in *Eledone cirrhosa*.

In those species where the outer coat of the egg capsule becomes a rigid "shell" soon after laying, the final volume of the egg can only be equal to that shell capacity. Apart from *Nautilus* (Arnold et al. 1993b), this situation is typical for the cirroctopod eggs (Fig. 8; Boletzky 1982) and for the eggs of the large sepiolid genus *Rossia* (Boletzky & Boletzky 1973). In all these instances, embryonic development covers several to many months, especially where temperatures are minimal, so that the eggs need a solid protection allowing the developing embryo to become a "sessile organism" (Boletzky 1994). How "complete" protection is under such conditions remains to be seen (Arnold et al. 1993a). All decabrachian eggs and egg masses (except those of *Rossia*) undergo a modification of their outline. It can be feeble (e.g. most sepiolid eggs have a leathery, slightly expandable outer coat, as shown in Fig. 4) or very marked, as in cuttlefish eggs (Fig. 3) or in most loliginid squid egg capsules (Figs 5b, 6). What invariably happens is a decrease in total wall thickness, so that the barrier for gas exchange is lowered and respiration facilitated (Wolf et al. 1985; see also Bouchaud & Daguzan 1989, Bouchaud 1991).

The processes involved in these structural changes are not yet known in detail. What is generally termed nidamental mucous secretion (Atkinson 1973), forming the mucinous capsule material, is a complex mixture of mucosubstances, only part of which is mucin (Sugiura & Kimura 1995). The progressive shrinkage of the nidamental jellies in decabrachian eggs is not simply due to the expansion of the chorionic volume. Cuttlefish eggs show very clearly the early loss of water from these envelopes, long before massive water uptake by the perivitellin fluid forces them to thin out (Jecklin 1934, Paulij 1991), or when no ovum is present (Boletzky 1975). This shrinkage of the jelly is somehow – in slow motion – the reverse of what happened at the moment of capsule formation, when the mucosubstances released by the oviducal and nidamental glands took up water rather quickly. This reverse process can be triggered artificially by placing a freshly laid egg capsule in alcohol; one then realizes that a dehydrated squid egg capsule fits transversely into the adult mantle cavity – an orientation probably necessary for the spiral rolling of the jelly sheaths (Jecklin 1934).

If the earliest task of jelly coats was to offer an appropriate substrate for spermatozoid locomotion, its last task is to offer an appropriate substrate for the locomotion of the hatchling when it leaves the capsule, using its hatching gland (Yung 1930, Denucé & Formisano 1982). Especially in loliginid squids, the purely ciliary locomotion of the hatchling, when moving through the tunnel opened by the hatching enzyme, requires a rather loose jelly consistency (Boletzky 1979). Experiments made with hatchlings introduced into capsules containing earlier embryonic stages show that the animals get stuck in these immature nidamental envelopes (S. v. Boletzky unpubl. obs.). On the other hand, similar experiments using the chorion membrane of a different species show that the hatching gland secretion is not species-specific; indeed squid hatchlings can open cuttlefish chorion and vice versa, but there is some group specificity, since octopod hatchlings cannot open decabrachian chorion, and vice versa (Boletzky 1992b).

Among the many unanswered questions relating to egg masses is the significance of bacteria found in various egg capsules (Herfurth 1936, Boletzky & Boletzky 1973, Biggs & Epel 1991), which apparently are not related to symbiotic luminous bacteria (e.g. in sepiolids, see McFall-Ngai 1994), and the role of polychaete annelids often found in the nidamental capsules of loliginids (Boletzky & Dohle 1967, Braun 1992).

Damage due to collection and/or maintenance conditions

Collection of cephalopod eggs and egg masses at sea is always burdened with a high risk of damaging the delicate structures, especially of floating egg masses. The example of *Thysanoteuthis* mentioned earlier shows that despite careful handling of an egg mass by experienced SCUBA divers, the slightest mechanical disturbance can change some details (double egg helix appearing single, in the case mentioned). The use of nets or trawls to catch pelagic floating egg masses will inevitably destroy the delicate structure of the jelly mass. What appears in preserved plankton samples as single eggs devoid of jelly may have been originally embedded in an extremely fragile gelatinous balloon. But even under ostensibly controlled aquarium conditions, intact egg masses (whether collected at sea or spawned in captivity) can be damaged by various noxious factors that are often difficult to assess. For example, disintegration of nidamental jellies in otherwise normal loliginid egg masses is frequently observed in the aquarium.

By far the most problematic phenomenon is progressively disturbed embryonic development in aquarium-raised egg masses, because the difference between normal and anomalous embryogenesis is not easily recognized. In particular, the viable embryos and hatchlings that have nevertheless undergone suboptimal conditions may be a source of errors in descriptions of a supposedly normal development (e.g. shape of yolk sac in some embryonic stages of *Loligo opalescens* described by Fields 1965), let alone the problems arising in rearing experiments using such individuals. Oxygen depletion (perhaps only in one part of an egg mass) and water temperatures too high or too low for the species, or salinity fluctuations can trigger slow degeneration or other abnormalities. Only an experienced observer is able to tell the difference between a healthy and a slightly disturbed embryo, for example, by looking at the yolk system (a slightly expanded yolk neck and inner yolk sac being one of the criteria generally used).

The final step of losing potentially important information is related to post-mortem maintenance, that is, inappropriate fixation and preservation of eggs and egg masses. Fortunately, the most simple methods generally provide the best compromise between the practically feasible and the scientifically desirable. Use of zinc formalin (F. G. Hochberg & R. E. Young pers. comm.), or neutralized 4% formalin (i.e. 1 volume of saturated 40% formaldehyde in 9 volumes of sea water buffered with calcium carbonate) is sufficient for fixation and short-term preservation of cephalopod embryos and their capsule material (see Roper & Sweeney 1983). If there is a rigid outer egg case, it should be perforated or partly broken to allow rapid penetration of the fixative. Embryonic material preserved in 4% formalin can later be rinsed in distilled water and post-fixed (e.g. with osmic acid for scanning electron microscopy). If formalin is not available (or prohibited for health reasons), cephalopod eggs and egg masses can be placed directly in 70% ethanol, a poor fixative for the embryonic material, leading to strong shrinkage of jelly capsules. Material so preserved can be rehydrated subsequently but will not recover its original volume. Ideally, embryos are dissected from the capsules and fixed in Bouin's solution (or in glutaraldehyde and/or osmic acid if electron microscopy is planned) along with some capsule material. A parallel sample of capsule material should be deep-frozen, or placed in 70% or 95% ethanol for subsequent molecular or chemical analyses.

Collection and observation routines

Under normal conditions of work aboard research vessels on oceanographic cruises, the last sentence of the preceding section can only appear illusory. No one will have time to dissect cephalopod embryos from eggs recovered from some sampling gear – unless a cephalopod specialist in person is present and ready to act. In order to save what can be saved, it is therefore wise to consider only simple procedures: if possible, dump the cephalopod eggs in a jar with 4% formalin and label it "Cephalopod eggs, Station ..., Cruise ..., Date . . .". When dispatching material obtained during a cruise, it is suggested that collections of cephalopod eggs be deposited with a museum that is used to handling such samples and will be able to get information to potentially interested researchers.

If a camera is at hand, it is always useful to take still photographs or a video footage of fresh egg mass material, making sure that some scale is on the picture (the photographers foot being better than nothing), and keeping a note (Station ..., Cruise ..., Date ...) about the event, especially if egg samples are obtained from stomach contents of predators (Nesis & Nigmatullin 1978). How misleading insufficient labelling can be is demonstrated by a description of an "Egg mass of *Eledone cirrhosa*" (Rees 1956: 179–80 and Pl. 9) from the collections of the Plymouth Laboratory, which must be the contents of the ovary of a mature female.

If the research vessel has an aquarium facility, egg masses (maintained in a suspended net or basket), or mature female cephalopods captured in good condition, may be kept alive in a shipboard tank. If such females spawn in captivity, the embryonic development can be observed from its very beginning (Boletzky 1987, Cheslin & Giragosov 1993). If no aquarium facilities are available, an egg mass of moderate size (preferably suspended from a floater) can be kept for a short period in a plastic bag containing well-oxygented sea water (Marthy 1972), or for a longer period if the sea water is changed daily. But even in such instances part of a larger sample should be preserved immediately after capture (and later perhaps some more material at periodic, for example, daily, intervals if that is feasible).

SCUBA divers collecting material in the vicinity of a marine laboratory may carry egg samples in a bucket or in a plastic bag (ideally in a cooler) holding enough sea water for transportation.

Finally, the beachcomber may encounter stranded cephalopod egg masses. Depending on how long the material has been exposed to air and sunlight, it may still be viable and worthy of being inspected to decide whether a sample should be saved (and again, a photograph or video footage may later provide useful information).

Discussion and conclusions

Cephalopod eggs and egg masses are curiosities that easily get lost in the course of a marine biological survey, because they are rather untidy and generally difficult to identify. It is therefore tempting to let them go down the drain, in the hope that they were of no special interest anyway. Even though much of the material thus cleaned out on shipboard may indeed be common stuff, there are at least two compelling reasons for saving it.

(1) Only specialists are able to tell the difference between known and hitherto unknown material, so all the material available has to get to them.

(2) Even the known material may provide insights into the reproductive patterns and biogeography of species, be it on a year-to-year basis or in relation to global changes of hydrological conditions.

As to the unknown material, nearly a decade ago, the hope was expressed "that ten years from now the embryonic development of *Spirula* and of *Vampyroteuthis*, especially with regard to cleavage patterns and shell sac differentiation, will be known" (Boletzky 1989). That is still a hope. In contrast, much has been achieved through laboratory and aquarium studies. If *Spirula* and *Vampyroteuthis* eggs are not found in the sea, perhaps artificial fertilization of mature ovarian eggs recovered from freshly caught mature females, using spermatozoids from fresh males, could fill the gap (Arnold & O'Dor 1990).

As to the known material, it is still true that the specialist who "knows" is rarely on the spot where the non-specialist "wants to know". So either the material is brought to the specialist, directly or via museum collections, or the non-specialist is given enough information to identify his/her material him/herself. It is noteworthy that since the monograph of Naef (1928) and the stage descriptions by Arnold (1965b), most of the detailed descriptions and tabulations of "normal embryonic stages" of cephalopods have been provided by authors who are not specialists in embryology (Harman & Gardiner 1927, Fields 1965, Natsukari 1970, Joll 1978, Ambrose 1981, O'Dor et al. 1982, Yamamoto 1982, Ito 1983, Forsythe & Hanlon 1985, Segawa 1987, Segawa et al. 1988, Baeg et al. 1992, Watanabe et al. 1996). These tabulations are so useful to the general biologist engaged in sorting and identification of samples, because they pay attention to what is recognizable by any interested observer. The long awaited "comprehensive presentation" of cephalopod eggs and egg masses alluded to in the Introduction will have to rely on such tables of normal embryonic stages combined with an illustration of the concomitant changes in the capsule material.

Acknowledgements

I thank Dr F. G. Hochberg (Natural History Museum, Santa Barbara, Southern California) and Professor P. R. Boyle (Zoology Department, University of Aberdeen, Scotland) for critical reading of the manuscript and for many helpful suggestions.

This article is dedicated to the late Professor Pierre Tardent (University of Zurich, Switzerland) in gratitude for his lasting encouragement and support for zoological research in marine biology.

References

Aldrich, F. A. & Lu, C. C. 1968. Report on the larva, eggs, and egg mass of *Rossia* sp. (Decapoda, Cephalopoda) from Bonavista Bay, Newfoundland. *Canadian Journal of Zoology* **46**, 369–71.

Ambrose, R. F. 1981. Observations on the embryonic development and early post-embryonic behavior of *Octopus bimaculatus*. *Veliger* **24**, 139–46.

Anderson, R. C. & Shimek, R. L. 1994. Field observations of *Rossia pacifica* (Berry, 1911) egg masses. *Veliger* **37**, 112–19.

Arakawa, K. Y. 1962. An ecological account on the breeding behavior of *Octopus luteus* (Sasaki). *Venus (Japanese Journal of Malacology)* **22**, 176–80.

Arnold, J. M. 1965a. Normal embryonic stages of the squid *Loligo pealii* (LeSueur). *Biological Bulletin* **128**, 24–30.

Arnold, J. M. 1965b. Observations on the mating behavior of the squid *Sepioteuthis sepioidea*. *Bulletin of Marine Science* **15**, 216–22.

Arnold, J. M. 1984. Cephalopods. In *The Mollusca. Vol. 7. Reproduction.* A. S. Tompa et al. (eds). London: Academic Press, 419–54.

Arnold, J. M., Awai, M. & Carlson, B. A. 1993a. Speculation and comments on predation of *Nautilus* on egg capsules of other *Nautilus*. *Journal of Cephalopod Biology* **2**(2), 47–50.

Arnold, J. M., Awai, M. & Carlson, B. A. 1993b. Nautilus embryology: egg capsule formation, deposition, and hatching. *Journal of Cephalopod Biology* **2**(2), 51–6.

Arnold, J. M. & Carlson, B. A. 1986. Living *Nautilus* embryos: preliminary observations. *Science* **232**, 73–6.

Arnold, J. M. & O'Dor, R. 1990. *In vitro* fertilization and embryonic development of oceanic squid. *Journal of Cephalopod Biology* **1**(2), 21–36.

Arnold, J. M., Singley, C. T. & Williams-Arnold, L. D. 1972. Embryonic development and post-hatching survival of the sepiolid squid *Euprymna scolopes* under laboratory conditions. *Veliger* **14**, 361–4.

Arnold, J. M. & Williams-Arnold, L. D. 1977. Cephalopoda: Decapoda. In *Reproduction of marine invertebrates, Vol. IV, Molluscs: gastropods and cephalopods* , A. C. Giese & J. S. Pearse (eds). New York: Academic Press, 243–90.

Atkinson, B. G. 1973. Squid nidamental gland extract: isolation of a factor inhibiting ciliary activity. *Journal of Experimental Zoology* **184**, 335–40.

Baeg, G. H., Sakurai, Y. & Shimazaki, K. 1992. Embryonic stages of *Loligo bleekeri* Keferstein (Mollusca: Cephalopoda). *Veliger* **35**, 234–41.

Batham, E. J. 1957. Care of eggs by *Octopus maorum*. *Transactions of the Royal Society of New Zealand* **84**, 629–38.

Biggs, J. & Epel, D. 1991. Egg capsule sheath of *Loligo opalescens* Berry: structure and association with bacteria. *Journal of Experimental Zoology* **259**, 263–7.

Björke, H., Hansen, K. & Sundt, R. C. 1997. Egg masses of the squid *Gonatus fabricii* (Cephalopoda, Gonatidae) caught with pelagic trawl off northern Norway. *Sarsia* **82**, 149–52.

Blainville, H. de 1837. Le Poulpe de l'Argonaute. *Annales d'Anatomie et de Physiologie*, **Mai 1837**, 1–30.

Bohadsch, J. B. 1752. De veris sepiarum ovis. Prague: F. I. Kirchner.

Boletzky, S. v. 1969. Zum Vergleich der Ontogenesen von *Octopus vulgaris*, *O. joubini* und *O. briareus*. *Revue Suisse de Zoologie* **76**, 715–26.

Boletzky, S. v. 1974a. Elevage de céphalopodes en aquarium. *Vie et Milieu* **24**(2A), 309–40.

Boletzky, S. v. 1974b. The "larvae" of Cephalopoda: a review. *Thalassia jugoslavica* **10**, 45–76.

Boletzky, S. v. 1975. Spawning behaviour independent of egg maturity in a cuttlefish (*Sepia officinalis* Linnaeus). *Veliger* **17**, 247–8.

Boletzky, S. v. 1978. Premières données sur le développement embryonnaire du Sepiolidé pélagique *Heteroteuthis* (Mollusca, Cephalopoda). *Haliotis* **9**, 81–4.

Boletzky, S. v. 1979. Ciliary locomotion in squid hatching. *Experientia* **35**, 1051–2.

Boletzky, S. v. 1981. Morphologie de l'oeuf et mode de ponte chez *Pteroctopus tetracirrhus* (Mollusca, Cephalopoda). *Vie et Milieu* **31**, 255–9.

Boletzky, S. v. 1982. On eggs and embryos of cirromorph octopods. *Malacologia* **22**, 197–204.

Boletzky, S. v. 1984. The embryonic development of the octopus *Scaeurgus unicirrhus* (Mollusca, Cephalopoda) – additional data and discussion. *Vie et Milieu* **34**, 87–93.

Boletzky, S. v. 1985. Céphalopodes. In *Peuplements profonds du Golfe de Gascogne*, L. Laubier & C. Monniot (eds). Brest: IFREMER Service Documentation-Publications, 401–8.

Boletzky, S. v. 1986. Encapsulation of cephalopod embryos: a search for functional correlations. *American Malacological Bulletin* **4**, 217–27.

Boletzky, S. v. 1987. On egg and capsule dimensions in *Loligo forbesi* (Mollusca: Cephalopoda): a note. *Vie et Milieu* **37**, 187–92.

Boletzky, S. v. 1989. Recent studies on spawning, embryonic development, and hatching in the Cephalopoda. *Advances in Marine Biology* **25**, 85–115.

Boletzky, S. v. 1992a. Evolutionary aspects of development, life style, and reproductive mode in incirrate octopods (Mollusca, Cephalopoda). *Revue Suisse de Zoologie* **99**, 755–70.

Boletzky, S. v. 1992b. Kopffüsser: Schlüpfprozesse. In *Meeresbiologische Exkursion – Beobachtung und Experiment*, P. Emschermann et al. (eds). Stuttgart: Gustav Fischer Verlag, 216–18.

Boletzky, S. v. 1994. Embryonic development of cephalopods at low temperatures. *Antarctic Science* **6**, 139–42.

Boletzky, S. v. & Boletzky, M. V. v. 1973. Observations on the embryonic and early post-embryonic development of *Rossia macrosoma* (Mollusca, Cephalopoda). *Helgoländer Wissenschaftliche Meeresuntersuchungen* **25**, 135–61.

Boletzky, S. v. & Dohle, W. 1967. Observations sur un capitellidé (*Capitella hermaphrodita* sp. n.) et d'autres polychètes habitant la ponte de *Loligo vulgaris*. *Vie et Milieu* **18**(1A), 79–98.

Boletzky, S. v. & Fioroni, P. 1990. Embryo inversions in incirrate octopods: the state of an enigma. *Journal of Cephalopod Biology* **1**(2), 37–57.

Bonnaud, L., Boucher-Rodoni, R. & Monnerot, M. 1997. Phylogeny of cephalopods inferred from mitochondrial DNA sequences. *Molecular Phylogenetics and Evolution* **7**, 44–54.

Bott, R. 1938. Kopula und Eiablage von *Sepia officinalis* L. *Zeitschrift für Morphologie und Oekologie der Tiere* **34**, 150–60.

Bouchaud, O. 1991. Energy consumption of the cuttlefish *Sepia officinalis* L. (Mollusca: Cephalopoda) during embryonic development, preliminary results. *Bulletin of Marine Science* **49**, 333–40.

Bouchaud, O. & Daguzan, J. 1989. Etude du développement de l'oeuf de *Sepia officinalis* L. (Céphalopode, *Sepiidae*) en conditions expérimentales. *Haliotis* **19**, 189–200.

Bouligand, Y. 1961. Le dispositif d'accrochage des oeufs de *Sepia elegans* sur *Alcyonium palmataum*. *Vie et Milieu* **12**, 589–93.

Boyle, P. R. (ed.). 1983. *Cephalopod life cycles, Vol. I, Species accounts*. London: Academic Press.

Bower, J. R. & Sakurai, Y. 1996. Laboratory observations on *Todarodes pacificus* (Cephalopoda: Ommastrephidae) egg masses. *American Malacological Bulletin* **13**, 65–71.

Braun, T. 1992. *Untersuchungen an* Capitella hermaphrodita (*Annelida: Polychaeta*) *aus der Laichgallerte von* Loligo vulgaris (*Mollusca: Cephalopoda*). Diploma dissertation, University of Bochum.

Brough, E. J. 1965. Egg-care, eggs and larvae in the midget octopus, *Robsonella australis* (Hoyle). *Transactions of the Royal Society of New Zealand* **6**, 5–19.

Cheslin, M. V. & Giragosov, V. Ye. 1993. The egg mass and embryonic development of the purple squid *Stenoteuthis oualaniensis* (the gigantic Arabian form) under experimental conditions. *Oceanology* **33**, 98–101.

Choe, S. 1966. On the eggs, rearing, habits of the fry, and growth of some Cephalopoda. *Bulletin of Marine Science* **16**, 330–48.

Choe, S. & Oshima, Y. 1961. On the embryonal development and the growth of the squid, *Sepioteuthis lessoniana* Lesson. Venus (*Japanese Journal of Malacology*) **21**, 462–75.

Clarke, M. R. 1966. A review of the systematics and ecology of oceanic squids. *Advances in Marine Biology* **4**, 91–300.

Cosgrove, J. A. 1993. *In situ* observations of nesting female *Octopus dofleini* (Wülker, 1910). *Journal of Cephalopod Biology* **2**(2), 33–45.

Cousteau, J.-Y. & Diolé, P. 1973. *Pieuvres – La fin d'un malentendu*. Paris: Flammarion.

De Leersnyder, M. & Lemaire, J. 1972. Sur la composition minérale du liquide périembryonnaire de l'oeuf de *Sepia officinalis* L. *Cahiers de Biologie Marine* **13**, 429–31.

Denucé, J. M. & Formisano, A. 1982. Circumstancial evidence for an active contribution of Hoyle's gland to enzymatic hatching of Cephalopod embryos. *Archives Internationales de Physiologie et de Biochimie* **90**(4), B185–6.

Dew, B. 1959a. Some observations on the development of two Australian octopuses. *Proceedings of the Royal Zoological Society of New South Wales 1957–1958* **1**, 44–52.

Dew, B. 1959b. Some observations on the development of the Australian squid *Sepioloidea lineolata* Quoy & Gaimard 1832. *Proceedings of the Royal Zoological Society of New South Wales 1957–1958* **1**, 53–5.

Drew, G. 1911. Sexual activities of the squid, *Loligo pealii* (Les.). *Journal of Morphology* **22**, 327–51.

Durward, R. D., Vessey, E., O'Dor, R. K. & Amaratunga, T. 1980. Reproduction in the squid, *Illex illecebrosus*: first observations in captivity and implications for the life cycle. *International Commission for the Northwest Atlantic Fisheries Selected Papers* **6**, 7–13.

Eibl-Eibesfeldt, I. & Scheer, G. 1962. Das Brutpflegeverhalten eines weiblichen *Octopus aegina* Gray. *Zeitschrift für Tierpsychologie* **19**, 257–61.

English, S. A. 1981. *The biology of two species of estuarine cephalopods from the Sydney region*. MS thesis, School of Biological Sciences, University of Sydney.

Férussac, A. E. de & d'Orbigny, A. 1835–1848. Histoire naturelle générale et particulière des Céphalopodes acétabulifères vivants et fossiles. Paris: J.-B. Baillière, Vol. 1, 361 pp., Vol. 2, Atlas, 144 Pls.

Fields, W. G. 1965. The structure, development, food relations, reproduction, and life history of the squid *Loligo opalescens* Berry. *California Fish and Game* **131**, 1–108.

Fioroni, P. 1978. Cephalopoda. In *Morphogenese der Tiere, Lieferung 2: G5-I*, F. Seidel (ed.). Jena: Gustav Fischer.

Fisher, W. K. 1923. Brooding-habits of a cephalopod. *Annals and Magazine of Natural History, Series 9*, **12**, 147–9.

Forsythe, J. W. & Hanlon, R. T. 1985. Aspects of egg development, post-hatching behavior, growth and reproductive biology of *Octopus burryi*, Voss 1950 (Mollusca: Cephalopoda). *Vie et Milieu* **35**, 273–82.

Forsythe, J. F. & Toll, R. B. 1991. Clarification of the western Atlantic ocean pygmy octopus complex: the identity and life history of *Octopus joubini* (Cephalopoda: Octopodinae). *Bulletin of Marine Science* **49**, 88–97

Gabe, S. H. 1975. Reproduction in the giant octopus of the North Pacific, *Octopus dofleini martini*. *Veliger* 18, 146–50.

Gomi, F., Yamamoto, M. & Nakazawa, T. 1986. Swelling of egg during development of the cuttlefish, *Sepiella japonica*. *Zoological Science* **3**, 641–5.

Gravely, F. H. 1908. IV. Notes on the spawning of *Eledone* and on the occurrence of *Eledone* with the suckers in double rows. *Manchester Memoirs* (*Memoirs and Proceedings of the Manchester Literary and Philosophical Society*) **53**(4), 1–14.

Grimpe, G. 1928. Pflege, Behandlung und Zucht der Cephalopoden für zoologische und physiologische Zwecke. *Handbuch der Biologischen Arbeitsmethoden, Abteilung IX* **5**, 331–402.

Guerra, A. 1992. *Mollusca Cephalopoda. Fauna Iberica, Vol. 1*. Madrid: Museo Nacional de Ciencias Naturales C.S.I.C.

Guerra, A. & Rocha, F. 1997. On a floating egg mass of the diamond shaped squid *Thysanoteuthis rhombus* (Cephalopoda: Thysanoteuthidae) in the western Mediterranean. *Iberus* **15**, 125–30.

Haaker, P. L. 1985. Observations of a dwarf octopus, *Octopus micropyrsus*. *Shells and Sea Life*, 1985 (January), 39–40.

Hall, J. R. 1970. Description of egg capsules and embryos of the squid, *Lolliguncula brevis* , from Tampa Bay, Florida. *Bulletin of Marine Science* **20**, 762–8.

Hamabe, M. 1961. Experimental studies on breeding habit and development of the squid, *Ommastrephes sloani pacificus* Steenstrup. *Zoological Magazine* (*Dobutsugaku Zasshi*) **70**, 25–34.

Hamabe, M. 1963. Spawning experiments of the common squid, *Ommastrephes sloani pacificus* Steenstrup, in an indoor aquarium. *Bulletin of the Japanese Society of Scientific Fisheries* **29**, 930–34.

Hamabe, M. 1973. Egg mass and newborns of *Tremoctopus violaceus* Delle Chiaje, caught in the harbour of Kasumi, Hyogo Prefecture. *Bulletin of the Tokai Regional Fisheries Research Laboratory* **72**, 1–5.

Hamada, T., Obata, I. & Okutani, T. (eds. Japanese Expert Consultation on Living Nautilus) 1980. Nautilus macromphalus *in captivity*. Tokyo: Tokai University Press.

Hanlon, R. T., Boletzky, S. v., Okutani, T., Perez-Gandaras, G., Sanchez, P., Sousa-Reis, C. & Vecchione, M. 1992. Suborder Myopsida Orbigny, 1845. In *"Larval" and juvenile cephalopods: a manual for their identification*, M. J. Sweeney et al. (eds). *Smithsonian Contributions to Zoology* **513**, 37–53.

Harman, M. T. & Gardiner, A. H. 1927. Development of the external form of a squid embryo. *Publications of the Puget Sound Biological Station* **5**, 181–203.

Hayashi, S., Uchiyama, I., Kasahara, Sh. & Minami, T. 1987. Vertical distribution of eggs of the squid, *Watasenia scintillans* and some species of fishes in Toyama Bay, the Japan Sea. *Bulletin of the Japan Sea Regional Research Laboratory* **37**, 163–74.

Herfurth, A. H. 1936. Beiträge sur Kenntnis der Bakteriensymbiose der Cephalopoden. *Zeitschrift für Morphologie und Ökologie der Tiere* **31**, 561–607.

Hochberg, F. G., Nixon, M. & Toll, R. B. 1992. Order Octopoda. In *"Larval" and juvenile cephalopods: a manual for their identification*, M. J. Sweeney et al. (eds). *Smithsonian Contributions to Zoology* **513**, 213–79.

Ikeda, Y., Sakurai, Y. & Shimazaki, K. 1993. Fertilizing capacity of squid (*Todarodes pacificus*) spermatozoa collected from various sperm storage sites, with special reference to the role of gelatinous substance from oviducal gland in fertilization and embryonic development. *Invertebrate Reproduction and Development* **23**, 39–44.

Ito, H. 1983. Some observations on the embryonic development of *Paroctopus conispadiceus* (Mollusca: Cephalopoda). *Bulletin of the Hokkaido Regional Fisheries Research Laboratory* **48**, 93–105.

Izuka, T., Segawa, S. & Okutani, T. 1996. Biochemical study of the population heterogeneity and distribution of the oval squid *Sepioteuthis lessoniana* complex in southwestern Japan. *American Malacological Bulletin* **12**, 129–35.

Jaeckel, S. G. A. 1958. Die Cephalopoden. *Die Tierwelt der Nord- und Ostsee* **9**(b3), 479–723.

Jatta, G. 1896. I Cefalopodi Viventi nel Golfo di Napoli (Sistematica). *Fauna und Flora des Golfes von Neapel* **23**, 268 pp., 31 Pls.

Jecklin, L. 1934. Beitrag zur Kenntnis der Laichgallerten und der Biologie der Embryonen decapoder Cephalopoden. *Revue Suisse de Zoologie* **41**, 593–673.

Joll, L. M. 1976. Mating, egg-laying and hatching of *Octopus tetricus* (Mollusca: Cephalopoda) in the laboratory. *Marine Biology* **36**, 327–33.

Joll, L. M. 1978. Observations on the embryonic development of *Octopus tetricus* (Mollusca: Cephalopoda). *Australian Journal of Marine and Freshwater Research* **29**, 19–30.

Joubin, L. 1888. Sur la ponte de l'Elédone et de la Sèche. *Archives de Zoologie Expérimentale et Générale* (*deuxième série*) **6**, 155–63.

Kölliker, A. 1844. *Entwickelungsgeschichte der Cephalopoden.* Zürich: Verlag Meyer und Zeller.

Lane, F. W. 1960. *Kingdom of the octopus.* New York: Sheridan House.

Laptikhovsky, V. V. & Murzov, S. A. 1990. Epipelagic egg mass of the squid *Sthenoteuthis pteropus* collected in the tropical eastern Atlantic. *Biologiya Moriya* **1990**(3), 62–3.

Larcombe, M. F. & Russell, B. C. 1971. Egg laying behaviour of the broad squid, *Sepioteuthis bilineata. New Zealand Journal of Marine and Freshwater Research* **5**, 3–11.

LaRoe, E. T. 1971. The culture and maintenance of the loliginid squids *Sepioteuthis sepioidea* and *Doryteuthis plei. Marine Biology* **9**, 9–25.

Lu, C. C. & Stranks, T. N. 1991. *Eledoni palari*, a new species of octopus (Cephalopoda: Octopodidae) from Australia. *Bulletin of Marine Science* **49**, 73–87.

Mangold, K. (ed.) 1989. Céphalopodes. In *Traité de zoologie, Vol. V/4* , P.-P. Grassé (ed.). Paris: Masson.

Mangold-Wirz, K. 1963. Biologie des céphalopodes benthiques et nectoniques de la Mer Catalane. *Vie et Milieu* (Suppl.) 13.

Mangold-Wirz, K., Boletzky, S. v. & Mesnil, B. 1976. Biologie de reproduction et distribution d'*Octopus salutii* Vérany (Cephalopoda, Octopoda). *Rapports de la Commission Internationale pour l'Exploration Scientifique de la Mer Méditerranée* **23**(8), 87–93.

Marthy, H. J. 1972. Tintenfischembryonen als Labortiere. *Mikrokosmos* **1972**(1), 29–31.

McFall-Ngai, M. 1994. Animal-bacterial interactions in the early life history of marine invertebrates: the *Euprymna scolopes*/Vibrio fischeri symbiosis. *American Zoologist* **34**, 554–61.

McGowan, J. A. 1954. Observations on the sexual behavior and spawning of the squid, *Loligo opalescens*, at La Jolla, California. *California Fish and Game* **40**, 47–54.

Misaki, H. & Okutani, T. 1976. Studies on early life history of Decapodan Mollusca. VI. An evidence of spawning of an oceanic squid, *Thysanoteuthis rhombus* Troschel, in the Japanese waters. *Venus* (*Japanese Journal of Malacology*) **35**, 211–13.

Möhres, F. P. 1964. *Welt unter Wasser.* Stuttgart: Belser Verlag.

Moynihan, M. & Rodaniche, A. F. 1982. The behavior and natural history of the Caribbean reef squid *Sepioteuthis sepioidea. Advances in Ethology* (*Supplements to Journal of Comparative Ethology*) **25**.

Naef, A. 1923. Die Cephalopoden (Systematik). *Fauna e Flora del Golfo di Napoli* **35**(I-1).

Naef, A. 1928. Die Cephalopoden (Embryologie). *Fauna e Flora del Golfo di Napoli* **35**(I-2).

Natsukari, Y. 1970. Egg-laying behavior, embryonic development and hatched larva of the pygmy cuttlefish, *Idiosepius pygmaeus paradoxus* Ortmann. *Bulletin of the Faculty of Fisheries Nagasaki University* **30**, 15–29.

Natsukari, Y. 1976. SCUBA diving observations on the spawning ground of the squid, *Doryteuthis kensaki* (Wakiya et Ishikawa, 1921)(Cephalopoda: Loliginidae). *Venus* (*Japanese Journal of Malacology*) **35**, 206–8.

Natsukari, Y. 1979. Discharged, unfertilized eggs of the cuttlefish, *Sepia* (*Metasepia*) *tullbergi. Bulletin of the Faculty of Fisheries*, Nagasaki University **47**, 27–8.

Nesis, K. N. 1977. The biology of paper nautiluses, *Argonauta boettgeri* and *A. hians* (Cephalopoda, Octopoda) in the western Pacific and the seas of the East Indian archipelago. *Zoologicheskii Zhurnal* **56**, 1004–14.

Nesis, K. N. 1982. *Cephalopods of the world.* Neptune City: TFH Publications.

Nesis, K. N. 1996. Mating, spawning, and death in oceanic cephalopods: a review. *Ruthenica* **6**, 23–64.

Nesis, K. N. & Nigmatullin, C. M. 1978. A record of egg-masses of the bottom octopus *Eledone caparti* (Octopodidae) in the stomachs of blue sharks. *Zoologicheskii Zhurnal* **57**, 1324–29.

Nigmatullin, C. M., Arkhipkin, A. I. & Sabirov, R. M. 1991. Structure of the reproductive system of the squid *Thysanoteuthis rhombus* (Cephalopoda: Oegopsida). *Journal of Zoology* **224**, 271–83.

Nigmatullin, C. M., Arkhipkin, A. I. & Sabirov, R. M. 1995. Age, growth and reproductive biology of diamond-shaped squid *Thysanoteuthis rhombus* (Oegopsida: Thysanoteuthidae). *Marine Ecology Progress Series* **124**, 73–87.

Norman, M. D. 1992a. Ocellate octopuses (Cephalopoda: Octopodidae) of the Great Barrier Reef, Australia: description of two new species and redescription of *Octopus polyzenia* Gray, 1849. *Memoirs of the Museum of Victoria* **53**, 309–44.

Norman, M. D. 1992b. *Systematics and biogeography of the shallow-water octopuses* (*Cephalopoda: Octopodinae*) *of the Great Barrier Reef*, Australia. Doctoral dissertation, University of Melbourne.

O'Dor, R. K. & Balch, N. 1985. Properties of *Illex illecebrosus* egg masses potentially influencing larval oceanographic distribution. *Northwest Atlantic Fisheries Organization Scientific Council Studies* **9**, 69–76.

O'Dor, R. K., Balch, N., Foy, E. A., Hirtle, R. W. M., Johnston, D. A. & Amaratunga, T. 1982. Embryonic development of the squid, *Illex illecebrosus* , and effect of temperature on development rates. *Journal of*

Northwest Atlantic Fishery Science **3**, 41–5.

Okada, Y. K. 1927. Versuche über die Wirkung der Dotterwegnahme am meroblastischen Ei (Ei von *Loligo bleekeri* Keferstein). *Zoologischer Anzeiger* **73**, 280–84.

Okiyama, M. 1965. Some considerations on the eggs and larvae of the common squid *Todarodes pacificus* Steenstrup. *Bulletin of the Japan Sea Regional Fisheries Research Laboratory* **15**, 39–53.

Okiyama, M. 1993. Why do gonatid squid *Berryteuthis magister* lose tentacles on maturation? *Nippon Suisan Gakkaishi* **59**, 61–5.

Okiyama, M. & Kasahara, S. 1975. Identification of the so-called "Common Squid Eggs" collected in the Japan Sea and adjacent waters. *Bulletin of the Japan Sea Regional Fisheries Research Laboratory* **26**, 35–40.

Okutani, T. 1978. Studies on early life history of decapodan Mollusca. VII. Eggs and newly hatched larvae of *Sepia latimanus* Quoy & Gaimard. *Venus (Japanese Journal of Malacology)* **37**, 245–48.

Okutani, T., Nakamura, I. & Seki, K. 1995. An unusual egg-brooding behavior of an oceanic squid in the Okhotsk Sea. *Venus (Japanese Journal of Malocology)* **54**, 237–9.

Orelli, M. v. 1960. Follikelfalten und Dotterstrukturen der Cephalopoden-Eier. *Verhandlungen der Naturforschenden Gesellschaft Basel* **71**, 272–82.

O'Shea, S. & Kubodera, T. 1996. Eggs and larvae of *Graneledone* sp. (Mollusca, Octopoda) from New Zealand. *Bulletin of the National Science Museum, Tokyo, Series A*, **22**, 153–64.

Overath, H. & Boletzky, S. v. 1974. Laboratory observations on spawning and embryonic development of a blue-ringed octopus. *Marine Biology* **27**, 333–7.

Owen, R. 1835. Cephalopoda. In *The Cyclopaedia of anatomy and physiology, Vol. 1* , R. B. Todd (ed.). London: Longman, Brown, Green, Longmans & Roberts, 517–62.

Paulij, W. 1991. *Hatching in decapod cephalopods.* PhD dissertation, Catholic University of Nijmegen, The Netherlands.

Pickford, G. E. 1949a. The distribution of the eggs of *Vampyroteuthis infernalis* Chun. *Journal of Marine Research* **8**, 73–83.

Pickford, G. E. 1949b. *Vampyroteuthis infernalis* Chun – an archaic dibranchiate cephalopod. II. External anatomy. *Dana Report* **32**, 1–133.

Pickford, G. E. 1950a. The Vampyromorpha (Cephalopoda) of the Bermuda Oceanographic Expeditions. *Zoologica (Scientific Contributions of the New York Zoologiocal Society)* **35**, 87–95.

Pickford, G. E. 1950b. A note on the eggs of *Octopus vulgaris* Lam. from the Western Atlantic: the identity of Tandy's eggs from the Dry Tortugas. *Proceedings of the Malacological Society of London* **28**, 88–92.

Portmann, A. 1933. Observations sur la vie embryonnaire de la pieuvre (*Octopus vulgaris* Lam.). *Archives de Zoologie Expérimentale et Générale* **76**, 24–36.

Portmann, A. 1937. Die Lageveränderungen der Embryonen von *Eledone* und *Tremoctopus. Revue Suisse de Zoologie* **44**, 359–61.

Quoy, J. R. C. & Gaimard, J. P. 1832. *Mollusques. Zoologie du voyage de l'Astrolabe sous les ordres du Capitaine Dumont d'Urville, pendant les années 1826-29*, Vol. 2, 321–661.

Racovitza, E. G. 1894. Notes de biologie. III. Moeurs et fécondation de la *Rossia macrosoma. Archives de Zoologie Expérimentale et Générale (3ème série)* **2**, 491–539.

Rees, W. J. 1956. Notes on the European species of *Eledone* with special reference to eggs and larvae. *Bulletin of the British Museum of Natural History, Zoology* **3**, 283–92.

Robson, G. C. 1926. Cephalopoda from NW African waters and the Biscayan region. *Bulletin de la Société des Sciences Naturelles du Maroc* **6**(7–8), 158–95.

Robson, G. C. 1929. *A monograph of the Recent Cephalopoda. Part I. Octopodinae.* London: British Museum.

Robson, G. C. 1932a. *A monograph of the Recent Cephalopoda. Part II. The Octopoda (excluding the Octopodinae).* London: British Museum.

Robson, G. C. 1932b. Notes on the Cephalopoda. No. 16. On the variation, eggs, and ovipository habits of Floridan octopods. *Annals and Magazine of Natural History* Series 10, **10**, 368–74.

Roper, C. F. E. 1965. A note on egg deposition by *Doryteuthis plei* (Blainville, 1823) and its comparison with other North American loliginid squids. *Bulletin of Marine Science* **15**, 589–98.

Roper, C. F. E. & Sweeney, M. J. 1983. Techniques for fixation, preservation, and curation of cephalopods. *Memoirs of the National Museum Victoria* **44**, 29–47.

Roper, C. F. E. & Voss, G. L. 1983. Guidelines for taxonomic descriptions of cephalopod species. *Memoirs of the National Museum Victoria* **44**, 49–63.

Sabirov, R. M., Arkhipkin, A. I., Tsygankov, V. Yu. & Shchetinnikov, A. S. 1987. Egg-laying and embryonal development of diamond-shaped squid *Thysanoteuthis rhombus* (Oegopsida, Thysanoteuthidae). *Zool-*

ogisheskij Zhurnal **64**, 1155–63.

Sauer, W. H. H., Roberts, M., Lipinski, M. R., Smale, M. J., Hanlon, R. T., Webber, D. M. & O'Dor, R. K. 1997. Choreography of the squid's "Nuptial Dance". *Biological Bulletin* **192**, 203–7.

Sauer, W. H. H. & Smale, M. J. 1993. Spawning behaviour of *Loligo vulgaris reynaudii* in shallow coastal waters of the South-eastern Cape, South Africa. In *Recent Advances in Fisheries Biology* , T. Okutani et al. (eds). Tokyo: Tokai University Press, 489–98.

Segawa, S. 1987. Life history of the oval squid, *Sepioteuthis lessoniana* in Kominato and adjacent waters, central Honshu, Japan. *Journal of the Tokyo University of Fisheries* **74**, 67–105.

Segawa, S., Yang, W. T., Marthy, H.-J. & Hanlon, R. T. 1988. Illustrated embryonic stages of the Eastern Atlantic squid *Loligo forbesi. Veliger* **30**, 230–43.

Segawa, S., Hirayama, S. & Okutani, T. 1993. Is *Sepioteuthis lessoniana* in Okinawa a single species? In *Recent advances in fisheries biology*, T. Okutani et al. (eds). Tokyo: Tokai University Press, 513–21.

Shaw, P. & Boyle, P. R. 1997. Multiple paternity within the brood of single females of *Loligo forbesi*, (Cephalopoda: Loliginidae), demonstrated with microsatellite DNA markers. *Marine Ecology Progress Series* **160**, 279–82.

Steenstrup, J. 1900. Heteroteuthis *Gray* med Bemaerkninger om Rossia-Sepiola-Familien i Almindelighed. *Kongelige Danske Videnskabernes Selskabs Skrifter 6 Raekk* **9**, 273–300.

Stranks, T. N. 1988. Redescription of *Octopus pallidus* (Cephalopoda: Octopodidae) from south-eastern Australia. *Malacologia* **29**, 275–87.

Sugiura, Y. & Kimura, S. 1995. Nidamental gland mucosubstance from the squid *Illex argentinus*: salt-soluble component as a mucin complex. *Fisheries Science* **61**, 1009–1011.

Suzuki, S., Misaki, H. & Okutani, T. 1979. Studies on early life history of decapodan Mollusca. VIII. A supplementary note on floating egg mass of *Thysanoteuthis rhombus* Troschel in Japan. The first underwater photography. *Venus (Japanese Journal of Malacology)* **38**, 153–5.

Tardent, P. 1962. Keeping *Loligo vulgaris* L. in the Naples Aquarium. (1st International Congress on Aquariology). *Bulletin de l'Institut Océanographique de Monaco* Special No. A, 41–6.

Tardent, P. 1993. *Meeresbiologie – Eine Einführung*. Stuttgart: Georg Thieme Verlag.

Thore, S. 1949. Investigations on the "Dana" Octopoda, I. *Dana Report* **33**, 1–85.

Thorson, G. 1946. Reproduction and larval development of Danish marine bottom invertebrates, with special reference to the planktonic larvae in the Sound (Oresund). *Meddelelser fra Kommissionen for Danmarks Fiskeri- og Havundersogelser* (*Serie Plankton*) 4, 1–523.

Tranter, D. J. & Augustine, O. 1973. Observations on the life history of the blue-ringed octopus *Hapalochlaena maculosa. Marine Biology* **18**, 115–28.

Van Heukelem, W. F. 1977. Laboratory maintenance, breeding, rearing, and biomedical research potential of the Yucatan octopus (*Octopus maya*). *Laboratory Animal Science* **27**, 852–9.

Vecchione, M. 1988. *In situ* observations on a large squid-spawning bed in the eastern Gulf of Mexico. *Malacologia* **29**, 135–41.

Verrill, A. E. 1885. Third catalogue of Mollusca, recently added to the fauna of the New England coast and the adjacent parts of the Atlantic, consisting mostly of deep-sea species, with notes on others previously recorded. *Transactions of the Connecticut Academy* **6**, 395–452.

Vevers, H. G. 1961. Observations on the laying and hatching of octopus eggs in the Society aquarium. *Proceedings of the Zoological Society of London* **137**, 311–15.

Watanabe, K., Sakurai, Y., Segawa, S. & Okutani, T. 1996. Development of the ommastrephid squid *Todarodes pacificus*, from fertilized egg to rhynchoteuthion paralarva. *American Malacological Bulletin* **13**, 73–88.

Wells, M. J. & Wells, J. 1977. Cephalopoda: Octopoda. In *Reproduction of marine invertebrates. Vol. IV. Molluscs: gastropods and cephalopods*, A. C. Giese & J. S. Pearse (eds). New York: Academic Press, 291–336.

Willey, A. 1902. *Contribution to the Natural History of the Pearly Nautilus*. A. Willey's Zoological Results, Part VI. (Cambridge: Cambridge University Press, 691–826.

Wolf, G., Verheyen, E., Vlaeminck, A., Lemaire, J. & Decleir, W. 1985. Respiration of *Sepia officinalis* during embryonic and early juvenile life. *Marine Biology* **90**, 35–9.

Wülker, G. 1913. Cephalopoden der Aru- und Kei-Inseln. Anhang: Revision der Gattung *Sepioteuthis. Abhandlungen der Senckenbergischen Naturforschenden Gesellschaft* **34**, 451–88.

Yamamoto, M. 1982. Normal stages in the development of the cuttlefish, *Sepiella japonica* Sasaki. *Zoological Magazine (Dobutsugaku Zasshi)* **91**, 146–57.

Young, J. Z. 1989. The angular acceleration receptor system of diverse cephalopods. *Philosophical Trans-*

actions of the Royal Society of London, Series B 325, 189–237.

Young, J. Z. 1997. The classification of octopods. *Vie et Milieu* **47**, 177 only.

Young, R. E. 1972. Brooding in a bathypelagic octopus. *Pacific Science* **26**, 400–4.

Young, R. E. 1995. Aspects of the natural history of pelagic cephalopods of the Hawaiian mesopelagic-boundary region. *Pacific Science* **49**, 143–55.

Young, R. E. & Harman, R. F. 1985. Early life history stages of enoploteuthin squids (Cephalopoda: Teuthoidea: Enoploteuthidae) from Hawaiian waters. *Vie et Milieu* **35**, 181–201.

Young, R. E., Harman, R. F. & Mangold, K. M. 1985a. The common occurrence of oegopsid squid eggs in near-surface oceanic waters. *Pacific Science* **39**, 359–66.

Young, R. E., Harman, R. F. & Mangold, K. M. 1985b. The eggs and larvae of *Brachioteuthis* sp. (Cephalopoda: Teuthoidea) from Hawaiian waters. *Vie et Milieu* **35**, 203–9.

Yung, K. C. 1930. Contribution à l'étude cytologique de l'ovogenèse, du développement et de quelques organes chez les Céphalopodes. *Annales de l'Institut Océanographique* **7**(8), 301–64.

Appendix

SCHEMATIC PRESENTATION OF EGGS AND EGG MASSES OF COLEOID CEPHALOPODS

CAPSULE SURFACE

hard <--> soft

COLLECTIVE CAPSULES

free (floating egg masses)

ommastrephid squids (e. g. *Illex*)
diamond squid (*Thysanoteuthis*)

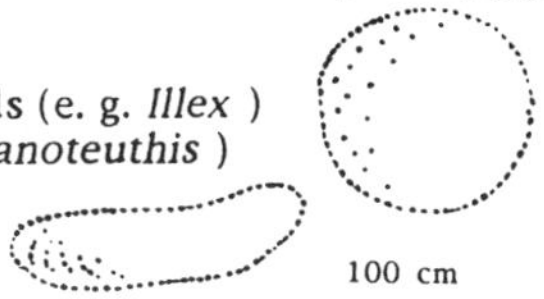

100 cm

attached

loliginid squids (capsules or "fingers" attached by fixating jelly, together forming a "mop", each finger containing many eggs (e.g. *Loligo*) or few eggs (*Sepioteuthis*)

10 cm

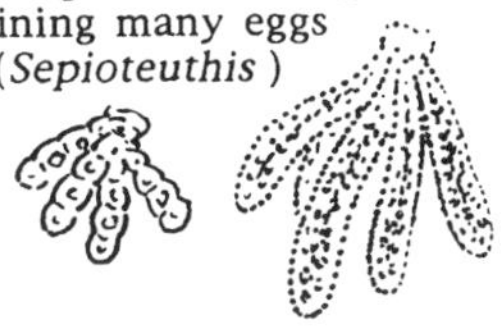

carried by female

gonatid squids

100 cm

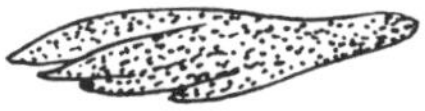

SINGLE EGGS without true capsule, with or without jelly coat around chorion

free (floating)

enoploteuthid and brachioteuthid squids
Vampyroteuthis

SINGLE EGGS

capsule attached, eggs generally laid in batches

pygmy cuttlefish (*Idiosepius*) 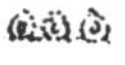 1 cm

sepiid cuttlefishes  1 cm

1 cm  sepiolid squids (except *Rossia*)

Rossia

1 cm finned octopods (eggs not laid in batches)

capsule reduced (cement embedding chorion stalk only)

finless octopods

CLUSTERS

(cement uniting several stalks)

1 cm

(cement uniting many stalks, thus forming the axis of an egg string also called a "festoon")

Oceanography and Marine Biology: an Annual Review 1998, **36**, 373–411

UCL Press

THE ECOLOGICAL IMPLICATIONS OF SMALL BODY SIZE AMONG CORAL-REEF FISHES

PHILIP L. MUNDAY & GEOFFREY P. JONES
Department of Marine Biology, James Cook University of North Queensland, Townsville, 4811, Australia

Abstract Small species (<100 mm total length) are a diverse and abundant component of coral-reef fish assemblages. However, difficulties identifying and estimating the abundance of small species has meant that they are often ignored or undersampled in community level studies. We use studies that sampled whole assemblages in order to examine size/diversity and size/abundance patterns within coral-reef fish assemblages. We then examine body size in relation to patterns of habitat use, species interactions (competition and predation) and life histories of coral-reef fishes.

Species diversity typically peaks in small size classes (<100 mm) and there are many low diversity larger size classes. In some assemblages, diversity also declines among the smallest species (<50 mm). The high diversity of small coral-reef fishes may partly be attributed to the complex structure of coral reefs. However, we find no evidence that diversity/size distributions are controlled by increased availability of niches or greater living space for small fishes. Regional-scale processes also influence local diversity of coral-reef fishes. In particular, high rates of speciation among small species might contribute to the observed diversity/size distributions.

Declines in abundance with increasing body size have been attributed to equal partitioning of energy among species of different body sizes ("energetic equivalence rule"). We find no support for the "energetic equivalence rule" in assemblages of coral-reef fishes. Although body size was generally a poor predictor of abundance for coral-reef fishes, the upper bound of the size/abundance distribution was very uniform among the assemblages we examined. This upper bound may prove useful for selecting species most likely to be resource-limited.

Small species of coral-reef fishes are more closely associated with the reef matrix than their larger counterparts. We propose that this is largely the result of high predation risk for small species. It is generally believed that smaller species and smaller size classes within species are subject to higher mortality rates. However, further studies are required to test this assumption.

The diversity and abundance of small reef fishes, and their restriction to specialized habitats may increase the likelihood of competitive interactions. There is ample evidence of intraspecific competition among individuals of similar size in small coral-reef fishes. There is less evidence of interspecific competition among coral-reef fishes. However, the outcome of interspecific interactions will, in part, depend on the relative size of species. Because body size can influence the intensity with which predation and competition act, consideration of population size-structure may help develop multi-factorial models of population dynamics of coral-reef fishes.

Introduction

Body size is one of the most obvious characteristics of an animal and a feature that has important ecological and evolutionary implications (Peters 1983, Calder 1984, LaBarbera 1989, Hanken & Wake 1993). For example, it is commonly observed that maximum diversity occurs in the smaller size classes of taxonomic and trophic groups and in naturally occurring assemblages of animals (Brown 1981, Griffiths 1992). In addition, small species are typically more abundant than large species (Damuth

1981, 1991). Despite their diversity and abundance, small species are often more susceptible to predation than larger species (Stanley 1979 in Brown & Maurer 1986, Werner & Gilliam 1984), they may have more restricted prey options than larger species (Brooks & Dobson 1965, Wasserman & Mitter 1978) and may be less successful competitors (Schoener 1983, Persson 1985, Morse 1980 in Brown & Maurer 1986). A range of mechanisms has been proposed to account for the diversity and abundance of small species despite the impacts of larger predators and competitors. Foremost among these are: (a) life-history strategies that promote short generation times and high intrinsic rates of increase (May 1978, Marzluff & Dial 1991), (b) habitat associations and behaviour that reduce predation (Werner 1984) and (c) resource partitioning that allows small species to coexist with larger competitors (Schoener 1974). In addition, Hutchinson (1959) proposed that small species are more diverse because small size permits animals to become specialized to small diversified elements of the environmental mosaic.

An extensive literature now exists on relationships between body size and species diversity, population abundance, assemblage energetics, species interactions, and life-history traits. However, attention in this literature has been given mostly to terrestrial organisms, particularly, insects, birds and mammals. The importance of body size has also been studied extensively in freshwater communities, most notably in relation to species interactions in size-structured populations of fishes (Werner & Gilliam 1984, chapters in Ebenman & Persson 1988). In contrast, relatively few studies have explicitly considered body size in relation to the diversity and abundance of marine fishes (Blackburn et al. 1993a, Blackburn & Lawton 1994, but see Miller 1979, 1996, Barlow 1981). The situation is even more apparent for coral-reef fishes where scant attention has been given to patterns of species diversity and abundance in relation to body size (Macpherson 1989). Similarly, the significance of body size in determining the outcome of interactions among species of coral-reef fishes has not been fully evaluated (Robertson 1998).

Although body size has rarely been explicitly considered, patterns of distribution and abundance of coral-reef fishes and the processes that help determine these patterns have been studied extensively (see chapters in Sale 1991a). Much of this research examines patterns and/or processes among species from a range of body sizes or through a range of ontogenetic stages. For example, the importance of recruitment of juvenile fishes to the adult population dynamics of coral-reef fishes has been an area of intense research activity over the past 10–20 yr (reviews by Doherty & Williams 1988, Caley et al. 1996). The influence of habitat structure, competition and predation on the community structure and population dynamics of coral-reef fishes have also been thoroughly investigated, with studies encompassing species that exhibit a range of body sizes and ontogenetic stages (Jones 1991, Hixon 1991). Here we review the literature relating to the distribution and abundance of coral-reef fishes in an attempt to elucidate patterns and processes that are influenced by body size. In particular, we examine the ecological consequences of small size in coral-reef fishes. We first establish the relationships between body size and patterns of diversity and abundance in local assemblages of coral-reef fishes and consider the relevance of hypotheses proposed to explain such patterns in other taxa. Because the high diversity and abundance of small species has often been linked to increased habitat opportunities offered by small body size, we then examine habitat associations of small coral-reef fishes. We also investigate the effects of predation and competition on small fish species and consider the importance of habitat use in these interactions. Finally we explore links between life-history strategies and the diversity and distributions of small reef fishes.

Body size is commonly expressed as mass or length. In fishes, length is measured from the tip of the snout to the end of the tail (total length) or from the tip of the snout to the base of the caudal fin (standard length). In ecological studies of coral-reef fishes, length is more frequently reported than mass. Therefore, throughout this review we primarily use length to describe body size. In addition, we follow the arbitrary classification used by Miller (1979, 1996) where small species are defined as those species having a mean maximum total length under 100 mm.

Diversity, abundance and body size

Although small species are common in many taxa, the tendency towards small body size is more apparent among fishes than in many other groups (Hanken & Wake 1993). The smallest known vertebrate is a tropical marine teleost, which matures at around 8 mm standard length (Winterbottom & Emery 1981), and at least 10% of all known teleosts are less than 100 mm total length (Miller 1979). Over two-thirds of small teleosts occur in the tropics (Miller 1979) and, in terms of taxa present, small species are particularly abundant on coral reefs (Choat & Bellwood 1991). For example, small species have been reported to constitute 72% and 68% of species collected from, or observed on, reefs in the Gulf of California and the Red Sea, respectively (Barlow 1981). Common families of coral-reef fishes that consist primarily of small species include, Tripterygiidae, Gobiidae, Callionymidae, Pseudochromidae, Microdesmidae, Blenniidae, Serranidae (Anthiinae), Apogonidae and Pomacentridae (Fig. 1). In the Caribbean, the Chaenopsidae is another diverse and abundant family of small fishes. In addition, some families (e.g. Labridae) have a few large species but most species within the family are small.

Research on communities of coral-reef fishes has usually been conducted on relatively small spatial scales, such as patch reefs (10's m across), whole reefs (100's m), or at single locations (1000's m). Studies at these scales are appropriate for examining the influence of body size on patterns of diversity and abundance because they encompass naturally occurring collections of interacting species. Furthermore, as recent studies on body size in community ecology have concentrated on local assemblages (Blackburn & Lawton 1994), direct comparisons between coral-reef fishes and other taxa are appropriate at these scales. Although many studies on assemblages of coral-reef fishes report patterns of diversity and abundance, rarely are the abundances of small species rigorously determined. Most published studies exclude the small and cryptic ichthyofauna because of difficulties identifying and accurately estimating the abundance of these species. Other studies use techniques and scales of sampling that are likely to undersample the diversity and abundance of small species (Sale & Douglas 1981, Brock 1982). To examine patterns of local diversity and abundance in relation to body size in coral-reef fishes we have chosen published studies in which a reasonable attempt was made to collect or estimate the abundance of all individuals in the assemblage regardless of body size. Published studies that met this criteria and provided raw data suitable for analysis, or for which we were able to obtain the raw data, are shown in Tables 1 and 2.

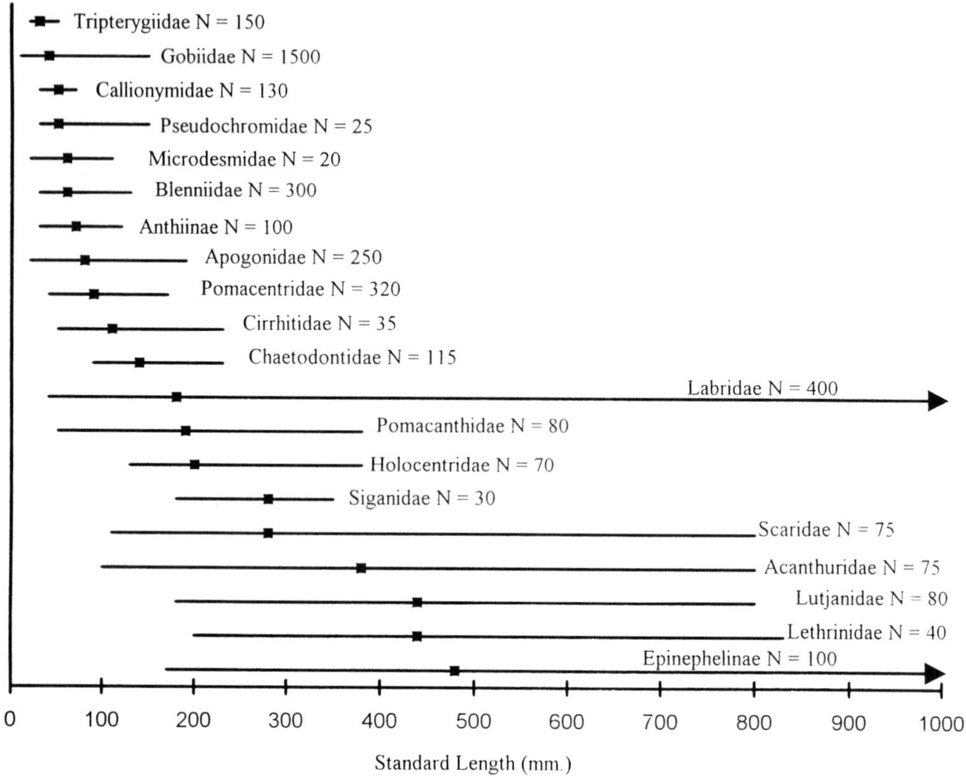

Figure 1 Length range and mean (■) of species within families of fishes common on coral reefs. N is the approximate number of species within each family. Derived from Randall et al. (1990), Woodland (1990), Choat & Bellwood (1991), Myers (1991), Randall & Heemstra (1991).

Body size and species richness

Most animal assemblages exhibit log normal or log skewed (right-skewed on logarithmic scale) species richness/body size distributions (Maurer et al. 1992, Loder et al. 1997). In these distributions, there are more small species than large species but the smallest species are not the most diverse. When grouped into 100-mm size classes, the number of species within local assemblages of coral-reef fishes examined here exhibit a monotonic decline in species richness with body size, or strongly right-skewed distributions (Fig. 2). Within these assemblages, small species (<100 mm) are always very diverse and in 75% of cases they are the most diverse group. A similar pattern was described by Barlow (1981) for an assemblage of coral-reef fishes in the Red Sea. Although species diversity peaks below 100 mm, the smallest species are not always the most diverse. In some assemblages, diversity declines below 50 mm total length (Barlow 1981). On a logarithmic scale the distributions of species within assemblages of coral-reef fishes range from approximately log normal to log-skewed (Fig. 3). The range of distributions observed here is not dissimilar from those observed in other marine (Fenchel 1993, Warwick & Clarke 1996) and terrestrial taxa (Loder et al. 1997).

Table 1 Regression analyses of log species richness on log standard length for local assemblages of coral-reef fishes. Species were grouped into 100 mm size classes and average size was used as the x variable in regression analysis. Slopes and coefficients of determination (r^2) are for species frequencies above the modal size class. Maximum reported standard lengths (Myers 1991) were used where body size of species was not published. Ordinary least squares regression was used for uniformity with other studies. Raw data from 10 stations per reef were pooled for Williams & Hatcher (1983).

Location	Source	Slope	r^2
Lameshur Bay, Caribbean	Randall 1963	−1.56	0.83
Rams Head, Caribbean	Randall 1963	−1.7	0.94
Lizard Island, Great Barrier Reef	Caley 1995a	−1.55	0.94
One Tree Island, Great Barrier Reef	Caley 1995a	−1.77	0.75
Pandora Reef, Great Barrier Reef	Williams & Hatcher 1983	−1.39	0.91
Rib Reef, Great Barrier Reef	Williams & Hatcher 1983	−1.61	0.8
Myrmidon Reef, Great Barrier Reef	Williams & Hatcher 1983	−1.57	0.73
Kaneohe Bay, Hawaii	Brock 1982	−1.01	0.77

Although the number of assemblages considered is few, it is apparent that small species constitute a very large percentage of the diversity in assemblages of coral-reef fishes.

Hutchinson (1959) proposed that the high diversity of small species results from their ability to utilize fine-grain aspects of the environment. Coral reefs offer a complex mosaic of habitat at a fine spatial scale and most small species are closely associated with the reef matrix (Warburton 1989). If habitat complexity at small spatial scales influences species richness/body size relationships of marine fishes, then the proportion of small species in assemblages should decline in environments with less fine-scale structural complexity. Hall (1996) described a distinct decline in total abundance and species richness below 100 mm body length for demersal fishes trawled from the North Sea. The relatively uniform and unstructured habitat utilized by these benthic fishes may offer less opportunities to be exploited by small species. In comparison, coral-reefs offer a highly complex benthic environment that may promote high diversity of small species. Although the comparison between assemblages of fishes collected by different methods in temperate and tropical waters is far from ideal, other observations support the notion that habitat structure helps determine species richness/body size distributions for coral-reef fishes. Positive correlations between species diversity and habitat complexity have been detected between sandstone and coral reefs in the same location (Öhman et al. 1997, Öhman & Rajasuriya in press). Furthermore, species diversity has often been positively correlated with habitat complexity among and within coral-reefs (Luckhurst & Luckhurst 1978, Gladfelter et al. 1980, Carpenter et al. 1981, Kaufman & Ebersole 1984, Roberts & Ormond 1987, Galzin et al. 1994, Chabanet et al. 1997) and on artificial reefs (Hixon & Beets 1993, Caley & St John 1996). Therefore, exploitation of the fine-scale aspects of the reef matrix has apparently enabled a large number of species to occupy coral reefs and may help explain why small species are very diverse on coral reefs.

In a further development of the habitat complexity argument, Hutchinson & MacArthur (1959) and others (May 1978, Southwood 1978) proposed that the way

Table 2 Regression analyses of log abundance on log body size (mass or length) for marine fishes. * Indicates statistics reported in literature. All other statistics are from analyses of published raw data. Maximum reported standard lengths (Myers 1991) were used where body size of species was not published. Ordinary least squares regression (OLS) was used for uniformity with other studies. r^2 = coefficient of determination, r = correlation coefficient, $Slope_{UB}$ = slope of the upper bound of the distribution calculated using the methodology of Blackburn et al. (1992), r^2_{UB} is the coefficient of determination for the upper bound slope.

Habitat	Source	Units	Number of Species	Slope (OLS)	r^2	r	$Slope_{UB}$	r^2_{UB}
Coral Reef Caribbean	Brown & Maurer 1986*	mass (g)	53	−0.05		−0.08		
Coral Reef Lameshur Bay	Randall 1963	mass (g)	87	−0.29	0.24		−0.64	0.92
Coral Reef Rams Head	Randall 1963	mass (g)	64	−0.3	0.24		−0.63	0.94
Coral Reef Pandora Reef	Williams & Hatcher 1983	mass (g)	93	−0.2	0.035		−1.4	0.97
Coral Reef Rib Reef	Williams & Hatcher 1983	mass (g)	175	−0.16	0.029		−1.24	0.97
Coral Reef Myrmidon Reef	Williams & Hatcher 1983	mass (g)	149	−0.17	0.035		−0.82	0.83
Coral Reef Lameshur Bay	Randall 1963	length (mm)	87	−0.66	0.14		−2.38	0.8
Coral Reef Rams Head	Randall 1963	length (mm)	64	0.13	0.007		−1.58	0.93

Table 2 *Continued*

Habitat	Source	Units	Number of Species	Slope (OLS)	r^2	r	Slope_{UB}	r^2_{UB}
Coral Reef Lizard Island	Caley 1995a	length (mm)	85	−0.39	0.022		−2.79	0.97
Coral Reef One Tree Island	Caley 1995a	length (mm)	73	−0.31	0.023		−2.27	0.78
Coral Reef Pandora Reef	Williams & Hatcher 1983	length (mm)	93	−0.3	0.016		−3.3	0.9
Coral Reef Rib Reef	Williams & Hatcher 1983	length (mm)	175	−0.25	0.014		−3.1	0.82
Coral Reef Myrmidon Reef	Williams & Hatcher 1983	length (mm)	149	−0.51	0.049		−2.58	0.78
Coral Reef Kaneohe Bay	Brock 1982	length (mm)	74	−0.33	0.029		−1.52	0.58
Tide Pool Gulf of California	Brown & Maurer 1986*	mass (g)	25	0.04		0.04		
Shallow Benthic Tropical	Macpherson 1989*	mass (g)	101	−0.7		−0.37		
Shallow Benthic Temperate	Macpherson 1989*	mass (g)	64	−0.41		−0.22		
Deep Sea	Blackburn & Lawton 1994*	mass (kg)	108	−0.16	0.011			

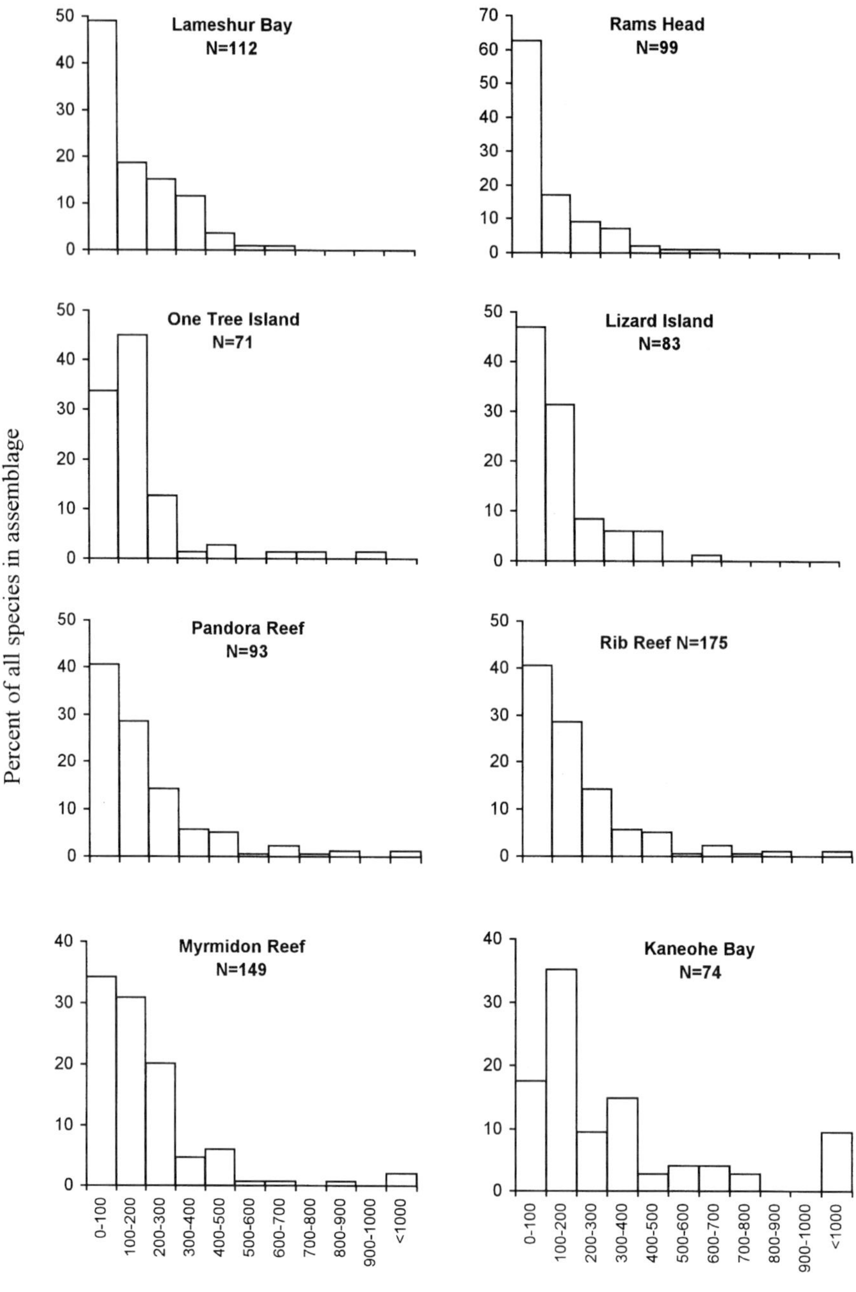

Figure 2 Body size distributions of species within local assemblages of coral-reef fishes. N = number of species. Derived from Randall (1963), Brock (1982), Williams & Hatcher (1983) (raw data supplied by D. Williams), Caley (1995a). Maximum reported standard lengths (Myers 1991) were used where body size of species was not published. For Kanehoe Bay the increase in abundance of species in the <1000 mm length class is due to a large number of moray eels in the collection.

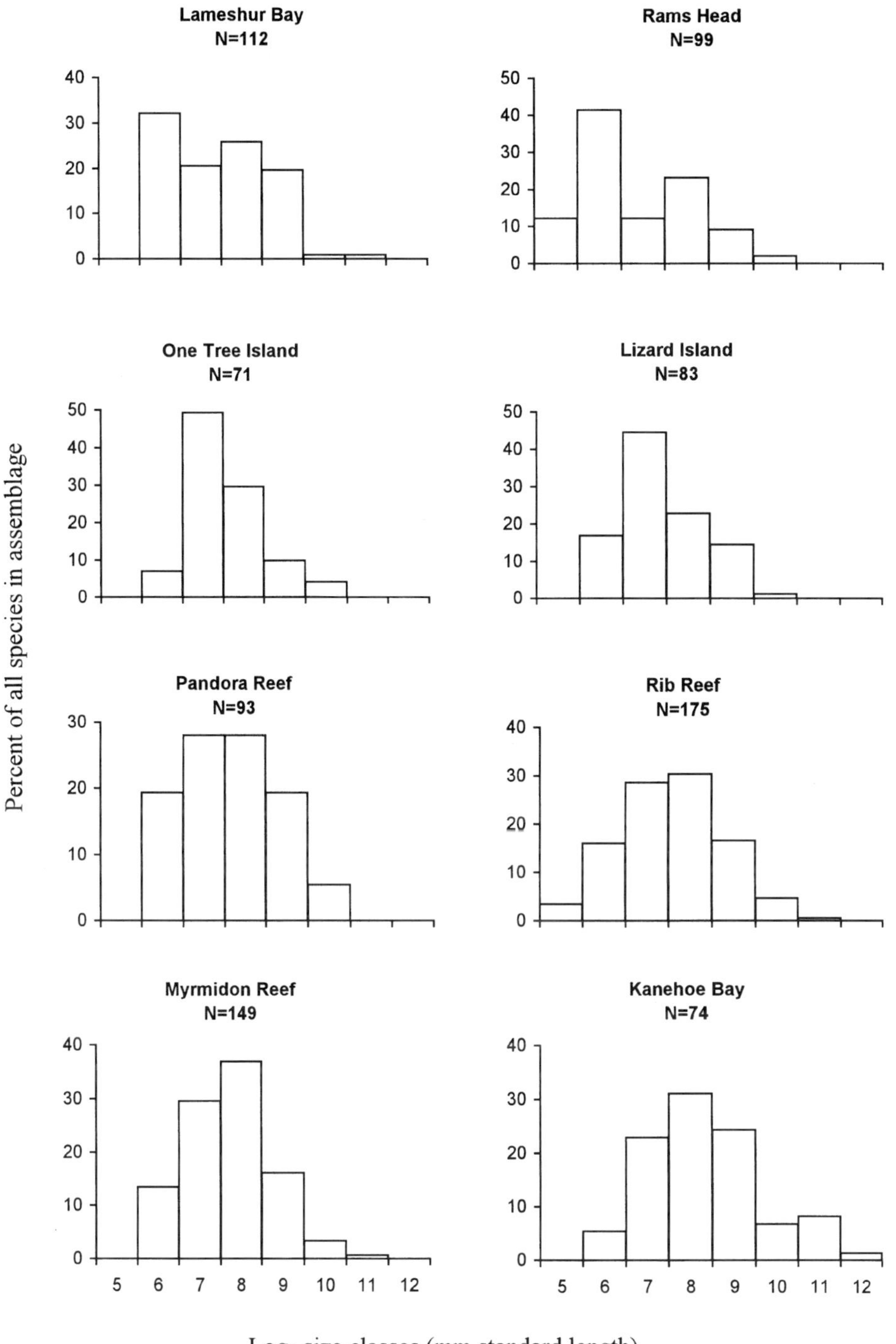

Figure 3 Body size distributions of species within local assemblages of coral-reef fishes classified by $\log_2$ size classes for comparison with other studies. N = number of species. Data as for Fig. 2.

animals perceive habitats scales with body size. To a large fish a coral reef would appear homogeneous but to a very small fish the reef would appear very complex and contain many habitat types. Therefore, species richness is expected to increase with decreasing body size, the argument being that there are many more ecological niches available at fine scales. Based on the notion that animals perceive their environment in two dimensions, the relationship between species richness (S) and body length (L) was formalized as $S = L^{-2}$ (May 1978). If this relationship holds, there should be 100 times more species for every decline in body size by a factor of ten (May 1978). However, the slopes of the body size/species richness curves for coral-reef fishes, when plotted on logarithmic scales, are shallower than -2 (Table 1, p. 377). This suggests that the model does not hold for coral-reef fishes. Indeed, empirical evidence that strongly supports the model is sparse in other taxa as well (Fenchel 1993). Although the relationship between species richness and body size for coral-reef fishes does not support the simple "niche diversity" model, the relationship appears uniform, with an average slope of approximately -1.5 (Table 1). This suggests that there may be some other common factor that is influencing the observed distribution of species in body size classes of coral-reef fishes.

Lawton (1986) and May (1986) suggested that increased species richness at small body size was determined by increased space available to small animals in addition to increased niche diversity at small spatial scales. Using the concept of fractals they argued that finer scale subdivision of the environment resulted in greater habitat space, which could be occupied by an increased number of individuals of smaller size. Greater species richness followed because species richness is generally positively correlated with the number of individuals present. By incorporating both niche diversity and increasing space arguments, May (1986) proposed that species richness should scale with body size within the bounds of L^{-D} and L^{-2D} where D is the fractal dimensions of the habitat. However, because of the range of slopes encompassed, this model is not precise enough to explain the relationship between species richness and body size. Using a fractal dimension (D) of approximately 1.15 reported for coral-reefs (Mark 1984), May's model predicts between 14 and 200 species in the 100 mm size class. More precise methods of using fractals to explain the relationship between species richness and body size have been used (Morse et al. 1985). However, these methods are of little use in assemblages of coral-reef fishes because they invoke the energetic equivalence rule for which there is little support in these assemblages (see Body size and population abundance section, p. 383). In addition, "habitat space" and "niche diversity" arguments assume that space is a limited resource, but there is little evidence to suggest that space is limiting for many species of coral-reef fishes (Doherty & Williams 1988, Jones 1991).

Local processes acting in ecological time are most commonly used to explain size–diversity distributions observed in local assemblages of animals (i.e., the hypotheses considered above). However, local-scale patterns of diversity in animal assemblages are also influenced by regional diversity and processes acting on historical timescales (Ricklefs 1987, Ricklefs & Schluter 1993, Caley 1995b, Caley & Schluter 1997). In assemblages of coral-reef fishes on the Great Barrier Reef, Caley (1997) found that local species richness increased with regional species richness and showed no signs of local species saturation. He concluded that diversity in these assemblages (Lizard Island and One Tree Island) was more strongly influenced by processes acting on regional spatial scales and historical timescales, than by local ecological interactions. Therefore, the number of small species in each of these local assemblages of coral-reef fishes is likely to be influenced by the number of small species in the surrounding regional assemblage.

The potential for processes acting on evolutionary scales to influence local body size/diversity distributions has been highlighted in theoretical models (Maurer et al. 1992, Brown et al. 1993). For example, body size/diversity distributions on large spatial scales are typically log-skewed (May 1986, Maurer et al. 1992), in a similar manner to local size/diversity distributions. Using simulation models, Maurer et al. (1992) demonstrated that these distributions could be produced by the combination of increased speciation rates for small species, increased extinction rates for large species and directional evolution in body size within extant species. Marzluff & Dial (1991) also proposed that high diversity among small species was the result of increased speciation rates and reduced extinction rates. If species diversity in local assemblages of coral-reef fishes is strongly influenced by regional diversity, the large number of small species in local assemblages might, at least in part, be a consequence of biased rates of speciation and extinction for small species.

Body size and population abundance

Population density and body size are negatively correlated in many data sets, some of which are naturally occurring assemblages (Marquet et al. 1990, 1995). More commonly, however, there is an approximately triangular or polygon-shaped relationship between size and abundance in animal assemblages (Lawton 1990, Cotgreave 1993, Blackburn et al. 1993a, Blackburn & Lawton 1994, Warwick & Clarke 1996). Within assemblages of coral-reef fishes the relationship between body size and abundance of each species is also triangular or polygonal (Fig. 4). As a result, body size is a poor predictor of abundance in assemblages of coral-reef fishes. This is indicated by the low correlation coefficients (r) and coefficients of determination (r^2) reported in Table 2. This pattern holds for marine fishes from other habitats for which analyses have been done (Table 2, p. 378). Therefore, it appears that the relationship between body size and abundance consistently reported for terrestrial animals and benthic marine invertebrates extends to coral-reef fishes and probably to marine fishes in general.

Although body size is generally a poor predictor of abundance in assemblages of coral-reef fishes, it is apparent that large species are, on average, less abundant than small to medium sized species (Fig 4). Many small species occur in high abundances but no large species are very abundant. As result, there is a weak negative relationship between body size and population abundance. In some animal assemblages the relationship between body mass (W) and population density is approximated by $W^{-0.75}$ when plotted logarithmically (Damuth 1981). Because basal metabolic rate scales as $W^{+0.75}$, Damuth (1981) proposed that energy use is independent of body size for species in these assemblages, a notion termed the "energetic equivalence rule"(Nee et al. 1991). In other words, each species within an assemblage uses the same amount of energy per year, regardless of size (Cotgreave 1993). Because small species use less energy per capita, per unit time, more small species than large species can be supported on a finite resource base (Cotgreave 1993). This implies that species abundances are limited by energetic requirements (Damuth 1981, 1987, 1991, Nee et al. 1991). Because mass is generally regarded as proportional to L^3, a slope of -2.25 would conform to

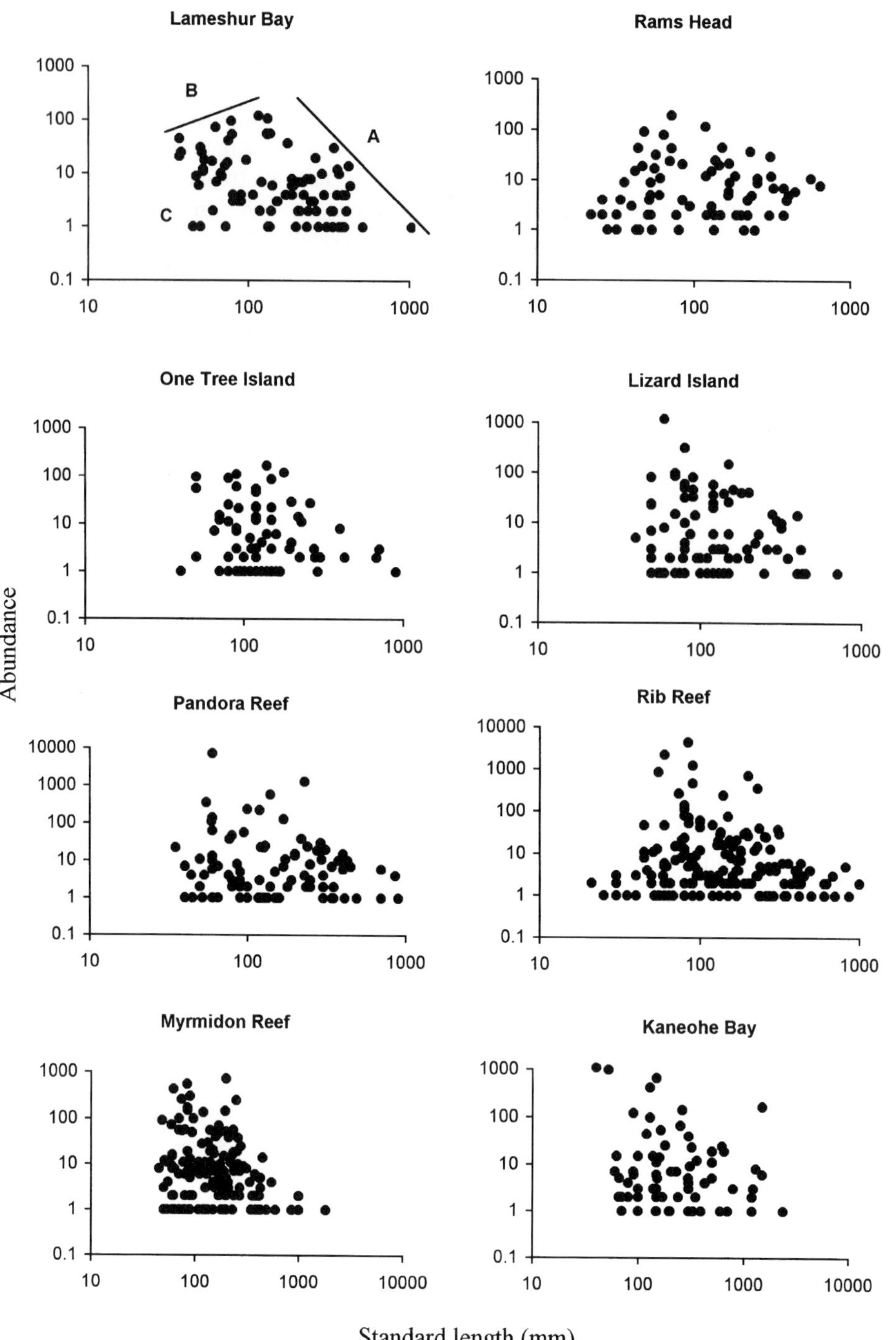

Figure 4 Body size versus abundance of each species within local assemblages of coral-reef fishes. A, B and C correspond to notations in the text. A delineates the upper boundary of the size/abundance distribution. B indicates the region of declining abundance with decreasing size. C indicates the minimum abundances of small species. Data as for Fig. 2.

the energetic equivalence rule if length was the unit of measure rather than mass. The slope of regressions of log abundance on log body mass for coral-reef fishes and other fishes is generally shallower than -0.75 for mass and -2.25 for length (Table 2, p. 378). This suggests that energy is not limiting the abundances of most species in these assemblages. Also, because of the large variation in population density at all but the largest and smallest body sizes, it is apparent that species of similar body size within these assemblages are using very different amounts of energy. Furthermore, studies of other animal assemblages have shown that log population abundance may scale with log body mass $\sim W^{-0.75}$ even where energy is not limiting (Marquet et al. 1990). Therefore, even for assemblages where the slope of the regression between log body mass and log population abundance is close to -0.75 (e.g. tropical shallow-benthic in Table 2) there is insufficient evidence to indicate energetic constraints within the assemblage.

Further evidence that energetic constraints may not limit the abundance of most species within assemblages of coral-reef fishes comes from natural and experimental changes in food availability on reefs. Increased algal growth on reefs did not influence the abundances of herbivorous fishes on the Great Barrier Reef (Hart et al. 1996) and manipulations of food availability did not influence settlement or survival of coral-reef fishes in the Caribbean (Shulman 1984). Also, Robertson et al. (1981) found that reducing the amount of algae available did not influence growth or condition of the territorial herbivore, *Stegastes planifrons*. In contrast, Eckert (1985a), Jones (1986) and Forrester (1990) have shown that food availability can influence growth and maturity of coral-reef fishes, thereby influencing adult population size. Therefore, available energy may not limit total abundance of most species but may be important in determining breeding population size of some species.

One consistent feature of the relationship between body size and abundance in coral reef-fishes (Fig. 4, area A) and other taxa (Lawton 1990) is the distinct, negative slope of the upper-boundary following the most abundant species in the assemblage. It may only be species that reach high abundance at any particular place and time that are limited by energy (food) or other resources (e.g. space). Therefore, the upper bounds of species abundance distributions may be a useful indicator of where we should look for resource-limitation in coral-reef fishes. Indeed, the slope of the upper bound for each assemblage considered here does not differ greatly from predictions of energy-limitation hypotheses ($M^{-0.75}$ or $L^{-2.25}$, Table 2). However, Blackburn et al. (1993b) and Blackburn & Gaston (1994) caution that the upper bound of abundance/body size distributions may approach -0.75 (or -2.25 for length) for purely statistical reasons. Field experiments will be required to distinguish between these two explanations.

Despite the general trend for small species to be more abundant than large species, the species with the highest abundances are not always the smallest species. In all but one of the assemblages (Kanehoe Bay) of coral-reef fishes examined here, the smallest species were not the most abundant (Fig. 4, area B). Similar patterns have been observed in many other animal assemblages (Marquet et al. 1995) and have been attributed to underlying frequency distributions of species body size and species abundances (Blackburn et al. 1993b), inefficient sampling of small species (Blackburn et al. 1990, Currie 1993), or physiological constraints (Brown et al. 1993, Marquet et al. 1995). Physiological constraints imposed by small body size may explain decreased species richness and abundance among the smallest species of many taxa (Finlay &

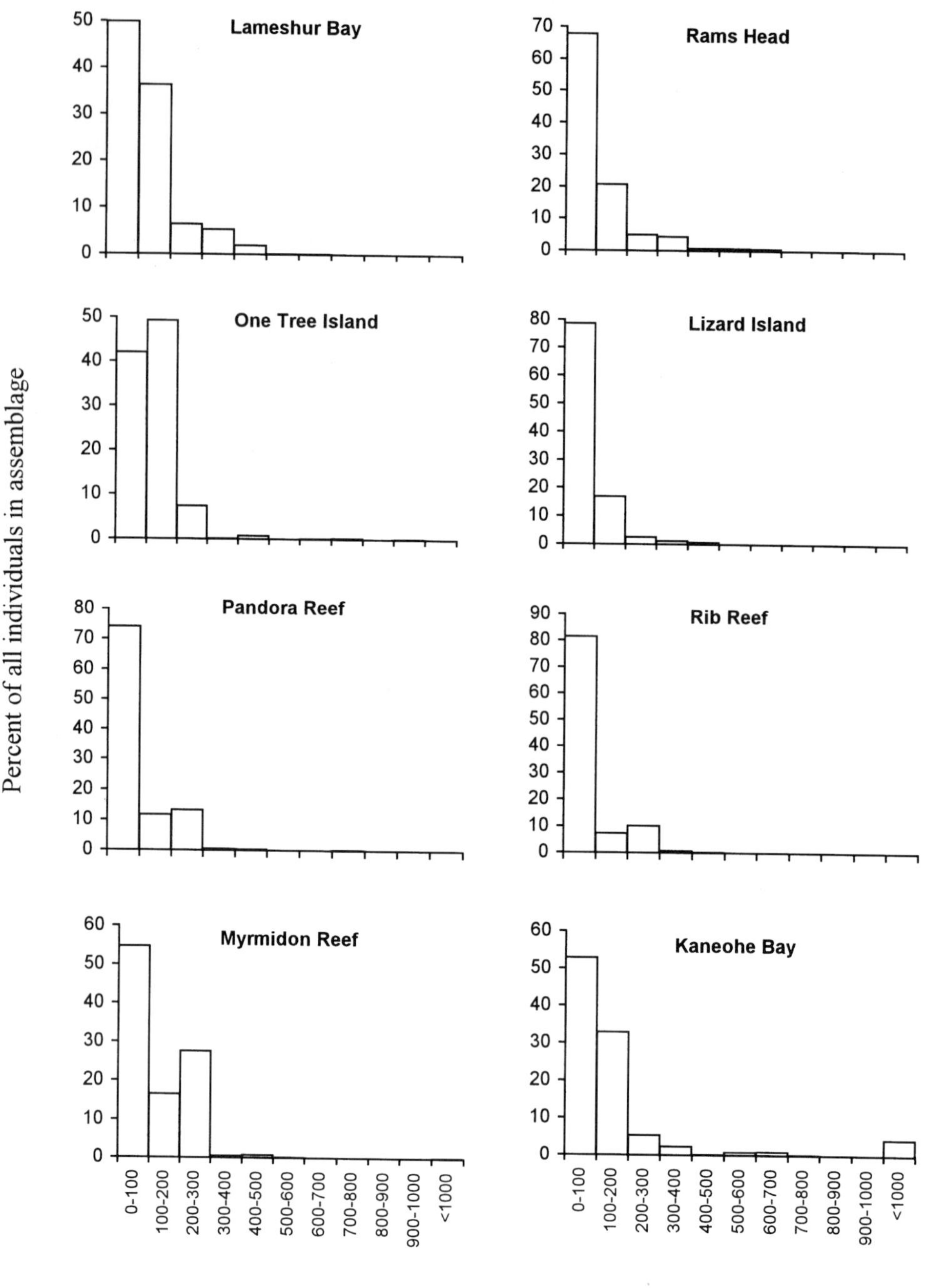

Figure 5 Total abundance of individuals grouped in 100-mm body size classes for local assemblages of coral-reef fishes. Data as for Fig. 2.

Fenchel 1996) and are likely to be particularly important for small coral-reef fishes (Miller 1996). Given that the smallest fishes on coral reefs extend to the lower limit of body size in vertebrates, it is likely that they are approaching the structural limits for miniaturization in vertebrates. Also, because there is a positive relationship between ovary size and fecundity in fishes (Blueweiss et al. 1978, Charnov 1983) reproductive potential may be constrained in very small species, thereby decreasing their abundances. Furthermore, because of their high metabolic rates, small species may be less able to direct energy towards reproduction (Brown et al. 1993). By combining allometric exponents for energy acquisition (0.75) and transformation of energy to reproduction (−0.25), Brown et al. (1993) predicted that energetic fitness declines in small species.

Although large species are usually less abundant than small species, small species are also frequently found in low abundances within assemblages of coral-reef fishes (Fig. 4, area C). This contrasts with many other animal assemblages where small species do not occur in the very low densities observed for much larger animals (Cotgreave 1993). Coral-reef fishes are widely believed to have open populations where much of the local reproductive output is uncoupled from recruitment to the population (review by Doherty & Williams 1988). Certainly on small spatial scales, it appears that populations of coral-reef fishes are divided into many sub-populations occupying disjunct habitat patches and these sub-populations are linked by larval dispersal (Sale 1991b, Caley et al. 1996). Therefore, at the small spatial scale of studies considered here, it is likely that some species of all sizes will occur in low abundance on particular reefs because of infrequent larval supply. Depending on the magnitude and variability of larval supply and the action of post-settlement processes, the presence of these species may be sporadic, or they might always be present in low numbers, or their abundance may be highly variable. A similar range of outcomes was observed by Sale et al. (1994) on small patch reefs monitored for over 10 yr.

Body size and total abundance

The total abundance of all species in different size classes is less frequently examined than the relationship between body size and population abundance. Where this has been investigated, a log normal distribution is usually observed (Hall 1996, Siemann et al. 1996, Warwick & Clarke 1996). Because body size, population abundance and species diversity are inter-related (Morse et al. 1988), total abundance will be influenced by the relationships between body size and species richness and body size and population abundance. Within assemblages of coral-reef fishes, there are more small species than mid to large size species (Fig. 2, p. 380) and small species are on average more abundant than large species (Fig. 4, p. 384). Consistent with expectations, total abundances within the assemblages of coral-reef fishes considered here peak in the smallest size classes (Fig. 5). The relationship between total abundance and species richness is well fitted to a power function with an average slope of 0.53 (Table 3). Therefore, $S = cI^{\sim 0.5}$ where S is species richness, I is the number of individuals within size classes and c is a case specific constant. A similar relationship was described by Sale (1980) among assemblages of coral-reef fishes. He found that among widely separated geographic locations the number of species increased with the number of individuals in a power function with a slope of approximately 0.5. Similarly, Siemann et al. (1996) have

Table 3 Regression analyses of log species richness on log total abundance of coral-reef fishes. Species were grouped into 100 mm size classes. Ordinary least squares regression was used for uniformity with other studies.

Location	Source	Slope	r^2
Lameshur Bay, Caribbean	Randall 1963	0.57	0.96
Rams Head, Caribbean	Randall 1963	0.83	0.97
Lizard Island, Great Barrier Reef	Caley 1995a	0.49	0.99
One Tree Island, Great Barrier Reef	Caley 1995a	0.57	0.98
Pandora Reef, Great Barrier Reef	Williams & Hatcher 1983	0.41	0.92
Rib Reef, Great Barrier Reef	Williams & Hatcher 1983	0.48	0.96
Myrmidon Reef, Great Barrier Reef	Williams & Hatcher 1983	0.52	0.95
Kaneohe Bay, Hawaii	Brock 1982	0.37	0.72
Mean		0.53	

reported that $S = cI^{\sim 0.5}$ in a grassland insect community where species richness and abundance peaked at intermediate body sizes. Curves that predict species richness from total abundance are not new, but the similarity in the relationship across spatial scales, taxa, and body size is intriguing.

Overall, relationships between body size and patterns of diversity and abundance observed in assemblages of coral-reef fishes are very similar to those described for terrestrial and other marine animals. There is some indication that small species are relatively more diverse on coral-reefs than in other habitats and this diversity may be associated with the finely-scaled habitat structure provided by the coral reef matrix and associated benthic organisms. Several theoretical models have been proposed to explain patterns of diversity and abundance in animal communities. However, our analyses show that these models do not adequately describe the structure of coral-reef fish assemblages. These models typically assume that communities are limited by the availability of a single resource, such as space or energy, and this resource is divided equally among size classes. Although populations of coral-reef fishes are probably resource-limited at least some of the time, there is little evidence to support resource-limitation across whole assemblages most of the time. Therefore, these models appear inadequate on both empirical and theoretical grounds. In a similar manner to studies on the population dynamics of coral-reef fishes (Caley et al. 1996), it may be from multifactorial models that the most useful explanations for relationships between species richness, abundance and body size will emerge.

Habitat associations and body size

Warburton (1989) found that body size differed significantly among habitat types for Indo–Pacific fishes. In general, small species occupied more restricted and sheltered habitats and microhabitats than large species. In addition, it is commonly observed that small species and small individuals stay in closer proximity to the reef than larger fishes (Fishelson et al. 1974, Allen 1975, Forrester 1991, Hobson 1991). Similar size-related patterns of habitat use have been described in temperate marine fishes (Choat

& Ayling 1987), freshwater fishes (Page & Swofford 1984) and fishes on artificial reefs (Anderson et al. 1989). As a consequence of their size, small species can use habitats not available to larger species. For example, a variety of blennies use holes in the reef for shelter (Lindquist 1985, Clarke 1989, 1994, Greenfield & Johnson 1990). Some gobies live only among the branches of specific corals (Patton 1994, Munday et al. 1997), on gorgonian corals (Randall et al. 1990), or on the surface of sponges (Tyler & Bohlke 1972). Clingfishes (Gobiesocidae) are symbionts of crinoids and sea urchins (Allen & Stark 1973, Randall et al. 1990) as are certain apogonids (Allen 1972). Damselfishes of the genus *Amphiprion* and *Premnas* have a symbiotic relationship with anemones (Fautin & Allen 1992) and other damselfishes, such as some *Dascyllus* and *Plectroglyphidodon* species are found in close association with branching corals (Sale 1971, Robertson & Lassig 1980). Some small fishes, such as the highly modified, eel-like fishes, belonging to the family Carapidae, live in the body cavities of echinoderms, tunicates and bivalves (Myers 1991). Strong habitat associations such as listed above, some of which are commensal or mutualistic, are not commonly observed for larger species of coral-reef fish.

These observations suggest that small species of fishes tend to be more specialized in terms of habitat use than large species and correspond with other findings, that, across a range of taxa, specialists tend to be smaller than generalists (Wasserman & Mitter 1978, Price 1984, Gaston & Lawton 1988). However, the scale of the interaction between animal and habitat must be considered. For example, the vertical rugosity of the reef measured in millimetres may be important in describing the distribution and abundance of small blennies, but for much larger fishes an order of magnitude greater rugosity may be important. Unfortunately, habitat variables are usually measured on only one spatial scale, even when comparing habitat use among species with vastly different body sizes. Studies that measure habitat variables scaled to body size are required to determine if small species are generally more specialized than large species. Although many small species are habitat specialists, the degree of habitat specialization varies widely, even among closely related species. Congeneric and confamilial differences in habitat specificity have been described among assemblages of anemonefishes (Fautin 1986, Fautin & Allen 1992), other pomacentrids (Clarke 1977, Itzkowitz 1977, Robertson & Lassig 1980, Öhman et al. 1998), apogonids (Chave 1978) and coral gobies (Munday et al. 1997).

In general, coral-reef fishes leave the planktonic stage and settle to reef environments at a very small size (mostly between 7 mm and 12 mm) (Leis 1991, Victor 1991). Thus, for most species there is a vast difference in size between newly-settled fishes and adults. Ontogenetic changes in diet (Horn 1989, Choat 1991), mortality rates (Eckert 1987, Hixon 1991) and competitive interactions (Jones 1987a,b) have been reported among coral-reef fishes. Changes in habitat use might be expected to coincide with such size-related changes in individual fitness or physiological requirements. Indeed, patterns of habitat use are associated with body size in many species. For example, depth-size relationships have been observed for some pomacentrids (Clarke 1977), other pomacentrids change microhabitats with size (Waldner & Robertson 1980, Lirman 1994) whereas others exhibit habitat specialization as juveniles but are less habitat-specific as adults (Helfman 1978). Juvenile French Grunts (*Haemulon flavolineatum*) settle onto sand and seagrass beds and migrate to nearby reefs within a few weeks (Shulman & Ogden 1987). Also, small individuals occupy different habitats from adults in some species of surgeonfishes (Robertson et al. 1979, Goodwin &

Kosaki 1989), wrasses (Goodwin & Kosaki 1989, Green 1996), parrotfishes (McAfee & Morgan 1996) and goatfishes (McCormick 1995). Furthermore, on some reefs, the size of individuals from a range of species increases from the protected back reef to the outer reef flat (Chabanet & Letourneur 1995).

Ontogenetic changes in habitat use do not occur in all species. Species with strong habitat associations, such as *Dascyllus aruanus* (Sweatman 1983), *Paragobiodon* (Kuwamura et al. 1994) and *Amphiprion* spp. (Leis 1991, Fautin 1992), settle directly to adult habitats. Settlement to these habitats may enhance growth and/or survival (Jones 1988, Forrester 1990, Booth 1995). *Dascyllus aruanus* preferentially recruits to corals with adults present (Sweatman 1983, 1985, 1988) and similar patterns have been observed in other species that use only a few microhabitat types (Doherty 1982, Eckert 1985b, Jones 1987b, Fowler 1990, Booth 1991, 1992, 1995, Öhman et al. 1998). The presence of large conspecifics could indicate suitable habitat for such species (Levin 1993).

Some species may settle to adult habitats but avoid settling among large conspecifics. For example, small, newly-settled *Pomacentrus chrysurus* avoid habitat patches with large conspecifics (Öhman et al. 1998) while some blennies recruit to the edges of adult territories (Nursall 1977). Intraspecific interactions are particularly frequent among small territorial herbivorous fishes (Nursall 1977, Sale et al. 1980, Doherty 1982, 1983) and in these species, interactions between new settlers and larger residents may have important implications for securing territories (Sale et al. 1980, Hunte & Côte 1989). Similarly, agonistic behaviour by resident anemonefishes may reduce the success of juveniles attempting to recruit to occupied anemones (Fautin 1992, Elliott et al. 1995).

The area of habitat occupied may also be related to body size. Home range size is often positively correlated with body size for site-attached animals (Gaston & Blackburn 1996) and a similar pattern has been observed among species of coral-reef fishes (Sale 1978a). However, Sale (1978a) noted that the range sizes of coral-reef fishes are often much smaller than for similar sized terrestrial animals. Most small coral-reef fishes are highly site attached and many defend individual or group territories (Low 1971, Nursall 1977, Sale 1978b, 1980, Sale et al. 1980). Small species such as most gobies, blennies and many territorial pomacentrids have home ranges encompassing only a few square metres (Low 1971, Lassig 1976, 1977, Nursall 1977, Sale 1978a,b, 1980, Forrester 1995). Although small coral-reef fishes are generally believed to be highly site attached, individuals at some life stages may travel considerable distances outside their primary home range (Marraro & Nursall 1983, Shibunto et al. 1993, Hattori 1994). Furthermore, Lewis (1997) found that movement among patch reefs was commonplace for some small species, particularly apogonids (Apogonidae).

Predation and body size

Within-species mortality rates typically decline with increasing body size in fishes and this decline is usually attributed to decreased predation rates on larger fishes (Werner 1984, Mittelbach & Chesson 1992). For most coral-reef fishes, size changes dramatically with ontogeny and they may best be described as having size-structured populations (*sensu* Ebenman & Persson 1988). Individuals recruit to the benthic population at a very small size and by maturity they have often increased in size by at least an order of magnitude. For many species it appears that mortality is very high following recruitment and decreases through time and, presumably, with increasing size (Eckert 1987, Shulman & Ogden 1987, Sale & Ferrell 1988, Hixon 1991). Although it is widely accepted that within species mortality rates of reef fishes vary with body size, Caley (1998) found that mortality rates of some coral-reef fishes is age invariant in the first year after settlement. He stressed the need for more studies on age- or size-specific mortality rates in coral-reef fishes in order to determine the influence of mortality on population dynamics and life histories.

Analysis of mortality rates among species of different body sizes have rarely been attempted. We analyzed the available data on species size versus mortality rates for non-cryptic, reef associated fishes and found that natural mortality rates decrease with increasing species size (Fig. 6, filled circles, mortality = $29.55 * \text{cm}^{-0.986}$, $r^2 = 0.765$). A similar among-species mortality rate (-1.03) was estimated for trawled demersal fishes off Namibia (Gordoa & Duarte 1992). Furthermore, when transformed for weight measurements ($m = 29.55 * w^{-0.328}$), the relationship we found does not appear markedly different from empirical observations of mortality rates of pelagic fishes plotted by Peterson & Wroblewski (1984, raw data in their Fig. 1). For most reef fishes the component of mortality attributable to predation is largely unknown, but is assumed to be considerable (Hixon 1991). Therefore, it appears that size is a major correlate of predation risk among species of reef fishes. Mortality rates of wrasses (Fig. 6, open circles),

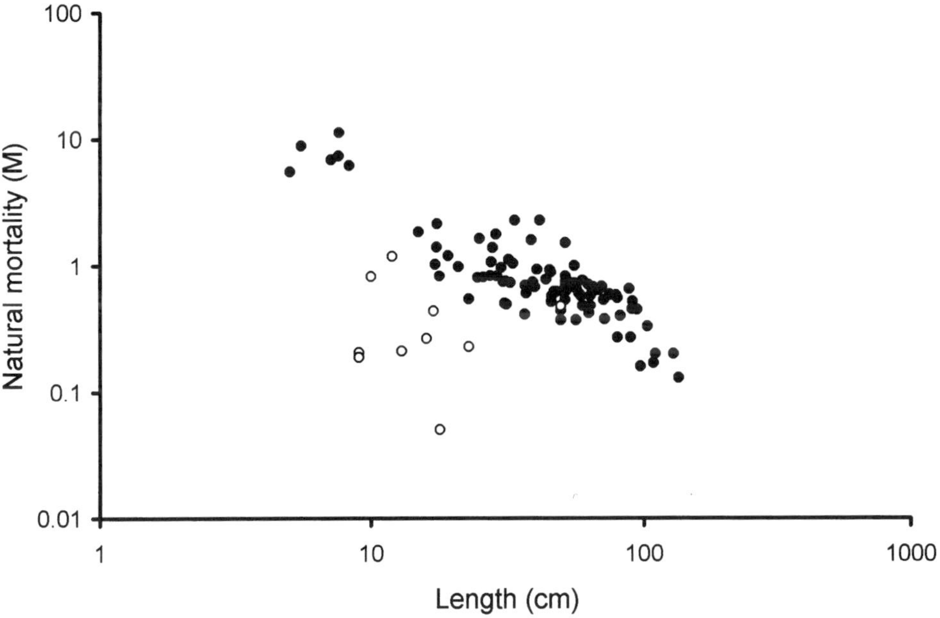

Figure 6 Instantaneous annual rates of mortality for coral-reef and reef-associated species. Filled circles are mortality estimates reported by Munro & Williams (1985) and Williams & Russ (1994). These estimates were mostly derived from Pauly's (1980) formula for mortality. Open circles are mortality estimates for coral-reef wrasses derived from empirical observations by Eckert (1987).

derived from empirical observations of Eckert (1987), are lower and more variable than mortality estimates for reef associated species (Fig. 6, filled circles). The cryptic behaviour of many small wrasses and protection from predation afforded by close association with the reef matrix may result in reduced mortality rates.

Predation pressure may be of particular importance in determining patterns of habitat use of small species and/or the smaller size classes of larger species (Jones 1984). This relationship was clearly demonstrated for freshwater fishes by Werner et al. (1983a,b). Small size classes of *Lepomis macrochirus* spent more time in vegetation when predators were present, even when foraging rates were much higher in more open habitats. In the absence of predators all size classes foraged as predicted by a profitability model.

Among coral-reef fishes, mortality is typically high in the first few days following settlement (Carr & Hixon 1995). Coral-reef fishes often settle to high complexity habitats (Shulman 1984, Öhman et al. 1998), a strategy that might reduce predation during the early benthic stage. The importance of complex habitats may also extend into the juvenile and adult phases. Jones (1988) found that survival of *Dascyllus aruanus* and *Pomacentrus amboinensis* during the first year of life was higher on structurally complex *Pocillopora* corals than on *Porites*, which offers less protection. Similarly, survival of juvenile *Pomacentrus moluccensis* in the presence of predators was much higher on reefs built from high complexity corals than low complexity corals (Beukers & Jones 1998). Hixon & Beets (1993) demonstrated that holes were used as a refuge from predation on artificial reefs. Furthermore, fish preferred shelter holes close to their body shape and predation was reduced on reefs with holes near the body size of resident fish. The importance of refuge availability in determining the diversity and abundance of small reef fishes was further elucidated by Caley & St John (1996). They found that survival of newly recruited fishes was greater on artificial reefs containing refuges from predation. Corresponding with an increase in abundance on these reefs was an increase in species richness. These experiments reinforce the numerous observations that small species and small size classes are often found in habitats offering the most protection from predation (Sale 1971, Fishelson et al. 1974, Allen 1975, Smith 1978, Jones 1984, Ebeling & Laur 1985, Sih 1992). However, relationships between refuge availability and prey densities are likely to change through time because fishes may outgrow juvenile refuges or become susceptible to a different suite of predators as they grow (Hixon 1991).

As dusk approaches on a coral reef there is a size assorted movement of diurnal fishes to shelter (Hobson 1991). Small species may be more vulnerable to predation in low light levels (Hobson 1991) and thus seek shelter first. Shulman (1985) observed numerous agonistic interactions among species using shelter sites. In these interactions, small individuals and small species were typically disadvantaged and frequently evicted from shelter sites by larger fish (Shulman 1985). Selection of shelter holes that closely match body size could be advantageous not only to reduce predation risk, as demonstrated by Hixon & Beets (1993), but also to reduce aggressive interactions with larger conspecifics and other species. Robertson & Sheldon (1979) found that *Thalassoma bifasciatum* defended shelter holes but failed to find any evidence that nocturnal shelter sites were limiting for *T. bifasciatum* or *Stegastes planifrons*. In this case, it appears that shelter defence probably reduces time exposed to predation rather than being a response to a limited resource. Even though *Thalassoma bifasciatum* and *Stegastes planifrons* are of similar size, they use different sized shelter holes. Therefore, not only size but also behavioural and morphological defences help determine nocturnal shelter use in small fishes (Robertson & Sheldon 1979).

The relative size of predator and prey may be important in predator–prey interactions. For example, Litvak & Leggett (1992) found that large larval fishes were preferentially eaten by predators. In this case, small size was a refuge from predation. Kingsford (1992) found that predator size varied across reef zones and that different sized predators tended to target different sized prey. Assemblages of small fishes in different areas of the reef might, therefore, be affected differently by predation depending on the types and abundances of predators present. Spatial variation in mortality rates of reef fishes has been demonstrated in several locations (Eckert 1985a, Aldenhoven 1986, Caley 1995b, Connell 1996, Beukers & Jones 1998). Much of this mortality has been attributed to the densities of predators (Connell 1996) but may also be influenced by the size of predators. Although predation risk may be influenced by body size, McCormick & Kerrigan (1996) did not detect differences in predation risk among newly-settled *Upeneus tragula* of different sizes. However, the size range of the *Upeneus* (27–32 mm standard length) was small in comparison with the size of predators (*Synodus variagatus* (130–143 mm)). In addition, *Synodus* ambush prey that stray too close, rather than targeting particular individuals (McCormick & Kerrigan 1996).

The use of fine-scale habitat features as a refuge may reduce predation by large predators but very small species may become prey for other small fishes. Members of the families Pseudochromidae and Apogonidae are typically less than 100 mm standard length and prey on small fishes as well as invertebrates. In addition, use of habitat as a refuge may be costly in other ways. For example, it may reduce the opportunities to mate (Endler 1980, Forsgren 1992) and to feed (Werner et al. 1983a, Dill & Fraser 1984, Forrester 1990).

Competition and body size

The occurrence of numerous small species on coral reefs, many with similar resource requirements (e.g. shelter, diet), would suggest that competition may be prevalent among small reef fishes. Also, if the response to predation is to concentrate small species and individuals into common refuges, the level of competition may be exacerbated. MacArthur & Levins (1967) proposed that there must be a limit to similarity in use of resources if species are to coexist where resources are limiting. Character displacement has often been proposed as a response to interspecific competition for limited resources that can promote coexistence (Wilson 1975, Slatkin 1980). For example, size differences between species could promote coexistence for a limiting resource such as shelter. Smith (1978) found that species of fishes sheltering in the reef differed in size by approximately equal proportions (5–12%) across the range of body sizes. He proposed that at least a 5% difference in size was needed to eliminate competition for shelter holes and other size-related resources. However, as noted above, shelter holes do not appear to be a limited resource for some reef fishes. In addition, size ratios within the range usually observed in assemblages of animals invariably occur where animal sizes are distributed lognormally and variances of these distributions are small (Eadie et al. 1987). Because species richness/body size distributions of coral-reef fishes are log-skewed, the size ratios observed by Smith (1978) may simply be statistical artefacts.

For some small, habitat-specialist species, it appears that habitat space may limit abundances. Fricke (1980) correlated coral colony size with group size for *Dascyllus marginatus* and Sale (1972) found that *D. aruanus* populations were related to coral colony size at some sites. However, where branching corals were more abundant, there was no such relationship. For anemonefishes, where social interactions limit the number of adults per colony, the amount of suitable habitat appears to limit abundance (Ross 1978, Fricke 1979). Some of the most habitat-specific fishes on coral reefs are coral gobies from the genera *Gobiodon* (Munday et al. 1997) and *Paragobiodon* (Lassig 1976, Kuwamura et al. 1994). Munday et al. (1997) found that the abundance of some species of *Gobiodon* at Lizard Island was correlated with the abundance of preferred coral species across reef zones. Furthermore, *Gobiodon* numbers declined significantly when the abundance of suitable corals declined following a crown-of-thorns starfish outbreak, but were not affected at sites where coral numbers did not decline. Lassig (1976) and Kuwamura et al. (1994) have shown that the number of *Paragobiodon* in coral heads is positively correlated with colony size. Furthermore, when the abundance of large coral colonies decreased, the number of large gobies decreased and the occupation rate of small colonies increased (Kuwamura et al. 1994). Therefore, it appears that population dynamics of coral-dwelling gobies is closely linked to host colony dynamics. For another small, habitat specialist, *Acanthemblemaria spinosa*, Buchheim & Hixon (1992) experimentally demonstrated that living space was a limited resource. In contrast, Robertson et al. (1981) did not detect any changes in abundance or fitness of *Stegastes planifrons* when space was experimentally manipulated.

Where habitat space is limiting, we might expect inter- and intraspecific competition to occur. Clarke (1992) demonstrated that sympatric species of hole-dwelling blennies compete for space in artificial habitats. *Acanthemblemaria spinosa*, with a higher metabolic rate, out-competed *A. aspera* for preferred spaces. However, *A. aspera* was able to persist in locations with lower food abundance because of lower metabolic demands. When *A. spinosa* was not present, *A. aspera* preferentially occupied holes usually occupied by *A. spinosa*. The natural distributions of these two species reflect this competitive interaction (Clarke 1989, 1994). Recently, Clarke (1996) has shown that the abundance of both species track changes in habitat availability in the wild. In addition, the relative abundances of the two species reversed, in accordance with predictions of the laboratory experiments, following a dramatic decline in habitats suitable for use by *A. spinosa*.

In addition to hole-dwelling blennies, interspecific interactions are known to influence patterns of habitat use and/or abundances in a variety of coral-reef damselfishes (Ebersole 1985, Robertson 1995, 1996), surgeonfishes (Robertson & Gaines 1986) and temperate water fishes (Hixon 1980, Schmitt & Holbrook 1990). For example, in a comprehensive examination of the interaction between seven sympatric species of damselfishes in the Caribbean, Robertson (1984, 1995, 1996) has shown that species size and aggressiveness are important in determining patterns of habitat use and species abundances. Robertson (1995) concluded that the largest species of *Stegastes*, *S. diencaeus*, was competitively dominant and that space holding ability among *Stegastes* species was size dependent. Larger individuals of *S. diencaeus* and *S. dorsopunicans* frequently evicted smaller conspecific and heterospecifics when their territory size was constrained (Robertson 1995). Neither species, however, was able to evict larger heterospecific or conspecific neighbours. Following removals of the most abundant, and second largest species, *S. planifrons*, the abundance of smaller and less aggressive species increased (Robertson 1996). The abundance of *S. partitus*, which is about half the size of *S. planifrons*, doubled after the removal of *S. planifrons*. *S. partitus* also increased its microhabitat use into areas previously used mostly by *S. planifrons*. The abundance of another small species, *S. variabilis*, also doubled. The combined increase in abundance of *S. partitus* and *S. variabilis* was ~ 70% of the abundance of *S. planifrons* prior to the removals. More importantly, there was a strong correlation between the prior density of *S. planifrons* on individual reefs and the combined increase in density of *S. partitus* and *S. variabilis* on these reefs (Robertson 1996). In addition, the largest damselfish in the assemblage, *Microspathodon chrysurus*, aggressively dominates and has a negative effect on the body mass and fat deposits of *Stegastes planifrons* (Robertson 1984). Clearly, size-related competitive asymmetries influence both patterns of habitat use and the abundance of damselfishes in this assemblage. However, Robertson (1995) notes that species effects may limit size effects where the smaller species is very aggressive, as is the case with interactions between *Microspathodon chrysurus* and *Stegastes planifrons*.

In anemonefishes, another group of small, habitat specialists, it appears that distributions of species among host anemones are controlled by a variety of factors including imprinting of larvae on their host anemone before they enter the pelagic environment (Arvedlund & Nielsen 1996), learned host preferences, competition for limited space and stochastic processes (Fautin 1986, 1992). The competitive dominant, *Premnas biaculeatus*, occurs with only one actinian species (Fautin 1986). Other species occur with more actinian species in decreasing order of competitive dominance (Fautin 1986). In field and laboratory experiments Fautin (1986) found that a combination of size, species and prior residence was important in determining competitive superiority. Only *P. biaculeatus* consistently won competitive interactions against similar sized anemonefishes and even evicted larger individuals of other species from their host anemones (Fautin 1986). Therefore, in anemonefishes competitive ability seems to be related to the degree of specialization, in addition to body size.

As many small species of coral-reef fishes are habitat specialists and some appear to be limited by habitat availability, it may be among these species that effects of inter- and intraspecific competition are most likely to be detected. However, rigorous experimental studies that have examined the potential for effects of both intra- and interspecific competition on important demographic parameters have not always supported this contention. In pairwise comparisons of closely related and ecologically similar species, there is often very little evidence of interspecific competition when properly controlled density manipulations are carried out (Doherty 1982, 1983, Jones 1987a, 1988, Roberts 1987). Although competition is more likely to occur among small species of near similar size, similarity in size is also likely to lead to competitive equality. Competitive dominance by larger species may only occur when there is a moderate size difference among the competing species. Where there is a large size difference, competition is less likely, and when it does occur, small species may persist because they have a spatial refuge from larger competing species (Robertson 1984). Rather than controlling for differences in body size, future experimental studies will need to incorporate relative size differences as a potentially important factor.

There is considerably more experimental evidence for intraspecific competition in small reef fish species. The results suggest that competition within species is more prevalent among individuals of similar sizes, and that the effects are more often evident in terms of reduced growth, rather than increased mortality. For example, Jones

(1987a) found that growth rate of juvenile *Pomacentrus amboinensis* was influenced by cohort density but not adult density in the first year of growth. However, in the second year of growth, adult densities significantly affected growth rates of the larger juveniles. At high adult densities, growth rates were depressed. Therefore, the influence of intraspecific competition on growth may become more prominent as individuals approach maturity. Jones (1987a) also determined that cohort density and adult density influenced the rate of maturity and the total number of individuals reaching maturity. Numbers maturing did not increase linearly with cohort density and when adults were present the number of individuals reaching maturity was reduced by 50%. Direct effects of increased density on mortality rates have been demonstrated in some very small reef fishes (Forrester 1995). Negative effects of adults on juvenile recruitment have also been observed in some species (e.g. Sweatman 1985, Jones 1987b, Forrester 1995) and may be more common in species with a restricted adult size. Also, as slower growing juveniles appear to suffer increased mortality (Forrester 1991, Jones 1997), indirect effects of competition stemming from effects on growth rates may be evident in longer term studies.

Life histories of small coral-reef fishes

The size of an organism is generally considered to be one of the most important life-history traits and one that may dramatically influence other life-history characteristics (Begon & Mortimer 1986, Roff 1992, Stearns 1992). Small species are typically expected to be short lived, and exhibit early maturity, high reproductive output and high intrinsic rates of increase (Begon & Mortimer 1986). Across a range of organisms there is a positive correlation between size and longevity (Calder 1984). For example, Blueweiss et al. (1978) presented a positive relationship between maximum life span and body weight in a variety of animals including fishes. However, there is no clear relationship between size and longevity when the data used in this analysis are examined for fish alone. Many small species of fish are short lived but it is becoming increasingly obvious that some small species are also long lived (Miller 1996). In particular, many small coral-reef fishes are relatively long lived (Choat & Bellwood 1991). For example, *Pomacentrus moluccensis* may live for over 17 yr (Doherty & Fowler 1994), *Amphiprion akindynos* for more than 10 yr, *Dascyllus aruanus* for more than 7 yr (Sale et al. 1994), some chaetodontids live for over 30 yr (Nangle & Jones unpubl. data), and the small acanthuroids, *Zebrasoma scopas* and *Ctenochaetus striatus*, are estimated to live at least 30 yr (Choat & Axe 1996). Comparisons of longevity in relation to body size among coral-reef fishes may be strongly influenced by phylogenetic history. For example, species in the family Acanthuridae reach much greater ages than species from the family Scaridae (Choat & Axe 1996, Choat et al. 1996) despite larger body sizes among the Scaridae.

Relationships between longevity and size might also be influenced by patterns of habitat use and predation risk. The potential importance of predation in modifying life-history traits of fishes was demonstrated by Reznick et al. (1990, 1996). They found that life histories of guppies in high predation localities evolved towards earlier maturity, higher reproductive effort and smaller offspring. Small temperate water fishes that are highly susceptible to predation also exhibit these traits (Miller 1979, 1996, Table 4).

Table 4 Life-history traits, life style and presumed predation risk of temperate water gobies (family Gobiidae).

Species	Max. size (mm)	Maturity (yrs)	Longevity (yrs)	Life style	Presumed predation risk	Source
Pseudaphya ferreri	35	<1	1	nektonic	high	Miller 1973
Aphia minuta	50	<1	1	nektonic	high	Miller 1973, 1979, 1984 Iglesias et al. 1997
Crystallogobius linearis	50	<1	1	nektonic	high	Miller 1973, 1979, 1984
Sigamia geneionema	115	<1	1	open sand	high	Sano 1997
Pomatoschistus minutus	75	0.6–1	1.3–1.6	open sand	high	Miller 1979, 1984, Magnhagen 1993
P. microps	55	0.6–1	1.6–2.1	open sand	high	Miller 1979, 1984, Magnhagen 1993
P. norvegicus	60	<1	2.7	epibenthic		Gibson & Ezzi 1981
Neogobius melanostomus	170	2–3	3	epibenthic		Miller 1984
N. syrman	190	1	3	epibenthic		Miller 1984
Gobius niger	150	2	5	cryptobenthic	low	Magnhagen 1993
G. couchi	75	2	6	cryptobenthic		Miller & El-Tawil 1974
G. paganellus	90	2–3	10	cryptobenthic	low	Miller 1979, 1984
Thorogobius ephippiatus	110	3–4	9	cryptobenthic	low	Miller 1979, 1984
Lepidogobius lepidus	90	1–2	>7	burrows	low-moderate	Grossman 1979, Miller 1996
Gobius cobitis	228		10	cryptobenthic	low	Gibson 1970
Lesueurigobius friesii	95	1–2	11	cryptic burrower	low	Gibson & Ezzi 1978, Miller 1979, 1984
Typhlogobius californiensis	80		12	symbiotic/cryptic	low	Miller 1984

In contrast, cryptobenthic fishes, which are more protected from predation, are longer lived and reproductive effort is spread over a number of years (Miller 1979, 1996, Table 4). Furthermore, Magnhagen (1993) demonstrated that short-lived gobies expose themselves to predation risk while breeding but longer-lived species would not breed in the presence of a predator. In order to reduce phylogenetic confounding we have confined the comparison in Table 4 to within the family Gobiidae. Temperate water fishes were used in the analysis because insufficient data are available for tropical species from a single family that uses broadly different habitats. However, from the small amount of data available on life histories in tropical gobies it appears that a similar trend is emerging. *Coryphopterus personatus* hovers above the substratum near corals and sponges around reefs in the Caribbean (Cole & Robertson 1988), a life style that Miller (1979) described as nektonic and offering little cover from predation. This species reaches only 34 mm total length (Cole & Robertson 1988), matures in approximately 1 month and has an adult half life of 1–2 months (Robertson & Kautman 1998). Another small goby, *Istigobius decoratus*, reaches approximately 80 mm total length and lives on open sand near reefs. It is presumably under high predation pressure and lives for less than 1 yr at Lizard Island on the Great Barrier Reef (J. Kritzer unpubl. data). In contrast, coral-dwelling gobies that live among the branches of corals are much longer lived (Lassig 1976, 1977, Kuwamura et al. 1994, 1996, Patton 1994), even though they reach only 75% of the maximum size of *I. decoratus*. Detailed information on life-history traits from a wide range of species are needed in order to separate the influence of body size, predation risk, habitat use and phylogenetic history on life histories of coral-reef fishes.

Small size may influence a number of other life-history traits in coral-reef fishes. For example, most small coral-reef fishes are demersal spawners with extended care of eggs (Barlow 1981). In contrast, larger species are typically pelagic spawners with no parental care of eggs (Barlow 1981, Thresher 1984). In Fig. 1 (p. 376) we show the size range of common families of coral-reef fishes. Seven of the first nine families are egg brooders (Apogonidae) or demersal spawners. Callionymidae and Serranidae (Anthiinae) are the only taxa containing mostly small species that are not demersal spawners. However, for the Anthiinae, phylogenetic constraints could apply since all other members of the family Serranidae are pelagic spawners (Thresher 1984). Similarly, many species of wrasse (Labridae) are small but they are pelagic spawners like larger species in the family. In contrast to the pattern of demersal spawning among small taxa, all families with larger species in Fig. 1 (from Cirrhitidae to Epinephelinae) are pelagic spawners. Most have buoyant eggs with no parental care (Thresher 1984, 1991). Siganidae are pelagic spawners with no parental care, but the eggs are heavier than sea water and settle to the substratum (Thresher 1991). The spawning mode of the Holocentridae is unknown. The only notable exception to pelagic spawning and no parental care among large coral-reef fishes are the Balistidae. Balistids lay demersal eggs that are cared for by the female (Thresher 1984, 1991).

Egg sizes are similar across a wide range of body sizes in fishes (Blueweiss et al. 1978). As a result, small fishes produce far fewer eggs per spawning than large fishes (Weatherly 1972, Thresher 1984). Demersal spawning and egg care by small species may increase early survival and, thereby, offset lower overall fecundity (Barlow 1981). In addition to caring for the eggs, incubation periods for demersally spawned eggs are longer than pelagic eggs (Thresher 1991). Therefore, larvae from demersal spawners are generally larger and more developed at hatching (Barlow 1981, Thresher

1991). A trade-off between fecundity and larval mortality has also been proposed from small bodied invertebrates where there is a general trend towards non-planktotrophic larvae (Jablonski & Lutz 1983). Although fecundity of fishes is strongly correlated with body size in females, reproductive success is not limited to female size in small fishes. Because paternal care is common among small coral-reef fishes (Barlow 1981), male body size may also be important. For example, clutch size and survival has been correlated with male size in *Acanthemblemaria crockeri* (Hastings 1988), *Ophioblennius atlanticus* (Côte & Hunte 1989) and *Paragobiodon echinocephalus* (Kuwamura et al. 1993). However, male size does not influence egg survival in all species with paternal care (Sunobe et al. 1995).

The demersal spawning mode of small reef fishes has implications for patterns of diversity and distribution of coral-reef fishes. Demersally spawned eggs are retained in the reef environment until hatching. In contrast, pelagic eggs are buoyant and are subject to hydrodynamic processes that may disperse the eggs from reef environments (Barlow 1981, Thresher 1991). Also, because newly-hatched larvae of demersal spawners are larger and more developed than newly-hatched larvae from pelagic spawners, they may be more able to influence their trajectory (Leis 1993). As a consequence, demersal spawning may increase opportunities for local colonization but may reduce opportunities for widespread colonization. Pre-settlement fishes from a range of demersal spawning families tend to be more frequently found near reefs, whereas those of pelagic spawners are frequently found well offshore (Leis & Miller 1976, Leis 1982, Kingsford 1988, Kingsford & Choat 1989). In a review of data on larval fish assemblages, Leis (1993) cautions that the dichotomy between the spatial distributions of larvae from demersal and pelagic spawners is not universal and might also be influenced by sampling biases. He concludes, however, that at least in some locations there are real differences in the distributions of larvae from demersal and pelagic spawners. For example, near Lizard Island on the Great Barrier Reef, the larvae of all demersally spawned species sampled declined in abundance with increasing distance into the Coral Sea (Leis 1993). In contrast, larvae of pelagic spawners had variable distributions including species with peak abundance near reefs and species with peak abundance offshore (Leis 1993). The tendency for larvae from demersal spawners to remain near-shore might influence the distributions of small coral-reef fishes. Indeed, coral-reef fishes with trans-Pacific distributions are typically pelagic spawners and the number of demersal spawners declines substantially from west to east across the Pacific (Leis 1984, Thresher 1991).

Barlow (1981) proposed that reduced dispersal characteristics of larvae as a result of demersal spawning may restrict gene flow among populations of small coral-reef fishes. This may lead to greater rates of speciation among small species than among larger pelagic spawners (Rosenblatt 1963, in Thresher 1991). Recently, Doherty et al. (1995) linked larval characteristics with gene flow among populations of coral-reef fishes separated by 1000 km on the Great Barrier Reef. Using allozyme electrophoresis to estimate genetic heterogeneity of populations, they concluded that gene flow between widespread locations is influenced by the mean larval duration of the species. Of the seven species studied, *Acanthochromis polyacanthus*, which does not have a pelagic larval phase, exhibited the greatest heterogeneity in genetic characters between regions. Four demersal spawning pomacentrids with intermediate larval durations also exhibited heterogeneity of genetic characters between regions. However, the species with the longest larval durations, *Ctenochaetus striatus* and *Pterocaesio chrysozona*, did not show significant differences between the regions. These two species were also the only pelagic spawners in the analysis and because pelagic spawners generally have more widespread larval distributions, this might be an important component to the gene flow argument. In the Caribbean, Shulman & Bermingham (1995) found that the degree of genetic heterogeneity among widespread populations of coral-reef fishes was not adequately explained by pelagic larval duration alone, but was confounded with traits such as spawning mode.

Behaviour of larvae has been implicated in dispersal patterns of larval coral-reef fishes (Leis 1993, Cowen & Sponaugle 1997) and, thus, in levels of gene flow among widely separated populations (Shulman & Bermingham 1995). For example, large larval fishes are able to swim further than small larvae and this difference may further influence the dispersal characteristics of species (Stobutzki & Bellwood 1997). Larval behaviour of some marine invertebrate has also been implicated in genetic differences on scales much less than predicted by their dispersal potential (Palumbi 1994). The relative importance of larval behaviour on dispersal patterns of coral-reef fishes remains largely unlnown.

Further evidence for the potential of restricted gene flow in small species comes from studies of communities of larval fishes within atoll lagoons. Leis (1994) found that species able to complete their life cycles within atoll lagoons were primarily small, substratum-associated fishes. The great majority of these were benthic spawners or brooders (apogonids). In contrast, few pelagic spawners appeared able to complete their life cycles within lagoons (Leis 1994). In general, it appears that the larvae of small species are less likely to disperse long distances or across ocean expanses as a consequence of their spawning mode. Also, at least some small species appear able to exist as isolated populations on small spatial scales. These characteristics favour high speciation rates among demersal spawners, which are characteristically small.

Small species generally have short generation times and this is also believed to enhance speciation rates (Marzluff & Dial 1991). Many small fishes mature within the first year (Miller 1979, 1996) and even relatively long-lived small species of fish may mature early. Age or size at maturity is generally believed to be intrinsically linked to growth rates and longevity in fishes (Roff 1984, Charnov & Berrigan 1990). However, Miller (1996) noted that some small cryptobenthic fishes mature earlier than predicted from their longevity. Consequently, some small cryptic fishes have relatively short generation times despite being relatively long lived. Potentially, short generation times in conjunction with decreased gene flow among widespread populations could enhance speciation rates among small coral-reef fishes. As discussed in previous sections, increased speciation among small species may help explain the species body/size distributions observed in assemblages of coral-reef fishes.

Conclusions

Diversity and abundance peaks in small size classes (<100mm) within local assemblages of coral-reef fishes. However, difficulties identifying and accurately estimating the abundance of small species have meant they are often ignored or undersampled in community level studies. An enormous volume of research has been conducted on the population ecology of coral-reef fishes during the past 20 yr. Much of this research has concentrated on small to moderate sized species and, particularly, on one family, the Pomacentridae. Consequently, there have been calls for more studies on larger-bodied reef species (Polunin & Roberts 1996). To this we add that more research is needed on families of small reef fishes other than the Pomacentridae. Also, in order to gain an unbiased understanding of reef fish communities, more attention is required on the large number of very small and cryptic species within reef fish assemblages.

Analysis of patterns of diversity and abundance in relation to body size within assemblages of animals have proved useful in generating hypotheses about processes structuring communities. We have extended these comparisons to assemblages of coral-reef fishes. Diversity/size and abundance/size distributions within local assemblages of coral-reef fishes are very similar to those observed in many other taxa. The factors generating these patterns remain unclear, however, there is some potential for testing the various hypotheses proposed. In particular, species at the upper boundary of the abundance/size distribution might be more frequently resource-limited than other species. If this is the case, intraspecific competition may be more intense in species at the upper bound of the abundance/size distribution. Also, interacting guild members at the upper bound of the abundance/size distribution would be most likely to demonstrate demographic effects of interspecific competition.

We propose that life-history traits and larval ecology of small species of coral-reef fishes has contributed significantly to their high diversity. Further studies on the genetic structure of widely separated populations of demersal and pelagicaly spawned species of coral-reef fishes will be useful in testing this notion. Also, more information is required on the behaviour of larval reef fishes. It is widely accepted that larvae of coral-reef fishes disperse long distances. However, for some small species, long-range dispersal might be relatively rare. Detailed analysis of species geographic distributions in relation to body size, generation time and spawning mode will also be useful for examining the relationship between body size, spawning mode and diversity.

Body size has been successfully integrated into models that describe how competition and predation influence the population dynamics of freshwater fishes. Similarly, the intensity and outcome of predator–prey and competitive interactions in assemblages of coral-reef fishes can be influenced by the absolute and relative sizes of the participating individuals. Predation is most likely to occur among individuals and species of different body sizes. Predation effects are also likely to be strongest during the early benthic stage. In contrast, competition is most likely among individuals and species of similar body sizes. Integrating size-structure into studies of coral-reef fishes may assist in attempts to develop multifactorial models of population dynamics of coral-reef fishes.

Acknowledgements

Discussions with J. Brown, J. Caley, H. Choat and R. Robertson contributed to the development of this review. D. Williams generously supplied us with raw data used in previously published studies. B. Kerrigan, R. Robertson and J. Kritzer supplied unpublished data on life histories of reef fishes. J. Brown, J. Caley, C. N. Johnson, H. Sweatman, N. Gust and K. Buckley gave constructive comments on the manuscript.

References

Aldenhoven, J. M. 1986. Local variation in mortality rates and life-expectancy estimates of the coral-reef fish *Centropyge bicolor*. *Marine Biology* **92**, 237–44.

Allen, G. R. 1972. Observations on a commensal relationship between *Siphamia fuscolineatus* (Apogonidae) and the crown-of-thorns starfish, *Acanthaster planci*. *Copeia* 1972, 595–7.

Allen, G. R. 1975. *Anemonefishes*. Neptune City: TFH Publications.

Allen, G. R. & Stark II, W. A. 1973. Notes on the ecology, zoogeography, and coloration of the gobiesocid clingfishes, *Lepadichthys caritus* Briggs and *Diademichthys lineatus* (Sauvage). *Proceedings of the Linnean Society of New South Wales* **98**, 95–7.

Anderson, T., DeMartini, E. E. & Roberts, D. A. 1989. The relationship between habitat structure, body size and distribution of fishes at a temperate artificial reef. *Bulletin of Marine Science* **44**, 681–97.

Arvedlund, M. & Nielsen, L. E. 1996. Do the anemonefish *Amphiprion ocellaris* (Pisces: Pomacentridae) imprint themselves to their host sea anemone *Heteractis magnifica* (Anthozoa: Actinidae)? *Ethology* **102**, 197–211.

Barlow, G. W. 1981. Patterns of parental investment, dispersal and size among coral-reef fishes. *Environmental Biology of Fishes* **6**, 65–85.

Begon, M. & Mortimer, M. 1986. *Population ecology: a unified study of animals and plants*. Cambridge: Blackwell Scientific.

Beukers, J. S. & Jones, G. P. 1998. Habitat complexity modifies the impact of piscivores on a coral reef fish population. *Oecologia* **114**, 50–59.

Blackburn, T. M., Brown, V L., Doube, B. M., Greenwood, J. J. D., Lawton, J. H. & Stork, N. E. 1993a. The relationship between abundance and body size in natural animal assemblages. *Journal of Animal Ecology* **62**, 519–28.

Blackburn, T. M. & Gaston, K. J. 1994. Body size: the limits to biomass and energy use. *Oikos* **69**, 336–9.

Blackburn, T. M., Harvey, P. H. & Pagel, M. D. 1990. Species number, population density and body size relationships in natural communities. *Journal of Animal Ecology* **59**, 335–45.

Blackburn, T. M. & Lawton, J. H. 1994. Population abundance and body size in animal assemblages. *Philosophical Transactions of the Royal Society of London* **343**, 33–9.

Blackburn, T. M., Lawton, J. H. & Perry, J. E. 1992. A method of estimating the upper bounds of plots of body size and distribution in natural animal assemblages. *Oikos* **65**, 107–12.

Blackburn, T. M., Lawton, J. H. & Pimm, S L. 1993b. Non-metabolic explanations for the relationship between body size and animal abundance. *Journal of Animal Ecology* **62**, 694–702.

Blueweiss, L., Fox, H., Kudzma, V., Nakashima, D., Peters, R. & Sams, S. 1978. Relationships between body size and some life history traits. *Oecologia* **37**, 257–72.

Booth, D. J. 1991. The effects of the sampling frequency on estimates of recruitment of the domino damselfish *Dascyllus albisella*. *Journal of Experimental Marine Biology and Ecology* **145**, 149–59.

Booth, D. J. 1992. Larval settlement patterns and preferences by domino damselfish *Dascyllus albisella*. *Journal of Experimental Marine Biology and Ecology* **155**, 85–104.

Booth, D. J. 1995. Juvenile groups in a coral-reef damselfish: density-dependent effects on individual fitness and population demography. *Ecology* **76**, 91–106.

Brock, R. E. 1982. A critique of the visual census method for assessing coral reef fish populations. *Bulletin of Marine Science* **32**, 269–76.

Brooks, J. L. & Dobson, S. I. 1965. Predation, body size and composition of plankton. *Science* **150**, 28–35.

Brown, J. H. 1981. Two decades of homage to Santa Rosalia: towards a general theory of diversity. *American Naturalist* **21**, 877–88.

Brown, J. H., Marquet, P. A. & Taper, M. L. 1993. Evolution of body size: consequences of an energetic definition of fitness. *American Naturalist* **142**, 573–84.

Brown, J. H. & Maurer, B. A. 1986. Body size, ecological dominance and Cope's rule. *Nature* **324**, 248–50.

Buchheim, J. R. & Hixon, M. A. 1992. Competition for shelter in the coral-reef fish *Acanthemblemaria spinosa* Metzelar. *Journal of Experimental Marine Biology and Ecology* **164**, 45–54.

Calder III, W. A. 1984. *Size, function and life history*. Cambridge: Harvard University Press.

Caley, M. J. 1995a. Community dynamics of tropical reef fishes: local patterns between latitudes. *Marine Ecology Progress Series* **129**, 7–18.

Caley, M. J. 1995b. Reef fish community structure and dynamics: an interaction between local and larger-scale processes? *Marine Ecology Progress Series* **129**, 19–29.

Caley, M. J. 1997. Are local patterns of reef fish diversity related to patterns of diversity at a larger scale? *Proceedings of the 8th International Coral Reef Symposium* **1**, 993–8.

Caley, M. J. 1998. Age-specific mortality rates in reef fishes: evidence and implications. *Australian Journal of Ecology* **23**, 241–5.

Caley, M. J., Carr, M. H., Hixon, M. A., Hughes, T. P., Jones, G. P. & Menge, B. A. 1996. Recruitment and the local dynamics of open marine populations. *Annual Review of Ecology and Systematics* **27**, 477–500.

Caley, M. J. & Schluter, D. 1997. The relationship between local and regional diversity. *Ecology* **78**, 70–80.

Caley, M. J. & St John, J. 1996. Refuge availability structures assemblages of tropical reef fishes. *Journal of Animal Ecology* **65**, 414–28.

Carpenter, K. E., Miclat, R. I., Albaladejo, V. D. & Corpuz, V. T. 1981. The influence of substrate structure on the local abundance and diversity of Philippine reef fishes. *Proceedings of the 4th International Coral Reef Symposium* **2**, 497–502.

Carr, M. H. & Hixon, M. A. 1995. Predation effects on early post-settlement survivorship of coral-reef fishes. *Marine Ecology Progress Series* **124**, 31–42.

Chabanet, P. & Letourneur, Y. 1995. Spatial pattern of size distribution of four species on Reunion coral reef flats. *Hydrobiologia* **300/301**, 299–308.

Chabanet, P., Ralambondrainy, M., Faure, G. & Galzin R. 1997. Relationships between coral reef substrata and fish. *Coral Reefs* **16**, 93–102.

Charnov, E. L. 1983. *Life history invariants: some explorations of symmetry in evolutionary ecology*. Oxford: Oxford University Press.

Charnov, E. L. & Berrigan, D. 1990. Dimensionless numbers and life history evolution: age of maturity versus the adult lifespan. *Evolutionary Ecology* **4**, 273–5.

Chave, E. H. 1978. General ecology of six species of Hawaiian cardinalfishes. *Pacific Science* **32**, 245–70.

Choat, J. H. 1991. The biology of herbivorous fishes on coral reefs. In *The ecology of fishes on coral reefs*, P. F. Sale (ed). San Diego: Academic Press, 120–55.

Choat, J. H. & Axe, L. M. 1996. Growth and longevity in acanthurid fishes; an analysis of otolith increments. *Marine Ecology Progress Series* **134**, 15–26.

Choat, J. H., Axe, L. M. & Lou, D. G. 1996. Growth and longevity in fishes of the family Scaridae. *Marine Ecology Progress Series* **145**, 33–41.

Choat, J. H. & Ayling, A. M. 1987. The relationship between habitat structure and fish faunas on New Zealand reefs. *Journal of Experimental Marine Biology and Ecology* **110**, 257–84.

Choat, J. H. & Bellwood, D. R. 1991. Reef fishes: their history and evolution. In *The ecology of fishes on coral reefs*, P. F. Sale (ed). San Diego: Academic Press, 39–66.

Clarke, R. D. 1977. Habitat distribution and species diversity of chaetodontid and pomacentrid fishes near Bimini, Bahamas. *Marine Biology* **40**, 277–89.

Clarke, R. D. 1989. Population fluctuation, competition and microhabitat distribution of two species of tube blennies, *Acanthemblemaria* (Teleostei: Chaenopsidae). *Bulletin of Marine Science* **44**, 1174–85.

Clarke, R. D. 1992. Effects of microhabitat and metabolic rate on food intake, growth and fecundity of two competing coral reef fishes. *Coral Reefs* **11**, 199–205.

Clarke, R. D. 1994. Habitat partitioning by chaenopsid blennies in Belize and Virgin Islands. *Copeia* 1994, 398–405.

Clarke, R. D. 1996. Population shifts in two competing fish species on a degrading coral reef. *Marine Ecology Progress Series* **137**, 51–8.

Cole, K. S. & Robertson, D. R. 1988. Protogyny in the Caribbean reef goby, *Coryphopterus personatus*: gonad ontogeny and social influences on sex change. *Bulletin of Marine Science* **42**, 317–33.

Connell, S. D. 1996. Variations in mortality of a coral-reef fish: links with predator abundance. *Marine Biology* **126**, 347–52.

Côte, I. M. & Hunte, W. 1989. Male and female mate choice in the redlip blenny: why bigger is better. *Animal Behaviour* **38**, 78–88.

Cotgreave, P. 1993. The relationship between body size and population abundance in animals. *Trends in Ecology and Evolution* **8**, 244–8.

Cowen, R. K. & Sponaugle, S. 1997. Relationships between early life history traits and recruitment among coral reef fishes. In *Early life history and recruitment in fish populations*, R. C. Chambers & E. A. Trippel (eds). London: Chapman & Hall, 423–49.

Currie, D. J. 1993. What shape is the relationship between body size and population density? *Oikos* **66**, 353–8.

Damuth, J. 1981. Population density and body size in mammals. *Nature* **290**, 699–700.

Damuth, J. 1987. Interspecific allometry of population density in mammals and other animals: independence of body mass and population energy use. *Biological Journal of the Linnean Society* **31**, 193–246.

Damuth, J. 1991. Of size and abundance. *Nature* **351**, 268–70.

Dill, L. M. & Fraser, A. H. G. 1984. Risk of predation and the feeding behaviour of juvenile coho salmon (*Oncorhynchus kisutch*). *Behavioural Ecology and Sociobiology* **16**, 65–71.

Doherty, P. J. 1982. Some effects of density on the juveniles of two species of tropical territorial damselfish. *Journal of Experimental Marine Biology and Ecology* **65**, 249–61.

Doherty, P. J. 1983. Tropical territorial damselfishes: is density limited by aggression or recruitment? *Ecology* **64**, 176–90.

Doherty, P. J. & Fowler, T. 1994. An empirical test of recruitment limitation in a coral reef fish. *Science* **263**, 935–9.

Doherty, P. J., Planes, S. & Mather, P. 1995. Gene flow and larval duration in seven species of fish from the Great Barrier Reef. *Ecology* **76**, 2373–91.

Doherty, P. J. & Williams, D. M. 1988. The replenishment of coral reef fish populations. *Oceanography and Marine Biology: an Annual Review* **26**, 487–551.

Eadie, J. McA., Broekhoven, L. & Colgan, P. 1987. Size ratios and artifacts: Hutchinson's rule revisited. *American Naturalist* **129**, 1–17.

Ebeling, A. W. & Lauer, D. R. 1985. The influence of plant cover on surfperch abundance at an offshore temperate reef. *Environmental Biology of Fishes* **12**, 169–79.

Ebenman, B. & Persson, L. (eds). 1988. *Size-structured populations.* Berlin: Springer-Verlag.

Ebersole, J. P. 1985. Niche separation of two damselfish species by aggression and differential microhabitat utilization. *Ecology* **66**, 14–20.

Eckert, G. J. 1985a. Settlement of coral reef fishes to different natural substrata and at different depths. *Proceedings of the 5th International Coral Reef Symposium* **5**, 385–90.

Eckert, G. J. 1985b. *Population studies of labrid fishes on the southern Great Barrier Reef.* PhD thesis, University of Sydney, Australia.

Eckert, G. J. 1987. Estimates of adult and juvenile mortality for labrid fishes at One Tree Reef, Great Barrier Reef. *Marine Biology* **95**, 167–71.

Elliott, J. K., Elliot, J. M. & Marsical, R. N. 1995. Host selection, location, and association behaviours of anemonefishes in field settlement experiments. *Marine Biology* **122**, 377–89.

Endler, J. A. 1980. Natural selection of colour patterns in *Poecilia reticulata. Evolution* **34**, 76–91.

Fautin, D. G. 1986. Why do anemonefishes inhabit only some host actinians? *Environmental Biology of Fishes* **15**, 171–80.

Fautin, D. G. 1992. Anemonefish recruitment: the roles of order and chance. *Symbiosis* **14**, 143–60.

Fautin, D. G. & Allen, G. R. 1992. *Field guide to anemonefishes and their host anemones.* Perth: Western Australian Museum.

Fenchel, T. 1993. There are more small species than large species? *Oikos* **68**, 375–8.

Finlay, B. J. & Fenchel, T. 1996. Global diversity and body size. *Nature* **383**, 132–3.

Fishelson, L., Popper, D. & Avidor, A. 1974. Biosociology and ecology of pomacentrid fishes around Sinai Peninsula (northern Red Sea). *Journal of Fish Biology* **6**, 119–33.

Forrester, G. E. 1990. Factors influencing the juvenile demography of a coral reef fish. *Ecology* **71**, 1666–81.

Forrester, G. E. 1991. Social rank, individual size and group composition as determinants of food consumption by humbug damselfish, *Dascyllus aruanus. Animal Behaviour* **42**, 701–11.

Forrester, G. E. 1995. Strong density-dependent survival and recruitment regulate the abundance of a coral reef fish. *Oecologia* **103**, 275–82.

Forsgren, E. 1992. Predation risk affects mate choice in a gobiid fish. *American Naturalist* **140**, 1041–9.

Fowler, A. J. 1990. Spatial and temporal patterns of distribution and abundance of chaetodontid fishes at One Tree Reef, southern GBR. *Marine Ecology Progress Series* **64**, 39–53.

Fricke, H. W. 1979. Mating system, resource defence and sex change in the anemonefish *Amphiprion akallo-*

pisos. Zeitschrift fur Tierpsychologie **50**, 313–26.

Fricke, H. W. 1980. Control of different mating systems in a coral reef fish by one environmental factor. *Animal Behaviour* **28**, 561–9.

Galzin, R., Planes, S., Dufour, V., Salvat, B. 1994. Variation in diversity of coral reef fish between French Polynesian atolls. *Coral Reefs* **13**, 175–80.

Gaston, K. J. & Blackburn, T. M. 1996. Range size–body size relationships: evidence of scale dependence. *Oikos* **75**, 479–85.

Gaston, K. J. & Lawton, J. H. 1988. Patterns in body size, population dynamics, and regional distribution of bracken herbivores. *American Naturalist* **132**, 662–80.

Gibson, R. N. 1970. Observations on the biology of the giant goby *Gobius cobitis* Pallas. *Journal of Fish Biology* **2**, 281–8.

Gibson, R. N. & Ezzi, I. A. 1978. The biology of a Scottish population of Fries' goby, *Lesueurigobius friesii. Journal of Fish Biology* **12**, 371–89.

Gibson, R. N. & Ezzi, I. A. 1981. The biology of the Norway goby, *Pomatoschistus norvegicus* (Collett), on the west coast of Scotland. *Journal of Fish Biology* **19**, 697–714.

Gladfelter, W. B., Ogden, J. C. & Gladfelter, E. H. 1980. Similarity and diversity among coral reef fish communities: a comparison between tropical western Atlantic (Virgin Islands) and tropical central Pacific (Marshall Islands) patch reefs. *Ecology* **61**, 1156–68.

Goodwin, J. R. & Kosaki, R. K. 1989. Reef fish assemblages on submerged larva flows of three different stages. *Pacific Science* **43**, 289–301.

Gordoa, A. & Duarte, C. M. 1992. Size-dependent density of the demersal fish off Namibia: patterns within and among species. *Canadian Journal of Fisheries and Aquatic Science* **49**, 1990–3.

Green, A. L. 1996. Spatial, temporal and ontogenetic patterns of habitat use by coral reef fishes (Family Labridae). *Marine Ecology Progress Series* **133**, 1–11.

Greenfield, D. W. & Johnson, R. K. 1990. Community structure of Western Caribbean blennioid fishes. *Copeia* 1990, 433–48.

Griffiths, D. 1992. Size, abundance, and energy use in communities. *Journal of Animal Ecology* **61**, 307–15.

Grossman, G. D. 1979. Demographic characteristics of an intertidal bay goby (*Lepidogobius lepidus*). *Environmental Biology of Fishes* **4**, 207–18.

Hall, S. J. 1996. Global diversity and body size. *Science* **383**, 133 only.

Hanken, J. & Wake, D. B. 1993. Miniaturization of body size: the organismal consequences and evolutionary significance. *Annual Review of Ecology and Systematics* **24**, 501–19.

Hart, A. M., Klumpp, D. W. & Russ, G. R. 1996. Response of herbivorous fishes to crown-of-thorns starfish *Acanthaster planci* outbreaks. II. Density and biomass of selected species of herbivorous fish and fish-habitat correlations. *Marine Ecology Progress Series* **132**, 21–30.

Hastings, P. A. 1988. Correlates of males reproductive success in the browncheek blenny, *Acanthemblemaria crockeri* (Blennioidea: Chaenopsidae). *Behavioural Ecology and Sociobiology* **22**, 95–102.

Hattori, A. 1994. Inter-group movement and mate acquisition tactics of the protandrous anemonefish, *Amphiprion clarkii*, on a coral reef, Okinawa. *Japanese Journal of Ichthyology* **41**, 159–65.

Helfman, G. S. 1978. Patterns of community structure in fishes: summary and overview. *Environmental Biology of Fishes* **3**, 129–48.

Hixon, M. A. 1980. Competitive interactions between California reef fishes of the genus *Embiotoca. Ecology* **61**, 918–31.

Hixon, M. A. 1991. Predation as a process structuring coral reef communities. In *The ecology of fishes on coral reefs*, P. F. Sale (ed). San Diego: Academic Press, 475–508.

Hixon, M. A. & Beets, J. P. 1993. Predation, prey refuges, and the structure of coral-reef fish assemblages. *Ecological Monographs* **63**, 77–101.

Hobson, E. S. 1991. Trophic relationships of fishes specialised to feed on zooplankters above coral reefs. In *The ecology of fishes on coral reefs*, P. F. Sale (ed). San Diego: Academic Press, 69–95.

Horn, M. H. 1989. Biology of marine herbivorous fishes. *Oceanography and Marine Biology: an Annual Review* **27**, 167–272.

Hunte, W. & Côte, I. M. 1989. Recruitment in the redlip blenny *Ophioblennius atlanticus*: is space limiting? *Coral Reefs* **8**, 45–50.

Hutchinson, G. E. 1959. Homage to Santa Rosalina, or why there are so many kinds of animals. *American Naturalist* **93**, 145–59.

Hutchinson, G. E. & MacArthur, R. H. 1959. A theoretical ecological model of size distributions among species of animals. *American Naturalist* **869**, 117–25.

Iglesias, M., Brothers, E. B. & Morales-Nin, B. 1997. Validation of daily increment deposition in otoliths. Age and growth determination of *Aphia minuta* (Pisces: Gobiidae) from the northwest Mediterranean. *Marine Biology* **129**, 270–87.

Itzkowitz, M. 1977. Spatial organisation of the Jamaican damselfish community. *Journal of Experimental Marine Biology and Ecology* **28**, 217–41.

Jablonski, D. & Lutz, R. A. 1983. Larval ecology of marine benthic invertebrates: palaeobiological implications. *Biological Review* **58**, 21–89.

Jones, G. P. 1984. The influence of habitat and behavioural interactions on the local distribution of the wrasse, *Pseudolabrus celidotus*. *Environmental Biology of Fishes* **10**, 43–58.

Jones, G. P. 1986. Food availability affects growth in a coral reef fish. *Oecologia* **70**, 136–9.

Jones, G. P. 1987a. Competitive interactions among adults and juveniles in a coral reef fish. *Ecology* **68**, 1534–47.

Jones, G. P. 1987b. Some interactions between residents and recruits in two coral reef fish. *Journal of Experimental Marine Biology and Ecology* **114**, 169–82.

Jones, G. P. 1988. Experimental evaluation of the effects of habitat structure and competitive interactions on the juveniles of two coral reef fishes. *Journal of Experimental Marine Biology and Ecology* **123**, 115–26.

Jones, G. P. 1991. Postrecruitment processes in the ecology of coral reef fish populations: a multifactorial perspective. In *The ecology of fishes on coral reefs*, P. F. Sale (ed). San Diego: Academic Press, 294–328.

Jones, G. P. 1997. Relationships between recruitment and postrecruitment processes in lagoonal populations of two coral reef fishes. *Journal of Experimental Marine Biology and Ecology* **213**, 231–46.

Kaufman, L. S. & Ebersole, J. P. 1984. Microtopography and the organisation of two assemblages of coral reef fishes in the West Indies. *Journal of Experimental Marine Biology and Ecology* **78**, 253–68.

Kingsford, M. J. 1988. The early life history of fish in coastal waters of northern New Zealand: a review. *New Zealand Journal of Marine and Freshwater Research* **22**, 463–79.

Kingsford, M. J. 1992. Spatial and temporal variation in predation of reef fishes by coral trout (*Plectropomus leopardus*, Serranidae). *Coral Reefs* **11**, 193–8.

Kingsford, M. J. & Choat, J. H. 1989. Horizontal distribution patterns of presettlement reef fish: are they influenced by proximity of reefs? *Marine Biology* **101**, 285–97.

Kuwamura, T., Nakashima, Y. & Yogo, Y. 1996. Plasticity in size and age at maturity in a monogamous fish: effect of host coral size and frequency dependence. *Behavioural Ecology and Sociobiology* **38**, 365–70.

Kuwamura, T., Yogo, Y. & Nakashima, Y. 1993. Size-assortive monogamy and parental egg care in a coral goby *Paragobiodon echinocephalus*. *Ethology* **95**, 65–75.

Kuwamura, T., Yogo, Y. & Nakashima, Y. 1994. Population dynamics of goby *Paragobiodon echinocephalus* and host coral *Stylophora pistillata*. *Marine Ecology Progress Series* **103**, 17–23.

LaBarbera, M. 1989. Analyzing body size as a factor in ecology and evolution. *Annual Review of Ecology and Systematics* **20**, 97–117.

Lassig, B. 1976. Field observations on the reproductive behaviour of *Paragobiodon* spp. (Osteichthyes: Gobiidae) at Heron Island Great Barrier Reef. *Marine Behaviour and Physiology* **3**, 283–93.

Lassig, B. 1977. Socioecological strategies adopted by obligate coral-dwelling fishes. *Proceedings of the 3rd International Coral Reef Symposium* **1**, 565–70.

Lawton, J. H. 1986. Surface availability and insect community structure: the effects of architecture and fractal dimension of plants. In *Insect and plant surface*, B. Juniper & R. Southwood (eds). London: Edward Arnold, 317–32.

Lawton, J. H. 1990. Species richness and population dynamics of animal assemblages. Patterns in body size: abundance space. *Philosophical Transactions of the Royal Society of London* **330**, 283–91.

Leis, J. M. 1982. Nearshore distributional gradients of larval fish (15 taxa) and planktonic crustaceans (6 taxa) in Hawaii. *Marine Biology* **72**, 89–97.

Leis, J. M. 1984. Larval fish dispersal and the East Pacific barrier. *Oceanographie Tropicale* **19**, 181–92.

Leis, J. M. 1991. The pelagic stage of reef fishes: the larval biology of coral reef fishes. In *The ecology of fishes on coral reefs*, P. F. Sale (ed). San Diego: Academic Press, 183–230.

Leis, J. M. 1993. Larval fish assemblages near Indo-Pacific coral reefs. *Bulletin of Marine Science* **53**, 362–92.

Leis, J. M. 1994. Coral Sea atoll lagoons: closed nurseries for the larvae of a few coral reef fishes. *Bulletin of Marine Science* **54**, 206–27.

Leis, J. M. & Miller, J. M. 1976. Offshore distributional patterns of Hawaiian fish larvae. *Marine Biology* **36**, 359–67.

Levin, P. S. 1993. Habitat structure, conspecific presence and spatial variation in the recruitment of a temperate reef fish. *Oecologia* **94**, 176–85.

Lewis, A. R. 1997. Recruitment and post recruitment immigration affect the local population size of coral reef fishes. *Coral Reefs* **16**, 139–49.

Lindquist, D. G. 1985. Depth zonation, microhabitat, and morphology of three species of *Acanthemblemaria* (Pisces: Blennioidea) in the Gulf of California, Mexico. *Marine Ecology* **6**, 329–44.

Lirman, D. 1994. Ontogenetic shifts in habitat preferences in the three-spot damselfish, *Stegastes planifrons* (Cuvier), in Roatan Island, Honduras. *Journal of Experimental Marine Biology and Ecology* **180**, 71–81.

Litvak, M. K. & Leggett, W. C. 1992. Age and size-selective predation on larval fishes: the bigger-is-better hypothesis revisited. *Marine Ecology Progress Series* **81**, 13–24.

Loder, N., Blackburn, T. M. & Gaston, K. J. 1997. The slippery slope: towards an understanding of the body size frequency distribution. *Oikos* **78**, 195–201.

Low, R. M. 1971. Interspecific territoriality in a pomacentrid reef fish, *Pomacentrus flavicauda* Whitley. *Ecology* **52**, 648–54.

Luckhurst, B. E. & Luckhurst, K. 1978. Analysis of the influence of substrate variables on coral reef fish communities. *Marine Biology* **49**, 317–23.

MacArthur, R. H. & Levins, R. 1967. The limiting similarity, convergence and divergence of coexisting species. *American Naturalist* **101**, 377–85.

Macpherson, E. 1989. Influence of geographical distributions, body size and diet on population density of benthic fishes off Namibia (South West Africa). *Marine Ecology Progress Series* **50**, 295–99.

Magnhagen, C. 1993. Conflicting demands in gobies: when to eat, reproduce, and avoid predators. *Marine Behaviour and Physiology* **23**, 79–90.

Mark, D. M. 1984. Fractal dimension of a coral reef at ecological scales: a discussion. *Marine Ecology Progress Series* **14**, 293–4.

Marquet, P. A., Navarrete, S. A. & Castilla, J. C. 1990. Scaling population density to body size in rocky intertidal communities. *Science* **250**, 1125–27.

Marquet, P. A., Navarrete, S. A. & Castilla, J. C. 1995. Body size, population density, and the energetic equivalence rule. *Journal of Animal Ecology* **64**, 325–32.

Marraro, C. H. & Nursall, J. R. 1983. The reproductive periodicity and behaviour of *Ophioblennius atlanticus* (Pisces: Blenniidae) at Barbados. *Canadian Journal of Zoology* **61**, 317–25.

Marzluff, J. M. & Dial, K. P. 1991. Life history correlates of taxonomic diversity. *Ecology* **72**, 428–39.

Maurer, B. A., Brown, J. H. & Rusler, R. D. 1992. The micro and macro in body size evolution. *Evolution* **46**, 939–53.

May, R. M. 1978. The dynamics and diversity of insect faunas. In *Diversity of insect faunas*, L. A. Mound & N. Waloff (eds). Oxford: Blackwell Scientific, 188–204.

May, R. M. 1986. The search for patterns in the balance of nature: advances and retreats. *Ecology* **67**, 1115–26.

McAfee, S. G. & Morgan, S. G. 1996. Resource use by five sympatric parrotfishes in the San Blas Archipelago, Panama. *Marine Biology* **125**, 427–37.

McCormick, M. I. 1995. Fish feeding on mobile benthic invertebrates: influence of spatial variability in habitat associations. *Marine Biology* **121**, 627–37.

McCormick, M. I. & Kerrigan, B. A. 1996. Predation and its influence on the condition of a newly settled tropical demersal fish. *Marine and Freshwater Research* **47**, 557–62.

Miller, P. J. 1973. The species of *Pseudaphya* (Teleostei: Gobiidae) and the evolution of aphyiine gobies. *Journal of Fish Biology* **5**, 353–65.

Miller, P. J. 1979. Adaptiveness and implications of small size in teleosts. *Symposium of the Zoological Society of London* **44**, 263–306.

Miller, P. J. 1984. The tokology of gobioid fishes. In *Fish reproduction*, G. W. Potts & R. J. Wootton (eds). London: Academic Press, 119–53.

Miller, P. J. 1996. The functional ecology of small fish: some opportunities and consequences. In *Miniature vertebrates: the implications of small body size*, P. J. Miller (ed). Oxford: The Zoological Society of London, 175–95.

Miller, P. J. & El-Tawil, M. Y. 1974. A multidisciplinary approach to a new species of *Gobius* (Teleostei: Gobiidae) from southern Cornwall. *Journal of Fish Biology* **174**, 539–74.

Mittelbach, G. G. & Chesson, P. L. 1992. Predation risk: indirect effects on fish populations. In *Predation: direct and indirect impacts on aquatic communities*, W. C. Kerfoot & A. Sih (eds). Hanover: University of New England Press, 315–32.

Morse, D. R., Lawton, J. H., Dodson, M. M. & Williamson, M. H. 1985. Fractal dimension of vegetation and the distribution of arthropod body lengths. *Nature* **314**, 731–3.

Morse, D. R., Stork, N. E. & Lawton, J. H. 1988. Species number, species abundance and body length relationships of arboreal beetles in Bornean lowland rain forest trees. *Ecological Entomology* **13**, 25–37.

Munday, P. L., Jones, G. P. & Caley, M. J. 1997. Habitat specialisation and the distribution and abundance of coral-dwelling gobies. *Marine Ecology Progress Series* **152**, 227–39.

Munro, J. L. & Williams, D. McB. 1985. Assessment and management of coral reef fisheries: biological, environmental and socioeconomic aspects. *Proceedings of the 5th International Coral Reef Symposium* **4**, 545–81.

Myers, R. F. 1991. *Micronesian reef fishes*. Guam: Coral Graphics.

Nee, S., Read, A. F., Greenwood, J. J. D. & Harvey, P. H. 1991. The relationship between abundance and body size in British birds. *Nature* **351**, 312–13.

Nursall, J. R. 1977. Territoriality in redlip blennies (*Ophioblennius atlanticus* – Pisces: Blenniidae). *Journal of the Zoological Society of London* **182**, 205–22.

Öhman, M. C., Rajasuriya, A. & Olafsson, E. 1997. Reef fish assemblages in north-western Sri Lanka: distribution patterns and influences of fishing practices. *Environmental Biology of Fishes* **49**, 45-61.

Öhman, M. C., Munday, P. L., Jones, G. P. & Caley, M. J. 1998. Settlement strategies and the distribution patterns of coral-reef fishes. *Journal of Experimental Marine Biology and Ecology* **225**, 219–38.

Öhman, M. C. & Rajasuriya, A. (in press). Relationships between habitat structure and fish assemblages on coral and sandstone reefs. *Environmental Biology of Fishes.*

Page, L. M. & Swofford, D. L. 1984. Morphological correlates of ecological specialisation in darters. *Environmental Biology of Fishes* **11**, 139–59.

Palumbi, S. R. 1994. Genetic divergence, reproductive isolation, and marine speciation. *Annual Review of Ecology and Systematics* **25**, 547–72.

Patton, W. K. 1994. Distribution and ecology of animals associated with branching corals (*Acropora* spp.) from the Great Barrier Reef, Australia. *Bulletin of Marine Science* **55**, 193–211.

Pauly, D. 1980. On the inter-relationships between natural mortality, growth parameters, and mean environmental temperature in 175 fish stocks. *Journal du Conseil, Conseil permanent international pour l'Exploration de la Mer* **39**, 175–92.

Persson, L. 1985. Assymetrical competition: are larger animals competitively superior? *American Naturalist* **126**, 261–6.

Peters, R. H. 1983. *The ecological implications of body size*. Cambridge: Cambridge University Press.

Peterson, I. & Wroblewski, J. S. 1984. Mortality rate of fishes in the pelagic ecosystem. *Canadian Journal of Fisheries and Aquatic Science* **41**, 1117–20.

Polunin, N. V. C. & Roberts, C. M. (eds) 1996. *Reef fisheries*. London: Chapman & Hall.

Price, P. W. 1984. Communities of specialists: vacant niches in ecological and evolutionary time. In *Ecological communities*, D. R. Strong (ed). New Jersey: Princeton University Press, 510–23.

Randall, J. E. 1963. An analysis of the fish populations of artificial and natural reefs in the Virgin Islands. *Caribbean Journal of Science* **3**, 31–47.

Randall, J. E., Allen, G. R. & Steene, R. C. 1990. *Fishes of the Great Barrier Reef and Coral Sea*. Bathurst: Crawford House Press.

Randall, J. E. & Heemstra, P. C. 1991. Revision of the Indo-Pacific groupers (Perciformes: Serranidae: Epinephelinae), with descriptions of five new species. *Indo-Pacific Fishes* **20**, 1–332.

Reznick, D. N., Bryga, H. & Endler, J. A. 1990. Experimentally induced life-history evolution in natural population. *Nature* **346**, 357–9.

Reznick, D. N., Butler, M. J., Rodd, F. H. & Ross, P. 1996. Life-history evolution in guppies (*Poecilia reticulata*). 6. Differential mortality as a mechanism for natural selection. *Evolution* **50**, 1651–60.

Ricklefs, R. E. 1987. Community diversity: relative roles of local and regional processes. *Science* **235**, 167–71.

Ricklefs, R. E. & Schluter, D. 1993. Species diversity: regional and historical influences. In *Species diversity in ecological communities: historical and geographical perspectives*, R. E. Ricklefs & D. Schluter (eds). Chicago: University of Chicago Press, 350–63.

Roberts, C. M. 1987. Experimental analysis of resource sharing between herbivorous damselfish and blennies on the Great Barrier Reef. *Journal of Experimental Marine Biology and Ecology* **111**, 61–75.

Roberts, C. M. & Ormond, R. F. G. 1987. Habitat complexity and coral reef fish diversity and abundance on Red Sea fringing reefs. *Marine Ecology Progress Series* **41**, 1–8.

Robertson, D. R. 1984. Cohabitation of competing territorial damselfishes on a Caribbean coral reef. *Ecology* **65**, 1121–35.

Robertson, D. R. 1995. Competitive ability and the potential for lotteries among territorial reef fish. *Oecologia* **103**, 180–90.
Robertson, D. R. 1996. Interspecific competition controls abundance and habitat use of territorial Caribbean damselfishes. *Ecology* **77**, 885–99.
Robertson, D. R. 1998. Implications of body-size for interspecific interactions and asemblage organization among coral-reef fishes. *Australian Journal of Ecology* **23**, 252–7.
Robertson, D. R. & Gaines, S. D. 1986. Interference competition structures habitat use in a local assemblage of coral reef surgeonfishes. *Ecology* **67**, 1372–83.
Robertson, D. R., Hoffman, S. G. & Sheldon, J. M. 1981. Availability of space for the territorial Caribbean damselfish *Eupomacentrus planifrons*. *Ecology* **62**, 1162–9.
Robertson, D. R. & Kaufman, K. W. 1998. Assessing early recruitment dynamics and its demographic consequences among tropical reef fishes: accommodating variation in recruitment seasonality and longevity. *Australian Journal of Ecology* **23**, 226–33.
Robertson, D. R. & Lassig, B. 1980. Spatial distribution patterns and coexistence of a group of territorial damselfishes from the Great Barrier Reef. *Bulletin of Marine Science* **30**, 187–203.
Robertson, D. R., Polunin, N. V. C. & Leighton, K. 1979. The behavioural ecology of three Indian Ocean surgeonfishes (*Acanthurus lineatus*, *A. leucosternum*, and *Zebrasoma scopas*): their feeding strategies, and social and mating systems. *Environmental Biology of Fishes* **4**, 125–70.
Robertson, D. R. & Sheldon, J. M. 1979. Competitive interactions and the availability of sleeping sites for a diurnal coral reef fish. *Journal of Experimental Marine Biology and Ecology* **40**, 285–98.
Roff, D. A. 1984. The evolution of life history parameters in teleosts. *Canadian Journal of Fisheries and Aquatic Science* **41**, 989–1000.
Roff, D. A. 1992. *The evolution of life histories*. New York: Chapman & Hall.
Ross, R. M. 1978. Territorial behaviour and ecology of the anemonefish *Amphiprion melanopus*. *Zeitschrift fur Tierpsychologie* **46**, 71–83.
Sale, P. F. 1971. Extremely limited home range in a coral reef fish, *Dascyllus aruanus* (Pisces; Pomacentridae). *Copeia* 324–27.
Sale, P. F. 1972. Influence of corals in the dispersion of the pomacentrid fish, *Dascyllus aruanus*. *Ecology* **53**, 741–4.
Sale, P. F. 1978a. Coexistence of coral reef fishes – a lottery for living space. *Environmental Biology of Fishes* **3**, 85–102.
Sale, P. F. 1978b. Reef fishes and other vertebrates: a comparison of social structures. In *Contrasts in behaviour*, F. J. Lighter (ed). New York: John Wiley, 314–46.
Sale, P. F. 1980. The ecology of fishes on coral reefs. *Oceanography and Marine Biology: an Annual Review* **18**, 367–421.
Sale, P. F. (ed) 1991a. *The ecology of fishes on coral reefs*. San Diego: Academic Press.
Sale, P. F. 1991b. Reef fish communities: open nonequilibrial systems. In *The ecology of fishes on coral reefs*, P. F. Sale (ed). San Diego: Academic Press, 564–98.
Sale, P. F., Doherty, P. J. & Douglas, W. A. 1980. Juvenile recruitment strategies and the coexistence of territorial pomacentrid fishes. *Bulletin of Marine Science* **30**, 147–58.
Sale, P. F. & Douglas, W. A. 1981. Precision and accuracy of visual census technique for fish assemblages on coral patch reefs. *Environmental Biology of Fishes* **6**, 333–9.
Sale, P. F. & Ferrell, D. J. 1988. Early survivorship of juvenile coral reef fishes. *Coral Reefs* **7**, 117–24.
Sale, P. F., Guy, J. A. & Steel, W. J. 1994. Ecological structure of assemblages of coral reef fishes on isolated patch reefs. *Oecologia* **98**, 83–99.
Sano, M. 1997. Temporal variation in density dependence: recruitment and postrecruitment demography of a temperate zone sand goby. *Journal of Experimental Marine Biology and Ecology* **214**, 67–84.
Schmitt, R. J. & Holbrook, S. J. 1990. Population responses of surfperch released from competition. *Ecology* **71**, 1653–65.
Schoener, T. W. 1974. Resource partitioning in ecological communities. *Science* **185**, 27–39.
Schoener, T. W. 1983. Field experiments in interspecific competition. *American Naturalist* **121**, 240–85.
Shibunto, T., Gushima, K. & Kakuda, S. 1993. Female spawning migrations of the protogynous wrasse, *Halichoeres marginatus*. *Japanese Journal of Ichthyology* **39**, 357–61.
Shulman, M. J. 1984. Resource limitation and recruitment patterns in a coral reef fish assemblage. *Journal of Experimental Marine Biology and Ecology* **74**, 85–109.
Shulman, M. J. 1985. Coral reef fish assemblages: intra and interspecific competition for shelter sites. *Environmental Biology of Fishes* **13**, 81–92.

Schulman, M. J. & Bermingham, E. 1995. Early life history, ocean currents, and the population genetics of Caribbean reef fishes. *Evolution* **49**, 897–910.
Shulman, M. J. & Ogden, J. C. 1987. What controls tropical reef fish populations: recruitment or benthic mortality? An example in the Caribbean reef fish *Haemulon flavolineatum*. *Marine Ecology Progress Series* **39**, 233–42.
Siemann, E., Tilman, D. & Haarstad, J. 1996. Insect species diversity, abundance and body size relationships. *Nature* **380**, 704–6.
Sih, A. 1992. Predators and prey lifestyles: an evolutionary and ecological overview. In *Predation: direct and indirect impacts on aquatic communities*, A. Sih (ed). Hanover: University of New England Press, 203–24.
Slatkin, M. 1980. Ecological character displacement. *Ecology* **61**, 163–77.
Smith, C. L. 1978. Coral reef fish communities: a compromise view. *Environmental Biology of Fishes* **3**, 109–28.
Southwood, T. R. E. 1978. The components of diversity. In *Diversity of insect faunas*, L. A. Mound & N. Woloff (eds). Oxford: Blackwell Scientific, 19–40.
Stearns, S. C. 1992. *The evolution of life histories*. Oxford: Oxford University Press.
Stobutzki, I. C. & Bellwood, D. R. 1997. Sustained swimming abilities of the late pelagic stages of coral reef fishes. *Marine Ecology Progress Series* **149**, 35–41.
Sunobe, T., Ohta, T. & Nakazono, A. 1995. Mating system and spawning cycle in the blenny, *Istiblennius enosimae*, at Kagoshima, Japan. *Environmental Biology of Fishes* **43**, 195–9.
Sweatman, H. P. A. 1983. Influence of conspecifics on choice of settlement sites by larvae of two pomacentrid fishes (*Dascyllus aruanus* and *D. reticulatus*). *Marine Biology* **75**, 225–29.
Sweatman, H. P. A. 1985. The influence of adults of some coral reef fishes on larval recruitment. *Ecological Monographs* **55**, 469–85.
Sweatman, H. P. A. 1988. Field evidence that settling coral reef larvae detect resident fishes using dissolved chemical cues. *Journal of Experimental Marine Biology and Ecology* **124**, 163–74.
Thresher, R. E. 1984. *Reproduction in reef fishes*. Neptune City: TFH Publications.
Thresher, R. E. 1991. Geographic variability in the ecology of coral reef fishes: evidence, evolution, and possible implications. In *The ecology of fishes on coral reefs*, P. F. Sale (ed). San Diego: Academic Press, 401–36.
Tyler, J. C. & Bohlke, J. E. 1972. Records of sponge dwelling fishes, primarily of the Caribbean. *Bulletin of Marine Science* **22**, 601–42.
Victor, B. C. 1991. Settlement strategies and biogeography of reef fishes. In *The ecology of fishes on coral reefs*, P. F. Sale (ed). San Diego: Academic Press, 231–60.
Waldner, R. E. & Robertson, D. R. 1980. Patterns of habitat partitioning by eight species of territorial Caribbean damselfishes (Pisces: Pomacentridae). *Bulletin of Marine Science* **30**, 171–86.
Warburton, K. 1989. Ecological and phylogenetic constraints on body size in Indo-Pacific fishes. *Environmental Biology of Fishes* **24**, 13–22.
Warwick, R. M. & Clarke, K. R. 1996. Relationships between body-size, species abundance and diversity in marine benthic assemblages: facts or artefacts? *Journal of Experimental Marine Biology and Ecology* **202**, 63–71.
Wasserman, S. S. & Mitter, C. 1978. The relationship of body size to breadth of diet in some Lepidoptera. *Ecological Entomology* **3**, 155–60.
Weatherly, A. H. 1972. *Growth and ecology of fish populations*. London: Academic Press.
Werner, E. E. 1984. The mechanisms of species interactions and community organisation in fish. In *Ecological communities*, D. R. Strong (ed). New Jersey: Princeton University Press, 360–82.
Werner, E. E. & Gilliam, J. F. 1984. The ontogenetic niche and species interactions in size structured populations. *Annual Review of Ecology and Systematics* **15**, 393–425.
Werner, E. E., Gilliam, J. F., Hall, D. J. & Mittelbach, G. G. 1983a. An experimental test of the effects of predation risk on habitat use in fish. *Ecology* **64**, 1540–48.
Werner, E. E., Mittelbach, G. G., Hall, D. J. & Gilliam, J. F. 1983b. Experimental tests of optimal habitat use in fish: the role of relative habitat profitability. *Ecology* **64**, 1525–39.
Williams, D. McB. & Hatcher, A. I. 1983. Structure of fish communities on outer slopes of inshore, mid-shelf and outer shelf reefs of the Great Barrier Reef. *Marine Ecology Progress Series* **10**, 239–50.
Williams, D. McB. & Russ, G. R. 1994. *Review of data on fishes of commercial and recreational fishing interest in the Great Barrier Reef*. Research Publication No. 33. Townsville: Great Barrier Reef Marine Park Authority.

Wilson, D. S. 1975. The adequacy of body size as a niche difference. *American Naturalist* **109**, 769–84.

Winterbottom, R. & Emery, A. R. 1981. A new genus and two new species of gobiid fishes (Perciformes) from the Chagos Archipelago, Central Indian Ocean. *Environmental Biology of Fishes* **6**, 139–49.

Woodland, D. J. 1990. Revision of the fish family Siganidae with descriptions of two new species and comments on distribution and biology. *Indo-Pacific Fishes* **19**, 1–136.

Oceanography and Marine Biology: an Annual Review 1998, **36**, 413–436
© A. D. Ansell, R. N. Gibson and Margaret Barnes, Editors
UCL Press

AUTHOR INDEX

References to complete articles are given in **bold type**; references to page numbers are given in normal type; references to bibliographical lists are given in *italics*.

Oceanography and Marine Biology: an Annual Review 1998, **36**, 437–452

UCL Press

SYSTEMATIC INDEX

References to complete articles are given in **bold type**; references to sections of articles are given in *italics*; references to pages are given in normal type.

Oceanography and Marine Biology: an Annual Review 1998, **36**, 453–459
© A. D. Ansell, R. N. Gibson and Margaret Barnes, Editors
UCL Press

SUBJECT INDEX

References to complete articles are given in **bold type**; references to sections of articles are given in *italics*; references to pages are given in normal type.